ADVANCES IN CHEMICAL PHYSICS

VOLUME XXXVI

EDITORIAL BOARD

Advances in
CHEMICAL PHYSICS

EDITED BY

I. PRIGOGINE

University of Brussels
Brussels, Belgium
and
University of Texas
Austin, Texas

AND

STUART A. RICE

Department of Chemistry
and
The James Franck Institute
The University of Chicago
Chicago, Illinois

VOLUME XXXVI

AN INTERSCIENCE ® PUBLICATION

JOHN WILEY AND SONS

NEW YORK • LONDON • SYDNEY • TORONTO

Library of Congress Catalog Number: 58-9935

ISBN 0-471-02274-8

Printed in the United States of America

10 9 8 7 6 5 4 3 2 1

CONTRIBUTORS TO VOLUME XXXVI

L. S. CEDERBAUM, Institut für Theoretische Physik, Physik—Department der Technischen Universität München, Germany

L. G. CHRISTOPHOROU, Department of Physics, University of Tennessee, Knoxville, Tennessee

A. P. CLARK, Department of Physics, University of Stirling, Stirling, Scotland

A. S. DICKINSON, Department of Atomic Physics, School of Physics, The University, Newcastle-upon-Tyne, England

W. DOMCKE, Institut für Theoretische Physik, Physik—Department der Technischen Universität München, Germany

M. GRANT, Department of Physics, University of Tennessee, Knoxville, Tennessee

V. KVASNICKA, Department of Mathematics, Faculty of Chemistry, Slovak Technical University, Bratislava, Czechoslovakia

D. L. McCORKLE, Health Physics Division, Oak Ridge National Laboratory, Oak Ridge, Tennessee

I. C. PERCIVAL, Department of Applied Mathematics, Queen Mary College, London, England

D. RICHARDS, Faculty of Mathematics, The Open University, Milton Keynes, England

D. G. TRUHLAR, Department of Chemistry, University of Minnesota, Minneapolis, Minnesota

R. E. WYATT, Department of Chemistry, University of Texas, Austin, Texas

INTRODUCTION

Few of us can any longer keep up with the flood of scientific literature, even in specialized subfields. Any attempt to do more, and be broadly educated with respect to a large domain of science, has the appearance of tilting at windmills. Yet the synthesis of ideas drawn from different subjects into new, powerful, general concepts is as valuable as ever; thus the desire to remain educated persists in all scientists. This series, *Advances in Chemical Physics*, is devoted to helping the reader obtain general information about a wide variety of topics in chemical physics, which field we interpret very broadly. Our intent is to have experts present comprehensive analyses of subjects of interest and to encourage the expression of individual points of view. We hope that this approach to the presentation of an overview of a subject will both stimulate new research and serve as a personalized learning text for beginners in a field.

ILYA PRIGOGINE

STUART A. RICE

CONTENTS

ADVANCES IN CHEMICAL PHYSICS

VOLUME XXXVI

SEMICLASSICAL THEORY
OF BOUND STATES

IAN C. PERCIVAL

Department of Applied Mathematics
Queen Mary College
Mile End Road
London E1 4NS
England

CONTENTS

I. INTRODUCTION AND EARLY THEORIES

A. Introduction

Before 1925 the determination of the properties of quantized systems from the properties of the corresponding classical systems was the central problem of the old quantum theory and of chemistry and nonrelativistic physics. Bohr's correspondence principle helped to guide the old quantum theorists toward quantum mechanics, but the discovery of quantum mechanics only partly solved the original problem by showing how to obtain a matrix equation or a wave equation from a classical Hamiltonian.

In quantum mechanics a quantal formulation is obtained from a classical formulation, but a quantal solution is not obtained from a classical solution.

Even now no general relation between solutions is known. An asymptotic relation is known for the separable systems, whose motion in either form of mechanics can be transformed by a time-independent transformation into independent one-dimensional motions. This is the field of traditional JWKB theory.

This chapter reviews the semiclassical theory of bound states, which may or may not be separable.

The development of semiclassical theory of bound states was impeded for many decades by our ignorance of the behavior of classical systems of many degrees of freedom. The problem was recognized by Einstein in 1917.[1] The situation has now changed drastically as a result of the Kolmogorov–Arnol'd–Moser (KAM) theorem and many numerical integrations of classical trajectories.

Following these classical results, as shown in Section VII, semiclassical mechanics differs from quantum mechanics because it has three basic types of energy spectrum instead of two. In quantum mechanics there is a continuous spectrum of free or collision states and a discrete spectrum of bound states. In semiclassical mechanics the discrete bound-state spectrum breaks into two distinct parts, the *regular* and the *irregular* spectrum. The latter exists only for two or more degrees of freedom.

Little is known about the irregular spectrum, but the most important properties of the regular spectrum are understood. They are the subject of Sections II to V. In the regular spectrum there is a definite and well-defined classical equivalent of the bound quantal state: it is the *invariant toroid*. These sections are devoted to its properties and its relation to the semiclassical wave function. For simplicity the subject is introduced through systems of one degree of freedom in Sections II and III, and generalized in Sections IV and V, leading to the Einstein–Brillouin–Keller (EBK) quantization rule and the Maslov index.

Like Maslov we adopt a phase-space picture of classical motion, but we go even further in using the same picture for semiclassical wave functions. The resulting freedom in changing representations allows us to overcome the usual problems of phase changes at caustics.

There is a practical need for semiclassical quantization of high bound states where the motion cannot be adequately approximated by the motion of a separable system. For example, with lasers it has become practical to investigate the higher vibrational states of polyatomic molecules. There are also applications to the physics of solids, nuclei, and particles. For such high states the quantal matrix and perturbation methods usually used for the approximate solution of wave equations are no longer effective: the number of stationary states required for a basis set, and the number of oscillations in the wave function of each state, are too large. However, it is just under these conditions that semiclassical theory is at its best; it is complementary to the traditional methods of quantum theory. In Section VI we describe four different methods that have been used to obtain bound-state energy levels using EBK quantization, thus demonstrating its efficacy by example.

For one degree of freedom an invariant toroid is nothing more than a closed classical trajectory in phase space. For more degrees of freedom the invariant toroid is not a trajectory, and there is no correspondence between individual closed trajectories and semiclassical wave functions. Nevertheless, there are connections between closed trajectories and the properties of the bound-state energy spectrum, which we review in Section VIII, indicating some of the pitfalls.

It is no longer possible to be both comprehensive and comprehensible in one article on semiclassical mechanics. Many important developments have had to be omitted from this review, some of which are significant for bound states. An understanding of the form of a wave function in the short-wavelength limit and the energy spectrum of a system with more than one minimum in its potential function requires the theory of caustics and of uniform approximations. These theories have been advanced considerably in recent years, as may be seen from the review articles of Berry and Mount and more recent contributions.[2] Furthermore, it has not been possible to do justice to theories of complex classical trajectories[3] nor to probability distributions for many-particle systems,[4] nor to semiclassical expansions.[5] Semiclassical scattering is reviewed in a chapter in this volume. Other recent contributions to semiclassical mechanics are given in Ref. 6.

B. Early Quantization Rules

1. Old Quantum Theory: Schwarzschild, Epstein, and Einstein

Let a system of N degrees of freedom have coordinates q_k and canonically conjugate momenta p_k. Let the N action integrals I_k be defined by the integrals

$$I_k = \frac{1}{2\pi} \oint p_k \, dq_k \qquad (k = 1, \ldots, N) \tag{1.1}$$

where each integral is taken around one cycle of the motion of the coordinate q_k. The Sommerfeld–Wilson quantization rule then states that for each stationary state of the system

$$I_k = n_k \hbar \qquad (k = 1, \ldots, N) \tag{1.2}$$

where $2\pi\hbar$ is Planck's constant and n_k is an integer.

The condition depends on the choice of coordinates; if it is valid for one choice it cannot be valid for all. Schwarzschild and Epstein[7] proposed that the coordinates be chosen so that the action function S_E, which is a solution of the classical Hamilton–Jacobi equation, can be expressed as a sum over N action functions, each dependent on one coordinate alone. This proposal is limited to separable systems, and the theory is not invariant under the canonical transformations of classical mechanics.

Jammer has written a good review[8] of this early work but does not mention the paper of Einstein,[1] who was naturally dissatisfied with the lack of invariance of the Schwarzschild–Epstein theory.

Einstein based his theory on the invariant differential sum

$$\sum_{k=1}^{N} p_k \, dq_k \tag{1.3}$$

Einstein's quantization rule is given by invariant line integrals of the form

$$I = \frac{1}{2\pi} \oint \sum_k p_k \, dq_k = n\hbar \tag{1.4}$$

along closed curves in coordinate space that have *no need* to be classical trajectories. In order to obtain N independent quantum conditions, Einstein was led to consider the invariance of the integrals under continuous transformations, and the topology (connectivity) of the momenta p_k as functions of the coordinates q_k.

For the simple case of a particle of one degree of freedom oscillating in a potential well, there are two momenta of opposite sign for almost all classically accessible positions, and the momentum may be considered as a continuous two-valued function of these positions, or alternatively as a single-valued function on two sheets, by analogy with the Riemann theory of complex variables.

For more degrees of freedom we quote Einstein in a fairly free translation:

"We now come to an absolutely essential point—. We follow the unrestricted motion of a single system for an unlimited period of time, and think of the corresponding trajectory traced out in q-space. Two possibilities arise:

1. There exists a part of q-space such that the trajectory comes arbitrarily close to every point in this N-dimensional subspace in the course of time.
2. The trajectory can be kept within a continuum of less than N dimensions. A special case of this is the exact closed trajectory.

Case 1 is general. Cases 2 are generated from 1 by specialization. As an example of Case 1 think of a particle under the action of a central force, the motion of which is described by 2 coordinates which determine the position of the particle within the plane of its orbit (for example polar coordinates r and ϕ). Case 2 occurs, for example, when the central force is attractive and exactly proportional to $1/r^2$ and when the relativistic corrections to the Kepler motion are neglected. In that case the orbit is closed and the points on it form a continuum of only one dimension. When considered in three-dimensional space, the motion under the central force is always of Type 2, because the orbit can be put into a continuum of two dimensions. From the three-dimensional viewpoint, one has to regard the central motion as a special case of a motion which is defined by a complicated law of force (for example the motion studied by Epstein in the theory of the Stark effect.)

The following is concerned with the general Case 1. Look at an element $d\tau$ in the q-space. The trajectories that we have considered pass through this element any number of times. Each such passage gives rise to a momentum vector p_k. A priori 2 fundamentally different types of trajectories are possible.

Type 1a): the vectors p_k repeat themselves so that only a finite number of them belong to each $d\tau$. In this case for the trajectories considered the p_k represent one- or many-valued functions of the q_k.

Type 1b): An infinite number of p_k vectors pass through the place considered. In this case p_k cannot be expressed as a function of the q_k.

One notices that the formulation of the (Einstein) quantum conditions is not possible for the case (b). On the other hand classical statistical mechanics holds only for the case (b); so only in this case is the microcanonical ensemble equivalent to the time ensemble."

Einstein's quantization rule is canonically invariant, and for separable systems it reduces to the Epstein–Schwarzschild condition. But its general validity requires averages over microcanonical and time ensembles to be different, that is, for the ergodic hypothesis of statistical mechanics to be false. At the time that Einstein wrote his paper, some form of ergodic theorem was believed to be generally valid for nonseparable systems.

Recent work has shown this belief to be unfounded.

2. Wave Functions: Brillouin and Keller

Since 1925, microscopic physics has been based on quantum mechanics, and the old quantum theory has been reinterpreted as an asymptotic limit of quantum mechanics as $\hbar \to 0$, as a part of "semiclassical" mechanics. For bound states this is the limit of high quantum numbers. If the wave function of a system is separable in some set of coordinates, then one-dimensional JWKB theory may be used, although modern textbooks frequently obscure its semiclassical basis.

Brillouin[9] sketched and Keller[9] developed a semiclassical theory of the wave functions of bound systems of many degrees of freedom.

According to this theory, a time-independent wave function should be expressed asymptotically as a finite sum of terms of the form

$$\psi \sim \sum_{r=1}^{R} B_r \exp(i\hbar^{-1}S_r) \qquad (\hbar \to 0) \tag{1.5}$$

where S_r is a classical action function of the coordinates and B_r is a normalization factor that nearly everywhere varies slowly by comparison with $S_r/\hbar$. Each term in the sum corresponds to one sheet of Einstein's many-sheeted momentum functions, which are given by the gradient of S_r on each sheet r.

At any classically accessible point in coordinate space, a stationary wave function is made up of a superposition of up to R travelling waves each with vector wave number of the form $\mathbf{p}_r/\hbar$, where $\mathbf{p}_r$ is the vector momentum on sheet r.

The boundary conditions on the action function provide the quantization rule

$$I_k = (n_k + \alpha_k/4)\hbar \qquad (k = 1, \dots, N) \tag{1.6}$$

which is the same as Einstein's except for the additional terms involving the integers α_k; these terms arise from the diffraction of quantal waves at classical turning points. The integer α_k depends only on the form of the classical action function; it is not a quantum number. With this correction Einstein's quantization rule becomes the EBK rule, which is now derived from the asymptotic limit of quantum mechanics.

Although Brillouin and Keller do not mention it, their theory cannot be applied to the classical systems of Einstein's Case 1b, for the same reason that Einstein's quantization rule cannot be applied to that case: the classical trajectories defined by the action function S_k cannot fill the $(2N-1)$-dimensional energy shell in phase space.

Maslov[10] propounded a general mathematical theory of semiclassical quantization, putting the work of Brillouin and Keller on a firm foundation. The integer α_k is named the Maslov index.

II. CLASSICAL MECHANICS FOR ONE DEGREE OF FREEDOM

A. The Oscillator

The classical mechanics of bounded motion is presented in a form suitable for semiclassical quantization. A phase-space picture is used. The theory is based on the properties of those action functions that are solutions of the time-independent Hamilton–Jacobi equation, supplemented by the *Maslov index function*,[11] which is a topological property of the motion, and accounts for the behavior of the system at turning points. These lead to the Maslov index.[10]

A "representation" terminology is used; it is analogous to that of quantum mechanics.

This section deals with closed trajectories of bounded systems with one degree of freedom. The Maslov index function and Maslov index are introduced by analogy with the properties of action functions and canonical action integrals. This theory is elementary, but makes a useful introduction to the more difficult general theory of bounded systems with many degrees of freedom.

Consider a one-dimensional oscillator with Hamiltonian $H(q,p)$. To fix ideas we can suppose that $H(q,p)$ has the form

$$H(q,p) = V(q) + T(p) = V(q) + \frac{1}{2m} p^2 \qquad (2.1)$$

but this is not necessary; we require only that the value of q oscillate between two values. The motion may be represented by a trajectory in phase space such as that illustrated in Fig. 2.1. The sense in which the trajectory is described is shown by the arrows.

The equations of motion in phase space are Hamilton's equations

$$\dot{q} = \frac{\partial H}{\partial p}, \qquad \dot{p} = -\frac{\partial H}{\partial q} \qquad (2.2)$$

They define the coordinate q and momentum p as functions of time and implicitly define p as a two-valued function of q. The two values lie on

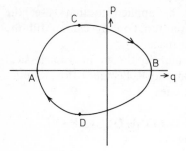

Fig. 2.1.   Phase-space diagram of an oscillator.

different "sheets" analogous to the Riemann sheets of complex variable theory.[1] The upper and lower parts of the trajectory in phase space correspond to the two sheets. They are joined at the turning points A and B, which we call the q-turning points. At these points p is a singular function of q and dp/dq becomes infinite.

The points $\mathbf{X} = (q,p)$ on the trajectory, like the momentum, are two-valued functions of q. However, we may also consider the points $\mathbf{X}$ as functions of p, which are also two-valued. The corresponding sheets are joined at the "p-turning" points C and D where q is a singular function of p and dq/dp becomes infinite. The coordinate q is a well-behaved function of p at the q-turning points A and B, whereas the momentum p is a well-behaved function of q at the p-turning points C and D.

When q is treated as the independent variable, we refer to "q-representation" and when p is the independent variable we refer to "p-representation," for which p is the coordinate and $-q$ is the conjugate momentum.

In q-representation the trajectory may be considered as the graph of p as a two-valued function of q, with ACB and ADB the two sheets of this function. In p-representation the trajectory is the graph of q as a function of p, with two sheets CAD and CBD.

B.   Action Function for the Oscillator

In q-representation the time-independent Hamilton–Jacobi (HJ) equation for the action function $S(q)$ is

$$H\left(q, \frac{dS}{dq}\right) - E = 0 \qquad (2.3)$$

where E is the energy. In the terminology of the textbooks,[12] $S(q)$ is the characteristic function.

Equation 2.3 is the classical analog of the quantal time-independent Schrödinger equation in q-representation for the stationary states of a bound system. In quantum mechanics much is made of the necessity to impose the correct boundary conditions for the wave function: without them one cannot obtain the correct discrete spectrum for the energy E. By contrast, there is usually less discussion of the boundary conditions for the action function, which is the solution of the HJ equation; we need them because they are necessary for the classical basis of semiclassical wave functions.

For a given energy E, a solution $S(q)$ of the HJ equation defines part of a trajectory in phase space, made up of points $(q, dS/dq)$. However, the equation can only be used to define the trajectory on one q-sheet, between the turning points A and B. At these points $S(q)$ is singular and the HJ equation cannot be used.

Clearly, the trajectory has a natural continuation into the second sheet. Normally $S(q)$ is continued by supposing it to be continuous at A and B, and sometimes its square-root singularity at the turning points is simply ignored. We use a different method based on the canonical (contact) transformation to momentum representation. This has the same result as the normal method for action functions, but is capable of generalization to other functions that are *not* continuous on passing from one sheet to another.

By the theory of canonical transformations the action function $\bar{S}(p)$ in momentum representation is defined as

$$\bar{S}(p) = S(q) - pq \tag{2.4}$$

It is the classical analog on the q-sheet ACB of the unitary transformation from q- to p-representation in quantum theory. It is a well-behaved function of p everywhere on this sheet except the point C, which is a p-turning point and where it is singular.

The momentum function $\bar{S}(p)$ satisfies the HJ equation

$$H\left(-\frac{d\bar{S}}{dp}, p\right) - E = 0 \tag{2.5}$$

everywhere on ACB except the point C. However, this equation may be used to continue $\bar{S}(p)$ into the complete p sheet CBD, past the q-turning point at B, and the relation (2.4) can then be used to define $S(q)$ in the segment BD of the second q-sheet. Continuation around the trajectory may be completed by successive transformations between q- and p-representation.

It is sometimes useful to distinguish the values of the function $S(q)$ between the two different q-sheets, by using a subscript r so that $S_r(q)$ has a unique value, $S_1(q)$ being the value on sheet 1: ACB and $S_2(q)$ being the value on sheet 2: BDA. The integer $\bar{r}$ is used to label the p-sheets.

A notation that is often more convenient uses the points $\mathbf{X}$ in phase space as the independent variable, it being understood that the action functions are defined only for phase points on the trajectory.

Let $\mathbf{X} = (q,p)$ be a point on the phase-space trajectory and let

$$S_q(\mathbf{X}) = S_r'(q), \qquad S_p(\mathbf{X}) = \bar{S}_{\bar{r}}(p) \tag{2.6}$$

In this notation the suffices q and p on S are not variables but labels; they denote the classical representation for the action function. In the function $S_q(\mathbf{X})$, with $\mathbf{X} \equiv (q,p)$, the coordinate q acts as the continuously varying argument and the momentum p serves to distinguish the sheet. There is only one value of $\mathbf{X}$ for each point on the trajectory, so that $S_q(\mathbf{X})$ and $S_p(\mathbf{X})$ are single-valued functions, provided $\mathbf{X}$ is not allowed to pass around the trajectory.

However, if $\mathbf{X}$ is allowed to pass once or more around the trajectory, $S_q(\mathbf{X})$ and $S_p(\mathbf{X})$ are seen to be multivalued, but this is a different kind of multivalued function than that considered previously. One passage in a negative sense results in an increment $[S]$ in the action function. We follow Landau and Lifshitz[13] in defining I to be the action integral or canonical action:

$$I = \frac{1}{2\pi} [S] = \frac{1}{2\pi} \oint p \, dq \tag{2.7}$$

In this notation the conjugate canonical angle variables are often angles and range over an interval of length 2π. The canonical action may be used in place of the energy to label the trajectories of the oscillator.

C. Maslov Index Function and Maslov Index

The index function keeps a record of the singularities in the action functions $S_q(\mathbf{X})$ and $S_p(\mathbf{X})$ that occur in passing from one sheet to another. The derivative dp/dq changes sign by passing through infinity whenever the q-sheet changes and by passing through zero whenever the p-sheet changes. The signature of dp/dq is the function

$$\text{Sgn}\left(\frac{dp}{dq}\right) = \begin{cases} +1 & \dfrac{dp}{dq} > 0 \\[2mm] -1 & \dfrac{dp}{dq} < 0 \end{cases} \tag{2.8}$$

and will be left undefined when dp/dq is zero. This signature is defined as a function of $\mathbf{X}$ everywhere on the trajectory except for q-turning points and p-turning points.

The signature of dp/dq is used to define the *Maslov index function*,[11] an integer function of X which, like the action function, has different forms in different representations. Let $\sigma_q(X)$ and $\sigma_p(X)$ be the respective index functions in q- and p-representation. As for the action function, the relation between q- and p-representations allows $\sigma_q(X)$ and $\sigma_p(X)$ to be continued from one sheet to another. They are defined up to an arbitrary additive integer constant by the properties

$$\sigma_q(X) \text{ is an integer constant on any } q\text{-sheet} \qquad (2.9a)$$

$$\sigma_p(X) \text{ is an integer constant on any } p\text{-sheet} \qquad (2.9b)$$

$$\sigma_p(X) = \sigma_q(X) - \mathrm{Sgn}\left(\frac{dp}{dq}\right) \qquad (2.9c)$$

where in (2.9c) the relation holds only at a point X on the trajectory that is neither a q-turning point nor a p-turning point.

The additive integer constant may be fixed by choosing a point X^0 on the trajectory for which

$$\sigma_q(X^0) = 0 \qquad (2.10)$$

Normally this point is chosen so that $S_q(X^0) = 0$ also.

In the example of the oscillator whose phase-space trajectory is illustrated in Fig. 2.1, let C be the point X^0. Then by (2.9a) $\sigma_q = 0$ throughout the q-sheet ACB. In the segment CB of the trajectory, p is a decreasing function of q, so by (2.9c), $\sigma_p = 1$ on CB. But by (2.9b) this must be the value of σ_p throughout the p-sheet CBD. In the segment BD of the trajectory, p is an increasing function of q, so by (2.9c), $\sigma_q = 2$ on BD, and thus in the entire q-sheet BDA. Thus $\sigma_q(X)$ is a discontinuous function of X at the q-turning points, with a discontinuity of magnitude 2. Continuing further around the trajectory in a negative sense, the direction of increasing time, the p-sheet DAB may be used to continue σ_q again into the original q-sheet ACB, where it is now found to have the value 4. Thus $\sigma_q(X)$ is multivalued integer function that increases by 4 on passing once around the trajectory in the direction of increasing time. The values of σ_q and σ_p on the segments of trajectory separated by turning points are shown in Table I.

TABLE I
Index Functions σ_q and σ_p for the Oscillator of Fig. 2.1

Segment	A	C	B	D	A	C
σ_q		0	0	2	2	4
σ_p		−1	1	1	3	3

The constant $[\sigma]$ denotes the change in $\sigma_q(\mathbf{X})$ on passing once around the trajectory. Clearly, from the definitions, the change in $\sigma_p(\mathbf{X})$ is the same. To avoid ambiguity the initial and final point is always chosen not to be a turning point. If $\mathbf{X}$ returns to its original value after passing around the trajectory in either direction any number of times, then the initial and final values of $\sigma_q(\mathbf{X})$ and $\sigma_p(\mathbf{X})$ will differ by an integer multiple of $[\sigma]$.

By analogy with the canonical action I, we define the canonical Maslov index α in terms of σ by the relation

$$\alpha = \frac{[\sigma]}{2} \tag{2.11}$$

For the oscillator of Fig. 2.1, $[\sigma]=4$ and $\alpha=2$.

There is a general theory of canonical transformations for the action function S, whereby any nontrivial function of q and p can be used as the coordinate of a new representation. We do not consider the general theory of the index function σ, although it does lead to some interesting topological problems.

D. Vibration and Rotation

In Figure 3.1 of the next section is illustrated the phase-space trajectory of a particle moving between two potential wells with a low hill between them. There are two q-turning points and six p-turning points, two of them convex outward from the region contained by the curve. The action and index functions may be continued across turning points by the same method as for the simpler oscillator of the previous section.

The change σ_q in σ_p is $[\sigma]=4$ and the Maslov index is $\alpha=2$, as for the simple oscillator. It is clear that the minimum of $p(q)$ near A and the maximum near B may be removed by continuous deformation without changing the value of the Maslov index. Any closed trajectory in phase space that is described in the negative sense and that can be continuously deformed to an arbitrarily small circle around a point has a Maslov index of two. In the continuous transformations the trajectory is not allowed to intersect itself.

This result is quite independent of the form of the Hamiltonian and indeed of the existence of a Hamiltonian. The Maslov index is a topological property of the trajectory.

In the above cases the motion is called a *vibration*, oscillation or libration.

However, suppose q is a coordinate that represents the same physical state of the system when $q=q^1$ and $q=q^1+l$, where l is the smallest positive quantity with this property. Typically, q is an angle and $l = 2\pi$, as in the case of a rotation about an axis. Then Fig. 3.2 (below) represents a possible phase-space trajectory of the motion, which is a rotation. The trajectory $XYZX$ has no q-turning points and two p-turning points. If we assume that A is $\mathbf{X}^0$, then the index functions σ_q and σ_p are obtained using relations (2.9) and shown in Table II. Clearly in this case $[\sigma] = 0$ and the Maslov index α is also zero. For any rotation the canonical Maslov index is zero. The phase space is a cylinder and the trajectory is wrapped once around it.

TABLE II
Index Functions for the One-Dimensional Rotor of Fig. 3.2 (Below)

Segment	A	B	C	A	B
σ_q	0	0	0	0	
σ_p	-1	1	-1	-1	

The trajectory cannot be continuously deformed into a small circle around a point. The way in which the trajectory is embedded into the phase space is topologically different for vibration and rotation.

In each case the Maslov index α is integral. This is because at each q-turning point σ_q changes by ± 2 and the number of turning points must be even, so that $[\sigma]$ is divisible by 4. For one degree of freedom, rotation cannot occur unless the state of the system is periodic in q. However, vibration can occur whether or not the state is periodic in q. There are systems, such as the plane vertical pendulum in a gravitational field, and hindered rotation in molecules, that exhibit both vibration and rotation. These topologically distinct forms of motion are divided by a singular motion that is not periodic. In the case of the pendulum this is the motion in which the bob asymptotically approaches a point vertically above the point of suspension. Such singular motions have probability zero and are ignored here. The vertical pendulum is considered by Born.[14]

III. SEMICLASSICAL MECHANICS FOR ONE DEGREE OF FREEDOM

We now show how the Maslov index function and canonical Maslov index appear naturally in the semiclassical theory, which is introduced at first without reference to a specific Hamiltonian or Schrödinger equation.

We define a truncated semiclassical wave function to be a wave function that is nonzero over a limited range $\mathcal{R}$ of q-space and that is a function of coordinate q and Planck's constant $\hbar$ of the form

$$\psi(q, \mathcal{R}) = B(q)\exp i\left[\frac{S(q)}{\hbar} - \frac{\sigma_q \pi}{4}\right] \quad (q \text{ in } \mathcal{R})$$

$$\psi(q, \mathcal{R}) = 0 \qquad\qquad\qquad\qquad \text{otherwise} \tag{3.1}$$

where $B(q)$ and $S(q)$ are analytic functions of q in $\mathcal{R}$ and $B(q)$ is not everywhere zero. The σ_q is an integer constant. The real function $S(q)$ has the dimensions of action and is called an action function. The integer σ_q is essential to the theory of phase increments at caustics.

The wave function $\psi(q, \mathcal{R})$ may be transformed to momentum representation by using a Fourier transform, but unlike the quantal transformation the semiclassical transformation is *defined* as the stationary phase value of the Fourier integral. This is the basis of semiclassical transformation theory.

We first restrict ourselves to those regions $\mathcal{R}$ for which

$$\frac{dp}{dq} = S''(q) \neq 0 \qquad (\text{all } q \text{ in } \mathcal{R}) \tag{3.2}$$

where $p(q)$ is the classical momentum function defined by this $S(q)$. Then p is an increasing or decreasing function of q; p and q are single-valued functions of one another. Let $\overline{\mathcal{R}}$ be the range in p corresponding to the range $\mathcal{R}$ in q.

Then the truncated semiclassical wave function in p-representation is

$$\overline{\psi}\left(p, \overline{\mathcal{R}}\right) \approx \frac{1}{(2\pi\hbar)^{1/2}} \int_{\mathcal{R}} dq\, B(q)\exp i\left[\frac{S(q)}{\hbar} - \frac{\sigma_q \pi}{4} - qp\right] \tag{3.3}$$

$$(\text{stationary})$$

where the stationary phase integral is obtained. For each value of p in $\overline{\mathcal{R}}$ there is only one value of q in $\mathcal{R}$ for which the exponent is stationary, the value given by the corresponding classical value of q that satisfies the relation

$$p = \frac{dS(q)}{dq} \tag{3.4}$$

There is no stationary value of q for any p outside $\overline{\mathcal{R}}$. Therefore, using the stationary phase integral (A1) (see Appendix) we have

$$\bar{\psi}\left(p,\overline{\mathfrak{R}}\right)=\bar{B}\left(p\right)\exp i\left[\frac{\bar{S}\left(p\right)}{\hbar}-\frac{\bar{\sigma}_p\pi}{4}\right] \quad \left(p\ \text{in}\ \overline{\mathfrak{R}}\right)$$

$$\bar{\psi}\left(p,\overline{\mathfrak{R}}\right)=0 \hspace{4cm} \text{otherwise,}$$

(3.5)

where

$$\bar{S}\left(p\right)=S\left(q\right)-qp$$

$$\bar{B}\left(p\right)=\frac{B\left(q\right)}{\left|dp/dq\right|^{1/2}}$$

(3.6)

$$\sigma_p=\sigma_q-\text{Sgn}\left(\frac{dp}{dq}\right)=\sigma_q-\text{Sgn}\left[S''\left(q\right)\right]$$

where all functions are defined for p in $\overline{\mathfrak{R}}$ and q in $\mathfrak{R}$ only.

The transformation of truncated semiclassical wave functions depends only on the transformation of the action and Maslov index functions of classical mechanics. It is convenient to consider the wave functions as functions of the phase points $\mathbf{X}$ on the classical curve defined by the action function $S(q)$. To distinguish the position and momentum wave functions we write

$$\psi_q(\mathbf{X})=\psi(q), \qquad \psi_p(\mathbf{X})=\bar{\psi}(p)$$

(3.7)

These are only defined on the segment of the curve for which q is in $\mathfrak{R}$ and p is in $\overline{\mathfrak{R}}$.

A. Wave Functions in Phase Space

The truncation of the wave function is artificial, but there is a natural continuation. For a bounded system there are natural boundaries for $\psi_q(\mathbf{X})$; in Fig. 2.1 these are at A and B. Similarly, there are natural boundaries for $\psi_p(\mathbf{X})$, but at different points on the curve. The conjugate representation is used to continue the semiclassical wave function through its natural boundary. We are able to define a semiclassical wave function $\psi_q(\mathbf{X})$ on a curve in phase space that has natural boundaries in q-representation by going over to p-representation. For such a case the curve will correspond to more than one sheet r.

A semiclassical phase-space wave function on the curve $\mathcal{C}$ is defined by the conditions

PW1. For each part of $\mathcal{C}$ that is entirely in a q-sheet, $\psi_q(\mathbf{X})$ has the form (3.1), where q is the position coordinate of $\mathbf{X}$.

PW2. For each part of $\mathcal{C}$ that is entirely in a p-sheet, $\psi_p(\mathbf{X})$ has the form (3.5), where p is the momentum coordinate of $\mathbf{X}$.

PW3. For each part of $\mathcal{C}$ that satisfies both conditions above, $\psi_q(\mathbf{X})$ and $\psi_p(\mathbf{X})$ are related by (3.6).

It is particularly important for quantization conditions that at a q-turning point the action function $S_q(\mathbf{X})$ be continuous, but the index function $\sigma_q(\mathbf{X})$ have an increment of ± 2.

We may think of a semiclassical wave propagating along the curve $\mathcal{C}$. Where $\mathcal{C}$ has a q-turning point the wave has a q-caustic; where it has a p-turning point the wave has a p-caustic.

The most important curves are those without end points. If they tend to infinity, they correspond to scattering processes, which we do not consider here. If they are closed they correspond to bound states. Wave functions on closed curves are required to satisfy the additional condition:

PW4. The wave functions $\psi_q(\mathbf{X})$ and $\psi_p(\mathbf{X})$ are single valued.

Coordinate and momentum wave functions are naturally defined in terms of phase-space wave functions as sums over the sheets. If r labels the q-sheets and $\mathbf{X}_r$ are the phase points on the curve $\mathcal{C}$ corresponding to the point q, then

$$\psi(q) = \sum_r \psi_q(\mathbf{X}_r) \tag{3.8}$$

For those values of q that are not reached by the trajectory $\psi(q)$ is zero. Similarly, if $\bar{r}$ labels the p-sheets, then

$$\bar{\psi}(p) = \sum_{\bar{r}} \psi_p(\mathbf{X}_{\bar{r}}) \tag{3.9}$$

For the second case of vibration, considered in Section II, the relation between the wave functions is illustrated in Fig. 3.1. In that example the value of the q-wave function at q_1 is

$$\psi(q_1) = \psi_q(A) + \psi_q(B) \tag{3.10}$$

and the value of the p-wave function at p_2 is

$$\bar{\psi}(p_2) = \psi_p(C) + \psi_p(D) + \psi_p(E) + \psi_p(F) \tag{3.11}$$

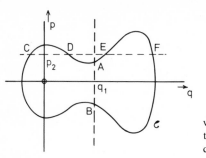

Fig. 3.1 Trajectory and wave functions for vibration across two wells with a hill between them. A, B, C, D, E, F label phase points on the curve $\mathcal{C}$.

Now consider the case in which q is a variable that is naturally periodic of period l, for example an angle, and the curve corresponds to a rotation. This is illustrated in Fig. 3.2. There is only one q-sheet and normally an even number of p-sheets. The wave functions at q_1 and p_2 are

$$\psi(q_1) = \psi_q(A)$$

$$\bar{\psi}(p_2) = \psi_p(B) + \psi_p(C) \tag{3.12}$$

Where there is more than one q-sheet the number of terms $\psi_q(\mathbf{X}_r)$ that contribute to $\psi(q)$ is greater than 1, and there is interference between the contributions from the different sheets. Thus for the trajectory of Fig. 3.1 two progressive waves interfere to produce the wave function $\psi(q)$ at q_1; similarly, four progressive waves contribute to $\bar{\psi}(p)$ at p_2.

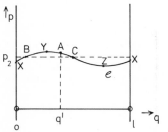

Fig. 3.2. Trajectory for rotation with a small potential.

B. Quantization Conditions

The quantization conditions follow from PW1 to PW4. In passing once around the curve neither S nor σ is single valued, but condition PW4 requires $\psi_q(\mathbf{X})$ and $\psi_p(\mathbf{X})$ to be single valued. For consistency the increment in the exponent of (3.1) must be an integer multiple of $2\pi i$, that is, in

q-representation

$$\frac{[S_q]}{\hbar} - \frac{[\sigma_q]\pi}{4} = 2\pi \tag{3.13}$$

Using the definition of action integral I and Maslov index α,

$$I = \frac{[S_q(\mathbf{X})]}{2\pi}, \qquad \alpha = \frac{[\sigma_q]}{2} \tag{3.14}$$

we obtain

$$I = \left(n + \frac{\alpha}{4}\right)\hbar \qquad (n = 0, 1, 2, \ldots) \tag{3.15}$$

For vibration $\alpha = 2$ and for rotation $\alpha = 0$, giving the correct quantization conditions for motion in one dimension.

C. Hamiltonian Theory and the Wave Equation

So far the semiclassical theory has had no need of a Hamiltonian function and a wave equation. The connection with them is made in this section.

Suppose the Hamiltonian function has the form

$$H(q,p) = V(q) + T(p) \tag{3.16}$$

The wave equations for the exact wave functions in coordinate and momentum space have the equivalent forms

$$\left[V(q) + T\left(-i\hbar\frac{d}{dq}\right) - E\right]\psi(q) = 0 \tag{3.17}$$

$$\left[V\left(i\hbar\frac{d}{dp}\right) + T(p) - E\right]\bar{\psi}(p) = 0 \tag{3.18}$$

which are related by Fourier transformation.

We use both forms to specify a semiclassical mechanics, thus avoiding some of the problems at caustics.

A semiclassical solution of the wave equation is defined to be a $\psi(q)$ and a $\bar{\psi}(p)$ with the following properties

SC1. $\psi(q)$ has the form

$$\psi(q) = \sum_r B_r(q)\exp i\left[\frac{S_r(q)}{\hbar} - \frac{\sigma_{qr}\pi}{4}\right] \tag{3.19}$$

and satisfies (3.17) to zeroth order in $\hbar$ except at q-caustics. σ_{qr} is a constant integer for each r.

SC2. $\bar{\psi}(p)$ has the form

$$\bar{\psi}(p) = \sum_{\bar{r}} \bar{B}_{\bar{r}}(p) \exp i \left[\frac{\bar{S}_{\bar{r}}(p)}{\hbar} - \frac{\sigma_{p\bar{r}}\pi}{4} \right] \qquad (3.20)$$

and satisfies (3.18) to zeroth order in $\hbar$ except at p-caustics. $\sigma_{p\bar{r}}$ is a constant integer for each $\bar{r}$.

SC3. $\psi(q)$ and $\psi(p)$ are related by the usual Fourier transforms to zeroth order in $\hbar$, except at caustics.

The indices σ_q and σ_p are integers. By substitution $S(q)$ and $\bar{S}(p)$ are solutions of the Hamilton-Jacobi equations (2.3) and (2.5). Using the stationary phase integral (A1) of the Appendix, S and $\bar{S}$ are related by the classical transformation relation (2.4) whereas σ_q and σ_p are related by the transformation relation (2.9c).

The functions S and $\bar{S}$, σ_q and σ_p may be considered as before as functions of points on a trajectory in phase space. However, a truncated wave function cannot satisfy the conditions SC1 to SC3, so the trajectory must be closed.

Therefore if the classical action and index functions S, $\bar{S}$, σ_q, σ_p satisfy the Hamiltonian–Jacobi equations for a classical system and also the transformation relations PW1 to PW3, the corresponding ψ and $\bar{\psi}$ given by (3.19) and (3.20) are semiclassical solutions of the wave equations (3.17) and (3.18) for that system and the semiclassical quantization condition is given by (3.15).

To every closed trajectory that satisfies Hamilton's equations and the quantization conditions there corresponds a semiclassical wave function.

IV. CLASSICAL MECHANICS FOR MANY DEGREES OF FREEDOM; POLYATOMIC MOLECULES AND INVARIANT TOROIDS

To fix ideas we consider the properties of polyatomic and particularly triatomic molecules. We make the following simplifying assumptions:

SA1. There are no significant deviations from the Born-Oppenheimer approximation.

SA2. There is a single potential-energy surface V belonging to a single nondegenerate electronic state.

SA3. V has only one minimum and no other stationary points.

SA4. All effects of rotation can be neglected; the angular momentum is zero.

None of these assumptions is absolutely necessary for the application of classical and semiclassical methods to molecular problems, but they avoid complications that divert attention from the central features of the theory.

Consider the classical vibration of a diatomic molecule. For coordinate q of relative motion of the nuclei and conjugate momentum p the Hamiltonian is $H(q,p)$. For a given energy E between the equilibrium and dissociation energies, the equation

$$H(q,p) = E \qquad (4.1)$$

defines p as a two-valued function of q, which may be represented as a graph in the two-dimensional phase space of points $\mathbf{X} = (q,p)$. This graph is the phase-space trajectory and occupies the entire one-dimensional energy shell of points in phase space that satisfy (4.1).

For a polyatomic molecule of N degrees of freedom, the classical motion near equilibrium is close to that of N independent harmonic oscillators in normal coordinates q_k with conjugate momenta p_k and characteristic angular frequencies

$$\omega_k = \frac{2\pi}{T_k} \qquad (4.2)$$

where T_k is a characteristic period.

We temporarily neglect all anharmonic coupling and consider only two degrees of freedom, such as the stretch modes q_1, q_2 of a linear triatomic molecule. The phase space of points

$$\mathbf{X} = (\mathbf{q}, \mathbf{p}) = (q_1, q_2, p_1, p_2) \qquad (4.3)$$

is four dimensional. For energy E the energy shell defined by the equation

$$H(\mathbf{q}, \mathbf{p}) = E \qquad (4.4)$$

is three dimensional. Energy is conserved for each mode separately, so the trajectory in phase space is confined to a two-dimensional region Σ. This region is invariant in the sense that if the molecule starts with its coordinates and momenta (q,p) in Σ, then they remain in Σ for all time. The shape of Σ resembles a doughnut or torus (Fig. 4.1); it is called an invariant toroid, whatever the number N of degrees of freedom. Table III summarizes the dimensionality of various regions for 2 and for N degrees of freedom.

As in the case of one degree of freedom, the points $\mathbf{X}$ on the toroid are many-valued functions of $\mathbf{q}$ and of $\mathbf{p}$. For two or more degrees of freedom the phase space cannot be illustrated clearly on paper, but the $\mathbf{q}$-space can. As we might expect, the two-dimensional problem is complicated.

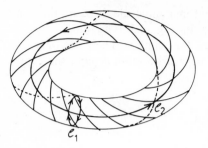

Fig. 4.1. Phase-space trajectory and invariant toroid for two degrees of freedom. $\mathcal{C}_1$ and $\mathcal{C}_2$ are curves for the definition of action integrals I_1 and I_2. The toroidal helix is the trajectory; normally it is not closed.

TABLE III
Dimensions for 1, 2, and N Degrees of Freedom

Degrees of freedom	1	2	N
Phase space	2	4	$2N$
Energy shell	1	3	$2N-1$
Invariant toroid	1	2	N
Trajectory	1	1	1

For the stretch modes of a triatomic molecule there are "box" toroids, as illustrated in Fig. 4.2 following Noid and Marcus.[15] If the toroid is thought of as an inner tube of a bicycle tire, and the "ground" is the Cartesian coordinate space of the normal modes q_1, q_2, then the box toroid is obtained by placing the tire vertically, with its axis horizontal, and then squashing it flat from above. There are clearly four q-sheets, corresponding to the four possible signs of p_1 and p_2 at a given q_1 and q_2. The turning edges overlap in pairs.

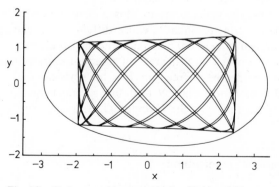

Fig. 4.2. Trajectory of box toroid, after Noid and Marcus.

For the rotation and vibration of a diatomic molecule there are the quite different "tube" toroids, as illustrated in Fig. 4.3. The tube toroid is obtained by laying it on the ground and flattening it. There are only two q-sheets, corresponding to inward and outward motion around the center of symmetry.

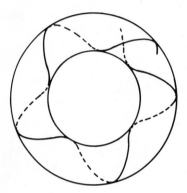

Fig. 4.3.   Tube toroid.

Normally, a trajectory in an invariant toroid approaches arbitrarily close to any point of that toroid. This is a definition of Einstein's Case 1 in terms of invariant toroids. Exceptionally, there is a Fermi resonance, which occurs when a nontrivial linear integer combination of the angular frequencies is zero,

$$\sum_{k=1}^{N} \nu_k \omega_k = 0 \qquad (\nu_k \quad \text{integers, not all zero}) \qquad (4.5)$$

In that case the trajectories are confined to toroids of lower dimension than N. For $N=2$ the trajectories are closed. This is Einstein's special Case 2. Approximate Fermi resonances cause difficulties in classical perturbation theory, known in celestial mechanics as the problem of small divisors.

An invariant toroid for N degrees of freedom corresponds to a closed trajectory for one degree of freedom. It is the classical analog of the wave function and is used in Section V to define a semiclassical wave function.

Let $S(\mathbf{q})$ be a solution of the Hamilton-Jacobi equation

$$H(\mathbf{q}, \nabla_q S) - E = 0 \qquad (4.6)$$

The N-dimensional momentum

$$\mathbf{p} = \nabla_q S(q) \qquad (4.7)$$

may be considered as a function of $\mathbf{q}$, defining an N-dimensional region of phase-space points $\mathbf{X} = (\mathbf{q}, \mathbf{p})$ related by (4.7). From Hamilton–Jacobi theory this region is made up of parts of classical trajectories.

As in the case of one degree of freedom, the solution of the Hamilton-Jacobi equation may be taken up to the q-turning-edge where $\mathbf{p}$ is a singular function of $\mathbf{q}$ and the ration of the volume elements in q- and p-space defined by

$$\frac{d\tau_p}{d\tau_q} = \mathrm{Det}\left\| \frac{\partial p_j}{\partial q_k} \right\| \tag{4.8}$$

becomes infinite. A q-sheet is such a region continued up to its turning edge by using the Hamilton–Jacobi equation.

By the theory of canonical transformations, the action function $\bar{S}(\mathbf{p})$ in momentum representation is defined to be

$$\bar{S}(\mathbf{p}) = S(\mathbf{q}) - \mathbf{p} \cdot \mathbf{q} \tag{4.9}$$

and it satisfies the Hamilton–Jacobi equation

$$H\left(-\nabla_p \bar{S}, \mathbf{p} \right) - E = 0 \tag{4.10}$$

The function $S(\mathbf{q})$ has a natural continuation obtained by transforming to momentum representation as in the case $N = 1$. Normally, the momentum action function is well-behaved at the q-turning edge. A p-sheet may be continued everywhere that $d\tau_q/d\tau_p$ is nonsingular.

We introduce $S_q(\mathbf{X})$, $S_p(\mathbf{X})$ to denote the action functions in the conjugate representations, in order to simplify the description of multivalued action functions under canonical transformation.

As for the case of one degree of freedom, the phase-space points $\mathbf{X}(\mathbf{q}, \mathbf{p}) = \mathbf{X}(\mathbf{q}, \nabla_q S(q))$ and $\mathbf{X}(\mathbf{q}, \mathbf{p}) = (-\nabla_p \bar{S}(p), \mathbf{p})$ lie in a smooth region of N dimensions, containing parts of classical trajectories. Unlike that case there is no guarantee for nonseparable systems that repeated continuation at caustics will form a closed N-dimensional region with no edges even for bound systems, nor to the author's knowledge that any closed region must have the form of a toroid. If it does, then, because the HJ equation is equivalent to Hamilton's equation, the toroid will consist of classical phase-space trajectories, and will be an invariant toroid.

In this section we make a further assumption:

SA5. Repeated continuation results in invariant toroids.

Unlike the other assumptions, this *is* essential to the remainder of the theory. The problem is considered in detail in Section 7; it is the problem of regular and irregular spectra.

On each q-sheet of the toroid we define a Maslov index function σ_q that is a constant integer on that sheet, and relate it, where it is meaningful, to the index function σ_p by the relation

$$\sigma_p(\mathbf{X}) = \sigma_q(\mathbf{X}) - \mathrm{Sgn}\left(\frac{\partial p_l}{\partial q_m}\right) \tag{4.11}$$

N independent closed curves $\mathcal{C}_k$ may be defined on an invariant toroid. For $N = 2$ they could be curves that pass around the toroid once the "long way" or the "short way" as illustrated in Fig. 4.1.

The action function $S_r(\mathbf{q})$ is single-valued on any one sheet; therefore, the variation of $S_r(\mathbf{q})$ around a curve $\mathcal{C}_0$ that lies entirely on that sheet is zero,

$$\left[S_r(\mathbf{q}) \right]_{\mathcal{C}_0} = \int_{\mathcal{C}_0} \mathbf{p} \cdot d\mathbf{q} = 0 \tag{4.12}$$

It follows by "adding contours" as in complex variable theory that any continuous deformation of $\mathcal{C}_k$ on the toroid leaves the value of the action integral

$$I_k = \frac{1}{2\pi} \left[S(q) \right]_{\mathcal{C}_k} = \frac{1}{2\pi} \int_{\mathcal{C}_k} \mathbf{p} \cdot d\mathbf{q} \tag{4.13}$$

unchanged. This invariance of the I_k under deformation of the $\mathcal{C}_k$ was noted by Einstein. It is also clear that the Maslov index

$$\alpha_k = \tfrac{1}{2} \left[\sigma_q \right]_{\mathcal{C}_k} \tag{4.14}$$

is similarly invariant.

The invariance of I_k under a continuous canonical transformation follows from the theory of canonical invariants.[12]

The Maslov index is also invariant, because it is an integer and can only change continuously. The only possibility is that it should stay constant. Such continuous transformations can be valuable in dealing with singular situations, such as the rotational symmetry of the diatomic molecule.

On each toroid the energy is constant, so the toroids therefore define a function

$$E(I_1, I_2, \ldots, I_N) = E(\mathbf{I}) \tag{4.15}$$

The Maslov indices and this energy function are all that is needed to determine the regular semiclassical energy spectrum.

V. SEMICLASSICAL MECHANICS FOR MANY DEGREES OF FREEDOM

Semiclassical mechanics for many degrees of freedom follows very closely the theory for one degree of freedom, the differences being due to the differences in structure of the classical solutions. The phase-space trajectories of the one-dimensional theory are replaced by invariant toroids of N dimensions in the $2N$-dimensional phase space. The truncated wave functions $\psi(\mathbf{q}, \mathcal{R})$ are nonzero in regions $\mathcal{R}$ of dimension N in q-space. Similarly for the corresponding wave function $\bar{\psi}(\mathbf{p}, \mathcal{R})$, which is related to $\psi(\mathbf{q}, \mathcal{R})$ by

$$\bar{\psi}(\mathbf{p}, \mathcal{R}) \approx \frac{1}{(2\pi\hbar)^{N/2}} \int_{\mathcal{R}} d\tau_q \psi(q, \mathcal{R}) \exp\left(-\frac{i}{\hbar}\mathbf{q}\cdot\mathbf{p}\right) \qquad \text{(stationary)} \quad (5.1)$$

provided that

$$\frac{d\tau_p}{d\tau_q} = \text{Det}\left\|\frac{\partial^2 S(\mathbf{q})}{\partial q_l \partial q_m}\right\|$$

$$= \text{Det}\left\|\frac{\partial p_l}{\partial q_m}\right\| \neq 0 \qquad \text{(all } \mathbf{q} \text{ in } \mathcal{R}) \qquad (5.2)$$

The relation (A4) of the Appendix is used.

The Maslov index functions are related by

$$\sigma_p = \sigma_q - \text{Sgn}\left(\frac{\partial p_l}{\partial q_m}\right) \qquad (5.3)$$

as in the classical theory.

Wave functions may be defined as functions of points $\mathbf{X} = (\mathbf{q}, \mathbf{p})$ on the invariant toroid and possess a natural continuation, as do the action and index functions of classical mechanics. A semiclassical wave function exists only where there is a closed invariant toroid.

The semiclassical wave function $\psi(\mathbf{q})$ is a sum over the phase-space wave functions defined on each sheet r for which there is an $\mathbf{X}_r$ corresponding to $\mathbf{q}$:

$$\psi(\mathbf{q}) = \sum_r \psi_q(\mathbf{X}_r) \qquad (5.4)$$

and similarly for $\bar{\psi}(\mathbf{p})$. Thus for uncoupled oscillators $\psi(\mathbf{q})$ is a sum of four terms, whereas for a central potential in two dimensions it is a sum of two terms. An interference pattern results from the superposition of the various progressive waves.

The quantization conditions are given by the action integrals I_k and Maslov indices α_k for the independent curves $\mathcal{C}_k$ on the toroid. The conditions are

$$I_k = (n_k + \alpha_k/4)\hbar \qquad (k = 1, \ldots, N; \ n_k = 0, 1, 2, \ldots) \qquad (5.5)$$

For N oscillators all $\alpha_k = 2$. For a particle in a three-dimensional central potential, using standard labels for quantum numbers, $\alpha_n = 2$ (or $\alpha_v = 2$) $\alpha_l = 2$, $\alpha_m = 0$.

The energy levels are completely determined by the Maslov indices and the energy function $E(\mathbf{I})$ of the canonical action variables I_k. They are given by

$$E_\mathbf{n} = E\left((\mathbf{n} + \alpha/4)\hbar\right) \qquad (5.6)$$

For a Hamiltonian function of the form

$$H(\mathbf{q}, \mathbf{p}) = V(\mathbf{q}) + T(\mathbf{p}) \qquad (5.7)$$

the wave equations for the exact wave functions $\psi(\mathbf{q})$, $\bar{\psi}(\mathbf{p})$ have the equivalent forms

$$\left[V(\mathbf{q}) + T(-i\hbar\nabla_q) - E \right]\psi(\mathbf{q}) = 0 \qquad (5.8)$$

$$\left[V(i\hbar\nabla_p) + T(\mathbf{p}) - E \right]\bar{\psi}(\mathbf{p}) = 0 \qquad (5.9)$$

which are related by Fourier transformation. A semiclassical solution is defined as for systems with one degree of freedom (Section III, SC1, SC2, SC3, 3.19, 3.20) with vectors $\mathbf{q}$ and $\mathbf{p}$.

By substitution in the Schrödinger equation and using the stationary phase integral (A4) of the Appendix, $S(\mathbf{q})$ and $\bar{S}(\mathbf{p})$ are related by the classical canonical transformation relation (4.9), whereas σ_q and σ_p are related by (4.11). The action functions $\mathbf{S}_q$ and $\bar{\mathbf{S}}_p$ and the index functions σ_q and σ_p may be considered as in the previous section as functions of the phase points $\mathbf{X}$ of an N-dimensional region of phase space, which will satisfy S1 to S3 only if it is closed.

Therefore, the classical action and index functions S_q, S_p, σ_q, σ_p satisfying the Hamilton–Jacobi equations (5.8) and (5.9) and the transformation relations define a semiclassical solution of the wave equation if the action integrals satisfy the quantization condition. It is helpful to note here that

$$\mathbf{p} = \nabla_q S_q(\mathbf{X}), \qquad S_q(\mathbf{X}) = \int_{\mathbf{X}_0}^{\mathbf{X}} \mathbf{p} \cdot d\mathbf{q} \qquad \text{(on the toroid)} \qquad (5.10)$$

so for contour $\mathcal{C}_k$

$$I_k = \frac{1}{2\pi} \left[S_q(\mathbf{X}) \right]_{\mathcal{C}_k} = \oint_{\mathcal{C}_k} \mathbf{p} \cdot d\mathbf{q} \tag{5.11}$$

Conversely, if a trajectory satisfies Hamilton's equation *and* Einstein's condition (1a), and there are N independent contours $\mathcal{C}_k$ in the N-dimensional region of phase space defined by the trajectory, which allow us to define Maslov indices and action integrals satisfying the quantization condition (5.5), then the trajectory defines a semiclassical wave function. These conditions are more severe than for one degree of freedom. In particular, there is no guarantee that Einstein's condition (1a) will be satisfied. This question is more fully discussed in Section VII.

VI. PRACTICAL EBK QUANTIZATION

A. Surfaces of Section

As shown in Section V, the problem of semiclassical quantization for a system of more than one degree of freedom is equivalent to the evaluation of the energy function $E(\mathbf{I})$ of the action variables at a set of discrete values given by the quantization condition

$$I_k = \hbar(n_k + \alpha_k/4) \tag{6.1}$$

The evaluation of the energy function in practice requires the direct or indirect calculation of some representation of the invariant toroids.

This problem is equivalent to the calculation of approximate constants of the motion in celestial mechanics, which has a long history; a standard modern work on the methods used in this field is that of Hagihara.[16] A method that can be used for stars and planets can also be used for molecules, but the field is too large to be covered in this review.

Four methods only are described; they have all been applied to model molecular potentials of more than one degree of freedom.

Eastes and Marcus[17] and Noid and Marcus[15] use stepwise numerical integration of trajectories to obtain invariant toroids and Poincaré's surfaces of section to derive the action integrals and semiclassical energy levels of systems of two degrees of freedom.

As shown in Table III, an invariant toroid for a system of two degrees of freedom lies in a three-dimensional energy shell. A surface of section in an energy shell is defined by a further condition, usually a simple one like $x = 0$ or $y = 0$. An invariant toroid intersects a surface of section in one or more closed curves $\mathcal{C}$. If we follow a trajectory on an invariant toroid, as illustrated in Fig. 4.1, and represent each intersection of the trajectory with a surface of section by a point on that surface, the points normally trace out a curve $\mathcal{C}$, with greater and greater precision as the number of intersections increases. This is the basis of the method of Marcus and his collaborators, who apply the method to a system with Hamiltonian of the form

$$H(x,y,p_x,p_y) = \tfrac{1}{2}\left(p_x^2 + p_y^2 + \omega_x^2 x^2 + \omega_y^2 y^2\right) + \lambda x\left(y^2 + \eta x^2\right) \qquad (6.2)$$

where ω_x, ω_y, λ, and η are constants. Figure 6.1 illustrates the intersections of five trajectories with a surface of section defined by $E = \text{constant}$, $y = 0$. The coordinates x and p_x are conjugate variables and $dy = 0$ on $\mathcal{C}$, so

$$2\pi I = \oint_{\mathcal{C}} \mathbf{p} \cdot d\mathbf{r} = \oint_{\mathcal{C}} p_x \, dx \qquad (6.3)$$

which is the area enclosed by $\mathcal{C}$ on the surface of section. This can be used directly to impose a quantization condition. Similarly, the surface of section defined by $x = 0$ contains independent curves and thus an independent quantization condition.

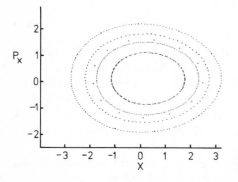

Fig. 6.1. Points on surface of section at $y = 0$ for five trajectories, after Noid and Marcus.

For a separable system the two surfaces of section would simply be the phase planes of the independent motions in the x- and the y-directions, and the curves would be the trajectories or energy shells for these independent systems. Introducing coupling between these systems destroys the separability, but the quantization conditions are the same.

Results obtained by Noid and Marcus are presented in Table IV.

The numerical integration of trajectories can be a lengthy procedure and it is difficult to obtain accurate integrals around curves $\mathcal{C}$; other methods have been used.

TABLE IV
Comparison of Semiclassical and Quantal Energy Levels
After Noid and Marcus[15] and Chapman, Garrett, and Miller[21]

System[a] ω_x^2, ω_y^2 λ, η		Quantum Numbers (n_x, n_y)	Quantum	NM	CGM	Uncoupled $\lambda = 0$
0.29375,	2.12581	0, 0	0.9916	0.9922	0.9920	1.0000
		1, 0	1.5159	1.5164	1.5164	1.5420
−0.1116,	0.08414	2, 0	2.0308	2.0313	2.0313	2.0840
		0, 1	2.4188	2.4198	2.4194	2.4580
0.36,	1.96	0, 0	0.9939	0.9942	0.9941	1.0000
−0.1,	0.1	1, 0	1.5809	1.5812	1.5812	1.6000
		2, 0	2.1612	2.1616	2.1615	2.2000
0.49,	1.69	0, 0	0.9955	0.9954	0.9955	1.0000
		1, 0	1.6870	1.6870	1.6870	1.7000
−0.1,	0.1	0, 1	2.2781	2.2785	2.2782	2.3000
		2, 0	2.3750	2.3751	2.3750	2.4000
		1, 1	2.9583	2.9588	2.9584	3.0000
		0, 2	3.5479	3.5480	3.5480	3.6000
0.81,	1.21	0, 0	0.9980	0.9978	0.9978	1.0000
		1, 0	1.8944	1.8944	1.8941	1.9000
−0.08,	0.1	0, 1	2.0890	2.0889	2.0890	2.1000
		2, 0	2.7899	2.7900	2.7896	2.8000

[a] The four parameters are those of (6.2).

B. Variational Principle

It is well known that in quantum mechanics the variational principle for the energy can be used to derive perturbation expansions for the bound-state wave functions and to obtain expressions for energy levels having errors that are of higher order than the error in the wave function.

An invariant toroid is the classical analog of a bound state of a system and can be used to approximate both wave functions and energy levels. The variational principle for the energy of an invariant toroid is analogous to the variational principle for the mean energy of a quantal state; this and further analogies are presented in Table V, which can be used as a guide to the classical theories.

Trkal[18] proposed a variational principle for action integrals without apparently realizing that the variations did not require to be constrained to the solutions of Hamilton's equations. Van Vleck[18] realized that the variations need not be so constrained and proposed a variational principle that is the same as the one presented here for one degree of freedom. However, in many dimensions time was used instead of the angle variables

TABLE V
Analogies Between Quantal and Classical Mechanics of
Systems with N Degrees of Freedom

	Quantal	Classical
1.	Bound-state wave function $\psi(\mathbf{q})$	Invariant toroid $\mathbf{X}(\boldsymbol{\theta}) \equiv [\mathbf{q}_{\text{tor}}(\boldsymbol{\theta}), \mathbf{p}_{\text{tor}}(\boldsymbol{\theta})]$
2.	Variational principle for mean energy for ψ	Variational principle for mean energy on $\mathbf{X}(\boldsymbol{\theta})$
3.	$\langle \psi \mid H \mid \psi \rangle$	$\int' d\tau_\theta\, H(\mathbf{q}(\boldsymbol{\theta}), \mathbf{p}(\boldsymbol{\theta})) = \langle H \rangle$
4.	Fixed normalization integral $\mathfrak{N} = \langle \psi \mid \psi \rangle$	Fixed action integrals $I_k = \int' d\tau_\theta\, \mathbf{p}(\boldsymbol{\theta}) \cdot \dfrac{\partial q(\boldsymbol{\theta})}{\partial \theta_k}$, $k = 1, 2, \ldots, N$
5.	Energy $E = \dfrac{\langle \psi \mid H \mid \psi \rangle}{\langle \psi \mid \psi \rangle}$	Angular frequencies $\omega_k = \dfrac{\partial \langle H \rangle}{\partial I_k}$, $k = 1, 2, \ldots, N$
6.	Dynamical operator A	Function of coordinates and momenta $A(\mathbf{q}, \mathbf{p})$ (sometimes called a phase function)
7.	Schrödinger equation for bound states $(E - H)\psi = 0$	Angle form of Hamilton's equations $\sum_k \omega_k \dfrac{\partial}{\partial \theta_k} \mathbf{q}(\boldsymbol{\theta}) = \dfrac{\partial H}{\partial \mathbf{p}}(\mathbf{q}(\boldsymbol{\theta}), \mathbf{p}(\boldsymbol{\theta}))$ $\sum_k \omega_k \dfrac{\partial}{\partial \theta_k} \mathbf{p}(\boldsymbol{\theta}) = -\dfrac{\partial H}{\partial \mathbf{q}}(\mathbf{q}(\boldsymbol{\theta}), \mathbf{p}(\boldsymbol{\theta}))$
8.	Quantal perturbation theory for Hamiltonian operator $H = H^0 + b_1 V$	Classical perturbation theory for Hamiltonian function $H(\mathbf{q}, \mathbf{p}) = H^0(\mathbf{q}, \mathbf{p}) + b_1 V(\mathbf{q}, \mathbf{p})$ (normally, V is a function of $\mathbf{q}$ alone)

θ_k, and the variations were implicitly restricted by an inadequate definition of action integral, which is ambiguous for approximate toroids. His principle was, therefore, incomplete. Percival[18] removed these restrictions.

To formulate the variational principle, a normalized integral of a function $f(\boldsymbol{\theta})$ over the entire space of the angle variables is defined by

$$\int' d\tau_\theta f(\boldsymbol{\theta}) = \frac{1}{(2\pi)^N} \int_{-\pi}^{\pi} d\theta_1 \int_{-\pi}^{\pi} d\theta_2, \ldots, \int_{-\pi}^{\pi} d\theta_N f(\theta_1, \theta_2, \ldots, \theta_N) \quad (6.4)$$

When a toroid Σ (which need *not* be invariant) is defined parametrically by a phase-space function

$$\mathbf{X}_\Sigma(\boldsymbol{\theta}) = \left[\mathbf{q}_\Sigma(\boldsymbol{\theta}), \mathbf{p}_\Sigma(\boldsymbol{\theta})\right] \tag{6.5}$$

the mean value of the energy on the toroid is

$$\langle E_\Sigma \rangle = \int' d\tau_\theta\, H\left(\mathbf{q}_\Sigma(\boldsymbol{\theta}), \mathbf{p}_\Sigma(\boldsymbol{\theta})\right) \tag{6.6}$$

and the mean of the kth action integral, defined for an arbitrary toroid, is

$$I_k(\Sigma) = \int' d\tau_\theta\, \mathbf{p}_\Sigma(\boldsymbol{\theta}) \cdot \frac{\mathbf{q}_\Sigma(\boldsymbol{\theta})}{\partial \theta_k} \tag{6.7}$$

This last definition is required because Einstein's definition in terms of $\int_{\mathcal{C}_k} \mathbf{p} \cdot d\mathbf{q}$ along curves around the toroid is no longer valid for arbitrary toroids; it depends on the curve.

For simplicity we now drop the arguments Σ and propose that the mean energy (6.6) should be stationary subject to the action integrals (6.7) remaining fixed; that is,

$$\Delta\Phi = \mathcal{O}(\Delta\mathbf{X})^2 \tag{6.8}$$

where

$$\Phi = \int d\tau_\theta \left(H(\mathbf{q}(\boldsymbol{\theta}), \mathbf{p}(\boldsymbol{\theta})) - \sum_{k=1}^N \omega_k \mathbf{p} \cdot \frac{\partial \mathbf{q}}{\partial \theta_k} \right) \tag{6.9}$$

for arbitrary variations

$$\Delta\mathbf{X}(\boldsymbol{\theta}) = \left[\Delta\mathbf{q}(\boldsymbol{\theta}), \Delta\mathbf{p}(\boldsymbol{\theta})\right] \tag{6.10}$$

The ω_k are here just Lagrange multipliers.

By the usual variational methods, equating coefficients of $\Delta\mathbf{q}$, $\Delta\mathbf{p}$ to zero we obtain the equations

$$\sum_k \omega_k \frac{\partial \mathbf{q}}{\partial \theta_k} = \frac{\partial H}{\partial \mathbf{p}} \tag{6.11a}$$

$$\sum_k \omega_k \frac{\partial \mathbf{p}}{\partial \theta_k} = -\frac{\partial H}{\partial \mathbf{q}} \tag{6.11b}$$

where

$$\frac{\partial}{\partial \mathbf{q}} = \left(\frac{\partial}{\partial q_1}, \frac{\partial}{\partial q_2}, \ldots, \frac{\partial}{\partial q_N} \right)$$

$$\frac{\partial}{\partial \mathbf{p}} = \left(\frac{\partial}{\partial p_1}, \frac{\partial}{\partial p_2}, \ldots, \frac{\partial}{\partial p_N} \right)$$

(6.12)

These equations are partial differential equations for a toroid and are called the angle Hamilton's equations. If the Lagrange multipliers ω_k are identified with angular frequencies and the relation

$$\theta_k = \omega_k t + \delta_k$$

(6.13)

is used, then the equations reduce to the more usual Hamilton's equations for a system on a trajectory that remains in the toroid. Thus the trajectory of any system with Hamiltonian function $H(\mathbf{q}, \mathbf{p})$ that starts at a point $(\mathbf{q}, \mathbf{p})$ on the toroid remains in the toroid for all time. It is in this sense that the toroid is invariant.

In obtaining the variational principle, the corresponding quantal theory has been used as a guide, but there are some essential differences between the two theories. The classical theory is nonlinear; this leads to difficulties when it comes to applying the theory. In particular, one is confronted with the problem of small divisors.

Percival and Pomphrey[19] used the variational principle for invariant toroids for two methods of obtaining the energy function $E(\mathbf{I})$. For a standard model system with Hamiltonian

$$H(\mathbf{r}, \mathbf{p}) = \tfrac{1}{2} \left(p_x^2 + p_y^2 + \lambda x^2 + \mu y^2 \right) + b_1 V^{\text{anh}}(x, y)$$

(6.14)

an application of the variational principle leads to the equations

$$\left[\lambda - (s_1 \omega_1 + s_2 \omega_2)^2 \right] x_{s_1 s_2} = F^{\text{anh}}_{x s_1 s_2}$$

$$\left[\mu - (s_1 \omega_1 + s_2 \omega_2)^2 \right] y_{s_1 s_2} = F^{\text{anh}}_{y s_1 s_2}$$

(6.15)

where $\mathbf{F}^{\text{anh}}_{s_1 s_2}$ are the Fourier components with respect to the angle variables θ_k of the anharmonic force

$$\mathbf{F}^{\text{anh}}(\mathbf{r}) = -b_1 \nabla V^{\text{anh}}(\mathbf{r})$$

(6.16)

At first these equations were iterated numerically to obtain $\mathbf{r}_{s_1 s_2}$ and thus by Fourier summation $\mathbf{r}(\theta_1, \theta_2)$. Choosing phase shifts δ_1 and δ_2 and using (6.13) to obtain θ_1 and θ_2 in terms of the time trajectories, $\mathbf{r}(t)$ (angle solutions) were obtained. These were checked against numerically integrated trajectories (Newton solutions) and it was found that the maximum of the magnitude of the difference between the angle and Newton solutions for 140 characteristic periods of the uncoupled motion was less than 10^{-5} units of distance. This was for a quartic anharmonic term. The calculation was repeated for a similar Hamiltonian with three degrees of freedom. In practice, the computing time required to obtain the toroid was less than that required for the numerical integration of the trajectories. A modification was made in which invariant toroids with quantized action integrals were obtained at each stage of the iteration. The results obtained by Marcus and Miller and their collaborators and all the results obtained by the second variational method were rederived.[20]

In the second variational method the equations were iterated analytically and the anharmonic term treated as a perturbation. For the two-dimensional potential

$$V_2 = \tfrac{1}{2}(\lambda x^2 + \mu y^2) + b_1 x^2 y^2 \qquad (6.17)$$

a perturbation series for the coordinates was obtained to first order in b_1. By using the stationary property of the functional Φ the resulting energy functions $E(\mathbf{I})$ were obtained to third order in b_1.

The energy function $E(\mathbf{I})$ to third order in b_1 is

$$E(I_1, I_2) = I_1 + I_2 \mu^{1/2} + \frac{b_1 I_1 I_2}{\mu^{1/2}} - b_1^2 \left(I_1^2 I_2 \frac{(3\mu - 2)}{4\mu^{3/2}(\mu - 1)} + I_1 I_2^2 \frac{(2\mu - 3)}{4\mu(\mu - 1)} \right)$$

$$+ b_1^3 \left[I_1^3 I_2 \frac{(5\mu^2 - 5\mu + 2)}{4\mu^{5/2}(\mu - 1)^2} + 3 I_1^2 I_2^2 \frac{(\mu^2 - 3\mu + 1)}{2\mu^2(\mu - 1)^2} \right]$$

$$+ I_1 I_2^3 \frac{(5 - 5\mu + 2\mu^2)}{4\mu^{3/2}(\mu - 1)^2} \qquad (6.18)$$

Approximate energy levels $E_{n_1 n_2}$ were obtained from (6.18) by using the relation (6.1).

The result was compared with the energy function obtained by second-order quantal perturbation theory,

$$E^{\text{Q.P.T.}}(I_1, I_2) = I_1 + I_2 \mu^{1/2} + \frac{b_1 I_1 I_2}{\mu^{1/2}} - b_1^2 \left(I_1^2 I_2 \frac{(3\mu - 2)}{4\mu^{3/2}(\mu - 1)} \right.$$

$$\left. + I_1 I_2^2 \frac{(2\mu - 3)}{4\mu(\mu - 1)} - \frac{3 I_1}{16\mu(\mu - 1)} + \frac{3 I_2}{16\mu^{1/2}(\mu - 1)} \right) \qquad (6.19)$$

where

$$I_k = \left(n_k + \tfrac{1}{2}\right)\hbar \qquad (k = 1, 2)$$

$$(n_k = 0, 1, 2, \ldots)$$

(6.20)

Each term of the semiclassical series is the dominant contribution, in powers of $\hbar$, to the corresponding term of the quantal perturbation series.

To investigate the accuracy of the semiclassical energy levels, they were compared with "exact" quantal energy levels found by diagonalizing the relevant Hamiltonian matrices, calculated using harmonic-oscillator wave functions as a basis.

The parameters μ and b_1 were chosen so that the number of bound states was typical of the stretch modes of a linear triatomic molecule, chosen to be carbon dioxide.

For each diagonalization a sufficient number of basis functions was chosen to ensure convergence of any tabulated energy levels to 0.01 cm^{-1} for the worst case. A comparison is shown in Table VI for the lowest 47 energy levels of V_2. Both numerical and analytic variational methods are clearly effective means of obtaining large numbers of reasonably accurate energy levels of systems with molecular-like potentials.

The numerical iteration scheme shows that toroids can be found for two and three degrees of freedom which are so accurate that it is difficult to determine any errors in them by stepwise integration of trajectories. The parametrization in terms of angle variables allows direct determination of action variables I_k, which is difficult using direct numerical integration of trajectories.

The analytic perturbation-variation method is easier in practice than quantal perturbation theory for a given order of perturbation, because it does not require the extensive summation procedures of the latter. Unlike the analogous quantal case, the stationary principle provides a powerful practical technique for obtaining higher-order energy levels from lower-order invariant toroids. The evaluation of the energy function $E(\mathbf{I})$ and the substitution of the quantization conditions $I_k = (v_k + \tfrac{1}{2})\hbar$ enables large numbers of energy levels to be obtained together.

The method is not as accurate as quantal methods for low quantum numbers, where the semiclassical expansion is not as good. It is straightforward to combine the method with low-order quantal perturbation theory, which may lead to improved results. Thus it is complementary to the usual quantal techniques.

TABLE VI

Exact and Semiclassical Energy Levels of the Two-Dimensional Hamiltonian.

$H = \frac{1}{2}(p_x^2 + p_y^2) + \frac{1}{2}(\lambda x^2 + \mu y^2) + b_1 x^2 y^2$; $\lambda = 1$, $\mu = 0.5$, $b_1 = -0.003$;

Dissociation Energy = 58817.9 cm^{-1}

Exact energy levels (cm^{-1})	Semiclassical energy levels (cm^{-1})	Exact energy levels (cm^{-1})	Semiclassical energy levels (cm^{-1})
1,203.29	1,203.29	8,245.59	8,245.64
2,198.35	2,198.35	8,358.65	8,358.66
2,611.77	2,611.78	8,544.56	8,544.53
3,193.39	3,193.38	8,778.52	8,778.38
3,600.78	3,600.78	8,933.52	8,933.52
4,020.23	4,020.26	9,163.02	9,162.96
4,188.40	4,188.38	9,210.16	9,210.21
4,589.70	4,589.70	9,334.91	9,334.92
5,003.16	5,003.18	9,533.03	9,533.01
5,183.39	5,183.36	9,654.02	9,654.08
5,428.70	5,428.73	9,748.58	9,748.62
5,578.54	5,578.54	9,915.74	9,915.74
5,985.96	5,985.98	10,157.83	10,157.78
6,178.35	6,178.31	10,174.46	10,174.51
6,405.53	6,405.56	10,310.96	10,310.98
6,567.30	6,567.29	10,521.42	10,521.40
6,837.15	6,837.19	10,612.43	10,612.49
6,968.63	6,968.63	10,718.58	10,718.61
7,173.27	7,173.22	10,897.80	10,897.80
7,382.19	7,382.20	11,062.43	11,062.50
7,556.01	7,555.96	11,138.46	11,138.51
7,807.87	7,807.90	11,152.63	11,152.56
7,951.15	7,951.15	11,286.82	11,286.83
8,168.15	8,168.10		

C. Hamilton-Jacobi Method

Chapman, Garrett, and Miller[21] solve the Hamilton-Jacobi equation for the action function F, which defines the invariant toroid, by a numerical version of the iterative procedure given by Born.[14] For either position or momentum representation, the action function is many-valued, so they use the *unperturbed angle* (θ_0) representation. For motion in one degree of freedom, the lines of constant θ_0 are the radii from the origin in the q-p phase plane (Fig. 2.1 and 3.1) and unless the trajectory is very distorted it will be a single-valued function of θ_0. The same applies to the toroids for more degrees of freedom. The action function $F(\theta_0, I)$ represents the invariant toroid corresponding to a bound quantal state when I satisfies the basic quantization conditions.

As in the method of the previous section, Fourier analysis is used, but in this case it is in the unperturbed angles θ_0 instead of the angle variables θ of the invariant toroid.

These workers applied the method to the potential of (6.2) with the results shown in Table VI.

VII. REGULAR AND IRREGULAR SPECTRA

A. Regular and Irregular Regions of Phase Space

In Sections II to VI the fundamental question raised by Einstein in his 1917 article has been put to one side. What is the nature of the motion of classical systems of N degrees of freedom when $N > 1$?

It has been assumed throughout these sections that the properties are not essentially different from those of separable systems, that there are invariant toroids of N dimensions in the $2N$-dimensional phase space.

A trajectory that lies on an invariant toroid will approach arbitrarily close to any point on that surface unless some nontrivial integral linear combination of the angular frequencies is zero,

$$\sum_{k=1}^{N} \nu_k \omega_k = 0 \qquad (\nu_k \text{ integers, not all zero}) \qquad (7.1)$$

Generally there is no such relation, and the trajectory is of Einstein Type 1a. When there is such a relation, the independent motions are "in resonance," the trajectory lies in a region of phase space of lower dimension than N and is of Einstein Type 2. The integer vector $(\nu_1, \nu_2, \ldots, \nu_N)$ is called the order of the resonance.

When there are $N-1$ relations like (7.1) the invariant toroid consists entirely of closed trajectories. For example, in two dimensions a single relation implies that a phase point returns to its original location after $|\nu_1|$ cycles of θ_1 and $|\nu_2|$ cycles of θ_2.

Because the frequencies ω_k are the same for every trajectory of an invariant toroid, each toroid consists entirely of Type 1a or Type 2 trajectories, so that the toroids themselves may be named Type 1a or Type 2. If an invariant toroid contains one closed trajectory, then all its trajectories are closed.

Although relations of the form (7.1) are an exception rather than the rule, arbitrarily close rational approximations to the ω_k result in resonances, and for most separable systems closed trajectories may be found arbitrarily close to every invariant toroid.

When a separable system is perturbed by a nonseparable perturbation, the independent motions are coupled together nonlinearly, and the coupling is particularly strong when the frequencies ω_k satisfy low-order relations of the type (7.1). Approximate relations of this type give rise to the "small denominators" of classical perturbation theory, and were the principal impediment to a general theory of invariant toroids for nonseparable systems.

The general mathematical theory of invariant toroids is due to Kolmogorov, Arnol'd, and Moser (KAM).[22] The second paper of Arnol'd is particularly clear. Reviews of their work have been given by Arnol'd and Avez, Galgani and Scotti, by Walker and Ford and by Ford,[23] the last two for the physicist interested in the foundations of statistical mechanics. They provide references to many other papers in this rapidly developing field. The relevance to molecules has been noted by Marcus, by Percival, and by Nordholm and Rice.[24]

Kolmogorov stated the theorem and sketched a proof, details being provided much later by Arnol'd and by Moser.

The KAM theorem says that if a bounded system is sufficiently close to being separable (or, more generally, integrable) then its phase space is almost always dominated by invariant toroids, given certain subsidiary conditions. However, the situation is not as clear as it might appear to be. There is a residual region, to which the theorem does not apply; although this residual region occupies a smaller volume of phase space than the region of invariant toroids, it does so in a very complicated fashion. Any invariant toroid has arbitrarily close to it trajectories from the residual region.

For systems of two degrees of freedom, the invariant toroids of two dimensions lie in a three-dimensional energy shell, so that a residual trajectory between two invariant toroids remains trapped between them. For more degrees of freedom there is no such trapping.

The KAM theorem shows that the invariant toroids required by EBK quantization are common for most nearly separable systems, but says nothing about the nature of the residual regions nor about systems that are far from being separable.

The many numerical integrations of classical trajectories that have now been carried out are consistent with the view that the residual regions contain unstable trajectories but that invariant toroids are common for many systems that are far from being separable.

The first numerical work in the field was by Fermi, Pasta, and Ulam[25] and directed towards an understanding of the mixing or ergodic properties of nonlinear systems. This was followed by calculations on particles in accelerators[26] and for molecular potentials,[27] but most of the investigations

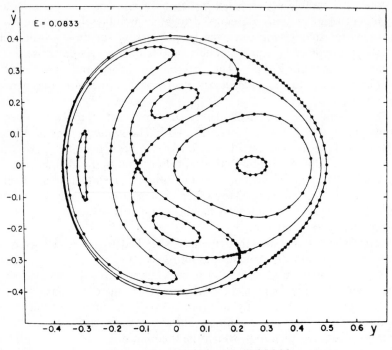

Results for $E = 0.08333$.

Fig. 7.1.   Surface of section for Hénon-Heiles potential at $E = \frac{1}{12}$.

were stimulated by dynamical astronomy, in particular the unexpected apparent discovery of a physically significant integral of the motion, besides energy and angular momentum, in the numerically integrated motion of a star in the smoothed gravitational potential of an axially symmetric galaxy.

The numerical experiments of Hénon and Heiles are among the most clearly presented and the work of Contopoulos is among the most exhaustive,[28] showing the complicated behavior of trajectories with simple cubic potentials.

Hénon and Heiles integrated trajectories of the system with Hamiltonian

$$H(\mathbf{r}, \mathbf{p}) = \tfrac{1}{2}\left(p_x^2 + p_y^2 + x^2 + y^2\right) + \left(x^2 y - \tfrac{1}{3} y^3\right) \tag{7.2}$$

with energies E ranging from the equilibrium energy $E = 0$ to the escape energy $E = \frac{1}{6}$. This system may be considered as a perturbation of the

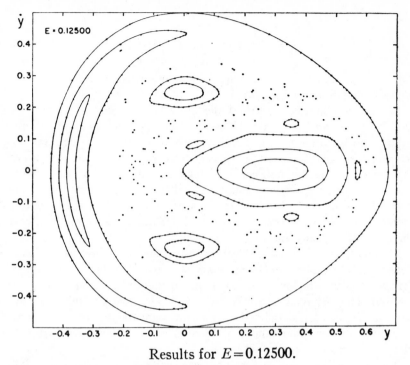

Results for $E = 0.12500$.

Fig. 7.2.   Surface of section for Hénon-Heiles potential at $E = \frac{1}{8}$.

system with separable quadratic Hamiltonian

$$H_0(\mathbf{q}, \mathbf{p}) = \tfrac{1}{2}\left(p_x^2 + p_y^2 + x^2 + y^2\right) \tag{7.3}$$

and the perturbation is relatively small for small E. This Hamiltonian H does not satisfy the conditions of the original KAM theory.

Nevertheless, they were able to use surfaces of section (see Section VI.A) to explore the energy shell for a range of energies. For trajectories that intersect this surface of section, and to the accuracy of their computations, they came to the following conclusions. Between $E = 0$ and $E = \frac{1}{12}$ the phase space occupied by these trajectories appears to consist entirely of invariant toroids (Fig. 7.1). Above $E = \frac{1}{12}$ "irregular" regions appear for which the trajectories do not lie on invariant toroids of two dimensions, but appear to wander throughout three-dimensional regions. For the energy $E = \frac{1}{8}$ (Fig. 7.2) the irregular regions already dominate, but "regular" regions consisting of invariant toroids still occupy a small but complicated region of phase space.

Contopoulos[28] has carried out very detailed studies of the system with

Hamiltonian

$$H = \tfrac{1}{2}\left(p_x^2 + p_y^2 + Ax^2 + By^2\right) - xy^2 \qquad (7.4)$$

These papers illustrate clearly the very complicated behavior of systems with simple polynomial potentials.

From Contopoulos' work and the more recent study of Hamiltonian (7.2) by Ford and Lunsford,[29] Hénon and Heiles' result is an oversimplification, and for finite nonseparable perturbations, however small, there are irregular regions in the phase space. These regions can be due to very high-order resonances between very high multiples of the basic frequencies. They then require very small intervals for the numerical integration. Arnol'd[30] proved for small perturbations, and Contopoulos found for larger ones, that the size of an irregular region decreases exponentially with the order of the resonance.

Irregular regions due to high-order resonances are therefore difficult to detect. As more accurate integrations of classical orbits have been performed, irregular regions associated with higher-order resonances are discovered. This behavior is consistent with the hypothesis[22] that very small irregular regions are associated with a significant fraction of resonances of arbitrarily higher order, and that there is an irregular region arbitrarily close to every invariant toroid.

Regular and irregular regions of phase space are associated with stable and unstable periodic closed trajectories. Normally stable periodic trajectories are surrounded by invariant toroids. Stable trajectories and associated invariant toroids occur for arbitrarily high energies E, well above escape energy.

Unstable periodic trajectories are surrounded by irregular regions. The physical instability of the trajectories in these regions leads to numerical instability in the integrations, which makes numerical experiment very difficult, so the regions are not fully understood. Very accurate integrations of Walker and Ford[31] show some long-term correlations in the unstable trajectories. The ergodic behavior over infinite times is not known: time averages and phase-space averages over the irregular regions may or may not be the same.

Numerical experiments do not constitute proofs, but on the basis of these experiments we are able to supplement the proofs of the KAM theory by plausible hypotheses.

The theory and hypotheses are combined in the following list of properties of regions of phase space containing only bounded trajectories of a nonseparable system of N degrees freedom with an analytic Hamiltonian function. For simplicity, we suppose that there are no constants of motion, such as momentum, which are derived from the symmetry of the Hamiltonian function, except for the energy E.

C1. The solutions of the classical equations of motion have a very complicated structure.

C2. There are *no* constants of the motion defined throughout an energy shell besides the energy E (strictly, no isolating integrals of the motion).

C3. The motion is not ergodic: time averages are not equal to averages over an energy shell.

C4. Almost all of the phase space of the bounded trajectories may be divided into regions of two types.

a. Regular regions, made up entirely of invariant toroids of N dimensions (Einstein case 1a).

b. Irregular regions, made up almost entirely of unstable trajectories (Einstein case 1b).

C5. At those energies for which the system is nearly separable, the regular regions normally occupy most of the phase space, but irregular regions remain.

C6. At those energies for which the system is far from being separable, irregular regions may occupy most of the phase space, but regular regions remain.

C7. The transition between C5 and C6 may or may not be rapid.

C8. There are irregular regions arbitrarily close to any point of a regular region. The trajectories of the regular regions are therefore not strictly stable.

C9. For systems of two degrees of freedom, predominantly regular regions surround stable periodic trajectories, irregular regions surround unstable periodic trajectories.

C10. The larger irregular regions are associated with low-order resonances. The trajectories of these regions are very unstable: those trajectories that result from a very small initial perturbation of a given trajectory soon wander through large part of an irregular region of dimension greater than N.

C11. The smaller irregular regions are associated with high-order resonances. The instabilities of the trajectories of these regions may take a long time to appear, and very precise calculation is required to find them.

It is the property C8 that causes the most complication. It is because of this property that we do not refer to the regions as "stable" and "unstable." If higher-order resonances are neglected, say by suppressing the high frequencies of the motion, the situation is much simpler. Properties C1, C2, and C8 are replaced by the properties:

Simplified properties of nonseparable classical systems

CS1. The structure of the solutions is greatly simplified.

CS2. There may be constants of the motion besides the energy E.

Where the system is nearly separable, the irregular regions are small or even absent, except for some special cases. If they are absent then the simplified system has N physically significant constants of the motion (isolating integrals of the motion).

CS8. The regular regions each fill a finite volume of phase space, and there is a finite volume around almost every point of every regular region containing no unstable trajectories.

This is similar to the simplified picture presented by Hénon and Heiles.

According to Contopoulos,[28] the irregular regions are associated with unstable orbits. According to Sinai,[32] who proved ergodicity for hard spheres, instability is associated with a continuous frequency spectrum. Numerical evidence from Contopoulos and others suggests that trajectories in irregular regions, even though possibly not ergodic, approach all points of a $(2N-1)$-dimensional region of phase space, so that it would appear that the frequency spectrum of a trajectory $(\mathbf{q}(t), \mathbf{p}(t))$ must be continuous.

On applying weak time-dependent perturbations to a system with a phase point in a regular region, the system responds at its resonant modal frequencies, with decreasing effect as the order of the resonance increases. After a weak short-lived perturbation has ceased, the new trajectory almost always stays on a toroid in the neighborhood of the old one.

If the system were on an irregular trajectory, it would respond over a continuous range of frequencies and rapidly diverge from its original trajectory.

B. Regular and Irregular Spectra

Percival[24] has applied the correspondence principle to the properties of regular and irregular regions of phase space.

In the asymptotic limit as $\hbar \to 0$ the quantal systems have all the complications of the classical systems given by C1 to C11, together with the additional problems of quantization. This limit is not of such physical importance as the properties of systems with high quantum numbers, where $\hbar$ is finite, but small when compared with the most important action variables of the classical motion.

In the limit $\hbar \to 0$ all regular and irregular regions must be considered, and there are an infinite number of them, but for finite $\hbar$ those regions that are much smaller in volume than $(2\pi\hbar)^N$ can almost all be neglected, and only a finite number of regions of either type remain.

A similar simplification can be made by putting a limit on the frequency. The higher frequencies of the quantized motion are given by relations of the type

$$\omega_{\gamma'\gamma} = \frac{E_{\gamma'} - E_{\gamma}}{\hbar} \tag{7.5}$$

and are not adequately represented by high multiples of classical frequencies. Such high multiples may be neglected without introducing significant additional error into the process of quantization. There remain a finite number of possible classical resonances, and these can produce no more than a finite number of regular and irregular regions of phase space.

Therefore the physical problem of quantization, with finite but small $\hbar$, should commence with the simplified classical model of a nonseparable system, in which the complex properties (C1, C2, C8) are replaced by the simpler properties (CS1, CS2, CS8).

The KAM theorem, which proves the existence of invariant toroids occupying finite volumes of phase space, shows that EBK quantization is valid in those parts of phase space. This fact, together with the hypotheses suggested by numerical experiment, implies the following properties of quantized nonseparable systems with N degrees of freedom. It is assumed throughout that $\hbar$ is small but finite.

A high quantum level of the discrete energy spectrum of a bound quantal system belongs to either (R), a regular energy spectrum, or (I), an irregular energy spectrum.

The regular energy spectrum and its associated states have the following properties:

R1. A quantal state may be labeled by the vector quantum number

$$\mathbf{n} = (n_1, n_2, \ldots, n_N)$$

R2. A state with quantum number $\mathbf{n}$ corresponds to those phase-space trajectories of the corresponding classical system that lie in an N-dimensional invariant toroid with action constants I_k given by quantum conditions

$$I_k = (n_k + \alpha_k/4)\hbar$$

R3. The quantal state must resonate at frequencies close to those of the corresponding classical motion. Given two quantal states with one n_k differing by unity and the others the same, the Planck relation for their energy difference is

$$\Delta E_k = \hbar \omega_k \tag{7.6}$$

where ω_k is a fundamental frequency on the corresponding toroid.

R4. A "neighboring state" to a state $\mathbf{n}^0$ with energy E^0 is a state with vector quantum number $\mathbf{n}$ close to $\mathbf{n}^0$, with energy difference no more than a small multiple of the maximum $|\Delta E_k|$.

R5. We now use the correspondence principle for weak perturbations, whose most well-known form is the correspondence principle for intensity of radiation. Under weak external perturbations the state $\mathbf{n}^0$ is much more strongly coupled to neighboring states than to other states, with the coupling tending to decrease rapidly with $|\mathbf{n} - \mathbf{n}^0|$. The apparent exception of optical transitions in atoms from high to low $\mathbf{n}$ when l is small is due to the singularity in the Coulomb potential, which contravenes the requirement of analyticity.

Not all states close in energy to a given state are neighboring states, not even all those belonging to the same regular spectrum. However, experiments that are able to select a few high-n states of a regular spectrum have a high probability of selecting neighboring states, so it is possible to observe the regularity. If transitions with energy differences ΔE^1 and ΔE^2 are observed, approximate multiples and integer combinations of them should also be observable. Bound integrable systems have a regular spectrum.

The correspondence principle predicts properties of an irregular spectrum in striking contrast to those of a regular spectrum:

I1. There is no unambiguous assignment of a vector quantum number to a state Ψ_0.

I2. The discrete bound-state quantal spectrum must tend to a continuous classical spectrum in the classical limit. The frequencies

$$\left[E(\overline{\Psi}) - E(\overline{\Psi}_0) \right] / \hbar = \omega \tag{7.7}$$

for fixed stationary state Ψ_0 and varying Ψ form a discrete distribution that tends to the continuous distribution in ω. The distribution of levels of the irregular spectrum could take on the appearance of a random distribution.

I3. By applying the correspondence principle for weak perturbations, there are no neighboring states in the sense of (R4) and (R5). Except for selection rules and accidents, a state of an irregular spectrum is coupled by a weak perturbation with intensities of similar magnitude to all those states of a similar energy that correspond to the same irregular region of classical phase space. The number of such states is very large, of order

$$n_j^{N-1} \tag{7.8}$$

where n_j is a typical quantum number.

I4. The energies of the irregular spectrum are more sensitive to a slowly changing or fixed perturbation than those of the regular spectrum.

In the Born–Oppenheimer approximation, the vibrational energy spectrum of a polyatomic molecule is observed to be regular near equilibrium, except for Fermi resonances, which are a residual form of irregular spectrum. Near the dissociation limit the spectrum should be mainly irregular, with a large number of weak optical lines in place of a small number of strong ones. A regular progression should be observed to terminate abruptly at a maximum energy below the dissociation limit. An irregular spectrum could easily be confused with a continuum under poor conditions of observation.

The instability of the irregular spectrum probably extends to instability under variations in $\hbar$. If that is the case the semiclassical limit does not exist for the irregular spectrum, except in a statistical sense.[33]

No unambiguous observations of individual states of irregular spectra are yet known to the author.

C. Numerical Calculations on Regularity

Pomphrey[34] has studied the energy spectrum of the Hénon-Heiles potential. It was important *not* to use semiclassical methods, although for part of the spectrum they are clearly more efficient, because of the risk of implicitly assuming what is to be proved. Pomphrey used the Rayleigh-Ritz method and an harmonic-oscillator basis, the eigenfunctions of the unperturbed Hamiltonian

$$H^0(\omega) = \tfrac{1}{2}(\mathbf{p}^2 + \omega^2 \mathbf{r}^2)$$

where ω is a free parameter. The Hénon-Heiles Hamiltonian takes the form

$$H = H^0(\omega) + \frac{\alpha r^3}{3}\sin 3\theta + \tfrac{1}{2}(\omega^2 - 1)r^2$$

Errors are introduced by the inevitable truncation of the basis set, but they were shown to be less than unity in the fourth decimal place by the following checks:

1. With a small value of the perturbation parameter α, the first few eigenvalues were found to agree with second-order quantal perturbation theory.

2. With fixed Hamiltonian, the basis set was changed by changing ω. The eigenvalues changed negligibly.

3. The size of the basis set was increased, leaving the eigenvalues of interest unchanged to within the fourth decimal place.

Eigenvalues were obtained for five different values of α from 0.086 to 0.090. For each value 71 eigenvalues were obtained that satisfied the above checks. The behavior of the spectrum under perturbation was examined by obtaining the magnitudes of the second differences Δ_i^2 of the energy levels with respect to the increments $\Delta\alpha = 0.001$ in α. They are

$$\Delta_i^2 = |E_i(\alpha + \Delta\alpha) - 2E_i(\alpha) + E_i(\alpha - \Delta\alpha)|$$

The values of Δ_i^2 are shown as a function of E_i in Fig. 7.3.

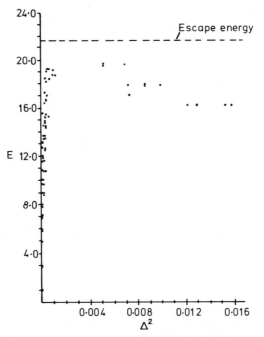

Fig. 7.3.   Variations in eigenvalues under a perturbation, after Pomphrey.

For energies less than $E = 16.0 \approx 0.74D$, all second differences are very small; D is the depth of the Hénon-Heiles well. For energies greater than $E = 16.0$, however, eigenvalues are found with corresponding Δ_i^2 values orders of magnitude larger. These eigenvalues are evidently very sensitive to small changes $\Delta\alpha$ in the perturbation. High-order terms in the perturbation expansion become important for these eigenvalues.

Thus two types of eigenvalue are distinguished by their behavior under a slowly changing perturbation.

From the results of Hénon and Heiles the relative area α_I of the surfaces of section covered by unstable trajectories is given approximately as a function of the energy E by

$$\alpha_I(E) = 0 \qquad\qquad E < 0.68D$$

$$= 3.125(E/D) - 2.125 \qquad E > 0.68D$$

Pomphrey[34] compared the classical integral

$$I(E) = \int^E \alpha_I(E)\,dE$$

which is the integrated area covered by unstable trajectories up to an energy E, with the quantal sum

$$S(E) = \frac{1}{D} \sum_i{}' n_I(E_i)\langle\Delta E_i\rangle \qquad (E_i < E)$$

In this sum, set $n_I(E_i) = 1$ if the eigenvalue E_i is very sensitive to the slight changes in the perturbation, and $n_I(E_i) = 0$ otherwise. $\langle\Delta E_i\rangle$ is a mean separation between an eigenvalue E_i and its two neighboring eigenvalues E_{i-1} and E_{i+1},

$$\langle\Delta E_i\rangle = (E_{i+1} - E_{i-1})/2$$

$S(E)$ represents the number of states of the irregular spectrum below energy E. Figure 7.4 shows a plot of $I(E)$ as a continuous curve and $S(E)$ as a series of points. We see that qualitatively the points (quantal spectrum results) follow the shape of the curve (classical trajectory results). The two sets of results cannot hope to agree quantitatively since the sum $S(E)$ corresponds to a volume of phase space of dimension four while the integral is over three dimensions. However, the energy at which the eigenvalues first become sensitive to the changing perturbation ($\approx 0.74D$) agrees well with the critical energy of Hénon and Heiles ($\approx 0.68D$).

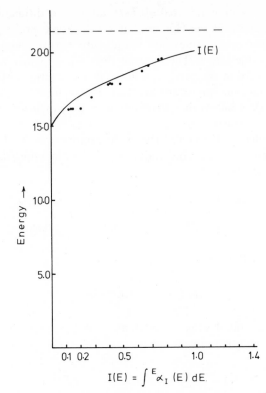

$$I(E) = \int^E \alpha_I (E) \, dE$$

Fig. 7.4.   Comparison of classical and quantal irregular regions.

Thus eigenvalues of the Hénon and Heiles nonintegrable Hamiltonian with energy less than a critical energy $E^c \approx 0.74D$ were found to be insensitive to a slight change in the perturbation. These eigenvalues belong to a regular spectrum. Above the critical energy, which is known to within a narrow band of energy, eigenvalues are found that are very sensitive to a small change in the perturbation. These eigenvalues belong to an irregular spectrum.

For this particular system the existence of two types of spectrum is confirmed insofar as the energy levels of the irregular spectrum are more sensitive to a fixed perturbation than those of the regular spectrum.

The classical trajectory calculations of Theile and Wilson and of Bunker and others[27] showed the features for molecular models similar to those for the potentials used by the astronomers. There are regular and irregular regions. In the irregular regions there is rapid vibrational relaxation and in the regular regions nearer equilibrium there is not.

Nordholm and Rice[24] have carried out quantal calculations on relaxation for the Hénon-Heiles and the Barbanis[35] potentials, the corresponding classical calculations being available. The Barbanis Hamiltonian is

$$H_B = \frac{1}{2\mu}(p_1^2 + p_2^2) + \frac{A}{2}(q_1^2 + q_2^2) - \lambda q_1^2 q_2 \qquad (7.9)$$

They use a quantal measure of ergodicity based on the overlap of the computed quantal wave functions and the wave functions of the unperturbed harmonic-oscillator Hamiltonian without the perturbing cubic term. With this measure they are able to distinguish regular (nonergodic) and irregular (ergodic) quantal behavior and correlate it with the corresponding classical case. For the most part they obtain agreement with the classical picture. It should be noted that they use the terms "regular" and "irregular" in the reverse sense to that of this chapter. They also obtained quantal estimates of the time scale of recurrences.

VIII. CLASSICAL TRAJECTORIES

A. Introduction

For motion of a system with one degree of freedom, a closed classical trajectory is identical to an invariant toroid. There is no problem of integrability or of regular and irregular energy spectra.

Recently there have been several more or less successful attempts to use closed trajectories insetad of invariant toroids for the semiclassical quantization of systems of many degrees of freedom. As we did for invariant toroids, we illustrate the theory first by means of a trivial system of one degree of freedom.[36]

Closed trajectories require somewhat more sophisticated quantum theory than invariant toroids.

The development of a quantum-mechanical system in time is given by a unitary transformation from the state $\psi(t_0)$ at time t_0 to the state $\psi(t)$ at another time t. Thus

$$\psi(t) = U(t,t_0)\psi(t_0) \qquad \text{(all } t_0, t) \qquad (8.1)$$

and

$$U(t,t_0) = \exp\left[-iH(t-t_0)\hbar\right] \qquad (8.2)$$

where U is the evolution operator. As defined here, the evolution operator is two-sided and relates the state at time t_0 to both earlier and later states $\psi(t)$.

The usual forward evolution operator $U^+(t,t_0)$, which relates $\psi(t_0)$ to the wave function at a later time t, is defined in terms of $U(t,t_0)$ by the equations

$$U^+(t,t_0) = U(t,t_0), \qquad t > t_0$$
$$= 0 \qquad\qquad t < t_0 \tag{8.3a}$$

The kernel of U^+ is the propagator. The backward evolution operator is similarly defined by the equations

$$U^-(t,t_0) = 0, \qquad\qquad t > t_0$$
$$= U(t,t_0), \qquad t < t_0 \tag{8.3b}$$

Thus

$$U(t,t_0) = U^+(t,t_0) + U^-(t,t_0) \tag{8.4}$$

and

$$U^-(t,t_0) = \left[U^+(t_0,t) \right]^\dagger \tag{8.5}$$

The Green operators are obtained by transforming the forward and backward evolution operators from time representation to energy representation by Fourier analysis. By taking Fourier transforms of operators there we obtain the spectral operator, and Green operators with constant factors

$$U \Rightarrow \delta(E - H) = \sum_\nu \delta(E - E_\nu) P_\nu \tag{8.6}$$

$$U^\pm \Rightarrow G^\pm(E)$$

where P_ν is the projection onto the space of eigenfunctions of H with eigenvalue E_ν. The operators are related by

$$G^- = (G^+)^\dagger \tag{8.7}$$

$$G^- - G^+ = 2\pi i \delta(E - H) \tag{8.8}$$

The difference between the Green operators, or the discontinuity on the real axis, is directly connected with the spectrum and the projection operators; through the projection operators it is related to wave functions in a particular representation. The trace of the spectral operator is the

(unsmoothed) spectral density

$$\operatorname{Tr}\delta(E-H)=\sum_{\nu}\delta(E-E_{\nu})=\rho(E)\qquad(8.9)$$

Semiclassical theories using closed classical trajectories for more than one degree of freedom have been based both on Green functions and on spectral operators.

For a particle of mass m on a ring (periodic boundary conditions) the spectral operator in position representation can be written as a sum over classical trajectories c joining x and x_0 and cycling around any number of times clockwise or counterclockwise:

$$\langle x|\delta(E-H)|x_0\rangle=\frac{1}{2\pi\hbar}\left(\frac{m}{2E}\right)^{1/2}\sum_{c=-\infty}^{\infty}\exp\left[\frac{i}{\hbar}S_{Ec}(x,x_0)\right]\qquad(8.10)$$

where $S_{Ec}(x,x_0)$ is the classical action for the path c. This not a purely formal result, as can be seen in Fig. 8.1, which shows partial sums for diagonal elements for 3 and for 11 paths. Peaks appear more clearly around the eigenvalues as the number of terms is increased.

This is a simple example of the Poisson summation formula.[36]

The spectral operator requires no analytic continuation but does require generalized functions or distributions. The Green operator requires analysis in the complex energy plane. The spectral operator is given by the

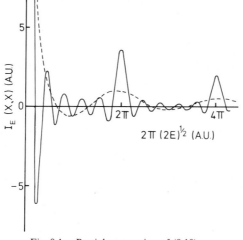

$2\pi(2E)^{1/2}$ (A.U.)

Fig. 8.1. Partial summation of (8.10).

difference between G^+ and G^-, or a sum over poles and cuts in this plane.

In a series of four papers Gutzwiller[37] used an asymptotic form of the Green function

$$G(E) \approx \sum_c B_c(E) \exp\left[\frac{iS_c(E)}{\hbar}\right] \qquad (\hbar \to 0) \qquad (8.11)$$

to obtain approximations to the energy spectrum and thus semiclassical quantization rules. $S(E)$ is the action function $\int p\, dq$, along a classical trajectory between initial and final points. It is thus distinct from the action function of the invariant toroids, which is not constrained to trajectories. The projection operators and spectrum are derived from the poles of the "response function," which is the trace of the Green function.

In every case the semiclassical response function is found to depend only on the closed periodic trajectories of the corresponding classical system. However, the resulting quantization rules are fundamentally different for separable and nonseparable systems. For the central motion (1970 paper) every closed trajectory belongs to a threefold infinity of similar closed trajectories of the same shape period and energy, by the spherical symmetry of the problem. The integration over all similar trajectories is carried out using the symmetry, leaving a sum over trajectories each of which is essentially different from the others. All trajectories may be characterized by the angular frequencies ω_r and ω_ψ of radial angular motion. As we know, each corresponds to a toroid. For the trajectories to be closed

$$\nu_r \omega_r + \nu_\psi \omega_\psi = 0 \qquad (8.12)$$

for some integers ν_r, ν_ψ. A pole, and thus an energy level F_j, is obtained as an infinite sum over a sequence of closed trajectories at energies that converge on E_j. In the case of separable systems, the quantization rule does not require closed trajectories at energy E_j. The rule gives the same results as EBK quantization.

By contrast, for the nonseparable systems (1971 paper) the evaluation of the response function depends critically on the existence of an isolated stable closed trajectory for the energy E_j, satisfying a quantization condition. This does not agree with EBK quantization, because closed trajectories on invariant toroids are not isolated. The question is reconsidered in the following sections.

For a special class of separable systems, Gutzwiller,[37] using the Green function, and Norcliffe, Percival, and Roberts,[38] using the spectral operator $\delta(E - H)$, show that both density matrices and spectrum for some quantal problems are given exactly by a sum over classical trajectories, provided a suitable representation is used. These correspondence identities are the subject of a review by Norcliffe.[38]

B.   Smoothed Densities and Complex Energies

Consider the regular spectrum of a system of two degrees of freedom. The classical frequencies ω_1, ω_2 are smooth functions of the action variables I_1 and I_2 and the energy differences

$$\Delta_1 E \approx \hbar\omega_1, \qquad \Delta_2 E = \hbar\omega_2 \tag{8.13}$$

are also smooth functions. Where the frequencies are near resonance, so that for integers ν_1 and ν_2

$$\nu_1\omega_1 + \nu_2\omega_2 \approx 0 \quad \text{and} \quad \nu_1\Delta_1 E + \nu_2\Delta_2 E \approx 0 \tag{8.14}$$

the trajectories on the toroids are nearly closed. Also, because of the "alignment" effect, there will be a bunching of energy levels at an energy interval given by

$$|\nu_1\Delta_1 E| = |\nu_2\Delta_2 E| \tag{8.15}$$

Thus nearly periodic trajectories are associated with the bunching of energy levels.

Such bunching is observed in nuclear spectra; it led to the theory of Balian and Bloch,[39] who overcome some of the difficulties encountered in the use of classical paths or trajectories. They introduce a *smoothed* density ρ_γ of the eigenvalues as in the Fermi-Thomas approximation and obtain a semiclassical expansion of ρ through the Green functions for complex values of the energy *off* the real axis. The distance from the real axis is related to the range of the smoothing.

The density of eigenvalues is

$$\rho(E) = \sum_m \delta(E - E_m) \tag{8.16}$$

and the smoothed density is

$$\rho_\gamma(E) = \sum_m f_\gamma(E - E_m) \tag{8.17}$$

where the smoothing function f_γ has a peak of width γ around the origin. If the smoothing function f_γ is Lorentzian, then ρ_γ is expressed in terms of the Green function $G(\mathbf{r}, \mathbf{r}'; z)$ by the relation

$$\rho_\gamma(E) = \frac{1}{\pi} \int d\tau \operatorname{Im} G(\mathbf{r}, \mathbf{r}'; E + i\gamma) \tag{8.18}$$

The Green function G is a bounded solution of

$$\left(-\hbar^2\nabla^2 + V(\mathbf{r}) - z\right)G(\mathbf{r},\mathbf{r}';z) = \delta(\mathbf{r}-\mathbf{r}') \tag{8.19}$$

It can be approximated by the semiclassical Green function (not to be confused with an unperturbed Green function)

$$G_0 = \left(\frac{A}{\hbar^2}\right)\exp\left(\frac{iS}{\hbar}\right) \tag{8.20}$$

where the action function $S(\mathbf{r},\mathbf{r}';\ z)$ is a solution of the complex Hamilton-Jacobi equation

$$(\nabla S)^2 + V(\mathbf{r}) - z = 0 \tag{8.21}$$

with the boundary condition

$$S(\mathbf{r},\mathbf{r}';z) \underset{\mathbf{r}\to\mathbf{r}'}{\sim} |\mathbf{r}-\mathbf{r}'|(z-V(\mathbf{r}'))^{1/2}, \qquad \operatorname{Im} S > 0 \tag{8.22}$$

The amplitude A can be obtained from S through the equation

$$\nabla(A^2\nabla^2 S) = 0 \qquad \left(A \underset{\mathbf{r}\to\mathbf{r}'}{\sim} (4\pi|\mathbf{r}-\mathbf{r}'|)^{-1}\right) \tag{8.23}$$

which expresses the conservation of flux. The use of a finite imaginary part for z eliminates the turning points and thus many of the difficulties of the Hamilton-Jacobi equation. The analysis depends on the assumption that (8.21) with boundary condition (8.22) has a unique solution for all real $\mathbf{r}$, $\mathbf{r}'$, and that for $\operatorname{Im}(z) = 0$ the limit

$$\lim_{\hbar\to 0}\left(\frac{\hbar}{i}\right)G(\mathbf{r},\mathbf{r},z) \tag{8.24}$$

exists and defines G_0 and thus S everywhere, where the S thus defined has a positive imaginary part, which produces a damping in the wave proportional to $\exp[-\operatorname{Im} S/\hbar]$. This damping is crucial to the theory, because long paths are strongly damped, so that they need not be considered in detail. It is the long paths that cause the problems on the real energy axis and the distinction between regular and irregular spectra.

The next step in the determination of G is to write the integral equation corresponding to (8.19) with the semiclassical approximation as the inhomogeneous "driving" term

$$G(\mathbf{r},\mathbf{r}') = G_0(\mathbf{r},\mathbf{r}') + \int d\tau\, G(\mathbf{r},\mathbf{r}'')\Gamma(\mathbf{r}'',\mathbf{r}') \tag{8.25}$$

where

$$\Gamma(\mathbf{r},\mathbf{r}') = (z - H)G_0(\mathbf{r},\mathbf{r}') + \delta^3(\mathbf{r} - \mathbf{r}')$$

$$= \left[\nabla^2 A(\mathbf{r},\mathbf{r}') + \delta^3(\mathbf{r} - \mathbf{r}') \right] \exp\left[\frac{iS(\mathbf{r},\mathbf{r}')}{\hbar} \right] \qquad (8.26)$$

The integral equation provides a multiple scattering expansion. The lowest-order term gives the Fermi-Thomas approximation to the smoothed density $\rho_\gamma(E)$ and the expansion yields successive corrections of order $\hbar^2$, $\hbar^2$, and so on.

If the potential is smooth, the dominant contributions for small γ arise from classical paths joining $\mathbf{r}$ and $\mathbf{r}'$, and reflections in singularities in the potential. Only the shortest paths contribute for finite γ. The limit $\gamma \to 0$ depends on the structure of the real classical solutions, and the problems of the previous section.

Balian and Bloch derive formal Laplace transform relations between the quantal Green function and a purely classical "path-generation function" $\Omega(\mathbf{r}',\mathbf{r}; s)$, which is defined by the integral equation

$$\Omega(\mathbf{r},\mathbf{r}'; s) = \frac{A(\mathbf{r},\mathbf{r}')}{s - S(\mathbf{r},\mathbf{r}')} + \frac{1}{2\pi i} \int_{-\infty}^{\infty} ds' \int d\tau'' \frac{\Omega(\mathbf{r},\mathbf{r}''; s')\nabla^2 A(\mathbf{r}'',\mathbf{r}')}{s - s' - S(\mathbf{r}'',\mathbf{r}')} \qquad (8.27)$$

The relations are

$$G(\mathbf{r},\mathbf{r}') = \frac{1}{2\pi i\hbar^2} \int_{-\infty}^{\infty} ds\, e^{is/\hbar}\Omega(\mathbf{r},\mathbf{r}'; s) \qquad (8.28)$$

$$\Omega(\mathbf{r},\mathbf{r}; z,s) = i \int_0^\infty d\hbar\, e^{-is/\hbar} G(\mathbf{r},\mathbf{r}'; z,\hbar) \qquad (8.29)$$

On the real axis the integrals over s' in (8.27) and s in (8.28) lead to problems but they can be shifted from the real axis and the results might then be considered as more than merely formal. The complex analysis gets rid of many of the problems at caustics.

The smoothed eigenvalue density is given by

$$\rho_\gamma(E) = \rho_\gamma^{FT}(E) - \frac{1}{2\pi^2\hbar^2} \sum_\alpha \mathrm{Re} \int_{\mathcal{C}_\alpha} ds\, e^{is/\hbar} \Xi(s, E + i\gamma) \qquad (8.30)$$

where

$$\Xi(s,z) = \int d\tau \left[\Omega(\mathbf{r}\mathbf{r}, s, z) - \Omega_0 \right] \qquad (8.31)$$

is the closed paths generating function. Ω_0 is the lowest-order term, which would give rise to an irrelevant divergence. The smoothed eigenvalue density is obtained as a sum over contributions each of which is associated with a closed complex trajectory. For small γ, in the semiclassical region where action functions are large compared to $\hbar$, each nearly real closed trajectory yields an oscillatory function of E in (8.30) and thus the bunching of energy levels discussed at the beginning of this section. One must also expect oscillatory behavior of the energy spectrum in the energy neighborhood of closed classical trajectories in the irregular spectrum.

C. Energy Levels in the Regular Spectrum

Berry and Mount, Marcus, Miller,[40] and others, including Gutzwiller, have expressed doubts about the method of quantization of nonseparable systems described in Section VIII.A. Berry and Tabor[41] have recently converted EBK quantization explicitly into a Poisson or "topological" sum over closed trajectories, thus extending Gutzwiller's method for separable systems to the case of the regular spectrum of nonseparable system, and extending Balian and Bloch's results for smoothed energy level identities to the unsmoothed case.

By (5.7) the density of states given by EBK quantization is

$$\rho(E) = \sum_n \delta\left(E - H\left((n + \alpha/4)\hbar\right)\right) \tag{8.32}$$

where for convenience the energy function $E(\mathbf{I})$ has been written as the Hamiltonian function $H(\mathbf{I})$ of the action variables alone. This is converted into a series of integrals over the positive quadrant in $\mathbf{I}$ space, using the Poisson summation formula

$$\rho(E) = \frac{1}{\hbar^n} \sum_\nu e^{-i\pi\alpha\cdot\nu/2} \int_{\substack{+\text{ve} \\ \text{quadrant}}} d\tau_I \, \delta\left(E - H(\mathbf{I})\right) \exp\frac{2\pi i}{\hbar}\nu\cdot\mathbf{I}. \tag{8.33}$$

The ν are the usual integer vectors. Retaining only $\nu = (0,0,\ldots,0)$ corresponds to replacing summations by integrations and reduces to the Fermi-Thomas result

$$\rho^{\mathrm{FT}}(E) = \frac{1}{(2\pi\hbar)^N} \int d\tau\,(\mathbf{q},\mathbf{p})\delta\left(E - H(\mathbf{q},\mathbf{p})\right) \tag{8.34}$$

relating the density of states to the volume of phase space in "quanta" of $(2\pi\hbar)^N$, where N is the number of degrees of freedom.

The terms $\nu \neq (0,0,\ldots,0)$ give oscillatory corrections to this Fermi-

Thomas background, as in the previous sections. Berry and Tabor obtain the result in terms of the integer q and integer vector μ where

$$\nu = \mu q \tag{8.35}$$

and q is the highest common factor of the components of ν, so that μ is a "prime vector." The result is

$$\rho(E) = \rho^{FT}(E) + \frac{2}{\hbar^{(N+1)/2}} \sum_{\mu}{}' \frac{1}{|\mu|^{(N-1)/2}|\omega|\sqrt{|K|}}$$

$$\times \sum_{q=1}^{\infty} \frac{\cos\left[q(W(\mu)/\hbar - \pi\alpha\cdot\mu) + \frac{1}{4}\pi\beta(\mu)\right]}{q^{(N-1)/2}} \tag{8.36}$$

where K is the scalar curvature of the energy shell in action space, $W(\mu)$ is the action around a closed trajectory, β is an index that plays a similar role on the energy shell to the canonical Maslov index on a toroid. $\sum'$ represents a restricted sum. No stability criteria for the trajectories appear. This summation formula has a number of deficiencies for special cases, which are considered by Berry and Tabor using uniformization procedures and complex toroids. They also relate the theory to the smoothed densities of the previous section.

They apply the theory with and without smoothing to the separable case of a particle of 1 proton mass moving in a Morse potential

$$V(r) = V_0\left[e^{-2\delta(r-r_0)} - 2e^{-\delta(r-r_0)}\right]$$

$$\left(V_0 = 0.2 \text{ eV}, \quad r_0 = 2.5 \text{ Å}, \quad \delta = 1 \text{ Å}^{-1}\right) \tag{8.37}$$

There are 166 bound states, not including degeneracy.

Starting with no smoothing ($\gamma = 0$) and including increasing numbers of bound-state trajectories the result is similar to Fig. 8.1. Introducing smoothing ($\gamma' = 0.001$ eV) without and with a uniform approximation gives progressive improvement in agreement with the exactly calculated quantal density of states, as shown in Fig. 8.2.

Results have also been obtained for the three-dimensional case.

Clearly, the smoothed sum over trajectories is an effective practical means of obtaining densities of states, and it will be interesting to see how it applies to the nonseparable case. For the regular spectrum the method does not appear to be as effective as the EBK methods of Section VI for obtaining energy levels.

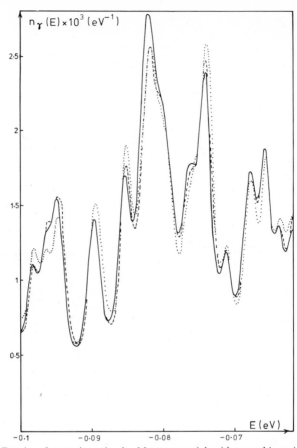

Fig. 8.2.   Density of states in a circular Morse potential, with smoothing width = .001 eV.
———, exact quantum calculation; · · · · · · ·, simple semiclassical formula; -------, uniform approximation.

APPENDIX: STATIONARY PHASE INTEGRALS.

For real $f(q)$, $g(q)$

$$\frac{1}{(2\pi\hbar)^{1/2}} \int dq\, g(q) \exp\frac{i}{\hbar} f(q) \approx \sum_{\mu} \frac{g_{\mu}}{|f''_{\mu}|^{1/2}} \exp\left[\frac{i}{\hbar} f_{\mu} + \frac{i\pi}{4}\sigma_{\mu}\right] \quad \text{(A1)}$$

where μ label the stationary points of $f(q)$, so that $f'(q_{\mu}) = 0$. The undefined quantities are

$$f_\mu = f(q_\mu), \qquad f_\mu'' = f''(q_\mu), \qquad g_\mu = g(q_\mu) \tag{A2}$$

$$\sigma_\mu = \mathrm{Sgn}(f_\mu'') = \begin{cases} +1 & f_\mu'' > 0 \\ -1 & f_\mu'' < 0 \end{cases} \tag{A3}$$

It is assumed that $f_\mu'' \neq 0$. For N-dimensional $\mathbf{q}$ with volume element $d\tau_q$ and for real $f(\mathbf{q})$, $g(\mathbf{q})$

$$\frac{1}{(2\pi\hbar)^{N/2}} \int d\tau_q \, g(\mathbf{q}) \exp \frac{i}{\hbar} f(\mathbf{q})$$

$$\approx \sum_\mu \frac{g_\mu}{|f_\mu''|^{1/2}} \exp\left[\frac{i}{\hbar} f_\mu + \frac{i\pi}{4} \sigma_\mu \right] \qquad \text{(stationary)}, \tag{A4}$$

where μ labels the stationary points of $f(q)$, so that $\nabla_q(q_\mu) = 0$. The undefined quantities are

$$f_\mu = f(q_\mu), \qquad f_\mu'' = \mathrm{Det}\left\| \frac{\partial^2 f}{\partial q_j \cdot \partial q_k} \right\|_{q=q_\mu}, \qquad g_\mu = g(q_\mu) \tag{A5}$$

$$\sigma_\mu = \mathrm{Sgn}\left(\frac{\partial^2 f}{\partial q_j \cdot \partial q_k} \right), \; \mathrm{Sgn}(A_{jk}) = \sum_j \mathrm{Sgn}(\lambda_j), \; \text{eigenvalues} \tag{A6}$$

It is assumed that $f_\mu'' \neq 0$.

Acknowledgements

I thank the U. K. Science Research Council for a grant and R. Balian, M. Berry, A. R. Edmonds, R. A. Marcus, W. H. Miller, K. S. J. Nordholm, N. Pomphrey, P. Pechukas, S. A. Rice, and D. Richards for helpful communications and for sending papers in advance of publication.

References

1. A. Einstein, *Verhand. Deut. Phys. Ges.*, **19**, 82 (1917).
2. M. V. Berry and K. E. Mount, *Reports Progr. Phys.*, **35**, 315 (1972); R. A. Marcus, *J. Chem. Phys.*, **56**, 311 (1972); H. Kreek, R. L. Ellis, and R. A. Marcus, *J. Chem. Phys.*, **61**, 4540 (1974); W. H. Miller, *Advan. Chem. Phys.*, **25**, 69 (1974); **30**, 77 (1975); M. S. Child, *Molec. Phys.*, **29**, 1421 (1975); J. N. L. Connor, *Ann. Rept. Progr. Chem. A*, **70**, 5 (1973); *Molec. Phys.*, **31**, 33 (1976); J. J. Duistermaat, *Commun. Pure Appl. Math.*, **27**, 207 (1974).
3. See Ref. 2. For early work see Ref. 12 of R. Balian and C. Bloch, *Ann. Phys.* (*N. Y.*) **85**, 514 (1974).
4. G. Starkschall and J. C. Light, *J. Chem. Phys.*, **61**, 3417 (1974) and references therein.

5. J. L. Dunham, *Phys. Rev.*, **41**, 721 (1932); P. N. Argyres, *Physics* **2**, 131 (1965); N. Fröman and P. O. Fröman, *Ann. Phys.* (*N. Y.*) **83**, 103 (1974).
6. See Ref. 2.
7. K. Schwarzschild, *Berliner Berichte*, 548 (1916); P. S. Epstein, *Ann. der Physik*, **50**, 489 (1916).
8. M. Jammer, *Conceptual Developments of Quantum Mechanics*, McGraw-Hill, New York, 1966 Chapters 2 and 3.
9. M. L. Brillouin, *J. de Physique* (*Ser.* 6) **7**, 353 (1926); J. B. Keller, *Ann. Phys.* (*N. Y.*) **4**, 180 (1958).
10. V. Maslov, *Théorie des Perturbations*, Dunod, Paris, 1972.
11. Our Maslov index function σ_q is related to the Maslov index $\mathrm{Ind}[l(A,B)]$ of Ref. 10 by

$$\sigma_q(B) - \sigma_q(A) = 2\,\mathrm{Ind}[l(A,B)]$$

    The factor 2 is introduced to ensure that σ_p, like σ_q, is an integer. The term Maslov index in this chapter refers to the index α for closed trajectories on toroids only, in which case it has the same meaning as in Ref. 10.
12. H. G. Goldstein, *Classical Mechanics*, Addison-Wesley, Reading, Mass., 1953; H. C. Corben and P. Stehle, *Classical Mechanics*, Wiley, New York, 1950.
13. L. Landau and E. M. Lifshitz, *Mechanics*, Pergamon, Oxford, 1960.
14. M. Born, *Mechanics of the Atom*, Ungar, New York, 1960, Chapter 2.
15. D. W. Noid and R. A. Marcus, *J. Chem. Phys.*, **62**, 2119 (1975).
16. Y. Hagihara, *Celestial Mechanics*, MIT Press, Cambridge, Mass., 1970.
17. W. Eastes and R. A. Marcus, *J. Chem. Phys.*, **61**, 4301 (1974).
18. V. Trkal, *Proc. Camb. Phil. Soc.*, **21**, 80 (1922); J. H. Van Vleck, *Phys. Rev.*, **22**, 547 (1923); I. C. Percival, *J. Phys. A: Math. Nucl. Gen.*, **7**, 794 (1974).
19. I. C. Percival and N. Pomphrey, *Molec. Phys.*, **31**, 97 (1976).
20. I. C. Percival and N. Pomphrey, *J. Phys. B* (*Atom. Molec. Phys.*), **9**, 3131 (1976).
21. S. Chapman, B. C. Garrett, and W. H. Miller, *J. Chem. Phys.*, **64** (1976).
22. A. N. Kolmogorov, *Dok. Akad. Nauk*, **98**, 527 (1954); A. N. Kolmogorov, in Gerretson and de Groot, Eds., *Proceedings of the International Congress of Mathematicians Ser. II*, *7th Congress*, North-Holland, Amsterdam, 1954, Vol. 1, p. 315 (1957) (in Russian). An English translation of this article forms Appendix D of R. Abraham, *Foundations of Mechanics*, W. A. Benjamin, New York, 1967. V. I. Arnol'd, *Iz. Akad. Nauk ser Matem*, **25**, 21 (1961); V. I. Arnol'd, *Usp. Matem. Nauk*, **18**, No. 5, 13 (1963) [English translation, Russian Math. Surv., **18**, No. 5, 9 (1963)]; V. I. Arnol'd, *Usp. Matem. Nauk*, **18**, No. 6, 91 (1963) [English translation, *Russian Math. Surv.*, **18**, No. 6, 85 (1963)]; J. Moser, *Nachr. Akad. Wiss. Göttingen*, No. 1, 1 (1962).
23. V. I. Arnol'd and A. Avez, *Ergodic Problems of Classical Mechanics*, New York, W. A. Benjamin, 1968; C. L. Siegel and J. K. Moser, *Lectures on Celestial Mechanics*, Springer, Berlin, 1971; J. Ford, *Lectures in Statistical Physics*, *Vol. II*, W. C. Schieve, Ed., Springer, New York, 1972; L. Galgani and A. Scotti, *Revista Nuovo Cimento*, **2**, 189 (1972); G. H. Walker and J. Ford, *Phys. Rev.*, **188**, 416 (1969).
24. R. A. Marcus, *Faraday Disc. Chem. Soc.*, **55**, 34 (1973); I. C. Percival, *J. Phys. B* (*Atom. Mol. Phys.*), **6**, L229 (1973); K. S. J. Nordholm and S. A. Rice, *J. Chem. Phys.*, **61**, 203 (1974); **61**, 768 (1974).
25. I. E. Fermi, J. Pasta, and S. Ulam, Los Alamos Report LA-1940 (1955), in Enrico Fermi, *Collected Works*, University of Chicago Press, Chicago, Vol. 2, p. 978.
26. K. R. Symon and A. M. Sessler, *Proceedings of the CERN Symposium on High Energy Accelerators and Pion Physics*, CERN, Geneva, 1956, Vol. 1, p. 44.

27. E. Thiele and D. J. Wilson, *J. Chem. Phys.*, **35**, 1256 (1961); D. L. Bunker, *J. Chem. Phys.*, **37**, 393 (1962).
28. M. Hénon and C. Heiles, *Astron. J.*, **69**, 73 (1964); G. Contopoulos, *Astrophys. J.*, **138**, 1297 (1963); G. Contopoulos, *Astron. J.*, **75**, 96 (1970); **76**, 147 (1971); J. Ford, *Advan. Chem. Phys.*, **24**, 155 (1973).
29. J. Ford and G. H. Lunsford, *Phys. Rev. A*, **1**, 59 (1970).
30. V. I. Arnol'd, *Russian Math. Surv.*, **18**, No. 6, 85 (1963).
31. G. H. Walker and J. Ford, *Phys. Rev.*, **188**, 416 (1969).
32. Ya. G. Sinai, *Izv. Akad. Nauk. SSSR Ser. Mat.*, **30**, 15 (1966): Amer. Math. Soc. Transl., **68**, 34 (1968).
33. I. C. Percival, Ref. 24; I. C. Percival and N. Pomphrey, Ref. 19.
34. N. Pomphrey, *J. Phys. B (Atom. Mol. Phys.)*, **7**, 1909 (1974).
35. B. Barbanis, *Astron. J.*, **71**, 415 (1966).
36. A. Norcliffe and I. C. Percival, *J. Phys. B (Atom. Mol. Phys.)*, **1**, 774 (1968); C. L. Pekeris, *Proc. Symp. Appl. Math.* **2**, 71 (1950).
37. M. C. Gutzwiller, *J. Math. Phys.*, **8**, 1979 (1967); **10**, 1004 (1969); **11**, 1791 (1970); **12**, 343 (1971); W. H. Miller, *J. Chem. Phys.*, **56**, 38 (1972).
38. A. Norcliffe, I. C. Percival, and M. J. Roberts, *J. Phys. B (Atom. Mol. Phys.)*, **2**, 578 (1969); A. Norcliffe, *Case Studies Atom. Phys.*, **4**, 1 (1973).
39. R. Balian and C. Bloch, *Ann. Phys. (N. Y.)*, **85a**, 514 (1974).
40. References 2 and 4; P. Pechukas, *J. Chem. Phys.*, **57**, 5577 (1972); W. H. Miller, Ref. 41.
41. M. V. Berry and M. Tabor, *Proc. Roy. Soc. (London) Ser. A*, **349**, 101 (1976); see also W. H. Miller, *J. Chem. Phys.*, **63**, 996 (1975).

THE CORRESPONDENCE PRINCIPLE
IN HEAVY-PARTICLE COLLISIONS

A. P. CLARK

Department of Physics,
University of Stirling,
Stirling, FK9 4LA, Scotland

A. S. DICKINSON

Department of Atomic Physics, School of Physics,
The University, Newcastle-Upon-Tyne,
NE1 7RU, England

D. RICHARDS

Faculty of Mathematics,
The Open University,
Milton Keynes, MK7 6AA, England

CONTENTS

I. INTRODUCTION

The simplest energy-transfer collision processes are those involving vibrationally and rotationally inelastic scattering. These collisions are important in such diverse areas as astrophysics, where, for example, He-H_2CO collisions have recently been studied (Garrison et al., 1975), and the absorption and dispersion of sound waves in gases (Herzfeld and Litovitz, 1959). Other areas where these processes are significant include the kinetics of gas-phase chemical reactions, infrared lasers, fluorescence spectroscopy, and pressure broadening of spectral lines (Levine and Bernstein, 1974, Chapter 5).

These collisions have attracted considerable interest over the past decade; reviews of the theoretical aspects have been written by Secrest (1973), Connor (1973), Lester (1975), and Balint-Kurti (1975).

Much of the impetus for theoretical studies has arisen from the availability of some reliable numerical solutions of the Schrödinger equation describing inelastic collisions, starting with the work of Secrest and Johnson (1966) and Allison and Dalgarno (1967). These solutions provided for the first time a benchmark for assessing the validity of many of the approximations that had been applied to atom-molecule collisions since the introduction of quantum mechanics. Further impetus has been provided by the advances in experimental technique that have made possible such measurements as the angular distributions for rotationally and vibrationally inelastic scattering (Toennies, 1974) and low temperature vibrational deexcitation rates for simple systems (e.g., Audibert et al., 1974).

This type of collision has the attractive property, for theoreticians, that numerically exact solutions can be obtained for realistic problems, such as the simultaneous rotational and vibrational excitation of H_2 by Li^+ (Schaefer and Lester, 1975). It is clearly impossible, however, at present to obtain similar reliable solutions to related problems, such as simultaneous rotational and vibrational excitation in methane-methane collisions.

Here we concentrate on the use of classical and semiclassical approximations for atom-molecule collisions. The best-known semiclassical method is the classical S-matrix theory introduced by Miller and Marcus. This has been reviewed extensively by Miller (1974, 1975), who also discusses the purely classical Monte Carlo trajectory methods. Classical S-matrix theory has been very successful for one-dimensional problems, including reactive scattering. Although it is likely to yield satisfactory results for realistic atom-molecule collisions, the computational effort required for multidimensional problems appears at least comparable with that of the full quantal solution, and partial averaging over some quantal states is usually necessary.

In calculating atom-molecule cross sections there are two main problems: determining the potential hypersurface and then deriving cross sections from it. The complexity of the first problem is related to the number of electrons involved, while that of the second is related to the number of nuclei. Consequently, the determination of potential surfaces is intrinsically more difficult. Reliable surfaces are expensive to calculate and available for only a few relatively simple systems. This appears likely to be the situation for some time to come. Since for many collisions only very approximate surfaces are available, there is a need for a simple, robust, approximate method giving cross sections accurate to within, say, 30%. The method should be tractable for the more complex molecule-molecule collisions and should be applicable to collisions involving large quantum numbers.

In this chapter we show that the use of classical perturbation theory in a classical S-matrix-type approach yields a simple approximation with acceptable accuracy for multidimensional inelastic collision problems. Both classically allowed and classically forbidden processes are described uniformly with a modest computing effort that is largely independent of the target quantum numbers. Only interactions that can lead to inelastic scattering are considered: quantal potential scattering is well described by uniform approximations based upon purely classical quantities (Berry and Mount, 1972), while reactive scattering cannot be treated using classical perturbation theory.

The initial thrust for the development of the approach described here came from studies of electron-atom collisions. The analysis of observations on radio-recombination lines emitted by highly excited hydrogen atoms required reliable inelastic level-to-level cross sections for electron-hydrogen collisions involving principal quantum numbers up to 250—larger quantum numbers than have been resolved in the laboratory for any system, to our knowledge. A suitable approximation, valid in the limit of large quantum numbers, was derived by Percival and Richards (1970) and it was termed the strong-coupling correspondence principle (SCCP). Their derivation of the SCCP was quite different from that of classical S-matrix theory. The SCCP has now been tested more thoroughly for molecular collisions than for electron-atom collisions. The application of the SCCP to charged-particle hydrogen collisions has been reviewed by Percival and Richards (1975).

This review covers the application of the SCCP to atom-molecule scattering. Its form is as follows. First, since any classical method involves the quantization of the internal states of the target and the use of classical approximations to matrix elements, we consider these briefly in Section II, illustrating how accurate the classical methods often are, even for low quantum numbers. Many of the classical results used here were standard in the days of the old quantum theory (Born, 1927), and their utility as accurate approximations in quantum mechanics has often been overlooked. Having established this foundation we review briefly some relevant collision theory in Section III and then derive the SCCP. Its connections with some other approximations are then established in Section IV. We illustrate the use of the SCCP by applying it in Section V to the well-studied problem of vibrational excitation in a collinear atom-diatomic molecule collision. This section shows that the SCCP is useful even for transitions involving very low quantum numbers. Rotational excitation of a rigid rotor is then examined in Section VI, with some emphasis on the implications of our findings for purely classical calculations in Section VII. Work in progress on the application of the SCCP to simultaneous vibrational and rotational transitions is considered briefly in Section VIII, and we end with our conclusions in Section IX.

II. THE CORRESPONDENCE PRINCIPLE

A. Introduction

Macroscopic systems obey the laws of classical mechanics; however, the transition from the quantum mechanics of atoms and molecules to the classical mechanics of macroscopic systems is not simple.

It is well known that quantum mechanics tends to classical mechanics in the limit $\hbar \to 0$. It is shown, for example, in Messiah (1964, p. 222), that in this limit the quantal flux of probability is the same as for a classical ensemble, so that there is a direct correspondence between classical and quantal statistics. However, this limit is subtle, as may be seen by considering Schrödinger's equation for a particle of mass m, energy E moving in a one-dimensional potential $V(x)$:

$$\hbar^2 \psi''(x) + 2m(E - V(x))\psi = 0. \tag{2.1}$$

On letting $\hbar \to 0$ the equation changes its form; quantal wave functions are always nonanalytic as $\hbar \to 0$.

Quantum mechanics can be formulated in terms of action functions and path integrals (Dirac, 1958, Section 32; Feynman, 1948; Feynman and Hibbs, 1965). When $\hbar$ is negligible in comparison with a typical action A of the system, the only significant contributions come from the classical paths. However, this is not a sufficient condition for the validity of classical mechanics since there can still be interference between classical paths with actions differing by $O(\hbar)$; this manifests itself in oscillations in the quantal probabilities. If ΔA is the observable difference in action between classical paths, these oscillations cannot be detected unless measurements are sufficiently precise to distinguish actions such that ΔA is of the order of $\hbar$, for example, high-resolution measurements of differential cross sections for potential scattering (Buck, 1975).

If R is a typical dimension of the system, it is generally true that $\hbar/A \to 0$ as $R \to \infty$; also, if the relative precision of measurements of any observable remains constant, then $\hbar/\Delta A \to 0$ so that quantal oscillations cannot be detected.

This limit of large actions is the basis of the elementary correspondence principles whereby every observable quantal quantity is supposed to have the same value as its classical analog, provided this analog exists. However, the application of this principle even in the simplest case can be ambiguous; see, for example, Section III.C.

Here we consider the application of this idea to the calculation of energy levels and matrix elements, as we shall need the results obtained when we consider collisions in Section III.

B. Semiclassical Energy Levels

Even for highly excited systems, the quantal energy levels are discrete. Thus in using classical mechanics to describe such systems, unless averages over many quantal states are required, the classical system must be quantised using the Bohr–Sommerfeld quantization rule. Here we briefly state the results for a bound N-dimensional system.

Let $H(\mathbf{p}, \mathbf{q})$ be the Hamiltonian of the N-dimensional system, the motion of which we suppose bound in configuration space. The solution, $S(\mathbf{q}, \mathbf{P})$, of the time-independent Hamilton–Jacobi equation,

$$H\left(\frac{\partial S}{\partial \mathbf{q}}, \mathbf{q}\right) = E, \tag{2.2}$$

where $\mathbf{P}$ are the N constants of integration (one of which is the energy E), then defines a canonical transformation from the old conjugate variables $(\mathbf{q}, \mathbf{p})$ to a new set $(\mathbf{Q}, \mathbf{P})$; see, for example, Goldstein (1959) or Born (1927),

$$\mathbf{p} = \frac{\partial S}{\partial \mathbf{q}} \qquad \mathbf{Q} = \frac{\partial S}{\partial \mathbf{P}}. \tag{2.3}$$

In principle, all information concerning the system is contained in the action function S. In practice, it is generally easier to solve the ordinary differential equations arising from Hamilton's equations than the partial differential equation (2.2).

For separable systems a new set of conjugate variables, the action-angle variables, may be defined by the equations

$$I_k = \frac{1}{2\pi} \oint \frac{\partial S}{\partial q_k} \, dq_k, \qquad k = 1, 2, \ldots, N, \tag{2.4a}$$

$$\theta_k = \frac{\partial S}{\partial I_k}, \qquad k = 1, 2, \ldots, N. \tag{2.4b}$$

The action variables I_k are constants of the motion and may be shown to be adiabatic invariants; the Hamiltonian is now $H(\mathbf{I})$, and the equation giving the angle variables $\boldsymbol{\theta}$ as functions of time is

$$\dot{\boldsymbol{\theta}} = \frac{\partial H}{\partial \mathbf{I}} = \text{const}, \qquad \boldsymbol{\theta} = \boldsymbol{\omega}^c t + \boldsymbol{\delta}, \qquad (2.5)$$

where $\boldsymbol{\delta}$ is a constant phase vector. The $\boldsymbol{\omega}^c$ vector is the fundamental frequency vector of the classical system. Thus the conjugate variables $(\boldsymbol{\theta}, \mathbf{I})$, and those obtained from them by canonical transformations of determinant ± 1 (Born, 1927), are of fundamental importance since the motion is multiply periodic in the angle variables $\boldsymbol{\theta}$; any dynamical variable $F(\mathbf{q}, \mathbf{p})$ may be written

$$F(\mathbf{q}, \mathbf{p}) = \sum_{\text{all } \mathbf{s}} F_{\mathbf{s}}(\mathbf{I}) \exp - i\mathbf{s} \cdot \boldsymbol{\theta}. \qquad (2.6)$$

The surfaces in the phase space $(\boldsymbol{\theta}, \mathbf{I})$ given by $\mathbf{I} = \text{constant}$ are N-dimensional tori, and the angle variables $\boldsymbol{\theta}$ are the angular coordinates on each torus.

In many systems of physical interest, for example, the rigid rotor, there are relations between the fundamental frequencies of the form

$$\sum_{k=1}^{N} n_{ik} \omega_k^c = 0, \qquad i = 1, 2, \ldots, P < N, \qquad (2.7)$$

where the n_{ik} are integers. In this case another set of action-angle variables may be found (Born, 1927) such that

$$\begin{aligned} \theta_k &= \text{constant}, \quad k = 1, \ldots, P, \\ \theta_k &= \omega_k^c t + \delta_k, \quad k = P+1, \ldots, N, \\ I_k &= \text{constant}, \quad k = 1, 2, \ldots, N. \end{aligned} \qquad (2.8)$$

In this case the motion in configuration space fills an $(N-P)$-dimensional region rather than the whole space.

In the early days of the "old quantum mechanics" it was found that setting

$$I_k = n_k \hbar \qquad (2.9)$$

(Bohr, 1918), where n_k is an integer, gave energy levels agreeing with experiment. There was no rigorous justification for this other than empirical evidence.

Subsequent work has shown that (2.9) gives agreement with quantum mechanics in the asymptotic limit $n_k \to \infty$, and that better agreement is obtained if we write the Bohr-Sommerfeld quantization condition as

$$I_k = (n_k + \gamma_k)\hbar \qquad n_k = 0, 1, 2, \ldots \qquad (2.10)$$

where γ_k depends upon the nature of the motion in the kth direction; for libration $\gamma = \frac{1}{2}$ and for rotation $\gamma = 0$. The asymptotic form of the quantal energy levels is then

$$E(\mathbf{n}) = E_{n_1 n_2 \cdots n_N} = H(I_1, I_2, \ldots, I_N) \qquad (2.11)$$

where I_k is obtained from (2.10).

In practice, it is found that the Bohr-Sommerfeld quantization rule (2.11) generally gives accurate estimates to quantal energy levels, and that for the important examples of the simple harmonic oscillator, the Coulomb potential, and the Morse potential it is exact.

The fundamental frequency vector $\boldsymbol{\omega}^c(\mathbf{I})$ is readily related to differences in energy levels using (2.11) and (2.5):

$$E(\mathbf{n}+\mathbf{s}) - E(\mathbf{n}) \cong \mathbf{s} \cdot \boldsymbol{\omega}^c(\mathbf{I})\hbar \qquad (2.12)$$

where $\mathbf{I} = (\hat{\mathbf{n}} + \boldsymbol{\gamma})\hbar$ and where $\hat{\mathbf{n}}$ is some mean of $\mathbf{n}$ and $\mathbf{n} + \mathbf{s}$; often we take $\hat{\mathbf{n}} = \mathbf{n} + \mathbf{s}/2$.

Although the Bohr–Sommerfeld quantization rule is generally a very good approximation, relations (2.10) and (2.11) for energy levels are unsatisfactory for three reasons. First, they represent only the first term in the asymptotic expansion of the energy levels. Second, they were derived implicitly assuming that there is only one classically accessible region. Finally, they are derived from a theory that assumes a separable Hamiltonian. Improvements have been made in all three of these directions.

Regarding the first and second points, considerable effort has been employed on one-dimensional systems using suitable approximations or expansions of the wave function.

Dunham (1932a) has obtained further terms in the asymptotic expansion of the energy levels for one-dimensional systems with only one classically accessible region. These correction terms have been applied to a vibrating rotor (Dunham, 1932b) and to a potential of the form $V(x) = x^{2p}$, for various integers p (Krieger et al., 1967). An alternative derivation of the Dunham correction is given by Argyres (1965).

The case of a number of classically accessible regions separated by barriers of finite height has been treated fairly extensively. Child (1974a) has reviewed the problem of three turning points and has obtained a generalization of the quantization condition (2.11) for the energy levels of quasibound states. The problem of four turning points has been treated by a number of authors (Dennison and Uhlenbeck, 1932; Fröman, 1966; Landau and Lifshitz, 1965; Child, 1974b).

When the Hamiltonian is nonseparable the quantal spectra may be regular or irregular (Percival, 1973). In the regular region the classical motion is topologically similar to that of separable systems and the correct quantization rules are given by Einstein (1917a) and Percival and Pomphrey (1976) (see also Percival, 1977).

C. Classical Radiation Theory and Bohr's Correspondence Principle

The principle of equating classical and quantal quantities can be illustrated using radiation theory; this is the basis of the old quantum theory, and it gives a physical foundation to the correspondence principles used later in this chapter.

According to classical radiation theory the intensity of dipole radiation emitted per unit time by a particle of charge e and position $\mathbf{r}$ is (Landau and Lifshitz, 1971, Section 7)

$$I = \frac{2e^2}{3c^3} |\ddot{\mathbf{r}}|^2. \tag{2.13}$$

If the motion of the particle is bounded and described by a separable Hamiltonian, the theory of the preceding section holds and we may use (2.6) to write

$$\mathbf{r} = \sum_{\text{all } \mathbf{s}} \mathbf{r}_{\mathbf{s}} \exp(-i\mathbf{s} \cdot \boldsymbol{\theta}), \qquad \boldsymbol{\theta} = \boldsymbol{\omega}^c t + \boldsymbol{\delta}. \tag{2.14}$$

If the total energy radiated in a period is much less than the total energy, the system radiates at each positive frequency $\mathbf{s} \cdot \boldsymbol{\omega}^c$ with intensity

$$I_{\mathbf{s}} = \frac{4e^2}{3c^3} (\mathbf{s} \cdot \boldsymbol{\omega}^c)^4 |\mathbf{r}_{\mathbf{s}}|^2. \tag{2.15}$$

The connection between this purely classical theory and the equivalent quantal theory is made by applying the correspondence principle. For energy levels this was first done by Bohr (1918); using the Einstein (1917b) relation between energy levels and the frequency of the radiation emitted in a transition from a level $E(\mathbf{n})$ to a level $E(\mathbf{n}')$,

$$\hbar\omega^q(\mathbf{n}, \mathbf{n}') = E(\mathbf{n}) - E(\mathbf{n}'), \tag{2.16}$$

he obtained (2.12) using the physical argument that classical mechanics must predict the observed frequencies when the initial and final orbits are close. Similarly, for intensities Bohr (1918) relates the intensity of the radiation with frequency $\omega^q(\mathbf{n}, \mathbf{n}')$ to the intensity calculated using ordinary classical radiation theory.

Thus on using the usual quantal theory of radiation (see, for example, Bethe and Salpeter, 1957, equation 59.10) we have, to within an arbitrary phase,

$$\langle \mathbf{n}' | \mathbf{r} | \mathbf{n} \rangle \cong \mathbf{r}_s, \qquad s = n - n', \qquad (2.17)$$

which is Heisenberg's form of the correspondence principle; this relates quantal matrix elements to classical Fourier components.

In both relations (2.12) and (2.17) two quantal states are connected by one classical orbit so that there is an ambiguity concerning the choice of the classical orbit. We discuss this difficulty further below.

D. Heisenberg's Form of the Correspondence Principle

The generalization of (2.17) to arbitrary functions and to systems of N degrees of freedom with separable Hamiltonians is straightforward; it is known as Heisenberg's form of the correspondence principle. For a system whose motion in configuration space is bound, we have, to within an arbitrary phase,

$$\langle \mathbf{n}' | F(\mathbf{q}) | \mathbf{n} \rangle \cong F_s(\mathbf{I}), \qquad s = n - n', \qquad (2.18)$$

where $F_s(\mathbf{I})$ is the Fourier component of $F(\mathbf{q})$ evaluated along a classical trajectory with action variables $\mathbf{I}$:

$$F_s(\mathbf{I}) = \left(\frac{1}{2\pi} \right)^N \int d\boldsymbol{\theta} \, F(\mathbf{q}(\boldsymbol{\theta}, \mathbf{I})) \exp i s \cdot \boldsymbol{\theta}. \qquad (2.19)$$

For a one-dimensional system with unbounded coordinate x we have

$$\langle E' | F(x) | E \rangle \cong F_{\Delta E}(\tilde{E}), \qquad \Delta E = E - E', \qquad (2.20)$$

where $F_{\Delta E}(\tilde{E})$ is the Fourier transform of the classical motion with energy $\tilde{E}$:

$$F_{\Delta E}(\tilde{E}) = \frac{\sqrt{vv'}}{2} \int_{-\infty}^{\infty} dt F\left(x(t, \tilde{E}) \right) \exp(i \Delta E t / \hbar) \qquad (2.21)$$

and where the normalization

$$\langle E | x \rangle \underset{x \to \infty}{=} \sin\left(\frac{\sqrt{2mE}\, x}{\hbar} + \delta \right) \qquad (2.22)$$

has been assumed.

The mean value of an operator provides a particular example of these correspondence principles.

In both (2.18) and (2.20) there is the same ambiguity as in (2.17) so that, in practice, an orbit that is some mean of the intial and final orbits is chosen. For example, we could take

$$I = \frac{(n+n')\hbar}{2}, \qquad \tilde{E} = \frac{E+E'}{2} \tag{2.23}$$

in (2.18) and (2.20), respectively. We return to this problem later in this section.

Heisenberg's form of the correspondence principle may be justified using WBK wave functions (Kramers, 1964, p. 416) or by considering wave packets (Landau and Lifshitz, 1965, p. 165), but neither of these methods gives the next term in the asymptotic expansion of the matrix element of which (2.18) or (2.20) is the first term. As far as the authors are aware, there is no proof of this correspondence principle for off-diagonal matrix elements that gives the next term in the asymptotic expansion.

In deriving Heisenberg's correspondence principle, it should be noted that two separate approximations are made. First, all quantum numbers must be large so that the quantal wave functions have many oscillations, and, second, the differences between all initial and final quantum numbers must be small so that there is interference between the initial and final states. It is also implicitly assumed that the operator F varies slowly by comparision with the wave functions. However, despite these reservations this correspondence principle is remarkably accurate when used with low quantum numbers.

The lack of a correction to Heisenberg's correspondence principle means that its accuracy has to be judged by comparison with exact matrix elements. The assumptions upon which it is based suggest that its accuracy decreases as $\min(n_k)$ decreases and as $\min(n_k/|n_k - n_k'|)$ increases; this seems to be true in practice. Also, its accuracy depends on the system to which it is applied. However, its accuracy can often be enhanced by careful choice of the classical orbit along which the Fourier components are evaluated.

In Table I we show (following Naccache, 1972) the comparison of various matrix elements of $\langle n|x|n+s\rangle$ for the Morse potential

$$V(x) = D\left\{1 - \exp\left[-a(x-x_e)\right]\right\}^2. \tag{2.24}$$

In this comparison two values of the classical action are chosen to illustrate the difference that this choice can make.

TABLE I

Comparison of Quantal and Correspondence Principle Values of $a^2 s^2 |\langle n|x|n+s \rangle|^2$ for a Morse Potential with About 50 Bound States. Two Different Values of the Classical Action Have Been Chosen, One Corresponding to That Given in (2.23). The Value in Paranthesis Is the Power of Ten by Which the Number Should Be Multiplied.

Transition $n - n+s$	Quantal value	Correspondence principle value	
		$I = (n + s/2)\hbar$	$I = \left[\dfrac{(n+\dot{s})!}{n!} \right]^{1/s} \hbar$
1–2	2.041 (−2)	1.523 (−2)	2.041 (−2)
3–4	4.166 (−2)	3.627 (−2)	4.167 (−2)
7–8	8.694 (−2)	8.108 (−2)	8.696 (−2)
15–16	1.904 (−1)	1.834 (−1)	1.905 (−1)
1–4	2.627 (−5)	1.686 (−5)	2.620 (−5)
1–12	1.053 (−13)	1.833(−13)	9.63 (−14)
3–6	1.398 (−4)	1.046 (−4)	1.397 (−4)
5–10	4.575 (−6)	3.504 (−6)	4.557 (−6)

Other examples are compared by Naccache (1972), Dickinson and Richards (1974), Dickinson and Shizgal (1975), Stwalley (1973), Tipping (1974), Percival and Richards (1975), Sharma and Hart (1975), and Fröman (1974).

An illustrative example of this correspondence principle is the mean value $\langle n|\exp(ia\alpha x)|n \rangle$ where $|n\rangle$ are eigenfunctions of a simple harmonic oscillator. We have

$$\langle n|\exp(ia\alpha x)|n \rangle = e^{-\alpha^2/2} L_n(\alpha^2), \qquad a^2 = \frac{2\omega}{\hbar}, \qquad (2.25)$$

$$\overset{c.p.}{\cong} J_0 \left(2\alpha\sqrt{n + \tfrac{1}{2}} \right), \qquad (2.26)$$

where (2.25) is given by Wilcox (1966) and (2.26) follows directly from (2.18) and the integral definition of the ordinary Bessel function. The functions $L_n(z)$ and $J_0(z)$ are respectively the Laguerre polynomial and the ordinary Bessel function.

For large α ($\gg \sqrt{n}$), $\exp(ia\alpha x)$ has more zeros in the classically accessible region than the quantal wave function; in this case any semiclassical approximation to the matrix element is likely to be wrong. In this example we have for $\alpha \gg \sqrt{n}$

$$J_0 \left(2\alpha\sqrt{n + \tfrac{1}{2}} \right) \cong \left(\frac{1}{\pi\alpha\sqrt{n + \tfrac{1}{2}}} \right)^{1/2} \cos\left(2\alpha\sqrt{n + \tfrac{1}{2}} - \frac{\pi}{4} \right), \qquad (2.27a)$$

$$e^{-\alpha^2/2} L_n(\alpha^2) \cong \alpha^{2n} e^{-\alpha^2/2}, \qquad (2.27b)$$

although for small α ($\ll \sqrt{n}$) the correspondence principle agrees well with the exact result and has error $O(\alpha^4)$.

It is clear that similar behavior would be found for the matrix element $\langle n'|\exp(ia\alpha x)|n\rangle$, $n' \neq n$.

This is an important example because matrix elements of this kind arise in scattering theory; here α would be related to the change in action of the target system and could be quite large. We return to this problem when it arises in Section III.B.

III. COLLISION THEORY

A. Introduction

In this section we review the theories applicable to the collision of a structureless particle with a system with internal structure. The extension to collisions between systems with internal structure is straightforward but leads to more complicated notation. Rearrangement collisions are not considered.

First we apply classical mechanics, then we consider the time-dependent Schrödinger equation, together with its approximate solutions, and finally we derive the strong-coupling correspondence principle, the approximate theory that we apply in Sections V to VIII.

B. Classical Mechanics

This is a simple method for obtaining cross sections, and a variety of workers have applied it to atom-molecule collisions; see, for example, LaBudde and Bernstein (1971) and references therein, and references in Section VII.

The method of obtaining classical cross sections using Newtonian mechanics is described fully elsewhere (LaBudde and Bernstein, 1971). Here we describe it briefly; to be specific, we consider the rotational excitation of diatomic molecules by atoms.

The rotor is assumed to be in a particular energy level and we require the classical differential cross section $d\sigma^c/d\Delta E$ and the classical probability distribution function, $P^c(\Delta E)$; $P^c(\Delta E)d\Delta E$ is the probability of the rotor making an energy change in the infinitesimal range $(\Delta E, \Delta E + d\Delta E)$. For the type of collision considered here these two quantities are related by an integral over the impact parameter;

$$\frac{d\sigma^c}{d\Delta E} = 2\pi \int_0^\infty dbbP^c(\Delta E). \qquad (3.1)$$

Generally it is assumed that the initial distribution in phase space is microcanonical, which for a rotor means that all orientations are equally likely.

For a given set of initial conditions of the rotor, Ω, say, and for an atom of given energy and impact parameter the final state is found by numerically integrating the equations of motion until the colliding pair is interacting negligibly. Then all dynamical variables will be functions of Ω; for example, the energy transfer would be $\Delta E^c(\Omega)$.

The probability distribution function of energy transfer is then

$$P^c(\Delta E) = \langle \delta(\Delta E - \Delta E^c(\Omega)) \rangle_\Omega \qquad (3.2)$$

where $\delta(x)$ is the Dirac delta function and the mean is over all initial conditions, Ω. When $\Delta E^c(\Omega)$ is obtained numerically, the integral (3.2) can be evaluated directly only by turning the initial-value problem into a boundary-value problem, and generally other techniques are preferable. There are two methods used in practice. First, the fraction of orbits giving energy transfers in specified ranges can be counted to give $P^c(\Delta E)$ as a histogram (see, for example, LaBudde and Bernstein, 1971). Second, the distribution function can be expressed in terms of a complete set of orthonormal functions:

$$P^c(\Delta E) = \sum a_k \phi_k(\Delta E) \qquad (3.3)$$

so that

$$a_k = \int d\Delta E \; P^c(\Delta E) \phi_k^*(\Delta E) \qquad (3.4a)$$

$$= \langle \phi_k^*(\Delta E^c(\Omega)) \rangle_\Omega. \qquad (3.4b)$$

There are three distinct sources of error in evaluating cross sections using this method. The first arises from inaccuracies due to the numerical solution of the equations of motion; in practice, this error can be made negligible by using a suitable integration routine and sufficient computer time. Second, if the histogram method is used there are statistical errors because only a finite sample of the initial conditions may be taken; these errors decrease as $N^{-1/2}$, where N is the number of sample points, but often by careful sampling faster convergence may be obtained (see, for example, LaBudde and Bernstein, 1971, or Hammersley and Handscomb, 1964). Alternatively, if the method of (3.3) and (3.4) is used, there are additional numerical errors in the integral (3.4b) and in the truncation of the infinite series (3.3).

The third source of error is the most important and, unlike those above, is impossible to control; it arises from the total neglect of all quantal effects. For atom-molecule scattering the two most important quantal effects are interference and the quantization of the energy levels of the target. Also, tunnelling can be important, especially for the soft long-range potentials in ion-molecule collisions. We consider the effects of interference in Section VII. Here we consider the problem of quantizing the (continuous) classical distribution function.

C. The Problem of Quantization

Since quantal energy levels are discrete and classical energies are continuous, it is not clear how to obtain cross sections for a transition between discrete states from the classical distribution function. If the final level is associated with a given band of energies detailed balance,

$$Eg_n\sigma(n \to n'; E) = E'g_{n'}\sigma(n' \to n; E') \tag{3.5}$$

where $E(E')$ is the initial (final) projectile energy and g_n is the degeneracy of the target in level n, is not satisfied; also, the band cannot be specified uniquely. If the final level is associated with a given classical energy transfer the cross section is zero, since the classical energy transfer is continuous.

Because of the problems with detailed balance inherent in using a band of energies, we prefer the latter choice and overcome its difficulty by taking a limit and weighting the classical probability by the ratio of the volumes of the classical and quantal phase space of the final state (Abrines and Percival, 1966).

The volume of phase space associated with a quantal state is as usual supposed to be $(2\pi\hbar)^N$, where N is the number of degrees of freedom of the system (see Landau and Lifshitz, 1968). If the quantal substates are equally populated then the total cross section for a transition between levels n and n' with energies E_n and $E_{n'}$ is

$$\sigma^q(n \to n') = \lim_{\delta E \to 0} \frac{d\sigma^c}{d\Delta E}(E_n \to E_{n'}) \cdot \delta E \left[\frac{(2\pi\hbar)^N g_{n'}}{K(E_{n'})\delta E} \right] \tag{3.6}$$

where the first term is the classical cross section and the term in square brackets is the ratio of the quantal and classical phase space volumes. The quantity $K(E)$ is the volume of classical phase space per unit energy interval:

$$K(E) = \int dV \delta(E - H) \tag{3.7}$$

where dV is the volume element of phase space and H is the target Hamiltonian. But if m is large the correspondence principle gives

$$(2\pi\hbar)^N g_m \cong K(E_m)\left|\frac{dE_m}{dm}\right| \tag{3.8}$$

so that

$$\sigma^q(n\rightarrow n') \cong \left.\left|\frac{dE_m}{dm}\right|\right._{m=n'} \frac{d\sigma^c}{d\Delta E}(E_n\rightarrow E_{n'}). \tag{3.9}$$

Accordingly, the total cross section behaves like a differential cross section with respect to n'. It is easy to see that (3.9) satisfies the detailed balance requirement (3.5).

The correspondence (3.9) is generally valid only if n, n', and $|n-n'|$ are all large. It is sometimes valid for small $|n-n'|$ (see Percival and Richards, 1975, Section 7), and sometimes it would be incorrect even when $|n-n'|$ is large, for example, in collisions with asymmetric tops. Thus care is necessary in its use. We shall see in Section IV that the correspondence (3.9) is the correct classical limit of quantum mechanics.

The use of a band of final energies to quantise the cross section has been considered by LaBudde and Bernstein (1973). Since quantum numbers are associated with classical action variables, they associate the quantal cross section with a mean of the classical distribution function of the action variable about the final state and assume that $\sigma^q(n\rightarrow n')$ behaves like a differential cross section with respect to n'. This gives, assuming $\alpha \ll n'$,

$$\sigma^q(n\rightarrow n') = \frac{1}{2\alpha}\int_{E_{n'-\alpha}}^{E_{n'+\alpha}} dE \frac{d\sigma^c}{d\Delta E}(E_n\rightarrow E)f(E) \tag{3.10}$$

where $f(E)$ is an arbitrary weighting function normalized to

$$\int_{E_{n-\alpha}}^{E_{n+\alpha}} dE f(E) = E_{n+\alpha} - E_{n-\alpha}. \tag{3.11}$$

LaBudde and Bernstein (1973) chose $f=1$ and considered the cases $\alpha=1$, $\frac{1}{2}$. By choosing $f(E)=(E_{n'+\alpha}-E_{n'-\alpha})\delta(E_{n'}-E)$ and taking the limit as $\alpha\rightarrow 0$, (3.9) is regained.

When $d\sigma^c/d\Delta E$ is a rapidly varying function of ΔE, σ^q as given by (3.10) will be strongly dependent upon α and will differ from the value given by (3.9). In these circumstances any classical result will be unreliable.

D. Time-Dependent Collision Theory

Here the target is treated quantally while the projectile is assumed to follow a classical path and to produce a time-dependent potential, $V(t)$, which induces transitions in the target. This assumption is valid when the interaction potential changes little in a distance comparable to the de Broglie wave length. This is generally true for heavy-particle collisions, and is often a good approximation for electron collisions.

A more important constraint on this method is that the incident particle is assumed to move along a classical path unperturbed by that part of the potential which causes transitions in the target. Thus energy and angular momentum are not conserved and detailed balance is not satisfied. Approximate adjustments can often be made to enforce these conditions (Vainshtein and Vinogradov, 1970; Dickinson and Richards, 1974). However, since energy is not conserved, this approximation should be most successful when the projectile energy is much greater than the energy transfer.

In this approximation Schrödinger's equation reduces to the infinite set of equations (see, for example, Schiff, 1955)

$$ i\hbar \frac{\partial S}{\partial t}(\mathbf{n'},\mathbf{n};t) = \sum_{\text{all } \mathbf{m}} \langle \mathbf{n'}|V|\mathbf{m}\rangle \exp\left(\frac{i(E_{\mathbf{n'}}-E_{\mathbf{m}})t}{\hbar}\right) S(\mathbf{m},\mathbf{n};t) \quad (3.12a) $$

$$ S(\mathbf{n'},\mathbf{n}; -\infty) = \delta_{\mathbf{n'n}} \quad (3.12b) $$

for the S-matrix elements, $S(\mathbf{n'},\mathbf{n}) = S(\mathbf{n'},\mathbf{n};\infty)$, for a transition from an initial state $\mathbf{n}$ to a final state $\mathbf{n'}$. Here $|\mathbf{n}\rangle$ are the eigenstates of the unperturbed target, with quantum numbers $\mathbf{n}$, and energy levels $E_{\mathbf{n}}$. The cross section is obtained from S by the usual integration over the impact parameter b:

$$ \sigma(\mathbf{n}\rightarrow\mathbf{n'}) = 2\pi \int_0^\infty b\,db\,|S(\mathbf{n'},\mathbf{n})|^2. \quad (3.13) $$

Equations (3.12) and (3.13) are the fundamental equations of time-dependent scattering theory. Exact solutions of (3.12) are impossible except in very special circumstances (see, for example, Section III.H).

A formal solution to (3.12a) and (3.12b) may be written down. This solution expresses the matrix S as an exponential of a matrix A, which itself is an infinite series; the nth term of this series contains nth-order commutators (Magnus, 1954; Robinson, 1963; Pechukas and Light, 1966). In practice, this solution is merely a formal device, although it provides a useful starting point for exponential-type approximations.

The numerical solution of a truncated set of the equations (3.12), however, is generally an easier task than the equivalent equations of time-independent scattering theory. The former is a first-order initial-value problem while the latter is a second-order boundary-value problem.

In heavy-particle collisions there are generally many states coupled together, so that the numerical integration of (3.12) involves handling large matrices. For this reason some effort has been spent in finding approximate solutions to these equations. We consider some of these approximations in Sections III.F to III.I. First we consider some devices for improving the accuracy of (3.12).

E.   The Choice of Trajectory

The transition probability obtained from (3.12) does not satisfy detailed balance. This difficulty may be overcome by choosing the energy of the projectile to be a mean of its initial and final energies and enforcing detailed balance on the resulting probability; see (6.20). This procedure is not unique, but it is found to improve the accuracy of the method considerably; it is equivalent to choosing a mean quantum number for the correspondence principle approximation to matrix elements, equations (2.18) and (2.20). The disadvantage of this simple improvement is that only one transition probability may be corrected at a time.

McCann and Flannery (1975a) have attempted to adjust the orbit of the projectile by allowing it to move in a field determined by the state of the target. Formally, energy is conserved in this scheme. This theory has been applied (McCann and Flannery, 1975a, b) to the rotational excitation of H_2 by H and has been compared to time-independent calculations using the effective-potential method of Rabitz (1975). At low incident energies, where one would expect this theory to be an improvement on (3.12), there are considerable differences between the two calculations; however, since subsidiary approximations are made, conclusions upon the accuracy of this method cannot be made from these studies.

A similar method has been developed and applied to vibrational excitation by Sorensen (1974). Another similar approximation due to Pechukas (1969) has been applied to vibrational excitation by Penner and Wallace (1974). None of these methods appear to give accuracies commensurate with the necessary extra computation.

F.   First-Order and Unitarized First-Order Approximations

The simplest approximate solution to (3.12) is to replace the S-matrix elements of the right-hand side by their initial values (3.12b) to give the first-order approximation,

$$S^{FO}(\mathbf{n'},\mathbf{n}) = -\frac{i}{\hbar} \int_{-\infty}^{\infty} dt \langle \mathbf{n'}|V|\mathbf{n}\rangle \exp\left(\frac{i(E_{\mathbf{n'}} - E_{\mathbf{n}})t}{\hbar}\right) \qquad (3.14)$$

which is valid when the potential V is small. Generally, however, this condition is not satisfied and the transition probability $|S^{FO}|^2$ does not satisfy unitarity. However, this approximation, together with estimates of the transition probability in regions where $|S^{FO}| > 1$, can provide reasonable estimates to cross sections; for example, this method has been successfully used for collisions of electrons with polar molecules (Dickinson and Richards, 1975).

Various unitarized approximations have been suggested (Seaton, 1961) that in one way or another produce a unitary S matrix from S^{FO}. Some of these methods are reviewed by Levine (1971); Levine and Balint–Kurti (1970) compare two unitarized approximations with first-order theories and with exact calculations for both vibrational and rotational excitation. They show that the unitarized approximations are a significant improvement on the first-order result, especially near threshold. Other calculations (Balint–Kurti and Levine, 1970; Bosanac and Balint–Kurti, 1975) support these conclusions.

Fisanick–Englot and Rabitz (1975) have derived a unitary approximation similar to the exponential approximation of Levine and Balint–Kurti, using assumptions the implication of which is difficult to assess. Saha et al. (1974) have applied a unitary approximation due to Takayanagi (1963) to the rotational excitation of CO by Ar and N_2 by Ne, and to TlF by Ar (Saha and Guha, 1975).

G.   The Sudden Approximation

If the potential operator in interaction representation,

$$V^I(t) = \exp\left(iH_0 \frac{t}{\hbar}\right) \cdot V(t) \cdot \exp\left(-iH_0 \frac{t}{\hbar}\right) \qquad (3.15)$$

where H_0 is the unperturbed Hamiltonian of the target, satisfies the commutation relation

$$\left[V^I(t_1), V^I(t_2)\right] = 0 \qquad \text{all} \quad t_1 \neq t_2 \qquad (3.16)$$

then (3.12) may be solved exactly to give

$$S(\mathbf{n}',\mathbf{n}) = \left\langle \mathbf{n}' \left| \exp\left(-\frac{i}{\hbar} \int_{-\infty}^{\infty} dt V^I(t) \right) \right| \mathbf{n} \right\rangle \tag{3.17}$$

which is the first term of the Magnus (1954) expansion.

If the interaction potential $V(t)$ is strongly peaked in time, that is, if the collision time is short in comparison with the characteristic time of the transition, then (3.16) is approximately true and the sudden approximation is valid; this is given by (3.17) with V^I replaced by V. For rotational excitation, except at the lowest incident energies, it is generally a good approximation. However, considerable computational effort is required for its evaluation. It should be mentioned that the infinite-order sudden approximation (Secrest, 1975), which is derived from time-independent scattering theory, is based partly on the same assumptions as the sudden approximation described above, and gives similar numerical results (Pack, 1975; Pattengill, 1975).

H.   The Equal Energy Spacing Approximation

The infinite set of coupled equations (3.12) may be solved if both the energy differences and matrix elements depend only on differences of quantum number:

$$\text{(a)} \qquad E_\alpha - E_\beta = \hbar(\alpha - \beta) \cdot \omega \tag{3.18a}$$

$$\text{(b)} \qquad \langle \alpha | V | \beta \rangle = V_{\alpha - \beta}(t) \tag{3.18b}$$

where the vector ω is a constant independent of α and β and $V_{\mathbf{n}}(t)$ depends only on $\mathbf{n}$. With these assumptions Presnyakov and Urnov (1970) have shown that the solution to (3.12) may be obtained by defining the generating function,

$$G(\boldsymbol{\theta},t) = \sum_{\mathbf{n}'} S(\mathbf{n}',\mathbf{n};t) \exp\left[-i(\mathbf{n}-\mathbf{n}') \cdot \boldsymbol{\theta} \right], \tag{3.19}$$

which, using (3.12), may be shown to satisfy

$$i\hbar \frac{\partial G}{\partial t} = G \sum_{\beta} V_\beta(t) \exp\left[-i\beta \cdot (\omega t + \boldsymbol{\theta}) \right], \tag{3.20}$$

$$G(\boldsymbol{\theta}, -\infty) = 1. \tag{3.21}$$

On solving for G and inverting the Fourier series (3.19) we obtain

$$S(\mathbf{n'}, \mathbf{n}) =$$

$$\left(\frac{1}{2\pi}\right)^N \int_0^{2\pi} d\boldsymbol{\theta} \exp\left\{i\left[\boldsymbol{\theta} \cdot (\mathbf{n'} - \mathbf{n}) - \frac{1}{\hbar}\int_{-\infty}^{\infty} dt \sum_{\beta} V_{\beta}(t) \exp\left[-i\boldsymbol{\beta} \cdot (\omega t + \boldsymbol{\theta})\right]\right]\right\}.$$

$$(3.22)$$

Since (3.22) involves an infinite series, further approximations are usually necessary. The more general case where ω is a function of t may be treated similarly (Presnyakov and Urnov, 1970).

This approximation has been applied to a simple harmonic oscillator perturbed by a time-dependent force (Presnyakov and Urnov, 1970) and to charged particle-hydrogen atom excitation (Presnyakov, 1974, private communication; see Lodge et al., 1976). In the former case the results are shown to agree with exact quantal solutions where the quantum numbers are large. In the latter case, although subsidiary approximations are made, good agreement is obtained with other results.

I.   The Strong-Coupling Correspondence Principle (SCCP)

This approximation is different from those considered above in that both the projectile and the target are described classically. With these simplifications the S-matrix elements may be expressed in closed form. This approximation was originally derived from the integral equation satisfied by the quantal evolution operator (see Percival and Richards, 1970). However, a simpler but equivalent derivation may be obtained using the equal-spacing approximation (3.22) in (3.12) (Richards, 1972); we outline this derivation here.

The SCCP rests on the three basic assumptions.

A1.   The equivalent purely classical model of the collision does not deviate far from its unperturbed motion.

A2.   All strongly coupled energy levels are equally spaced.

A3.   Heisenberg's form of the correspondence principle, equation (2.18), may be used.

In any application of the SCCP these assumptions need to be considered separately; in Sections V and VI we discuss them for vibrational and rotational excitation respectively. Generally, A2 and A3 are better approximations for excited states.

Clearly, the equal-spacing approximation assumed A2 and that the incident particle follows its unperturbed classical trajectory. We now show that by invoking A3 and A1 we may obtain the SCCP from (3.22).

The classical interaction potential may be expressed as a multiple Fourier series in the action-angle variables of the unperturbed target $(\mathbf{I}, \boldsymbol{\theta})$, see (2.6):

$$V^c(\mathbf{I}, \boldsymbol{\theta}; t) = \sum_{\text{all } \mathbf{s}} V^c_{\mathbf{s}}(\mathbf{I}, t) \exp(-i\mathbf{s} \cdot \boldsymbol{\theta}). \tag{3.23}$$

Using the correspondence principle, (2.12), we see that the ω defined in (3.18a) becomes in the limit of large quantum numbers, but small changes in quantum number, the fundamental frequency vector of the classical system, $\omega = \omega^c$. Also, for the same conditions, using Heisenberg's correspondence principle, (2.18), we have

$$\langle \alpha | V | \beta \rangle = V_{\alpha - \beta}(t) \cong V^c_{\beta - \alpha}(\mathbf{I}, t) \tag{3.24}$$

In deriving (3.22) it was assumed that the matrix elements depended only on $(\alpha - \beta)$, which we see from (3.24) is the same as assuming that the action variable $\mathbf{I}$ is constant for all relevant matrix elements. This is equivalent to assuming A1.

Substituting (3.24) into (3.22) and using (3.23) we find

$$S(\mathbf{n}', \mathbf{n}) = \left(\frac{1}{2\pi}\right)^N \int_0^{2\pi} d\boldsymbol{\theta} \exp\left\{ i\left[\boldsymbol{\theta} \cdot (\mathbf{n} - \mathbf{n}') - \frac{1}{\hbar} \int_{-\infty}^{\infty} dt V^c(\mathbf{I}, \boldsymbol{\theta} + \omega^c t; t) \right] \right\}. \tag{3.25}$$

This is the SCCP approximation.

It is readily seen that in the two limits of quantal first-order perturbation theory and the sudden approximation, (3.25) reduces to the equivalent of (3.14) and (3.17); see Percival and Richards (1975).

Often the target system is degenerate, for example, rotational excitation of a diatomic molecule, in which case the transition probability between levels rather than states is usually required. In this instance the quantal sum over degenerate levels is approximated by replacing it by an integral over a microcanonical ensemble of the target. Here we illustrate the approximation in the special case of a rigid rotor.

We require

$$P(j \to j') = \frac{1}{(2j+1)} \sum_m \sum_{m'} |S(j'm', jm)|^2 \tag{3.26}$$

where the S-matrix element is given by

$$S(j'm',jm) = \left(\frac{1}{2\pi}\right)^2 \int_0^{2\pi} d\theta_1\, d\theta_2 \exp\left\{ i\left[\Delta j\theta_1 + \Delta m\theta_2 - A(\theta_1,\theta_2) \right] \right\} \quad (3.27)$$

where A is the change in action of the target,

$$A(\theta_1,\theta_2) = \frac{1}{\hbar} \int_{-\infty}^{\infty} dt\, V^c \left(\mathbf{I}, \theta_1 + \omega_1 t, \theta_2; t \right). \quad (3.28)$$

Since the system is degenerate the theory of (2.8) may be used. Then θ_1 and θ_2 become the azimuthal angles of the rotor in the plane of rotation and in the plane perpendicular to the axis of quantization respectively (see Dickinson and Richards, 1974).

The sum (3.26) may be replaced by a sum over m and Δm, and then approximated using the relations

$$\sum_{\Delta m = -\infty}^{\infty} e^{i\Delta m\theta_2} = 2\pi\,\delta(\theta_2), \qquad \sum_{m=-l}^{l} \cong \left(j+\tfrac{1}{2}\right)\int_{-1}^{1} d(\cos\beta) \quad (3.29)$$

where β is the angle between the angular momentum vector and the quantization axis. With these approximations we find

$$\mathcal{P}(j\to j') = \frac{1}{2\pi} \int_0^{2\pi} d\theta_2 \int_{-1}^{1} \frac{d(\cos\beta)}{2} \left| \frac{1}{2\pi} \int_0^{2\pi} d\theta_1 \exp\left\{ i\left[\Delta j\theta_1 - A(\theta_1,\theta_2) \right] \right\} \right|^2.$$

$$(3.30)$$

Generally it is easier to evaluate (3.30) than the combination of (3.26) and (3.27).

IV. RELATIONS WITH OTHER APPROXIMATIONS

A. Introduction

Many semiclassical approximations have appeared in recent years. In this section we show how some are related to the SCCP derived in the last section and we also show that the SCCP has the correct form in the limit

$$n_k \gg |n_k - n_k'| \gg 1 \qquad k = 1,2,\ldots,N \quad (4.1)$$

which is a necessary criterion for classical theory to be valid.

B. Classical Limit of the Strong-Coupling Correspondence Principle

For convenience we consider a one-dimensional system only; the same analysis is valid for a totally degenerate system and the analysis for a separable higher-dimensional system follows similarly.

From (3.25) we define

$$g(\theta) = s\theta - \frac{1}{\hbar} \int_{-\infty}^{\infty} dt\, V^c(\theta + \omega^c t, I; t), \tag{4.2}$$

and assume that most of the contribution to the integral (3.25) comes from the M neighbourhoods of the stationary points of $g(\theta)$. Writing

$$g_k = g(\theta_k), \qquad g'(\theta_k) = 0, \qquad g_k'' = g''(\theta_k), \qquad k = 1, \ldots, M \tag{4.3}$$

then

$$S(n', n) = \frac{1}{2\pi} \sum_{k=1}^{M} e^{ig_k} \left[\frac{2\pi i}{g_k''} \right]^{1/2}. \tag{4.4}$$

If the differences between the actions of all the paths labelled by θ_k are large in comparison to $\hbar$,

$$|g_k - g_l| \gg 1, \qquad \text{all} \quad k \neq l, \tag{4.5}$$

then the interference terms may be neglected to give the probability

$$P(n \to n') = \frac{1}{2\pi} \sum_{k=1}^{M} \frac{1}{|g_k''|}. \tag{4.6}$$

The stationary points θ_k satisfy

$$g'(\theta) = 0 = s - \frac{1}{\hbar} \frac{d}{d\theta} \int_{-\infty}^{\infty} dt\, V^c(\theta + \omega^c t, I; t) \tag{4.7a}$$

or

$$s\omega^c \hbar = \Delta E^{\text{CPT}}(\theta_k), \qquad k = 1, 2, \ldots, M, \tag{4.7b}$$

where $\Delta E^{\text{CPT}}(\theta)$ is the classical energy transfer as obtained using classical perturbation theory. Since $\Delta E^{\text{CPT}}(\theta)$ is periodic in θ there are usually an even number of θ_k; in the classically inaccessible region, by definition, there are no real roots and the probability (4.6) is zero. Writing the probability in terms of the energy transfer we have

$$P(n \to n') = \frac{\hbar \omega^c}{2\pi} \sum_{k=1}^{M} \frac{1}{|d \Delta E^{CPT}/d\theta_k|}. \tag{4.8}$$

Now consider the distribution function of energy transfer $P^c(\Delta E)$ as given by classical mechanics:

$$P^c(\Delta E) = \frac{1}{2\pi} \int_0^{2\pi} d\theta \, \delta(\Delta E - \Delta E^c(\theta)) \tag{4.9a}$$

$$= \frac{1}{2\pi} \sum_{k=1}^{M'} \frac{1}{|d \Delta E^c/d\theta_k|}, \qquad \Delta E = \Delta E^c(\theta_k), \tag{4.9b}$$

where $\Delta E^c(\theta)$ is the exact classical energy transfer.

Using the density of states correspondence principle for transition probabilities analogous to (3.9), this gives

$$P^c(n \to n') = \left| \frac{dE'}{dn'} \right| P^c(\Delta E) = \hbar \omega^c(n') P^c(\Delta E) \qquad \Delta E = E_n - E_{n'} \tag{4.10a}$$

$$= \frac{\hbar \omega^c(n')}{2\pi} \sum_{k=1}^{M'} \frac{1}{|d \Delta E^c/d\theta_k|} \tag{4.10b}$$

using (2.12). When classical perturbation theory is valid $\omega^c(n) \cong \omega^c(n')$, and $\Delta E^c \cong \Delta E^{CPT}$ so that in this limit the SCCP reduces to classical mechanics and the density of states correspondence principle.

C. Connection with Other Semiclassical Theories

Levine and Johnson (1970) and Beigman et al. (1969) have obtained an expression for the S-matrix element in terms of the change in action of the target, $\Delta S(\beta)$:

$$S(\mathbf{n'}, \mathbf{n}) = \left(\frac{1}{2\pi} \right)^N \int d\beta \exp i \left\{ \beta \cdot (\mathbf{n} - \mathbf{n'}) + \frac{\Delta S(\beta)}{\hbar} \right\} \tag{4.11}$$

where β are the initial phases of the angle variables of the target. The difference between (4.11) and the SCCP is that the former uses exact classical trajectories; equation (4.11) is the first term in the asymptotic expansion of S in $\hbar$. We now show that (4.11) reduces to the SCCP when classical perturbation theory is valid.

Let the Hamiltonian of the target be H_0 so that the full Hamiltonian is

$$H = H_0(\mathbf{I}^0) + \lambda V(\boldsymbol{\theta}^0, \mathbf{I}^0; t) \qquad (4.12)$$

where $(\mathbf{I}^0, \boldsymbol{\theta}^0)$ are the action-angle variables of the unperturbed system, $\lambda = 0$ [previously denoted by $(\mathbf{I}, \boldsymbol{\theta})$]. The angle variables $\boldsymbol{\theta}^0$ have the initial phases $\boldsymbol{\beta}$ before the collision. The action $S(\boldsymbol{\theta}^0, \mathbf{I}; t)$ is the solution of the Hamilton–Jacobi equation

$$H_0\left(\frac{\partial S}{\partial \boldsymbol{\theta}^0}\right) + \lambda V\left(\boldsymbol{\theta}^0, \frac{\partial S}{\partial \boldsymbol{\theta}^0}; t\right) + \frac{\partial S}{\partial t} = 0 \qquad (4.13)$$

where $\mathbf{I}$ are now N nontrivial constants of integration that take on the initial values of $\mathbf{I}^0$. It follows (Landau and Lifshitz, 1969) that

$$\mathbf{I}^0 = \frac{\partial S}{\partial \boldsymbol{\theta}^0} \qquad \text{and} \qquad \boldsymbol{\beta} = \frac{\partial S}{\partial \mathbf{I}} = \text{constant.} \qquad (4.14)$$

Generally, the solution of (4.14) cannot be expressed in closed form, but a formal solution may be obtained as a power series in λ;

$$S = \boldsymbol{\theta}^0 \cdot \mathbf{I} - Et + \lambda S^{(1)} + \lambda^2 S^{(2)} + \cdots \qquad (4.15)$$

where E is the unperturbed energy. Substituting (4.15) into (4.13) and collecting terms we obtain for the zeroth- and first-order approximations,

$$H_0(\mathbf{I}) = E, \qquad (4.16a)$$

$$\boldsymbol{\omega}^c \cdot \frac{\partial S^{(1)}}{\partial \boldsymbol{\theta}^0} + V(\boldsymbol{\theta}^0, \mathbf{I}; t) + \frac{\partial S^{(1)}}{\partial t} = 0, \qquad (4.16b)$$

where $\boldsymbol{\omega}^c = \partial H_0 / \partial \mathbf{I}$ are the frequencies of the unperturbed system, which we assume nondegenerate. The solution to (4.16b) is

$$S^{(1)}(\boldsymbol{\theta}^0, \mathbf{I}; t) = -\int_{-\infty}^{t} d\tau\, V\left(\boldsymbol{\theta}^0 + \boldsymbol{\omega}^c(\tau - t), \mathbf{I}; \tau\right). \qquad (4.17)$$

Thus from equation (4.14), to first order in λ,

$$\mathbf{I}^0 = \mathbf{I} - \lambda \int_{-\infty}^{t} d\tau\, \frac{\partial}{\partial \boldsymbol{\beta}} V(\boldsymbol{\beta} + \boldsymbol{\omega}^c \tau, \mathbf{I}; \tau), \qquad (4.18a)$$

$$\boldsymbol{\theta}^0 = \boldsymbol{\omega}^c t + \boldsymbol{\beta} + \lambda \int_{-\infty}^{t} d\tau\, \frac{\partial V}{\partial \mathbf{I}}(\boldsymbol{\beta} + \boldsymbol{\omega}^c \tau, \mathbf{I}; \tau). \qquad (4.18b)$$

The change in action is thus

$$\Delta S(\boldsymbol{\beta}) = -\lambda \int_{-\infty}^{\infty} dt\, V(\boldsymbol{\beta} + \boldsymbol{\omega}^c t, \mathbf{I}; t) + O(\lambda^2), \tag{4.19}$$

which on substitution into (4.11) and putting $\lambda = 1$ gives the SCCP equation (3.25). For time-independent scattering theory the classical S-matrix approximation yields an integral representation similar to (4.11) (Marcus, 1971). The relation between this in the classical perturbation limit and the SCCP is discussed in Dickinson and Richards (1976a).

D.   The Modified Strong-Coupling Correspondence Principle

Generally, $\Delta S(\boldsymbol{\beta})$ in (4.11) must be evaluated numerically, but the approximation (4.19), used in the SCCP may often be evaluated analytically; frequently, first-order classical perturbation theory is a good description of the collision. Some account may be taken of second-order (in λ) effects in a way leading to another approximation to ΔS, which is simpler than the full theory. This approximation, due to Clark (1973), may be obtained by noting that the integrand of (4.18b) is periodic in $\boldsymbol{\beta}$ and may be approximated by its mean value. Then, following Clark (1973) we define

$$\overline{\mathbf{V}}(\mathbf{n}, t) = \left(\frac{1}{2\pi}\right)^N \hbar \int_0^{2\pi} d\boldsymbol{\theta}^0 \cdot \frac{\partial}{\partial \mathbf{I}^0} V(\boldsymbol{\theta}^0, \mathbf{I}^0; t) \tag{4.20a}$$

so that

$$\overline{V}_k(\mathbf{n}, t) \cong \langle n_1, \ldots, n_k + 1, \ldots, n_N | V | n_1, \ldots, n_k + 1, \ldots, n_N \rangle - \langle \mathbf{n} | V | \mathbf{n} \rangle. \tag{4.20b}$$

The last relation follows from the correspondence principle for matrix elements, (2.18). Thus from (4.18b) we have

$$\boldsymbol{\theta}^0 = \boldsymbol{\omega}^c t + \boldsymbol{\beta} + \frac{\lambda}{\hbar} \int_{-\infty}^{t} d\tau\, \overline{\mathbf{V}}(\mathbf{I}, \tau). \tag{4.21}$$

Thus an approximate second-order correction to ΔS is given by

$$\Delta S(\boldsymbol{\beta}) = -\lambda \int_{-\infty}^{\infty} dt\, V\left(\boldsymbol{\beta} + \boldsymbol{\omega}^c t + \frac{\lambda}{\hbar} \int_{-\infty}^{t} dt'\, \overline{\mathbf{V}}(\mathbf{I}, t'), \mathbf{I}; t\right) \tag{4.22}$$

which on substitution into (4.11) and putting $\lambda = 1$ gives the modified SCCP of Clark (1973).

V. VIBRATIONAL EXCITATION

A. Introduction

The simplest example to which the SCCP has been applied is the vibrational excitation of a molecule in a collinear collision with an atom. Although such a collision is clearly a rare occurence physically, a study of this model provides a good test for approximate theories since it is sufficiently simple for exact quantum solutions to be available.

Unlike other systems to which the SCCP has been applied, for example, rotational excitation of diatomic molecules (see Section VI) and the excitation of highly excited atoms (Percival and Richards, 1975), transitions between the vibrational states of a molecule require large energy transfers. Since the SCCP is based on classical perturbation theory, we would expect it to be intrinsically less reliable for vibrational excitation. Furthermore, typical vibrational quantum numbers are small so that the basic assumptions of the semiclassical method are invalid.

Despite these obvious disadvantages, the SCCP provides quite accurate transition probabilities for vibrational excitation, even for transitions involving low quantum numbers. Although the reasons for this are not fully understood, they are clearly related to the existence of a correspondence identity (Clark and Percival, 1975); however, at present there is no formal justification for using classical or semiclassical methods for systems in low quantum states or when the classical action is comparable to $\hbar$.

The existence of correspondence identities (Percival, 1969; Norcliffe, 1973), which include, for example, the prediction by Rutherford, using classical mechanics, of the correct differential cross section for scattering of a charged particle by a fixed charge, is always dependent on the particular representation used. The identity relevant to vibrational excitation is the "Feynman Identity." This exists for systems having a Lagrangian at most quadratic in the coordinates and velocities. For such systems, the quantum kernel or propagator is given exactly in terms of the classical action in position-time (or momentum-time) representation (Feynman and Hibbs, 1965). The identity does not exist in energy representation. The "Feynman Identity" is used as a basis for a formulation in which quantities such as bound-state spectra or transition probabilities due to a general time-dependent perturbation are expressed in terms of closed loops formed from classical paths in position-time (or momentum-time) space (Clark and Percival, 1975). So far this formulation has been applied only to systems having Lagrangians quadratic in position and velocities, where it was shown to yield the exact quantal results as expected (Clark and Percival, 1975). For general systems, its evaluation is difficult and further approximations are necessary. The existence of this identity, however, suggests that semiclassical methods for vibrational excitation are often more accurate than the formal theory would indicate.

The simple model of the collinear collision of an atom A with a diatomic molecule B-C has been extensively reviewed by Rapp and Kassal (1969); more recent developments are covered by the reviews of Secrest (1973) and Connor (1973). The system is shown in Fig. 1. All three atoms are constrained to move along the line defined by the molecular axis BC. The incoming atom A is reflected by the repulsive core of the potential $V_{AB}(\tilde{z})$, between atoms A and B only, and atoms B and C are bound by the potential $V_{BC}(\tilde{y})$, where $\tilde{y}$ and $\tilde{z}$ are the BC and AB separations, respectively. Rotational excitation of the diatomic molecule is thus not possible. Exact quantum results for this model have been used widely as a test of approximate theories and, because of its simplicity, the effects of changes in the inter- and intramolecular potentials may easily be investigated.

In order to apply time-dependent perturbation methods to this system, it is necessary to represent the effect of the incident atom on the target by a time-dependent field. There is no unique way of doing this since the interaction potential V_{AB} depends on the coordinates of the bound system. In the following applications, we assume that the oscillator motion has negligible effect on the incident particle trajectory.

B.   Linearly Perturbed Simple Harmonic Oscillator

We assume that the intramolecular potential $V_{BC}(\tilde{y})$ is that of a simple harmonic oscillator and that $V_{AB}(\tilde{z})$ is a purely repulsive exponential potential. It is more convenient to use the dimensionless coordinates $(x, y,$

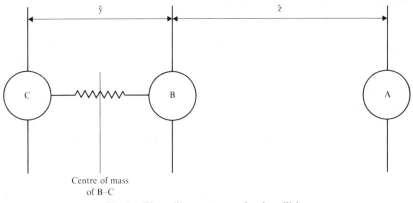

Fig. 1.   The collinear atom-molecule collision.

z) defined by Secrest and Johnson (1966). In these units we have

$$V_{BC}(y) = \tfrac{1}{2}y^2 \tag{5.1}$$

$$V_{AB}(z) = V_0 \exp\{-\alpha(x-y)\} \tag{5.2}$$

$$x = z + y \tag{5.3}$$

where V_0 is a constant and α depends on the collision system. In these coordinates, the collinear collision reduces to that of a particle of mass $m = m_A m_C / m_B(m_A + m_B + m_C)$ in collision with a harmonic oscillator of unit mass and force constant and with equilibrium position $y = 0$. Also $\hbar = 1$ and the energy is measured in units of $\hbar\omega$ where ω is the vibrational angular frequency of the molecule. The reduced time $\tau = \omega t$ is used.

If the amplitude of the vibrational motion is small, then we may expand (5.2) to give

$$V_{AB}(z) \cong V_0 \exp(-\alpha x)(1 + \alpha y). \tag{5.4}$$

This is the so-called linearized potential or Landau–Teller approximation (Landau and Teller, 1936). In order to express the incident particle motion as a time-dependent perturbation, we assume that the oscillator motion may be neglected and set $y = 0$ in (5.4). The classical equation of motion of the incident particle may then be solved to obtain $x(\tau)$. The collision problem is thus approximated to a harmonic oscillator perturbed by a linear time-dependent force $y\alpha V_0 \exp(-\alpha x(\tau))$.

The linearly perturbed harmonic-oscillator problem has been solved exactly quantum mechanically (Kerner, 1958; Treanor, 1965), and the probability of excitation of the harmonic oscillator from state n to state n' is

$$P^{\text{exact}}(n \to n') = \frac{n_<!}{n_>!} \exp(-\eta_0)\eta_0^s \left\{ L_{n_<}^s(\eta_0) \right\}^2 \tag{5.5}$$

where $s = |n - n'|$, $n_> (n_<)$ is the larger (smaller) of n and n', and $L_n^m(x)$ is the generalized Laguerre polynomial (Abramowitz and Stegun, 1965). The parameter η_0 is the classical energy transferred to the oscillator, in units of $\hbar\omega$, averaged over all initial phases of the oscillator, and is given by

$$\eta_0 = \tfrac{1}{2} \left| \int_{-\infty}^{\infty} \alpha V_0 \exp(-\alpha x(\tau) - i\tau) d\tau \right|^2 \tag{5.6}$$

where η_0 is independent of the initial excitation of the oscillator. It may be interpreted as the total energy transferred to a harmonic oscillator initially at rest. The quantum solution (5.5) depends only on the classical quantity η_0, this is due to the "Feynman Identity." In terms of the collision parameters, η_0 is given by (Rapp and Kassal, 1969)

$$\eta_0 = \frac{2\pi^2 m^2}{\alpha^2} \operatorname{cosech}^2 \left\{ \frac{\pi}{\alpha} \left(\frac{m}{2E_t} \right)^{1/2} \right\} \tag{5.7}$$

where E_t is the initial relative translational energy.

The SCCP transition probability is (Clark and Dickinson, 1971)

$$P^{CP}(n \to n') = J_s^2 \left\{ 2(\bar{n}\eta_0)^{1/2} \right\} \tag{5.8}$$

where $J_m(x)$ is the ordinary Bessel function and $\bar{n}$ is a "mean quantum number." We choose (see Section II, Table I)

$$\bar{n} = \left(\frac{n_>!}{n_<!} \right)^{1/s}, \qquad s \neq 0. \tag{5.9}$$

It may be shown that (5.8) is the leading term in the asymptotic expansion of (5.5) in powers of $s/n_<$ (Clark and Dickinson, 1971).

A direct comparison may be made by expanding (5.5) and (5.8) in powers of $\eta_0/E_{n_<}$, where $E_{n_<} = n_< + \frac{1}{2}$,

$$\frac{P^{CP}(n \to n')}{P^{exact}(n \to n')} = 1 + \frac{(s-1)}{6(2+s/E_{n_<})} \left(\frac{\eta_0}{E_{n_<}} \right) + O\left\{ \left(\frac{\eta_0}{E_{n_<}} \right)^2 \right\} \tag{5.10}$$

showing that the SCCP is valid for small $\eta_0/E_{n_<}$, that is, when classical perturbation theory is valid.

Although the derivation of the SCCP (Section III) assumed the bound system to be in a highly excited state, (5.10) shows that for this system it is valid for low quantum numbers. Again this is a consequence of the "Feynman Identity."

In Fig. 2 we show the $0 \to 1$ transition probabilities, as a function of η_0, comparing the SCCP result (5.8) with the exact quantal solution (5.5), quantal first-order time-dependent perturbation theory, and several quantum N-state computer solutions. There is satisfactory agreement between the quantal first-order result and $P^{exact}(0 \to 1)$ for $\eta_0 < 0.1$. $P^{CP}(0 \to 1)$, however, is valid for $\eta_0 < 0.4$ and has a similar range of validity as an eight-state computer solution. $P^{CP}(0 \to 2)$ and $P^{CP}(0 \to 3)$ and so on exhibit a similar range of validity.

In Fig. 3 comparisons for the $5 \to 6$ transition are shown. $P^{CP}(5 \to 6)$ is in

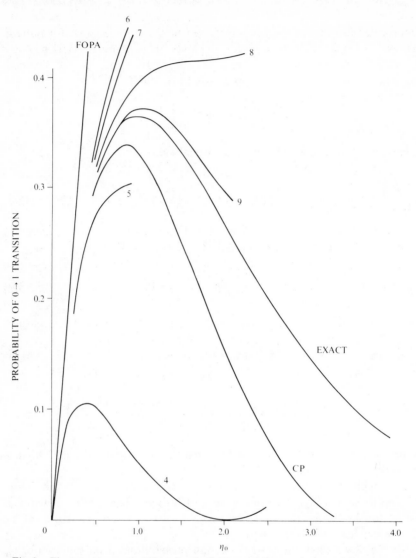

Fig. 2. The transition probabilities for the 0→1 transition are shown as a function of the phase-averaged classical energy transfer η_0. EXACT is the exact quantum-mechanical solution (5.5). CP is the correspondence principle approximation (5.8). FOPA is the first-order perturbation approximation. The curves numbered 4,5,...,9 are the 4,5,...,9 state computer solutions.

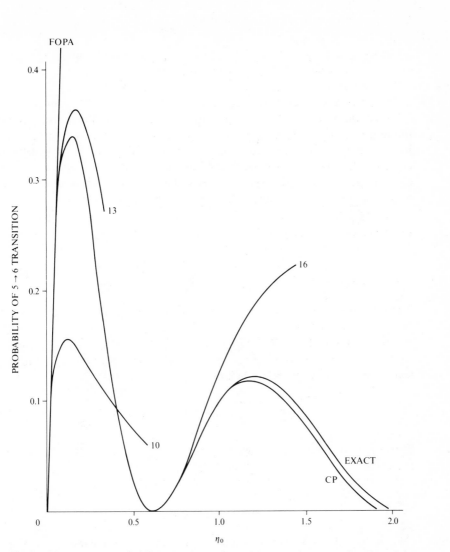

Fig. 3. The transition probabilities for the 5→6 transitions are shown as a function of the phase-averaged classical energy transfer η_0. EXACT is the exact quantum-mechanical solution (5.5). CP is the correspondence principle approximation (5.8). FOPA is the first-order perturbation approximation. The curves numbered 10, 13, 16 are the 10, 13, 16 state computer solutions.

good agreement with $P^{\text{exact}}(5 \to 6)$ for all η_0 shown and for a significant range of η_0 it is better than a sixteen-state computer solution.

The SCCP becomes accurate over a larger range of η_0 as $n_<$ becomes larger, as may be seen from Figs. 2 and 3. For the $20 \to 21$ transition, for example, the SCCP is accurate up to $\eta_0 \cong 7$—over the first eight oscillations in the transition probability. In the classical limit, that is, when $n_< \gg s$ and for $s \gg 1$, the condition for the validity of the SCCP is given by $(s/12)(\eta_0/E_{n_<}) \ll 1$, clearly showing the dependence on the validity of classical perturbation theory, $\eta_0/E_{n_<} \ll 1$. The classical perturbation theory result for the relative energy transferred to the oscillator is $\Delta E^{\text{CPT}}/E_{n_<} = -2\sqrt{\eta_0/E_{n_<}} \sin \delta$, where δ is the initial phase angle of the oscillator. The exact classical energy transfer, however, ΔE, is given by

$$\frac{\Delta E}{E_{n_<}} = \frac{\eta_0}{E_{n_<}} - 2\sqrt{\frac{\eta_0}{E_{n_<}}} \sin \delta. \tag{5.11}$$

The second term is just the perturbation theory result $\Delta E^{\text{CPT}}/E_{n_<}$ and will be dominant when $\eta_0/E_{n_<} \ll 1$.

The validity of the SCCP may be contrasted with the quantal first-order time-dependent perturbation theory result, $P^{\text{QFO}}(n \to n')$, which for an $s = 1$ transition yields (Clark and Dickinson, 1971)

$$\frac{P^{\text{QFO}}(n \to n \pm 1)}{P^{\text{exact}}(n \to n \pm 1)} = 1 + n_> \eta_0 + O(\eta_0^2) \tag{5.12}$$

showing that the range of validity of $P^{\text{QFO}}(n \to n \pm 1)$ decreases as $n_>$ increases.

Transitions $n \to n + s$ will be classically forbidden when

$$s > \eta_0 + 2\sqrt{\eta_0 E_n} . \tag{5.13}$$

This is the region, however, where the SCCP is most accurate. The SCCP will also be accurate in the classically allowed region providing classical perturbation theory is valid.

A general discussion of the validity of semiclassical methods for the linearly forced oscillator problem has been given by Pechukas and Child (1976).

C. Quadratically Perturbed Simple Harmonic Oscillator

As a further test of correspondence principle methods, we consider the problem of a harmonic oscillator perturbed by a potential $V(y) = y^2 f(\tau)$ where $f(\tau)$ is an arbitrary function of time. This corresponds to the case of a harmonic oscillator with time-varying frequency. Popov and Perelomov (1969) have obtained an exact quantum-mechanical solution for this system and the probability that the oscillator makes an $n \to n'$ transition is given by

$$P^{\text{exact}}(n \to n') = \frac{n_<!}{n_>!} \sqrt{1-\rho}\ |P^{s/2}_{(n+n')/2}\{\sqrt{1-\rho}\ \}|^2, \qquad (5.14)$$

$P_n^m(x)$ being the associated Legendre function (Abramowitz and Stegun, 1965) and $s = |n - n'| = 0, 2, 4, \ldots$ because of parity. The parameter ρ is obtained from the classical equation of motion for the perturbed oscillator, which, in the reduced units of Section V.B, is given by

$$\frac{d^2y}{d\tau^2} + (1 - 2f(\tau))y = 0. \qquad (5.15)$$

Then ρ is related to the classical phase-averaged energy transfer to the oscillator, $\langle \Delta E \rangle$, with initial energy of $\frac{1}{2}$, by (Husimi, 1953)

$$\rho = \frac{\langle \Delta E \rangle}{1 + \langle \Delta E \rangle}. \qquad (5.16)$$

Alternatively, as pointed out by Popov and Perelomov (1969), (5.15) may be interpreted as a Schrödinger equation. The parameter ρ then becomes the reflection coefficient for a particle of energy $\frac{1}{2}$, from the potential barrier defined by $f(\tau)$.

Equation (5.15) cannot be solved analytically for a general function $f(\tau)$ and thus, to discuss results quantitatively, we choose the particular form $f(\tau) = (a/2)\text{sech}^2(2\sqrt{a}\ \tau)$. With this function, the perturbation approximately represents a collision situation. If a is small, the magnitude of the perturbation is small, but its effect is spread over a long time, corresponding to adiabatic collisions. If a is large, the perturbation is large but strongly peaked in time, corresponding to sudden collisions. Equation (5.15) may now be solved (Gol'dman and Krivchenkov, 1961) giving

$$\rho = \frac{0.3669}{\sinh^2(\pi/2\sqrt{a}\ ) + 0.3669}. \qquad (5.17)$$

We now examine the validity of the SCCP and the modified SCCP (4.11 and 4.22) for this problem. The SCCP solution for the transition probabilities is (Clark and Dickinson, 1971)

$$P^{\text{CP}}(n \to n') = J^2_{s/2}(\bar{n}\sqrt{\rho'}\ ), \qquad s = 0, 2, 4, \ldots \qquad (5.18)$$

where $\bar{n}$ is defined by (5.9) and ρ' may be interpreted as a quantal first-order perturbation theory approximation to ρ (Clark and Dickinson, 1971), given by

$$\rho' = \frac{0.6169}{\sinh^2(\pi/2\sqrt{a}\,)}. \tag{5.19}$$

The modified SCCP may also be evaluated analytically (Clark, 1973) and is given by (5.18) with ρ' replaced by ρ'' where

$$\rho'' = \rho' \left| M\left(1 + \frac{i}{2\sqrt{a}}, 2, i\sqrt{a}\,\right)\right|^2 \tag{5.20}$$

and $M(a,b,x)$ is the confluent hypergeometric function (Abramowitz and Stegun, 1965).

A comparison between the SCCP result (5.18) and the exact result (5.14) yields for small ρ (Clark and Dickinson, 1971)

$$\frac{P^{CP}(n \to n')}{P^{exact}(n \to n')} = \left(\frac{\rho'}{\rho}\right)^{s/2}\left\{1 + \frac{n_<(n_< + s + 1) + (1 + s/2)}{s+2}(\rho - \rho')\right.$$

$$\left. - \frac{(s-2)}{6}\rho' + O(\rho^2)\right\}. \tag{5.21}$$

Thus (5.21) implies that the SCCP will be valid for small ρ providing (1) $s\rho' \ll 1$ and (2) $\rho' \cong \rho$, that is, providing classical perturbation theory is valid.

For the particular form of $f(\tau)$ chosen, (5.19) shows that although ρ' has the correct dependence on the parameter a it is about a factor of 2 larger than ρ. This is because of the breakdown of first-order classical perturbation theory caused by the assumption that the orbit of the bound system is unperturbed throughout the interaction. For small values of a, corresponding to weak interactions, the force is small in absolute magnitude although its effect lasts for a long period of time. To a good approximation, the amplitude of the oscillator will be unperturbed. The phase of the oscillator, however, will be perturbed and thus first-order classical perturbation theory will not be valid.

The modified SCCP includes a correction term that approximately takes into account the phase change of the oscillator. In the limit as $a \to 0$ we obtain

$$\rho'' \underset{a \to 0}{\to} \frac{0.3657}{\sinh^2(\pi/2\sqrt{a})}. \tag{5.22}$$

Thus ρ'' is a good approximation to ρ in this limit.

In Figs. 4 and 5 we show the $0 \to 2$ and $2 \to 4$ transition probabilities respectively as a function of the parameter a. It may be seen that $P^{\text{MODCP}}(n \to n')$ is a good approximation, at least up to the maximum in the transition probability, for both transitions, whereas $P^{\text{CP}}(n \to n')$ over-estimates $P^{\text{exact}}(n \to n')$ for small a.

This particular problem was chosen specifically to show that care must be taken, when applying the SCCP, to ensure that first-order classical perturbation theory is valid for the classical problem. The modified SCCP result reduces to the SCCP for the case of a linearly perturbed oscillator considered in Section V.B, the correction term being zero in this case.

Also, for rotational excitation of diatomic molecules by atoms, it is found that the correction term is too small, for most collisions, for such modifications to be significant.

D.   Collinear Collision of an Atom with a Morse Oscillator

In Section V.B and V.C we have concentrated on comparing the SCCP with exact quantum-mechanical solutions of time-dependent models. In this section, we compare the SCCP with time-independent close-coupling results for a collinear atom-diatomic molecule collision. This tests the additional approximation of representing the effect of the incident particle by a time-dependent field.

Secrest and Johnson (1966) have obtained close-coupling results for the harmonic-oscillator system described by (5.1) and (5.2). Comparisons between these exact results and the SCCP approximation have been made by Clark (1971). Several methods of deriving approximate time-dependent perturbing potentials were investigated including the impulse approxima-tion, $V_I(y, \tau)$, of Heidrich et al. (1971). Using $V_I(y, \tau)$ to obtain a phase-averaged classical energy transfer, η_0^I, and (5.5) to obtain transition proba-bilities (the ITFITS approximation), Heidrich et al. (1971) have shown that excellent agreement is obtained with the exact results of Secrest and Johnson (1966). Since $V_I(y, \tau)$ is linear in y, similarly good agreement was found between the SCCP [(5.8) using η_0^I] and the exact results, except for very strong collisions.

A more realistic representation of the intramolecular potential is the

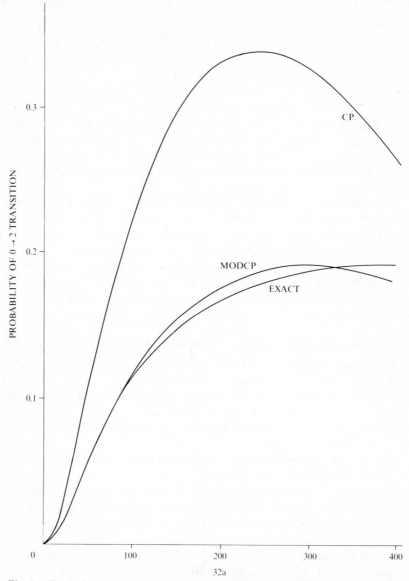

Fig. 4. Graph showing the 0→2 transition probability as a function of the parameter a, for a harmonic oscillator perturbed by a potential $y^2 f(\tau)$. EXACT is the exact quantal result, (5.14) and (5.17); CP is the correspondence principle probability, (5.18) and (5.19); MODCP is the modified correspondence principle probability, (5.18) and (5.20).

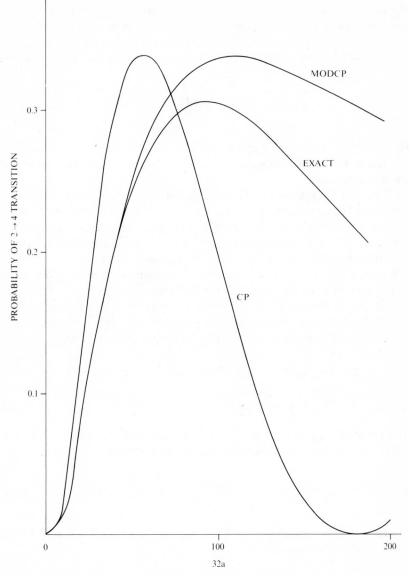

Fig. 5.   Graph showing the 2→4 transition probability as a function of the parameter a, for a harmonic oscillator perturbed by a potential $y^2 f(\tau)$. The notation is as in Fig. 4.

Morse potential and we consider in more detail the collinear collision problem with the molecule bound by this potential. Quantum-mechanical close-coupling transition probabilities have been computed for this system (Clark and Dickinson, 1973) with the interaction potential (5.2) and with

$$V_{BC}(y) = D_e \left[1 - \exp\left(-\frac{y}{\sqrt{2D_e}} \right) \right]^2. \tag{5.23}$$

This potential is expressed in the reduced units of Section V.B and D_e is the Morse well depth measured in units of $\hbar\omega_e$, ω_e being the equilibrium oscillation frequency.

Clark and Dickinson (1973) have shown that considerable differences can occur between the transition probabilities for this system and for the equivalent harmonic-oscillator model, even when only small anharmonicities of the Morse potential are present. Reasons for these differences have been discussed in detail by Schinke and Toennies (1975).

We apply the SCCP to this model and compare our results and those of several approximate theories with the exact quantum results.

In order to represent the system as a Morse oscillator perturbed by a time-dependent field, we assume that the incoming particle is scattered by the average static potential of the initial (n) and final (n') states,

$$U(x) = \tfrac{1}{2} V_0 \exp(-\alpha x) \left[\langle n|\exp(\alpha y)|n \rangle + \langle n'|\exp(\alpha y)|n' \rangle \right]$$

$$= \tfrac{1}{2} V_0 \exp(-\alpha x) \overline{U}. \tag{5.24}$$

The classical equation of motion for the incident particle may then be solved to give the time-dependent perturbing potential,

$$V(y,\tau) = \frac{\overline{E}}{2\overline{U}} \operatorname{sech}^2 \left\{ \frac{\alpha}{2} \left(\frac{\overline{E}}{m} \right)^{1/2} \tau \right\} \exp(\alpha y) \tag{5.25}$$

where $\overline{E}$ is the velocity-averaged translational energy (Rapp and Kassal, 1969) of the incoming particle,

$$\overline{E} = \tfrac{1}{4} \left\{ (E_T - E_n)^{1/2} + (E_T - E_{n'})^{1/2} \right\}^2, \tag{5.26}$$

E_T is the total energy of the system and E_m are the energy levels of the oscillator.

The SCCP transition probabilities were obtained to an accuracy of 1% by numerical integration. In Fig. 6, we compare $P^{\mathrm{MODCP}}(0 \rightarrow 1)$ (modified

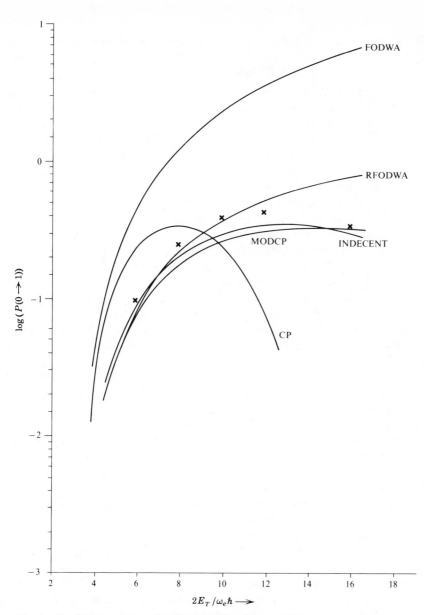

Fig. 6. The 0→1 transition probabilities for the collinear collision of an atom and a Morse oscillator ($m = \frac{1}{3}, \alpha = 0.314, D_e = 9.3$) as a function of the total energy in units of $\hbar\omega_e/2$. X are the exact transition probabilities. CP is the correspondence-principle approximation. MODCP is the modified correspondence-principle approximation. FODWA is the first-order distorted wave approximation. RFODWA is the revised first-order distorted-wave approximation. INDECENT is the classical trajectory forced oscillator approximation.

103

SCCP), $P^{CP}(0 \to 1)$ (unmodified SCCP), and several other approximate theories with the exact $0 \to 1$ transition probabilities for the collision parameters $m = \frac{1}{3}$, $\alpha = 0.314$, $D_e = 9.3$, which approximate an $H_2 + H$ collision.

The unmodified SCCP overestimates the exact transition probability $P^{exact}(0 \to 1)$ at low energies and underestimates it at higher energies. $P^{CP}(0 \to 1)$ does, however, agree with the first-order distorted-wave calculation $P^{FODWA}(0 \to 1)$ of Jackson and Mott (1932) at low energies. This agreement is to be expected since it has been shown by Rapp and Kassal (1969) that, providing the interaction potential changes slowly in the de Broglie wavelength of the incident particle, the FODWA reduces to the first-order time-dependent perturbation theory result (3.14) and equivalently to the first-order correspondence principle result by replacing the matrix element by a Fourier component.

The modified correspondence principle result $P^{MODCP}(0 \to 1)$ is in good agreement with $P^{exact}(0 \to 1)$ over the range of energies shown, which is the range covered by the close-coupling results, and agrees with the revised first-order distorted-wave result $P^{RFODWA}(0 \to 1)$ of Mies (1964) at low energies. The RFODWA differs from the FODWA in that it allows for differences between the diagonal matrix elements of the interaction potential. Both MODCP and RFODWA allow for perturbations of the bound-energy levels in a similar way.

Also shown are the "classical trajectory forced quantum oscillator" calculations, $P^{INDECENT}(0 \to 1)$ of Gentry and Giese (1975). In this method one calculates the classical energy transferred to the Morse oscillator, initially at rest, by the incident particle with "velocity-averaged" energy (5.26). Using this energy transfer the quantum transition probabilities are then obtained from the linearly forced harmonic oscillator result (5.5). One of the assumptions of this method is that the differences between the transition probabilities for the Morse and harmonic oscillators are mainly due to the different energy transfers. This has been justified by Schinke and Toennies (1975) for transitions among low vibrational states.

Approximate transition probabilities for this problem have also been obtained by McKenzie (1975) (not shown in Fig. 6). These are calculated from the solution of the close-coupling equations for time-dependent scattering (3.12) and are also in good agreement with $P^{exact}(0 \to 1)$. Classical S-matrix calculations have been made by Schinke and Toennies (1975). For the range of energies stated, these results agree with $P^{exact}(0 \to 1)$ to within 3% and are more accurate than any of the other approximate results discussed above. Similar agreement with quantum results has been obtained by Connor (1974), using classical S-matrix theory, for the $He + H_2$ system.

Approximate theories requiring a time-dependent model for the collinear collision are generally more accurate for smaller values of the mass parameter m. Large values of m occur for heteronuclear molecules in which the mass of atom B, m_B, is much less than m_C. For these collisions, the oscillator is perturbed to a much greater extent. It has been shown (Secrest, 1969; Bergeron et al., 1976) that, for large values of m, "chattering" can occur. In these cases, the simple time dependence of (5.25) will clearly be inadequate. These are, however, an artifact of the one-dimensional model. In three-dimensional models such collisions will be very unlikely and the use of a time-dependent trajectory will be more successful.

The value of $m = \frac{1}{3}$ was chosen specifically to demonstrate that large discrepancies can occur between the SCCP and MODCP theories. The accuracy of both the SCCP and MODCP theories increases with decreasing values of m; for $m = 0.006268$, approximately representing a $Br_2 + H_2$ collision, the SCCP and MODCP results are both accurate to within a few percent of the exact results in the region before the first maximum in the transition probability.

The FODWA, RFODWA, and INDECENT approximations will become progressively worse for transitions among higher states—the former two because more quantum channels will be strongly coupled and the latter because the anharmonicity of the molecule will become more apparent. The time-dependent close-coupling approach of McKenzie (1975) has the disadvantage that for transitions among higher states the number of channels required in the computation increases in a manner similar to time-independent close-coupling calculations.

The validity of classical S-matrix theory (Miller, 1975; Marcus, 1973) will, however, be independent of the initial quantum state. Although this theory is simple for one-dimensional problems, it becomes impracticable for three-dimensional calculations, since a simultaneous search of several variables must be made, and further approximations usually have to be introduced (Doll and Miller, 1972; Raczkowski and Miller, 1974).

The validity of the SCCP and MODCP theories, for this problem, is mainly dependent on the accuracy of approximating the classical dynamics of the incident particle trajectory. This will be more successful for three-dimensional collisions. The generalization of the correspondence principle to problems of several degrees of freedom is straightforward and requires relatively little extra computational effort.

VI. ROTATIONAL EXCITATION OF DIATOMIC MOLECULES BY ATOMS

A. Introduction

We now consider the application of the SCCP to rotational excitation. Clearly, at energies well below the threshold for vibrational excitation we should expect that neglect of vibration is not serious. As is usual we take a rigid rotor model for the molecular rotation. By employing an effective moment of inertia, appropriate to the transition under consideration, the centrifugal stretching of the rotor could be allowed for, if required. The standard quantum close-coupling (CC) formulation of this problem is that of Arthurs and Dalgarno (1960). In any comparisons with CC calculations we will, of course, use the same potential surface. It proves convenient in the SCCP to employ the same expansion of the interaction potential between the atom and the molecule as is used in the quantum description,

$$V(R,\theta) = \sum_{\lambda=0}^{N} F_\lambda(R) P_\lambda(\cos\theta) \qquad (6.1)$$

where R is the separation of the atom from the centre of mass of the molecule and θ is the angle between the molecular axis and the line joining the centre of mass of the molecule to the atom. For homonuclear molecules, symmetry requires $F_\lambda = 0$ when λ is odd. The number of terms, N, taken in the expansion (6.1) depends on the degree of anisotropy of the potential surface under consideration. Normally one or two anisotropic terms suffice, but up to nine have been used (Green and Thaddeus, 1974).

B.   The Change in Action of the Rotor

As discussed in Section III.D, to apply the SCCP it is necessary to represent the effect of the incident atom by a time-dependent field acting on the rotor. This is accomplished by assuming that the atom moves on the classical trajectory determined by $F_0(R)$ only. Classical trajectory calculations (e.g., Kistemaker and de Vries, 1975) show that this is generally quite a good approximation for finding the deflection of the atom. With this model we allow fully for the distortion of the orbit by the spherically symmetric component of the potential (6.1), which cannot induce rotational transitions.

We use a coordinate system with the origin at the centre of mass of the rotor, the x-y plane containing the orbit of the atom, and the y-axis towards the point of closest approach of the atom. Since the rotor traces out a disk it is convenient to choose unit vectors $\hat{\xi}$ and $\hat{\eta}$ in this disk and $\hat{\zeta}$ as a unit vector along the rotor angular momentum $\mathbf{j}$. Following Edmonds (1960) we take $0xyz$ and $0\xi\eta\zeta$ as initially coincident and obtain the final orientation by keeping $0xyz$ fixed and rotating $0\xi\eta\zeta$ through angles $\gamma\beta\alpha$ about the $0z$, $0y$, and $0z$ axes, respectively. The angles γ and α are now the angle variables conjugate to the total angular momentum of the rotor and its azimuthal component, respectively, while β is the angle between the angular momentum of the rotor and that of the atom. Central to our approach is the change in action $\hbar A$ of the rotor, given by classical perturbation theory (CPT), when the rotor is perturbed by a collision with an atom of energy E and impact parameter b. Using the standard techniques of angular momentum algebra (Edmonds, 1960) it is straightforward to show (Dickinson and Richards, 1974)

$$A(\alpha,\beta,\gamma;E,b)=4\pi \sum_{\lambda=1}^{N} (2\lambda+1)^{-1} \sum_{\mu,\nu=-\lambda}^{\lambda} N_{\lambda\mu}N_{\lambda\nu}\mathcal{D}_{\nu\mu}^{(\lambda)}(\gamma,\beta,\alpha)V_{\nu\mu}^{\lambda}(E,b).$$

(6.2)

Here $N_{\lambda\mu}=Y_{\lambda\mu}(\pi/2,0)$, $\mathcal{D}_{\nu\mu}^{(\lambda)}$ is the matrix element for finite rotations (Edmonds, 1960), and

$$V_{\nu\mu}^{\lambda}(E,b)=\frac{2}{\hbar}\int_0^\infty dt F_\lambda(R)\cos(\nu\omega^c t-\mu\phi),$$
(6.3)

where $(R,\phi+\pi/2)$ are the plane polar coordinates of the atom moving in the spherically symmetric potential F_0 and where ω^c is the classical frequency of the rotor.

Much of the simplification due to the use of CPT in evaluating A arises because the dependence on the orientation of the rotor, contained in the $\mathcal{D}$ functions, has been separated from the dynamical factors in the V^{λ} integrals so that once these integrals have been evaluated for a specified E, b, and ω^c the change in action A can be obtained trivially for any orientation. This separation is independent of the form of F_0. Classically the changes in j and its z component, m, can now be obtained from the derivatives of A with respect to the conjugate angle variables,

$$\Delta j=-\frac{\partial A}{\partial\gamma}, \qquad \Delta m=-\frac{\partial A}{\partial\alpha}.$$
(6.4)

With our choice of coordinates $\Delta l=-\Delta m$ where $\hbar l$ is the angular momentum of the atom.

C. Validity of Classical Perturbation Theory

Since the simplification resulting from the use of CPT is crucial to the efficiency with which SCCP calculations are performed, it is important to assess the validity of CPT by comparison with exact classical results. Because of the large number of independent parameters in an individual collision, the results of an exhaustive test of all possible variations would be almost impossible to assimilate. Furthermore, no experiments are yet possible to measure nonaveraged collisions. Cohen and Marcus (1970) have investigated the validity of CPT when the parameters affecting the collision are varied individually. They used a potential of the form (6.1) retaining only the F_0 and F_2 terms, these being of the Lennard–Jones (12,6) form. They found that over a wide range of relevant parameters CPT is accurate to within 10%, tending to be better for large incident energies, large rotational energies, and for rotors with large moment of inertia.

Classical perturbation theory has also been employed for the charge-dipole potential, $F_\lambda(r) = (eD/r^2)\delta_{\lambda 1}$, in (6.1). This application is of considerable interest for the scattering of polar molecules by electrons (Smith et al., 1975) and ions (Gentry, 1974 and references therein). Although the results of CPT for this potential have been derived previously, we give them here as an example of the general method.

Here

$$A(\alpha,\beta,\gamma;E,b) = \left\{ V_{11}^1 \cos^2(\beta/2)\cos(\alpha+\gamma) - V_{1-1}^1 \sin^2(\beta/2)\cos(\alpha-\gamma) \right\}$$

$$(6.5)$$

where

$$V_{1\pm1}^1(E,b) = \frac{2\omega^c eD}{\hbar v^2} \left[K_1(z) \pm K_0(z) \right], \qquad z = \frac{\omega^c b}{v}, \qquad (6.6)$$

$K_n(z)$ is the modified Bessel function (Abramowitz and Stegun, 1965, p. 374), and v is the speed of the incident particle.

It is now elementary to find Δj and Δm using (6.4), so for fixed β it is straightforward to solve for the values of α and γ giving a specified Δj, Δm transition. Such values are required in classical S-matrix theory.

Gentry (1974) has compared the sudden approximation (for fixed rotor orientation) in CPT with the solution to the full classical equations of motion. The parameters used were appropriate for $H^+ + HF$ collisions. The sudden approximation to CPT is not essential and simply results in a small z approximation to the V integrals in (6.6) being used. Since low rotor energies were involved, this was not a favorable case for the application of CPT.

Very poor results for individual orbits were obtained using sudden CPT until high incident particle energies (~ 1000 eV) were reached. Gentry (1974) also did a few calculations numerically solving the classical equations of motion for a rotor in a time-dependent field due to a charged particle passing on a straight-line path, and much smaller errors were obtained, suggesting that much of the error in his CPT calculations is due to the additional nonessential assumption of the sudden approximation. Gentry (1974) has also shown that averages over initial orientations of

CPT energy transfers are much more accurate than the calculations for individual orbits. This property of averages over CPT being more accurate than individual orbits has also been noted by Banks (private communication) for electron-hydrogen collisions.

D. Evaluation of the Scattering Amplitude

We consider initially the evaluation of the scattering amplitude S for a $\Delta j, \Delta m$ transition. Using (3.25) we find

$$S(\Delta j, \Delta m; \beta, E, b) = \frac{1}{4\pi^2} \int_0^{2\pi} \int_0^{2\pi} d\alpha \, d\gamma \exp i \{ \Delta j \gamma + \Delta m \alpha - A(\alpha, \beta, \gamma; E, b) \}.$$

$$(6.7)$$

As an illustration we consider the case examined in the previous section of scattering by a potential with a P_1 anisotropy only. Then using (6.5) and introducing new variables $\theta_\pm = \alpha \pm \gamma$ in (6.7) the double integral is separable yielding

$$S(\Delta j, \Delta m; \beta, E, b) = J_{\Delta k_+}(V_{11}^1 \cos^2 \beta/2) J_{\Delta k_-}(V_{1-1}^1 \sin^2 \beta/2),$$

$$\Delta k_\pm = (\Delta j \pm \Delta m)/2 \tag{6.8}$$

where the integral representation of the Bessel function (Abramowitz and Stegun, 1965, p. 360) has been used. This result was obtained from classical S-matrix theory with additional uniformization by Smith et al. (1975).

The separability of the double integral (6.7) is a special feature of the use of CPT and a P_1 anisotropy only; however, it is quite independent of the radial potentials and holds for any forms of F_0 and F_1.

Returning to the general case, usually the degeneracy-averaged $j \rightarrow j'$ probability is required. From (3.30) this may be written

$$\mathcal{P}(j \rightarrow j'; E, b) = \frac{1}{4\pi} \int_0^{2\pi} d\alpha \int_0^\pi \sin \beta \, d\beta \, |\bar{S}(\Delta j; \alpha, \beta, E, b)|^2 \tag{6.9}$$

where

$$\bar{S}(\Delta j; \alpha, \beta, E, b) = \frac{1}{2\pi} \int_0^{2\pi} \exp i(\Delta j \gamma - A) \, d\gamma \tag{6.10}$$

A being given by (6.2). For a P_1 anisotropy only

$$\bar{S}(\Delta j; \alpha, \beta, E, b) = \exp\left[-i\Delta j \left(\epsilon_1 + \frac{\pi}{2} \right) \right] J_{\Delta j}(R_1) \tag{6.11}$$

where ϵ_1 and R_1 are given by Dickinson and Richards (1974, Eq. 4.19). The averages over α and β in (6.9) must be performed numerically.

For a P_2 anisotropy only the change in action A again takes a simple form. Neglecting the term independent of γ, which cannot produce $\Delta j \neq 0$ transitions,

$$A = R_2 \cos(2\gamma + \epsilon_2) \tag{6.12}$$

where ϵ_2 and R_2, both independent of γ, are given by Dickinson and Richards (1974, Eq. 4.24). Then, because A depends on (2γ), only transitions with even Δj have nonzero amplitudes. For Δj odd complete destructive interference occurs between phases γ and $\pi + \gamma$, and for Δj even

$$\bar{S}(\Delta j) = \exp\left[-i\Delta j(\epsilon_2 + \pi/2)/2\right] J_{\Delta j/2}(R_2). \tag{6.13}$$

For most other anisotropies $\bar{S}$ must be evaluated numerically but some exceptions are discussed by Dickinson and Richards (1974, Section 4.2.3).

E. Limits of the Transition Probabilities for Rotational Excitation Evaluated Using the SCCP

For weak collisions where the trajectory integrals $V_{\nu\mu}^\lambda$ are small, the SCCP reduces (Section III.I) to the semiclassical limit of first-order time-dependent perturbation theory. However, for rotational excitation due to atoms, weak collisions are seldom important so we do not investigate this limit further here. Details of the comparison between the first-order SCCP and time-dependent theories are given in Dickinson and Richards (1974, Section 4.1).

In the sudden approximation, where it is assumed that the rotor is stationary throughout the collision ($\omega^c = 0$), the $V_{\nu\mu}^\lambda$ integrals (6.3) all become equal to $V_{0\mu}^\lambda$, thus simplifying the evaluation of the change in action (6.2). In general, the orientation averages must still be done numerically but, for a P_1 anisotropy only, one angular average can be done simply and the other may be expressed as a sum of Bessel functions,

$$\mathcal{P}(j \to j + \Delta j) = \frac{1}{V_{01}^1} \sum_{n=\Delta j}^\infty J_{2n+1}\left(2V_{01}^1\right). \tag{6.14}$$

An interesting example is provided by the special case of a $0 \to 2$ transition with a P_1 potential only. Jamieson (1976) has evaluated the transition probability in the quantum-mechanical sudden approximation:

$$P^q(0 \to 2) = 5j_2^2\left(V_{01}^1\right), \tag{6.15}$$

where $j_n(x)$ is a spherical Bessel function (Abramowitz and Stegun, 1965, p. 437). For values of V_{01}^1 up to about 4 (6.14), with the detailed balance correction (6.20), is within 20% of the quantal result (6.15). Larger differences occur as the strong collision limit, $V_{01}^1 \to \infty$, is approached. This is because of the breakdown of the correspondence principle approximation for matrix elements of highly oscillatory functions discussed in Section II.D.

F. The Body-Fixed Approximation

In this section we discuss the use of body-fixed coordinates (BF) and the approximations suggested by them. If we examine the general form of the potential (6.1) we note that there is axial symmetry about the atom-molecular line; consequently, the potential cannot induce changes in the component of the rotor angular momentum along this direction, either in a quantal or a classical treatment. However, since coordinate axes tied to this line define a noninertial frame of reference, additional terms arise in the Hamiltonian and the exact solution in these coordinates is as difficult as in the original coordinates. However, there are some computational advantages in the quantum problem (Launay, 1976). If we ignore the centrifugal term for the rotor but retain it for the atom, we obtain the approximate Hamiltonian

$$\bar{H} = H_0(j,\beta') + \frac{P_R^2}{2\mu} + \frac{Eb^2}{R^2} + V(R,\theta) \tag{6.16}$$

where the z-axis now lies along the atom-rotor line and the rotor angular momentum lies in the xz plane with polar angle β'. The quantal coupled-states approximation (McGuire and Kouri, 1974, see McGuire 1976 for a good discussion and extensive references) is based on $\bar{H}$. For this Hamiltonian $j\cos\beta'$ is a constant and the collision problem is now one dimensional, changes in j only being possible. To calculate the degeneracy-averaged cross section, the cross sections from this one-dimensional problem must be averaged over all initial orientations of j (or all initial values of m quantum mechanically). This separation of a two-dimensional problem into a series of one-dimensional problems produces significant simplifications.

Proceeding as in Section VI.B we find for the change in action

$$A^{\mathrm{BF}}(\beta',\gamma';E,b) = 4\pi \sum_{\lambda=1}^{N} (2\lambda+1)^{-1} \sum_{\nu=-\lambda}^{\lambda} N_{\lambda\nu} Y_{\lambda\nu}(\beta',0) V_{\nu 0}^{\lambda} e^{i\nu\gamma'} \tag{6.17}$$

where γ' is the polar angle of the rotor in the plane perpendicular to $\mathbf{j}$.

If one starts in the old coordinates and makes the approximation $V_{\nu\mu}^{\lambda} \cong V_{\nu 0}^{\lambda}$ then, on rotating the axes, the expression (6.17) is regained (Dickinson and Richards, 1976b). Thus we see directly in this classical perturbation-theory approach that the usual body-fixed axes approximation is valid when the $V_{\nu\mu}^{\lambda}$ integrals are independent of μ. The μ dependence arises from the polar angle of the incident atom and vanishes for backward scattering. We show in Table II the behaviour of $V_{2\pm 2}^{2}$ for various impact parameters when F_0 is a LJ (12,6) potential of well depth ϵ occurring at separation R_m and when $F_2 = C/R^n$. The atom energy is ϵ.

TABLE II

Percentage Deviations from Body-Fixed Limit $e_{\pm} = 100\,(1 - V_{20}^{2}/V_{2\pm 2}^{2})$

b/R_m		0.2	0.4	0.6	0.8	1.0	1.2
$n = 12$	e_+	1.1	2.2	3.4	4.8	7.0	11.7
	e_-	-1.3	-3.0	-5.4	-8.9	-15.2	-30
$n = 6$	e_+	4.2	8.1	12.2	16.9	23	
	e_-	-4.8	-10.6	-18.3	-30	-48	

For $b = 0$ the body-fixed approximation is exact since $\phi = 0$, but as the impact parameter increases so does the deviation from the body-fixed result. The body-fixed approximation is clearly more successful for the shorter-ranged potential. At $b/R_m = 0.8$ the angle of deflection is $\cong \pi/2$, so that the scattering is ultimately far from backward yet no gross errors are occurring. This may be attributed to a significant portion of the final deflection arising at distances where the anisotropic potential has effectively vanished. The opposite signs on $e_{\pm}$ are quite typical and suggest that some cancellation of the errors due to the body-fixed approximation is likely in calculating any orientation-averaged quantities.

The scattering amplitude in the SCCP approximation is now calculated for fixed rotor orientation β' using (3.25),

$$S^{\mathrm{BF}}(\Delta j; \beta', E, b) = \frac{1}{2\pi} \int_0^{2\pi} \exp i\left[\Delta j\gamma' - A^{\mathrm{BF}}(\beta', \gamma'; E, b)\right] d\gamma' \quad (6.18)$$

where A^{BF} is given by (6.17). Averaging over all initial orientations β',

$$\mathcal{P}^{\mathrm{BF}}(j \to j + \Delta j; E, b) = \frac{1}{2} \int_0^{\pi} \sin\beta'\, d\beta' |S^{\mathrm{BF}}(\Delta j; \beta', E, b)|^2. \quad (6.19)$$

Comparing this with (6.9) and (6.10), we see that the BF approximation has saved one integration as well as using a simpler change in action.

For a P_1 anisotropy only the average over β' is similar to that arising in the sudden approximation and (6.14) is regained with V_{01}^1 replaced by V_{10}^1. For a P_2 anisotropy only the integral over β' yields a power series in $(V_{20}^2)^2$ (Dickinson and Richards, 1976b).

G. Numerical Evaluation of Cross Sections Using the SCCP

In virtually all work to date on rotational excitation, interest has focussed on the degeneracy-averaged cross sections between rotor levels, $\sigma(j \rightarrow j')$. An exception arises in the calculation of spin-lattice relaxation times and Rayleigh scattering line widths, which depend primarily on reorientation cross sections (Shafer and Gordon, 1973 and references therein).

To satisfy the detailed balance condition (3.5) we modify the probability (6.9):

$$P(j \rightarrow j'; E, b) = \left(\frac{E'}{E} \right)^{1/2} \sqrt{\frac{2j'+1}{2j+1}} \; \mathscr{P}(j \rightarrow j'; \bar{E}, b) \qquad (6.20)$$

where $E(E')$ is the initial (final) translational energy and $\bar{E}$ is a mean energy: we obtain $\bar{E}$ using the arithmetic mean of the initial and final relative speeds. For the rotor frequency appearing in the $V_{\nu\mu}^\lambda$ integrals, (6.3), needed for $\mathscr{P}$, the arithmetic mean of the initial and final frequencies is used; this gives equal classical and quantal energy differences. Then the cross section,

$$\sigma(j \rightarrow j'; E) = 2\pi \int_0^\infty b \, db P(j \rightarrow j', E, b) \qquad (6.21)$$

satisfies the exact detailed balance requirement

$$E(2j+1)\sigma(j \rightarrow j'; E) = E'(2j'+1)\sigma(j' \rightarrow j; E'). \qquad (6.22)$$

A counting-of-states factor similar to that used in (6.20) has been employed by Zarur and Rabitz (1974) in an effective-potential calculation. The ratio of speeds factor is necessary in any time-dependent model of the collision.

The computational techniques to evaluate σ are discussed fully in Dickinson and Richards (1974). While two integrals with oscillatory integrands are required—that for $V_{\nu\mu}^\lambda$, (6.3), and that for the scattering amplitude, (6.10)—in neither case does the physics of the problem require the evaluation of the integrals in the highly oscillatory region. Much of the computing time is spent at present on the $V_{\nu\mu}^\lambda$ integrals, and both more efficient techniques for evaluating them exactly and approximations of useful accuracy are currently being examined. These improvements could lead to a reduction by a factor of about four in the computing time.

H. Comparison with Close-Coupling Calculations: Space-Fixed Results

In our earlier work (Dickinson and Richards, 1974, 1976a) it was stated erroneously that the most detailed comparison possible between SCCP results and close-coupling results (CC) was for

$$\bar{P}_b(j \to j'; \beta) = \frac{1}{2\pi} \int_0^{2\pi} |\bar{S}(\Delta j; \alpha, \beta, E, b)|^2 d\alpha. \tag{6.23}$$

However, as was shown in Section VI.D, comparisons with individual S-matrix elements are possible although no such comparisons are yet available. As with the comparisons of classical perturbation theory, there is the problem of selecting the most meaningful comparisons for a many-parameter dependent quantity.

Comparisons of (6.23) with its quantal equivalent are discussed in Dickinson and Richards (1974, 1976a). Here we concentrate on comparisons of cross sections. These comparisons have been made with the CC calculations for (a) Ne–N$_2$ (Bosanac and Balint–Kurti, 1975), (b) He–N$_2$ (Erlewein et al., 1968; Tabor, private communication), and (c) Ar–N$_2$ (Pack, 1975). For He–N$_2$ and Ne–N$_2$ the SCCP results are within 20% of the CC results for transitions including 0–2, 2–4, 4–6, 6–8, 0–4, 0–6, and 2–6 at various energies (Dickinson and Richards, 1974, 1976a).

We discuss the Ar–N$_2$ results more fully because there is a greater body of reliable quantum-mechanical results for this system than for the others; also, a variety of approximations have been evaluated. For Ar–N$_2$ Pack (1975) has performed 49-channel CC calculations including rotor levels up to $j = 14$. The body-fixed calculations of McGuire (1976), which are discussed more fully in the next section, suggest that satisfactory convergence of the close-coupling expansion had been achieved. In Table III we compare the CC results with the SCCP results (Dickinson and Richards, 1976a). Only cross sections for upward transitions are shown since the corresponding downward results may be obtained from the detailed balance relation (6.22).

TABLE III

Comparison of Close-Coupling (CC) and Coupled-States (CS) Results with Strong-Coupling Correspondence Principle (SCCP) and Body-Fixed Correspondence Principle (BFCP) Results for Ar–N_2 Collisions. All Cross Sections Are in Å^2, and the Initial Translational Energy Is 450 k_B when the Molecule Is in $j=10$ Level (k_B is Boltzmann's Constant).

| Transition | Method | | | |
	CC^a	$SCCP^b$	CS^c	$BFCP^d$
0→2	$(15.7)^e$	18.1	15.2	17.6
2→4	14.4	10.8	14.8	10.9
4→6	12.8	9.6	13.0	9.8
6→8	11.0	9.1	10.7	9.1
8→10	10.8	8.7	9.5	8.7
10→12	9.8	8.6	8.2	8.5
0→4	11.4	13.9	11.9	14.1
2→6	10.5	7.3	11.0	7.5
4→8	7.9	6.0	7.7	6.0
6→10	5.9	5.2	5.5	5.2
8→12	(2.08)	4.0	3.11	3.93
0→6	10.8	10.8	11.2	11.2
2→8	6.6	5.1	6.6	5.3
4→10	3.5	3.6	3.3	3.6
6→12	(0.81)	2.02	1.26	1.96
0→8	5.7	6.5	5.8	6.8
2→10	1.86	2.39	1.87	2.41
4→12	(0.32)	0.96	0.48	0.93
0→10	1.15	2.07	1.17	2.11

[a] Pack (1975).
[b] Dickinson and Richards (1976a).
[c] McGuire (1976).
[d] Dickinson and Richards (1976b).
[e] Close-coupling results in parentheses have uncertainties exceeding 4%.

For the $\Delta j = 2$ transition the SCCP error is about 25% for low initial levels j and decreases surprisingly nonuniformly as j increases. For $\Delta j = 4$ the errors are somewhat larger but there is clear evidence of improvement as j increases. For $\Delta j = 6$ and 8 the errors are only slightly larger than for $\Delta j = 2$ or 4, and it is only for $\Delta j = 12$ that the SCCP results are in error by 100%. It appears for strong collisions, $|\Delta j| \cong j$, that although classical perturbation theory is not valid the average over the trajectories predicted by CPT is similar to the average over the exact trajectories.

These results plus the fact that the SCCP satisfies unitarity and is forced to satisfy detailed balance suggest that it is a reasonably robust and reliable approximation.

In Fig. 7 we show the transition probability as a function of impact parameter for a typical transition in $Ar-N_2$. The CC "transition probability" is based on relating the total angular momentum $\hbar J$ of the system to the impact parameter. It is clear that at small impact parameters the SCCP underestimates significantly the transition probability. This is because the time-dependent theory allows probability to flow into energetically closed channels: in these very strong collisions CPT produces transitions up to $\Delta j = 15$. The difference at large impact parameters probably results because the quantum contribution arises from a range of orbital angular momenta for each value of J so that an averaging of our rapidly varying probability appears in the quantum result.

The most direct comparison available of computing times is for $Ne-N_2$ transitions where CC calculations, using the method of Gordon (1969), and SCCP calculations were performed on the same computer. For a 16-channel CC calculation the SCCP was two orders of magnitude faster. For an

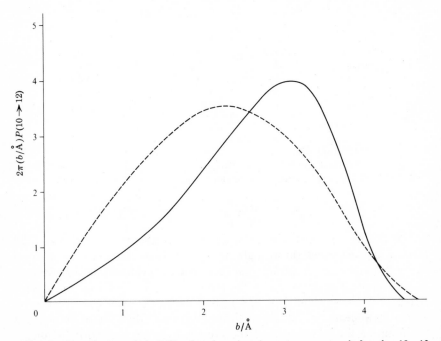

Fig. 7. Contributions $2\pi b\, P(b)$ plotted against impact parameter b for the $10\rightarrow12$ transition in $Ar-N_2$ at $E/k_B = 450°K$. The area under each curve gives the integral cross section $\sigma(10\rightarrow12)$. The CC results are denoted ------ and the SCCP results ———. The abbreviations are explained in the test.

M-channel CC calculation the time increases as M^3, while the SCCP time is largely independent of the quantum numbers involved.

I.   Comparison with Close-Coupling Calculations: Body-Fixed Results

Again we concentrate on the Ar–N$_2$ system discussed in the previous section. Results for other systems are discussed in Dickinson and Richards (1976b). For this system McGuire (1976) has made an extensive comparison of Pack's CC results with the coupled-states (CS) approximation, which makes the same physical approximations leading to the Hamiltonian (6.16). Because of the smaller number of coupled equations in the CS approximation, McGuire was able to establish explicitly that the CS expansion had converged with respect to the number of levels retained. These results, along with the body-fixed CP (BFCP) results, are shown in Table III.

It is clear that the level of agreement between the CS and the BFCP calculations is very similar to that between the CC and CP calculations. This is consistent with the comparability of the strength of coupling in the space-fixed and approximate body-fixed calculations. Furthermore, the agreement between the body-fixed and space-fixed results is about the same in both the quantal and the semiclassical calculations. For this system the additional error arising from the approximations made in the body-fixed frame is less than that associated with the use of correspondence principle methods.

As an example of a system where most of the cross section comes from long-range forces, we have considered the model Ar–TlF potential used by Tsien et al. (1973) for which close-coupling results are available for the 0–2 and 0–4 transitions. Because of the large reduced mass involved, many states are strongly coupled and all levels up to $j = 14$ were used in the CC expansion. A feature of this potential was the fairly strong long-range anisotropy so that over half of the 0–2 cross section is associated with the long-range forces. The value of the long-range anisotropy is at the limit suggested by McGuire (1976) for the success of the CS method. The results are shown in Table IV. Somewhat surprisingly, the BFCP result is in error only by about 20% for the 0→2 cross section but by about 100% for the much smaller 0→4 cross section.

We estimate that BFCP calculations can be performed at least twice as fast as SCCP calculations; to obtain the BFCP results quickly we have not yet used the full simplifications possible in the SCCP program. Any improved methods for calculating the V integrals (6.3), such as those discussed at the end of Section VI.G, will give a relatively greater increase in speed for BFCP calculations than SCCP calculations.

TABLE IV

Comparison of Ar–TlF Cross Sections; Comparison of Close Coupling (CC) Results, Tsien et al. (1973), with Strong-Coupling Correspondence-Principle Results (SCCP) and Body-Fixed Correspondence-Principle Results (BFCP) for Ar–TlF Collisions, Dickinson and Richards (1976b). All Cross Sections Are in $\overset{\circ}{A}^2$, Initial Translational Energy Is 1344 k_B Where k_B is Boltzmann's Constant.

Method	$\sigma(0\rightarrow2)$ $(\overset{\circ}{A}^2)$	$\sigma(0\rightarrow4)$ $(\overset{\circ}{A}^2)$
CC	58.9	16.4
SCCP	60.0	21.6
BFCP	72.9	32.0

J. Comparison with Other Approximations

A variety of approximate schemes are currently being used for atom-molecule rotational excitation. One approach is to reduce the dimensionality of the full coupled equations and then solve the resulting equations quantum mechanically. This type of approximation includes the coupled-states method, discussed in Section VI.F (McGuire, 1976), the effective Hamiltonian approximation (Rabitz, 1975) and the l-dominant approximation (DePristo and Alexander, 1975b). No classical or semiclassical analog of these latter approaches appears to have been investigated.

Other approaches retain the quantal description of both the internal and relative motion but approximate the coupling between them, e.g., the exponential-Born-distorted-wave (EBDW) approximation of Bosanac and Balint–Kurti (1975). Some comparisons of this with SCCP calculations for Ne–N$_2$ collisions are discussed in Dickinson and Richards (1976a), and these show the EBDW approximation to be more accurate but apparently significantly slower to calculate.

Another class of approximations retains the quantal description of the rotor but treats the relative motion classically, Section III.C. Decoupling approximations, such as those discussed above, may also be used (Lawley and Ross, 1965a,b). The resulting Schrödinger equation for the rotor may then be solved numerically, as in McCann and Flannery (1975a,b) and Jamieson et al. (1975), or further approximations may be introduced, as in Saha and Guha (1975). Some comparisons of the SCCP with this latter approximation are discussed in Dickinson and Richards (1976a).

A third group of approximations uses classical mechanics entirely for the dynamics. Either exact trajectories (LaBudde and Bernstein, 1973) can be used or classical perturbation theory can be employed and different methods are used to determine quantal level-to-level cross sections, as discussed in Section III.C. While the combination of exact classical trajectories and quantization using classical S-matrix theory (Miller, 1975) is very powerful, the root finding problem in selecting the trajectories normally requires further approximations such as partial-averaging. However, its use with the decoupling approximations that replace the two-dimensional S matrix by a number of one-dimensional S matrices might well be practicable.

In view of the uncertainties in our present knowledge of potential hypersurfaces, it hardly seems worthwhile to undertake extensive CC calculations for systems other than those for which accurate surfaces exist and some typical model systems. Existing CC calculations now provide quite an extensive body of results. Experience with the various time-dependent calculations indicates that classical trajectories are adequate for the relative motion unless results very close to threshold are required. Our results indicate that classical perturbation theory is adequate for describing the rotor motion even in low rotor states for many systems where the attractive forces are of Van der Waals' origin. There are well-founded reasons for expecting CPT to improve as the excitation of the rotor increases. By contrast, all methods involving a quantal description of the rotor become increasingly difficult to calculate in this limit.

K. Sensitivity of the Cross Section to the Potential Hypersurface

As mentioned in Section VI.A, considerable uncertainty surrounds most potential surfaces. Consequently, the results of even the best scattering calculations using these potentials may have a substantial error. For collinear atom-molecule vibrational excitation, Clark and Dickinson (1973, 1975) have shown, using CC methods, that transition probabilities are very sensitive to the part of the surface describing the oscillator motion. Alexander (1975) has shown, using the effective-potential method of Rabitz (1975), that vibrationally inelastic processes in three-dimensional collisions appear to be extremely sensitive to the specific choice of potential surface.

To estimate the sensitivity of rotational excitation cross sections to changes in the potential surface, we have evaluated cross sections after making physically reasonable alterations to the potential surface. These calculations have been done using the SCCP because this provides cheap reliable cross sections.

We concentrate again on the Ar–N_2 system discussed previously in Section VI.H. The potential used was of the form (6.1) with $F_0(R)/\epsilon = z^{-12} - 2z^{-6}$, $F_2(R)/\epsilon = a_{12}z^{-12} - a_6z^{-6}$, $z = R/R_m$, ϵ and R_m have their usual significance and the standard values were $\epsilon = 0.0103$ eV, $R_m = 3.93$ Å, $a_{12} = 0.5$, and $a_6 = 0.26$. De Pristo and Alexander (1975a) have also studied this system using the same values for ϵ and R_m but with $a_{12} = a_6 = 0.315$. Kistemaker and de Vries (1975) have used a pairwise additive potential for this system with a Morse form for the Ar–N interaction. Classical calculations using this potential yielded a rotational relaxation time in reasonable agreement with experiment but about half that calculated by Russell et al. (1972) using the potential of Pattengill et al. (1971).

The alterations we made are (i) increasing a_{12} by 20%, (ii) decreasing a_{12} by 20%, (iii) decreasing R_m by 10%, and (iv) decreasing ϵ by 10%. The percentage changes in almost all the possible inelastic $\sigma(10{\rightarrow}j)$ cross sections for each of the alterations are shown in Table V.

TABLE V

Sensitivity of the Ar–N_2 Cross Sections to Changes in the Potential Surface. Shown Are $100[\sigma^*(10{\rightarrow}j)/\sigma(10{\rightarrow}j) - 1]$, Where σ^* Is the Cross Section for the Potential Modified as Shown in the First Column and σ Is the Cross Section for the Standard Potential Defined in the Text. All Calculations Except the Last Row Had $E/\epsilon = 3.764$.

j	2	4	6	8	12	14
$a_{12} = 0.4$	-70	-45	-13	9	-4	-51
$a_{12} = 0.6$	50	9	-1	-4	-5	52
$R_m = 3.54$ Å	-45	-33	-22	-15	-28	-78
$\epsilon = 0.00927$ eV	-53	-39	-26	-17	-32	-83
$E/\epsilon = 4.182$						

We consider together the changes in the value of a_{12}. The modified values for a_{12} both lie within the range estimated by Pattengill et al. (1971).

We consider first the cross sections for large changes in energy. The main contribution to these cross sections comes from nearly head-on collisions where the greatest classical angular momentum transfer occurs. In our model the maximum classical angular momentum transfer is proportional to the strength of the potential anisotropy. Consequently, decreasing or increasing the value of a_{12} makes a marked difference to the volume in $(b, \alpha, \beta, \gamma)$ space in which the transition is classically accessible and thus to the volume in which the classical transition probability is nonzero. While quantal effects will doubtless blur this classical picture, the main outline should remain, so that significant differences, nonlinear in a_{12}, can be expected.

For small energy transfer the volume of space in which the transition is classically accessible is less sensitive to changes in the value of a_{12}, since the boundaries are determined mainly by competition between the short-range and long-range potentials. Thus the cross section for small energy transfer is relatively insensitive to the short-range anisotropy. This qualitative model is consistent with the results of our calculations.

We have examined also the effect of reducing each of ϵ and R_m by 10%. These changes are considerably larger than the uncertainties in the values of ϵ and R_m suggested by Pattengill et al. (1971). Such changes, however, which are comparable with those in the anisotropic potential, provide an opportunity of comparing the sensitivity of the cross sections to similar changes in the isotropic and anisotropic parts of the potential. Classically, the use of reduced units shows that for fixed E/ϵ the cross sections for transfer of angular momentum, measured in units of $\hbar B_z^{1/2}$, ($B_z = 2\mu\epsilon R_m^2/\hbar^2$), are proportional to R_m^2. For quantal cross sections for a change in the value of the angular momentum in absolute units such a scaling is no longer exact. Deviations are expected to be most marked for large absolute angular momentum transfer since the boundary in parameter space for classically accessible processes changes. This picture is consistent with our calculations.

When the value of ϵ was changed, the same physical scattering energy was retained. Thus the classical results, in units of $\hbar B_z^{1/2}$, will be altered because of the change in the reduced energy (E/ϵ). The higher energy collision will normally have less angular momentum transfer, so there are two effects leading to reduced cross sections for fixed angular momentum transfer. The previous discussion concerning the greater sensitivity of large Δj transitions again applies. In view of the slight energy dependence for the small Δj transitions obtained for this system in Dickinson and Richards (1976a) the changes of $\cong 20\%$ in these cross sections for small Δj transitions on a change of 10% in ϵ are surprising.

Our limited study suggests that the bigger cross sections are more sensitive to the isotropic than the anisotropic part of the potential. Since, in general, the isotropic part is much better known, this is encouraging for the calculation of reliable cross sections, at least for small Δj, but discouraging for the prospects of inverting cross sections to obtain information on potentials.

Clearly for the Ar–N_2 system the errors in the SCCP calculations discussed in Section VI.H are comparable with the uncertainties arising from our incomplete knowledge of the potential surface for this system. It appears likely that this situation is quite typical.

VII.   THE CLASSICAL LIMIT IN ROTATIONAL EXCITATION

A.   Introduction

Normally, purely classical trajectory calculations are easier to perform than either quantal or classical S-matrix calculations; this is especially true in the case of large quantum numbers. However, there is some doubt as to when purely classical mechanics can be expected to provide accurate results.

Classical mechanics is known to be valid when all relevant actions and their changes during the collision are much greater than $\hbar$; this is often not a necessary criterion since it has been demonstrated that under certain circumstances classical mechanics provides reliable results for transitions involving small changes in quantum number (see Percival and Richards, 1975, p. 69).

In Section IV we showed that the SCCP reduces to classical mechanics and the density of states correspondence principle when all relevant actions are large, but this analysis did not show how fast this limit is approached. Here we study this problem after reviewing other related work.

B.   Comparison of Classical and Quantal Cross Sections for Homonuclear Molecules

Purely classical and quantal calculations have been compared by various workers for particular systems. The $Li^+ + H_2$ system has been studied quantally by Lester and Schaefer (1973) and classically by LaBudde and Bernstein (1971, 1973), who used the two different methods of obtaining $\sigma^c(j \to j')$ from the classical differential cross section, $d\sigma^c/d\Delta E$ discussed in Section III, equations (3.9) and (3.10). Some of their results are shown in Table VI.

TABLE VI

Comparison of the Classical (LaBudde and Bernstein, 1973) and Quantal (Lester and Schaefer, 1973) Cross Sections for the $j = 0$ to $j = 2$ Transition in the $Li^+ + H_2$ System. The Cross Sections Are in $\overset{\circ}{A}^2$ and the Three Different Classical Results Are Described in the Text. The Errors in the Classical Results Are Statistical Errors.

E/eV	Classical results			Quantal result
	I	II	III	
0.2	3.54 ± 0.2	0.22 ± 0.2	0.002	3.69
0.5	17.35 ± 1.0	9.9 ± 1.0	—	15.16

In this table the classical results I and II are twice the results obtained from (3.10) with $\alpha = 1$ and $\frac{1}{2}$, respectively, and the results III are twice those obtained using the density of states correspondence principle, (3.9). The factor of two was introduced by LaBudde and Bernstein to allow for the symmetry of this system, which allows only even Δj transitions.

The large difference in these three results is a consequence of $d\sigma^c/d\Delta E$ being a rapidly varying function of ΔE. In these circumstances the good agreement between I and the quantal results should be considered as fortuitous.

Barg et al. (1976) have also studied the $Li^+ + H_2$ system, using the quantization rule corresponding to I above, and compared classical angular differential cross sections with the quantal calculations of Schaefer and Lester (1975). They found differences of between 7 and 100%, and noted that the classical results agreed to within 10% at large scattering angles (small impact parameters). They did not investigate the effects of applying different quantization methods, which is unfortunate since the work of LaBudde and Bernstein suggests that this system is particularly sensitive to the choice of method.

Augustin and Miller (1974) have studied the $He + H_2$ system using only the quantization rule (3.10) with $\alpha = 1$ and have found the classical and quantal results to agree to within 20% in the energy range 1000 to 5000 cm^{-1}.

Using the same quantization method, Pattengill et al. (1971) have considered the much heavier $Ar-N_2$ system and Pattengill (1975) has shown that the classical results are within 2 to 24% of the quantal results (Pack, 1975), depending on the transition considered. The success of the quantization rule (3.10) with $\alpha = 1$, which has no rigorous theoretical foundation, is surprising; clearly, further work is required.

C. Comparison of Classical and Quantal Cross Sections for Heteronuclear Molecules

All of the above comparisons deal with homonuclear molecules, and in each case some ad hoc allowance is made for the selection rule allowing only even Δj transitions. When there is no symmetry these ad hoc rules do not work and purely classical calculations cannot be adjusted to mimic the quantal distribution in ΔE.

There have been two papers dealing with the case where there is little or no symmetry. Augustin and Miller (1974) and Brumer (1974) have studied CO excited by He and H, respectively; each used the quantization rule (3.10) with $\alpha = \frac{1}{2}$; some of their results are given in Table VII. The results of Brumer (1974) for the He–HCN collision, a case of high symmetry, are also shown. We note that classical mechanics predicts the total inelastic cross section more accurately than the individual cross sections. This fact was first noted by Williams (1931).

Brumer (1974) also studied collisions of He + HCN, a system having only a slight asymmetry giving rise to large even Δj and small odd Δj cross sections (Green and Thaddeus, 1974). Purely classical calculations cannot reproduce this approximate selection rule; they overestimate the odd Δj and underestimate the even Δj cross sections (see Table VII). Brumer tried two ways of enforcing this selection rule but without very much success. This approximate selection rule, which survives all averaging, except for inelasticity, is clearly going to require a sophisticated semiclassical treatment, such as classical S-matrix theory or the SCCP.

TABLE VII

Table of $\sigma(0{\rightarrow}j)/\text{Å}^2$ for Various Systems. The Results for H + CO Are from Brumer (1974); the Errors in These Results Are the Statistical Errors of the Monte Carlo Method. The H + CO Results Are for an Incident Energy of 80.7 cm^{-1} While the He + HCN Results Are for a Total Energy of 60 cm^{-1}. The Results for He + HCN and He + CO Are from Augustin and Miller (1974) with an Incident Energy of 100 cm^{-1}; No Statistical Errors Are Given.

| | System | | | | | |
| j | H + CO | | He + HCN | | He + CO | |
	Classical	Quantal	Classical	Quantal	Classical	Quantal
1	17.6 ± 1.4	16.1	18 ± 3	2.0	7.04	6.16
2	6.4 ± 0.6	5.9	12 ± 2	20.2	8.64	13.04
3	3.4 ± 0.6	2.8	6.7 ± 1	1.6	7.68	4.08
4	0	1.5	2.8 ± 1	6.61	5.20	4.24
5	0	0.45	2.8 ± 1	0.62	2.80	1.04
$\Sigma\sigma(0{\rightarrow}j)$	27.4 ± 2.6	26.75	42.3 ± 8	31.03	31.36	28.56

D. The Classical Limit for Model Systems

The results described above suggest that a reliance on purely classical calculations is not justified. However, as it is difficult to draw general conclusions from these calculations, we have considered an idealized collision for which the transition probabilities may be obtained in analytic form. We consider that all of the essential physics remains so that the main features of a more realistic collision are faithfully reproduced.

The three cases considered are:

1. P_1 anisotropy only in the sudden or body-fixed approximation.
2. P_2 anisotropy only in the body-fixed approximation.
3. P_1 and P_2 anisotropy only in the body-fixed approximation.

In all cases we evaluate the quantum transition probability using the SCCP. First we consider our claim that all the essential physics is contained in our model.

Both the sudden and the body-fixed approximations have been evaluated and compared with exact close-coupling calculations in a variety of situations (Sections VI.H and VI.I) and both are found to reproduce reasonably well the general features of the transition probability. The sudden approximation is invalid at low energies, while the body-fixed approximation is valid only for short-range interactions. These two considerations are the only limits to the validity of our models.

The validity of the SCCP for rotational excitation is discussed in Section VI.H; here we note that it is generally accurate to within 20%.

In the approximations that we consider, it happens that the transition probability is a function of V, the maximum change in action of the rotor. Thus the numerical value of V is important. As a rough guide, V is proportional to $\sqrt{B_z}$ where $B_z = 2\mu\epsilon R_m^2/\hbar^2$; it is a slowly varying function of both the incident energy and the impact parameter if the latter is small ($b \leqslant R_m$), while for larger impact parameters it decreases rapidly. It also decreases as the rotor frequency increases, and so decreases with increasing quantum number for a given system. Typically, the maximum value of V varies between $\hbar$ (for a light system) and $10\hbar$ (for a heavy system).

1.   Classical Limit for P_1 Anisotropy
(Sudden or Body-Fixed Approximations)

For these collisions the SCCP transition probability is given by (6.14). Dickinson and Richards (1974) have shown that this has the asymptotic form for large V_1

$$P(j \rightarrow j') = \frac{1}{2V_1}\left[1 + \frac{\sin\left(2V_1 - \left(s + \frac{1}{4}\right)\pi\right)}{\sqrt{\pi V_1}}\right] + O\left(V_1^{-3/2}\right), \qquad s = |j - j'|.$$

(7.1)

Here $V_1 = |V_{01}^1|$ or $|V_{10}^1|$ [see (6.3)] in the sudden or body-fixed approximation, respectively. They show also (for the sudden approximation, the body-fixed analysis is similar) that

$$P^c(\Delta E) = \frac{H(V_1\omega^c\hbar - \Delta E)}{2V_1\omega^c\hbar}$$

(7.2)

where $H(x)$ is the Heaviside unit function. Using the density of states correspondence principle (3.9),

$$P^c(j \to j') = \frac{H(V_1 - s)}{2V_1} \qquad (7.3)$$

which is just the leading term of the SCCP result (7.1). The second term in (7.1) is due to interference between the two contributing classical paths (Dickinson and Richards, 1976a), so that these two terms constitute the "primitive semiclassical approximation" in the large V_1 limit.

If the classical result (7.2) is quantized using (3.10), we obtain

$$
\begin{aligned}
P^c(j \to j') &= 1/2V_1 & V_1 &\geqslant s + \alpha \\
&= \frac{V_1 + \alpha - s}{4\alpha V_1} & s + \alpha &\geqslant V_1 \geqslant s - \alpha \qquad (7.4) \\
&= 0 & V_1 &\leqslant s - \alpha.
\end{aligned}
$$

The classical transition probabilities (7.4) and (7.3) agree except in the neighbourhood of the classically inaccessible region, $V_1 < s$, where neither is valid.

In Fig. 8 $V_1 P^c(j \to j')$ is plotted against V_1 for $j' = j + 1$; for the values obtained from (7.4) we have chosen $\alpha = \frac{1}{2}$. It is seen that both classical results (7.3) and (7.4) can differ from the SCCP by more than 20% for $V_1 \leqslant 3$, the worst errors occurring where the probability is largest; but any averaging over a range of values of V_1 will improve the accuracy of the classical results; classical mechanics will be better for total than for differential cross sections. The classical result (7.4) is seen to mimic the SCCP result in the neighborhood of $V_1 = 1$ better than the density of states result, (7.3), but small changes in α can make large changes in the probability so that this result cannot be relied upon here.

The primitive semiclassical probability, (7.1), is seen to be significantly better than the purely classical results.

For small V_1 ($\cong s$) the classical probabilities, either with or without the interference term, are not to be trusted. Since V_1 is largest for small impact parameters, we would expect classical results to be best in the backward scattering region; this is consistent with the results of Barg et al. (1976). However, for light systems V_1 is generally small so that classical results cannot be trusted.

2. P_2 Anisotropy in the Body-Fixed Approximation

In this case the SCCP yields (Section VI.F) the transition probability as a power series in the action V_{20}^2 (equation 6.3).

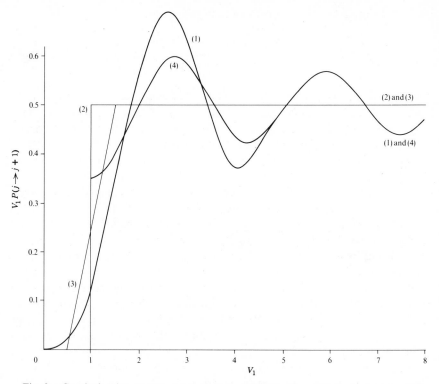

Fig. 8.   Graph showing $V_1 P(j \rightarrow j+1)$ plotted against V_1. Curve (1) is obtained from the SCCP, (6.14); curves (2) and (3) are the purely classical results of (7.3) and (7.4), respectively; curve (4) is the primitive semiclassical result, (7.1).

Using (4.7b) and (6.17) we obtain

$$\Delta E^c = \Delta E_{max} \sin^2 \beta' \cos 2\gamma', \qquad \Delta E_{max} = 3\omega^c \hbar V_2/2, \qquad (7.5)$$

where $V_2 = |V_{20}^2|$. We note that the maximum classically allowed $|\Delta j|$ of the target is $3V_2/2$. Using a generalization of (4.9a) we obtain

$$P^c(\Delta E) = \frac{K(m) H(\Delta E_{max} - |\Delta E|)}{\pi \Delta E_{max}(1 + |\Delta E|/\Delta E_{max})^{1/2}} \qquad (7.6a)$$

where $m = (\Delta E_{max} - |\Delta E|)/(\Delta E_{max} + |\Delta E|)$ and where $K(m)$ is a complete elliptic integral of the first kind (Abramowitz and Stegun, 1965, p. 590). Using the density of states correspondence principle (3.9) we obtain the classical probability for transitions between levels:

$$P^c\,(j\to j') = \frac{2K\,(m(s))\,H\,(3V_2 - 2s)}{3\pi V_2\,(1 + 2s/3V_2)^{1/2}}\,, \qquad s = |j - j'|, \qquad (7.6b)$$

where now $m(s) = (3V_2 - 2s)/(3V_2 + 2s)$. The expression for $P^c\,(j\to j')$ obtained from (3.10) cannot be obtained in closed form; however, we have shown the result of this quantization rule, for $\alpha = 1$, in Fig. 9. We have not obtained the primitive semiclassical approximation.

The classical probability (7.6b) is a decreasing function of $|\Delta j|$; in particular, $P^c\,(j\to j+1) > P^c\,(j\to j+2)$, which violates the quantal selection rule allowing only even Δj transitions. Furthermore, a stationary-phase evaluation of the SCCP shows that, for large V_2, $P^{SCCP}(j\to j+2p) \cong 2P^c\,(j\to j+2p)$; this is a direct consequence of the selection rule. In order to use classical mechanics to estimate $j\to j'$ cross sections, LaBudde and Bernstein (1973) multiplied their classical result by two: this arbitrary procedure is consistent with the above result.

In Fig. 9 we compare the SCCP result with twice the classical result of (7.6a) for the $\Delta j = 2$ transition and for various V_2. Also shown are the results of the quantization (3.10) with $\alpha = 1$, and the classical results for the forbidden transitions $\Delta j = 1, 3$, both multiplied by two for ease of comparison with the $\Delta j = 2$ result.

In general, the results show much the same behavior as in the P_1 case; for $V_2 < 4$ it is apparent that pure classical calculations are unreliable. The similarity of these results and those for the P_1 case suggest that for a single anisotropy our conclusions may be more general.

3. $P_1 + P_2$ Anisotropy in the Body-Fixed Approximation

In this case neither the SCCP nor the purely classical transition probabilities may be expressed in closed form, although the former may be expressed as an infinite series of Bessel functions, similar to Dickinson and Richards (1974, equation 4.28), and the latter as a one-dimensional integral. Of particular interest is the case of a relatively weak P_1 interaction; here the classical probability for odd Δj is dominated by contributions from the P_2 potential. This cannot happen in quantum mechanics or the SCCP.

In view of the results of Brumer (1974) and Augustin and Miller (1974) we felt it worthwhile to make some comparisons between classical and SCCP results.

Perturbation theory gives the classical energy transfer

$$\Delta E^c = \Delta E^{c1} + \Delta E^{c2} \qquad (7.7)$$

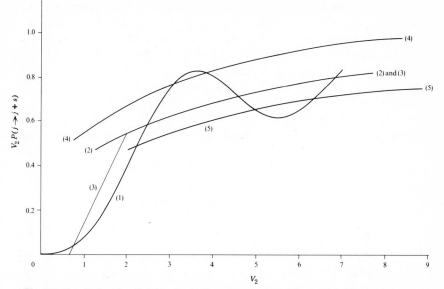

Fig. 9.   Graph showing $V_2 P(j \rightarrow j + s)$ plotted against V_2. Curve (1) is obtained from the SCCP with $s = 2$; curves (2) and (3) are twice the purely classical result with $s = 2$ using the density-of-states correspondence principle, (7.6), and the quantization method (3.10), respectively; curves (4) and (5) show twice the classical probability, (7.6), with $s = 1$ and 3, respectively.

where ΔE^{ck} is the energy transfer from the kth anisotropy. Thus the classical distribution function is given by the convolution

$$P^c(\Delta E) = \int_{-\infty}^{\infty} dx \, P^{(1)}(\Delta E - x) P^{(2)}(x) \qquad (7.8)$$

with obvious notation: $P^{(k)}$ is given by (7.2) and (7.6a) for $k = 1, 2$, respectively. Using the density of states correspondence principle (3.9) we obtain

$$P^c(j \rightarrow j + \Delta j) = \frac{1}{3 V_1 V_2 \pi} \int_{-\infty}^{\infty} dy \, \frac{H(V_1 - |s - y|) H(3V_2 - 2|y|) K(m(y))}{(1 + 2|y|/3V_2)^{1/2}}$$

$$(7.9)$$

where $m(y)$ is defined in (7.6b). Equation (7.9) takes on different forms according to the relative sizes of V_1 and V_2.

In Figs. 10 and 11 we have plotted the probability as a function of V_1, calculated using (7.9) and using the SCCP, for the $\Delta j = 1$ transition for $V_2 = 3$ and 5, respectively. We see that in both cases the classical result is in error for small V_1; here the selection rules ensure that $P^{SCCP} \alpha V_1^2$, but according to (7.9) P^c is nonzero when $V_1 = 0$.

For large $V_1 (\gg V_2)$ P^{SCCP} oscillates about P^c as in the previous cases, although this limit has not yet been reached in Fig. 11.

For $V_1 \cong V_2$ there is some structure in P^{SCCP} that is not given by the classical result. This is entirely due to interference. It may be shown that for

$$V_1 < V_1^{CRIT} = 3V_2 \sin\beta' - \frac{(9s)^{2/3}}{2}\left(\frac{V_2}{\sin\beta'}\right)^{1/3}, \qquad \sin\beta' > \frac{3^{1/4}}{2^{3/4}}\left(\frac{s}{V_2}\right)^{1/2}$$

$$(7.10)$$

there are four real classical paths that contribute to P^{SCCP} in the stationary-phase limit. For larger V_1 there are only two paths. At the boundary $V_1 \cong V_1^{CRIT}$ there is additional structure as the four paths coalesce into two paths.

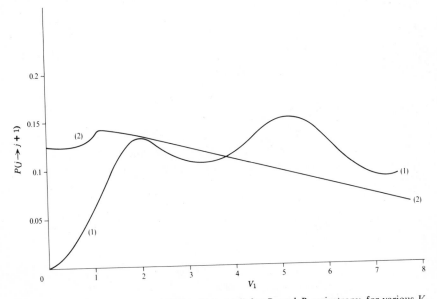

Fig. 10.   Graph showing the probability $P(j \to j+1)$ for P_1 and P_2 anisotropy, for various V_1 and $V_2 = 3.0$. Curve (1) is the SCCP probability and curve (2) the classical result of (7.9).

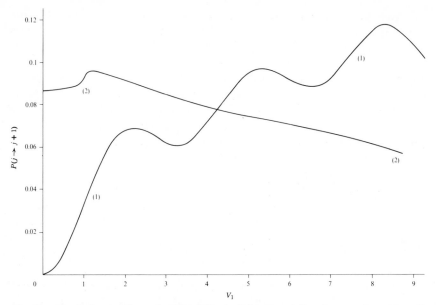

Fig. 11. Graph showing the probability $P(j \rightarrow j+1)$ for P_1 and P_2 anisotropy, for various V_1 and $V_2 = 5.0$. Curve (1) is the SCCP probability and curve (2) the classical result of (7.9).

Clearly, when contributory classical paths coalesce a purely classical result will be seriously in error; this will happen at particular impact parameters or scattering angles. The primitive semiclassical result would be a significant improvement, and would probably give the correct structure; a uniform approximation to this would be better still; the relevant theory under these circumstances is under active development. Connor (1976) uses catastrophe theory to classify the results based on the structure of the turning points. However, in the limit where classical perturbation theory is valid, the SCCP is the correct uniform approximation and is readily evaluated. The SCCP is easier to evaluate than the primitive semiclassical approximation, as the integrand is a well-behaved function.

E. Conclusion

In this section we first discussed some of the available comparisons between classical and quantal cross sections for rotational excitation. We then considered the relation between the classical and quantal transition probabilities as a function of impact parameter. Further work is necessary before quantitative results for cross sections may be obtained from our analysis.

Qualitatively, it is clear that averaging over the impact parameter will improve the agreement between classical and quantal transition probabilities and that the cross sections most sensitive to quantal effects are either those where most of the contribution comes from near the boundary between classically allowed and classically forbidden transitions, or those for odd Δj due to a potential that is almost an even function of $\cos \theta$.

For collisions involving homonuclear molecules where all the atoms are at least as heavy as, say, carbon, typical anisotropies lead to small Δj transitions having geometric cross sections at energies well above threshold and quantal effects are relatively small. For the large Δj cross sections, quantal effect are more important. For light homonuclear systems quantal effects may be expected. The evidence from cross section calculations is consistent with this for heavy systems. For light systems, however, the quantization procedure based on a "bin" of width 2 in j is quite successful. Further work is clearly needed to determine the extent of the applicability of this quantization method.

For heteronuclear systems with no near symmetry in the potential, the discussion above is expected to be valid. When near symmetry occurs, only quantization of classical results including interference between trajectories can be expected to be successful. Our SCCP calculations have shown the expected behavior in this limit.

VIII.   SIMULTANEOUS ROTATIONAL-VIBRATIONAL EXCITATION

Work is in progress to apply the SCCP to atom diatomic-molecule collisions allowing for simultaneous rotational and vibrational transitions. This work combines the developments described in Sections V and VI.

Previous studies of this topic have been covered in the recent reviews by Lester (1975) and Balint–Kurti (1975). Most of the theoretical work has been concentrated on the two systems $Li^+ + H_2$ and $He + H_2$. Ab initio potential surfaces exist for both these systems: $Li^+ + H_2$ (Lester, 1971; Kutzelnigg et al., 1973), $He + H_2$ (Krauss and Mies, 1965; Gordon and Secrest, 1970; Tsalpine and Kutzelnigg, 1973). The strong anisotropy of the $Li^+ + H_2$ surface, however, renders it less suitable for the application of SCCP methods.

The $He + H_2$ system has been studied extensively. Exact numerical solutions of the close-coupling equations have been obtained (Eastes and Secrest, 1972; McGuire and Micha, 1972). Because of the rapid increase in the number of coupled equations as more levels are included in the close-coupling basis, these calculations have been limited to very low-lying levels. However, two of the quantum approximations discussed in Section VI.I for rotational excitation have been applied to the problem; the coupled-states approximation by Alexander and McGuire (1976 and references therein) and the effective-potential method by Rabitz and Zarur (1975). Although these approximations reduce the number of coupled equations to be solved, cross sections have, at present, only been calculated for vibrational quantum numbers up to $v = 2$ and rotational quantum numbers up to $j = 12$. For the general validity of SCCP methods, these quantum numbers are relatively low. However, it is expected that the validity of the SCCP, shown for low j and v states in the previous sections, will also be apparent for rotational-vibrational excitation. These quantum-mechanical studies will thus provide a useful test for the SCCP for this problem.

The interaction potential will now take the general form

$$V(R,r_1,\theta) = \sum_{\lambda=0}^{\infty} F_\lambda(R,r_1) P_\lambda(\cos\theta) \qquad (8.1)$$

where r_1 is the separation of the bound nuclei. The extension of the rigid rotor formulation, described in Section VI, to include vibrational excitation thus requires an additional integration, over the vibrator angle variable w_1. Since r_1 is periodic in w_1 it may be expanded as a Fourier series.

The $V_{\nu\mu}^\lambda(E,b)$ trajectory integrals for the rigid rotor system will be generalized to

$$V_{\nu\mu}^{\lambda n}(E,b) = \frac{1}{\hbar} \int_0^{\infty} a_{n\lambda}(R)\cos(n\omega_1 t + \nu\omega t - \mu\phi)\, dt \qquad (8.2)$$

where $a_{m\lambda}(R)$ is the mth Fourier coefficient of $F_\lambda(R,r_1(w_1))$ and $\omega_1(\omega)$ is the vibrational (rotational) frequency. The number of $V_{\nu\mu}^{\lambda n}$ integrals now depends both on the number of terms necessary to represent adequately the functions F_λ as a Fourier series in w_1 and on the number of terms taken in the expansion of the potential (8.1).

A major difference between these calculations and those for the rigid rotor is that the spherically symmetric part of the potential $F_0(R,r_1)$ can now cause Δv transitions and hence cannot be ignored.

The vibrational frequency ω_1 is generally much larger than the rotational frequency ω. This suggests that use may be made of the sudden approximation (3.15) for determining the Δj part of the transition. Furthermore, for $\Delta v = 1$ transitions, the first-order SCCP approximation (3.12) may be applicable for the relatively low collision energies for which quantum calculations have been performed. Use may also be made of the body-fixed approximation discussed in Section VI.F.

Although the He + H$_2$ system will provide a useful test for the SCCP, this is not the major aim of this extension to rotational-vibrational excitation. Recent developments in experimental technique have enabled laser-induced fluorescence experiments to be successfully carried out. These experiments allow rate constants to be measured for moderate vibrational ($v \sim 5$) and high rotational ($j \sim 40$) quantum numbers. The systems that have been studied are $Li_2(B\,^1\Pi_u)$ (Ottinger et al., 1970; Ottinger and Poppe, 1971; Ennen and Ottinger, 1974) and $Na_2(B\,^1\Pi_u)$ (Bergmann and Demtröder, 1972) in collision with the rare gases.

Because of the high j and v quantum numbers involved in these measurements, the usual quantum-mechanical theories will clearly be inappropriate owing to the strong coupling between many states. The SCCP, however, will be valid for these transitions and calculations of cross sections and rate constants should yield important information on the potential surfaces of these systems.

IX. CONCLUSION

In this chapter we have considered methods used to study rotational and vibrational excitation of diatomic molecules in collisions with atoms. These collision processes have the advantage that in some circumstances the Schrödinger equation describing them may be reliably solved numerically.

However, purely quantal methods are generally not practicable since, except for excitation of low levels, the basis set is unmanageably large and other approximations are necessary. Those close-coupling approximations treating the projectile classically and the target quantally fail for the same reason; so also do the angular momentum decoupling methods. However, both of these methods significantly increase the range of problems accessible to a quantal approach. Nevertheless, it is clear that for larger quantum numbers and for more complex systems than those considered here alternative methods are necessary.

Rotational excitation is generally a classically allowed process so that for large quantum numbers, at least, purely classical methods would be expected to suffice. This is probable but the presently available evidence, reviewed in Section VII, is scant. Classical mechanics has only been tested for transitions involving low quantum numbers and small changes in quantum numbers—exactly the situation where it is most suspect. More work is necessary to establish those circumstances where it may safely be used.

Vibrational excitation, on the other hand, is generally a classically forbidden process so that quantal or semiclassical methods are necessary. For one-dimensional systems Schrödinger's equation may be satisfactorily solved numerically, but as the dimension of the system increases the computational effort involved quickly becomes prohibitive.

Clearly, to treat simultaneous rotational and vibrational excitation, it would be desirable to have an approximation that is uniformly valid for classically allowed and forbidden processes. Also, as the number of quantal states involved increases, it becomes necessary to use approximations based upon classical mechanics.

The method considered in detail in this chapter, the strong-coupling correspondence principle, like the classical S-matrix theory, describes both collision partners classically. Both of these methods incorporate quantal tunnelling and interference effects. These methods have the advantage that the computational effort is almost independent of the size of the quantum numbers, but have the disadvantage that their use for small quantum numbers is theoretically unjustified; however, empirical evidence suggests that serious errors are not engendered by using this type of approximation in this case.

The SCCP differs from other semiclassical theories in using classical perturbation theory to calculate the relevant changes in the dynamical variables due to the collision. This single approximation greatly simplifies the theory and reduces the computational effort considerably.

Neither multidimensional root-finding procedures nor complex classical trajectories are necessary. In practice, the use of classical perturbation theory in the applications discussed in Sections V and VI has not engendered serious error.

From these examples we infer that useful accuracies will be obtained for more complex collisions; from the general theory, we expect that for these processes the computational effort will be reasonable.

Acknowledgements

One of us (A.P.C.) thanks the Science Research Council for support. We thank Professor I. C. Percival for a critical reading of the manuscript.

References

Abramowitz, M., and I. A. Stegun. 1965. *Handbook of Mathematical Functions*, Dover, New York.
Abrines, R., and Percival, I. C., 1966, *Proc. Phys. Soc.*, **88**, 873–83.
Alexander, M. H. 1975. *Chem. Phys.*, **8**, 86.
Alexander, M. H., and P. McGuire. 1976. *J. Chem. Phys.*, **64**, 452.
Allison, A. C., and A. Dalgarno. 1967. *Proc. Phys. Soc. (London)*, **90**, 609.
Argyres, P. N. 1965. *Physics*, **2**, 131.
Arthurs, A. M., and A. Dalgarno. 1960. *Proc. Roy. Soc. (London)*, **A256**, 540.
Audibert, M. M., C. Joffrin, and J. Ducuing. 1974. *J. Chem. Phys.*, **61**, 4357.
Augustin, S. D., and W. H. Miller. 1974. *Chem. Phys. Lett.*, **28**, 149.

Balint-Kurti, G. G. 1975. In A. D. Buckingham and C. A. Coulson, Eds., *International Review of Science, Physics and Chemistry, Series 2*, Butterworths, London, Vol. 1, p. 285.

Balint-Kurti, G. G., and R. D. Levine. 1970. *Chem. Phys. Lett.*, **7**, 107.

Barg, G. D., G. M. Kendall, and J. P. Toennies. 1976, *Chem. Phys.*, **16**, 243.

Beigman, I. L., L. A. Vainshtein, and I. I. Sobel'man. 1969. *Zh. Eksp. Teor. Fiz.*, **57**, 1703 [English transl. *Soviet Phys.—JETP*, **30**, 920 (1970)].

Bergeron, G., X. Chapuisat, and J.-M. Launay. 1976. *Chem. Phys. Lett.*, **38**, 349.

Bergmann, K., and W. Demtröder. 1972. *J. Phys. B.*, **5**, 2098.

Berry, M. V., and K. E. Mount. 1972. *Reports Progr. Phys.*, **35**, 315.

Bethe, H. A., and E. E. Salpeter, 1957. *Handbuch der Physik*, Vol. 35, Atoms, 1, p. 88, Springer–Verlag, Berlin.

Bohr, N. 1918. *Kgl. Danske Vid. Selsk. Skr. Nat.-Math. afd* **8**, IV, 1; reprinted in B. L. Van der Waerden, Ed., *Sources of Quantum Mechanics*, Dover, New York, 1967.

Born, M. 1927. *Mechanics of the Atom*, Bell, London; reprinted edition: Ungar, New York, 1967.

Bosanac, S., and G. G. Balint-Kurti. 1975. *Mol. Phys.*, **29**, 1797.

Brumer, P. 1974. *Chem. Phys. Lett.*, **28**, 345.

Buck, U. 1975. *Advan. Chem. Phys.*, **30**, 313.

Child, M. S. 1974a. *Mol. Spect.*, **2**, 466.

Child, M. S. 1974b. *J. Mol. Spect.*, **53**, 280.

Clark, A. P. 1971. M. Sc Thesis, University of Stirling, unpublished.

Clark, A. P. 1973. *J. Phys. B.*, **6**, 1153.

Clark, A. P., and A. S. Dickinson. 1971. *J. Phys. B.*, **4**, L112.

Clark, A. P., and A. S. Dickinson. 1973. *J. Phys. B.*, **6**, 164.

Clark, A. P., and A. S. Dickinson. 1975. *J. Phys. B.*, **8**, L364.

Clark, A. P., and I. C. Percival. 1975. *J. Phys. B.*, **8**, 1939.

Cohen, A. O., and R. A. Marcus. 1970. *J. Chem. Phys.*, **52**, 3140.

Connor, J. N. L. 1973. *Ann. Rep. Chem. Soc.*, **A70**, 5.

Connor, J. N. L. 1974. *Mol. Phys.*, **28**, 1569.

Connor, J. N. L. 1976. *Mol. Phys.*, **31**, 33.

Dennison, D. M., and G. E. Uhlenbeck. 1932. *Phys. Rev.*, **41**, 313.

DePristo, A. E., and M. H. Alexander. 1975a. *J. Chem. Phys.*, **63**, 3552.

DePristo, A. E., and M. H. Alexander. 1975b. *J. Chem. Phys.*, **63**, 5327.

Dickinson, A. S., and D. Richards. 1974. *J. Phys. B.*, **7**, 1916.

Dickinson, A. S., and D. Richards. 1975. *J. Phys. B.*, **8**, 2846.

Dickinson, A. S., and D. Richards, 1976a. *J. Phys. B.*, **9**, 515.

Dickinson, A. S., and D. Richards. 1976b. *J. Phys. B.* (to be published in 1977, **10**, 323–343).

Dickinson, A. S., and B. Shizgal. 1975. *Mol. Phys.*, **30**, 1221.

Dirac, P. A. M. 1958. *Principles of Quantum Mechanics*, 4th ed., Oxford University Press, New York.

Doll, J. D., and W. H. Miller. 1972. *J. Chem. Phys.*, **57**, 5019.

Dunham, J. L. 1932a. *Phys. Rev.*, **41**, 713.

Dunham, J. L. 1932b. *Phys. Rev.*, **41**, 721.

Eastes, W., and D. Secrest. 1972. *J. Chem. Phys.*, **56**, 640.

Edmonds, A. R. 1960. *Angular Momentum in Quantum Mechanics*, Princeton University Press, Princeton.

Einstein, A. 1917a. *Verhand. Deut. Phys. Ges.*, **19**, 82.

Einstein, A. 1917b. *Phys. Z.*, **18**, 121; in B. L. van der Waerden, Ed., *Sources of Quantum Mechanics*, Dover, New York, 1967.

Ennen, G., and Ch. Ottinger. 1974. *Chem. Phys.*, **3**, 404.

Erlewein, W., M. von Seggern, and J. P. Toennies. 1968. *Z. Phys.*, **211**, 35.

Feynman, R. P. 1948. *Rev. Mod. Phys.*, **20**, 67.

Feynman, R. P., and A. R. Hibbs. 1965. *Quantum Mechanics and Path Integrals*, McGraw-Hill, New York.

Fisanick–Englot, G., and H. Rabitz. 1975. *J. Chem. Phys.*, **62**, 1409.

Fröman, N. 1966. *Arkiv Fys.*, **32**, 79.

Fröman, N. 1974. *Phys. Lett.*, **48A**, 137.

Garrison, B. J., W. A. Lester, W. H. Miller, and S. Green. 1975. *Astrophys. J.*, **200**, L175.

Gentry, W. R. 1974, *J. Chem. Phys.*, **60**, 2547.

Gentry, W. R., and C. F. Giese. 1975. *J. Chem. Phys.*, **63**, 3144.

Gol'dman, I. I., and V. D. Krivchenkov. 1961. *Problems in Quantum Mechanics*, Pergamon Press, London.

Goldstein, H. 1959. *Classical Mechanics*, Addison-Wesley, Reading, Mass.

Gordon, M. D., and D. Secrest. 1970. *J. Chem. Phys.*, **52**, 120.

Gordon, R. G. 1969. *J. Chem. Phys.*, **51**, 14.

Green, S., and P. Thaddeus. 1974. *Astrophys. J.*, **191**, 653.

Hammersley, J. M., and D. C. Handscomb. 1964. *Monte Carlo Methods*, Methuen, London.

Heidrich, F. E., K. R. Wilson, and D. Rapp. 1971. *J. Chem. Phys.*, **54**, 3885.

Herzfeld, K. F., and T. A. Litovitz. 1959. *Absorption and Dispersion of Ultrasonic Waves*, Academic Press, New York.

Husimi, K. 1953. *Progr. Theor. Phys.*, **9**, 381.

Jackson, J. M., and N. F. Mott. 1932. *Proc. Roy. Soc. (London)*, **A137**, 703.

Jamieson, M. J., P. M. Kalaghan, and A. Dalgarno. 1975. *J. Phys. B.*, **8**, 2140.

Jamieson, M. J. 1976. *Chem. Phys. Lett.*, **37**, 191.

Kerner, E. H. 1958. *Can. J. Phys.*, **36**, 371.

Kistemaker, P. G., and A. E. de Vries. 1975. *Chem. Phys.*, **7**, 371.

Kramers, H. A. 1964. *Quantum Mechanics*, Dover, New York.

Krauss, M., and Mies, F. H., 1965, *J. Chem. Phys.*, **42**, 2703.

Krieger, J. B., M. L. Lewis, and C. Rosenzweig. 1967. *J. Chem. Phys.*, **47**, 2942.

Kutzelnigg, W., V. Staemmler, and C. Hoheisel. 1973. *Chem. Phys.*, **1**, 27.

LaBudde, R. A., and R. B. Bernstein. 1971. *J. Chem. Phys.*, **55**, 5499.

LaBudde, R. A., and R. B. Bernstein. 1973. *J. Chem. Phys.*, **59**, 3687.

Landau, L. D., and E. M. Lifshitz. 1965. *Quantum Mechanics, Course in Theoretical Physics*, Vol. 3, Pergamon, Oxford.

Landau, L. D., and E. M. Lifshitz. 1968. *Statistical Physics, Course in Theoretical Physics*, Vol. 5, Pergamon, Oxford.

Landau, L. D., and E. M. Lifshitz. 1969. *Mechanics, Course in Theoretical Physics*, Vol. 1, Pergamon, Oxford.

Landau, L. D., and E. M. Lifshitz. 1971. *Classical Theory of Fields, Course in Theoretical Physics*, Vol. 2, Pergamon, Oxford.

Landau, L. D., and E. Teller. 1936. *Phys. Z.*, **10**, 34.

Launay, J.-M. 1976. *J. Phys. B*, **9**, 1823–38.

Lawley, K. P., and J. Ross. 1965a. *J. Chem. Phys.*, **43**, 2930.

Lawley, K. P., and J. Ross. 1965b. *J. Chem. Phys.*, **43**, 2943.

Lester, W. A. 1971. *J. Chem. Phys.*, **54**, 3171.

Lester, W. A. 1975. *Advan. Quantum Chem.*, **9**, 199.

Lester, W. A., and J. Schaefer. 1973. *J. Chem. Phys.*, **59**, 3676.

Levine, R. D. 1971. *Mol. Phys.*, **22**, 497.

Levine, R. D., and G. G. Balint–Kurti. 1970. *Chem. Phys. Lett.*, **6**, 101.

Levine, R. D., and R. B. Bernstein. 1974. *Molecular Reaction Dynamics*, Oxford University Press, New York.

Levine, R. D., and B. R. Johnson. 1970. *Chem. Phys. Lett.*, **7**, 404.

Lodge, J. G., I. C. Percival, and D. Richards. 1976. *J. Phys. B.*, **9**, 239.

McCann, K. J., and M. R. Flannery. 1975a. *Chem. Phys. Lett.*, **35**, 124.

McCann, K. J., and M. R. Flannery. 1975b. *J. Chem. Phys.*, **63**, 4695.

McGuire, P. 1976. *Chem. Phys.*, **13**, 81.

McGuire, P. and D. A. Micha. 1972. *Int. J. Quant. Chem.*, **6S**, 111.

McGuire, P. and Kouri, D. J., 1974, *J. Chem. Phys.*, **60**, 2488.

McKenzie, R. L. 1975. *J. Chem. Phys.*, **63**, 1655.

Magnus, W. 1954. *Commun. Pure Appl. Math.*, **7**, 649.

Marcus, R. A. 1971. *J. Chem. Phys.*, **54**, 3965.

Marcus, R. A. 1973. *Faraday Disc. Chem. Soc.*, **55**, 34.

Messiah, A. 1964. *Quantum Mechanics*, Vol. 1, North-Holland, Amsterdam.

Mies, F. H. 1964. *J. Chem. Phys.*, **40**, 523.

Miller, W. H. 1974. *Advan. Chem. Phys.*, **25**, 69.

Miller, W. H. 1975. *Advan. Chem. Phys.*, **30**, 77.

Naccache, P. F. 1972. *J. Phys. B*, **5**, 1308.

Norcliffe, A. 1973. *Case Studies in Atomic Physics*, Vol. 4, North-Holland, Amsterdam.

Ottinger, Ch., and D. Poppe. 1971. *Chem. Phys. Lett.*, **8**, 513.

Ottinger, Ch., R. Velasco, and R. N. Zare. 1970. *J. Chem. Phys.*, **52**, 1636.

Pack, R. T. 1975. *J. Chem. Phys.*, **62**, 3143.

Pattengill, M. D. 1975. *J. Chem. Phys.*, **62**, 3137.

Pattengill, M. D., R. A. LaBudde, R. B. Bernstein, and C. F. Curtiss. 1971. *J. Chem. Phys.*, **55**, 5517.

Pechukas, P. 1969. *Phys. Rev.*, **181**, 174.

Pechukas, P., and M. S. Child. 1976. *Mol. Phys.*, **31**, 973.

Pechukas, P., and J. C. Light. 1966. *J. Chem. Phys.*, **44**, 3897.

Penner, A. P., and R. Wallace. 1974. *Phys. Rev.* A, **9**, 1136.

Percival, I. C. 1969. In F. Bopp and H. Kleinpoppen (Eds.), *Physics of the One and Two Electron Atoms*, North-Holland, Amsterdam, p. 252.

Percival, I. C. 1973. *J. Phys. B.*, **6**, L229.

Percival, I. C. 1976. *Advan. Chem. Phys.*, **36**.

Percival, I. C. and N. Pomphrey. 1976. *Mol. Phys.*, **31**, 97.

Percival, I. C., and D. Richards. 1970. *J. Phys. B.*, **3**, 1035.

Percival, I. C., and D. Richards. 1975. In *Advances in Atomic and Molecular Physics*, Vol. 11, p. 2, Academic Press, New York.

Popov, V. S., and A. M. Perelomov. 1969. *Soviet Phys.—JETP*, **29**, 738.

Presnyakov, L. P., and A. M. Urnov. 1970. *J. Phys. B*, **3**, 1267.

Rabitz, H. 1975. *J. Chem. Phys.*, **63**, 5208.

Rabitz, H., and G. Zarur. 1975. *J. Chem. Phys.*, **62**, 1425.

Raczkowski, A. W., and W. H. Miller. 1974. *J. Chem. Phys.*, **61**, 5413.

Rapp, D., and T. Kassal. 1969. *Chem. Rev.*, **69**, 61.

Richards, D. 1972. *J. Phys. B*, **5**, L53.

Robinson, D. W. 1963. *Helv. Phys. Acta*, **36**, 140.

Russell, J. D., R. B. Bernstein, and C. F. Curtiss. 1972. *J. Chem. Phys.*, **57**, 3304.

Saha, S., E. Guha, and A. K. Barua. 1974. *J. Phys. B.*, **7**, 2264.

Saha, S., and E. Guha. 1975. *J. Phys. B.*, **8**, 2293.

Schaefer, J., and W. A. Lester. 1975. *J. Chem. Phys.*, **62**, 1913.
Schiff, L. I. 1955. *Quantum Mechanics*, 2nd Ed., McGraw-Hill, New York.
Schinke, R., and J. P. Toennies. 1975. *J. Chem. Phys.*, **62**, 4871.
Seaton, M. J. 1961. *Proc. Phys. Soc. (London)*, **77**, 174.
Secrest, D. 1969. *J. Chem. Phys.*, **51**, 421.
Secrest, D. 1973. *Ann. Rev. Phys. Chem.*, **24**, 379.
Secrest, D. 1975. *J. Chem. Phys.*, **62**, 710.
Secrest, D., and B. R. Johnson. 1966. *J. Chem. Phys.*, **45**, 4556.
Shafer, R., and R. G. Gordon. 1973. *J. Chem. Phys.*, **58**, 5422.
Sharma, R. D., and R. R. Hart. 1975. *J. Chem. Phys.*, **63**, 5383.
Smith, F. T., D. L. Huestis, D. Mukherjee, and W. H. Miller. 1975. *Phys. Rev. Lett.*, **35**, 1073.
Sorensen, G. B. 1974. *J. Chem. Phys.*, **61**, 3340.
Stwalley, W. C. 1973. *J. Chem. Phys.*, **58**, 3867.
Takayanagi, K. 1963. *Progr. Theor. Phys. Suppl.*, **25**, 1.
Tipping, R. H. 1974. *J. Mol. Spec.*, **53**, 402.
Toennies, J. P. 1974. *Chem. Soc. Rev.*, **3**, 407.
Treanor, C. E. 1965. *J. Chem. Phys.*, **43**, 532.
Tsalpine, B., and W. Kutzelnigg. 1973. *Chem. Phys. Lett.*, **23**, 173.
Tsien, T. P., G. A. Parker, and R. T. Pack. 1973. *J. Chem. Phys.*, **59**, 5373.
Vainshtein, L. A., and A. V. Vinogradov. 1970. *J. Phys. B*, **3**, 1090.
Wilcox, R. M. 1966. *J. Chem. Phys.*, **45**, 3312.
Williams, E. J. 1931. *Proc. Roy. Soc. (London)*, **A130**, 328.
Zarur, G., and H. Rabitz. 1974. *J. Chem. Phys.*, **60**, 2057.

H + H$_2$: POTENTIAL-ENERGY SURFACES AND ELASTIC AND INELASTIC SCATTERING*

DONALD G. TRUHLAR

Joint Institute for Laboratory Astrophysics
University of Colorado and National Bureau of Standards
Boulder, Colorado 80309

and

Department of Chemistry
University of Minnesota, Minneapolis, Minnesota 55455

ROBERT E. WYATT

Department of Chemistry
University of Texas, Austin, Texas 78712

CONTENTS

*Supported in part by the National Science Foundation through Grant No. MPS7506416 and by the Alfred P. Sloan Foundation through a research fellowship to one of the authors (D.G.T.).

I. INTRODUCTION

The $H + H_2$ reaction has long been considered an important prototype for chemical reactions with activation barriers. We have recently reviewed the history of H_3 kinetics, with emphasis on experimental and theoretical studies of the reactive dynamics in the ground electronic state (Truhlar and Wyatt, 1976). The first step in all such calculations is the Born-Oppenheimer approximation, i.e., the assumption of a potential-energy hypersurface (called the potential, the potential surface, or the surface for short). Thus there has been considerable work attempting to obtain this potential surface more accurately, especially by calculating the electronic energy of H_3 but also in some cases by attempting to fit analytic expressions to results of collision experiments. In particular, such work has emphasized the region around the saddle point because this region is thought to be the most important for determining the rate of reaction at low and medium temperatures (less than about $1000°K$). This work on the potential surface is reviewed here. We also review theoretical and experimental information about other parts of the ground-state potential surface, about potential surfaces for excited electronic states, about nonreactive dynamics on the ground-state surface, and about the dynamics of processes involving excited electronic states. This review and our previous one cited above, taken together, constitute a review of all work concerned with gas-phase collisions of H with H_2.

II. POTENTIAL-ENERGY SURFACES

A.   Ground Electronic State: Saddle Region and Short Range

The first major contribution to understanding the $H + H_2$ reaction was made by London (1929). He presented a formula, without derivation, for the energy of the lowest adiabatic electronic state of H_3 as a function of Coulomb and exchange integrals Q_{ij} and J_{ij} for each pair ij of atoms. These integrals depend on the internuclear separation of that pair. This is the generalization to H_3 of the valence-bond approach that Heitler and London had used to explain the binding energies of H_2 using the potential curve concept, which derives from Born and Oppenheimer's electronically adiabatic separation of the electronic and internuclear motions. The first importance of this formula is that it showed how the energy needed to cause the reaction could be much less than the energy needed to break an H_2 bond. Thus the activation energy of the $H + H_2$ reaction can be understood in terms of internuclear motion governed by one electronically adiabatic potential-energy surface which is the ground-state fixed-nuclei electronic energy of H_3, including internuclear repulsion. Thus future work on the reaction used this Born-Oppenheimer separation and involved the determination of a potential-energy surface as the first step. The second importance is that his formula later served as a basis for many semiempirical potential-energy surfaces that have been used to study the dynamics.

Eyring and Polanyi (1930) used the London formula in a semiempirical way to treat the reaction H + H$_2$, showing that the collinear potential-energy surface as a function of the two nearest-neighbor distances R_1 and R_2 had a saddle point that they correctly interpreted as a crucial factor for determining the energy of activation. The height of this saddle point, with respect to the energy of H infinitely far from H$_2$, is called the classical barrier height, or, for short, the barrier.

The first published derivation of the London equation was made by Slater (1931), who used valence-bond theory with only the two linearly independent covalent configurations that can be obtained from a basis of one $1s$ hydrogen orbital on each center. Orbital overlap integrals and multiple-exchange integrals were neglected.

The approximations involved in the London formula and the semiempirical way of using it (as developed by Erying and Polanyi) have been critically examined many times (see, e.g., Kassel, 1932; Coolidge and James, 1934; Van Vleck and Sherman, 1935; Hirschfelder et al. 1936, Glasstone et al. 1941, Hirschfelder 1941, Hirschfelder et al., 1954; Aroeste, 1964; Laidler and Polanyi, 1965; Parr and Truhlar, 1971). In particular, Coolidge and James (1934) critically examined these approximations with respect to the cancellation of errors that occurs in the London formula. We should remember their conclusion that the relation of quantum mechanics to the semiempirical method of London, Eyring, and Polanyi (LEP method) is "merely suggestive, rather than justificatory." In spite of this and in spite of London's attitude toward the many uses to which his formula would be put [Hirschfelder (1966), in reminiscing about this, commented, "London told me that he was appalled at the way chemists mangled his formula and still attached his name to the semiempirical results"], the first studies of H$_3$ reaction dynamics were made possible by the early semiempirical H$_3$ surfaces. Thus we will examine this work of this period in more detail.

The basis of the LEP method is to realize that the Heitler-London treatment of H$_2$, with neglect of orbital overlap integrals, expresses the ground-state energy of H$_2$ as the sum of Q_{ij} and J_{ij}. Thus a Morse curve for H$_2$ gives a semiempirical value for the sum of these integrals at each distance. Then it was assumed that Q_{ij} is always a constant fraction ρ of this sum. This scheme was originally motivated by examining Sugiura's (1927) calculated values of the integrals. These yielded a roughly constant

ρ of 0.14 at large distances. However, in practice ρ was often treated as an adjustable parameter. It was given various values at various times: 0.10 (Eyring and Polanyi, 1930; Eyring, 1931), 0.00 (Eyring and Polanyi, 1931), 0.14 (Eyring, Gershinowitz, and Sun, 1935), and 0.20 (Hirschfelder, Eyring, and Rosen, 1936). For any value of ρ the method predicts that the lowest-energy reaction path is linear. For low ρ the barrier is symmetric $(R_1 = R_2)$ but much too high. As ρ is increased the barrier is lowered. However, for any constant $\rho \gtrsim 0.07$ the symmetric configuration is a local minimum flanked by twin nonsymmetric saddle points. For example (Eyring, 1931), for $\rho = 0.035$ the barrier is symmetric with height 0.91 eV, but for $\rho = 0.10$ the local minimum has a nearest-neighbor distance R_1 of 1.76 a_0 and is 0.49 eV above the energy of $H + H_2$ but the twin saddle points are 0.07 eV higher. Yet Farkas's rate experiments (reviewed in Truhlar and Wyatt, 1976) were interpreted as leading to an activation energy of 0.17 to 0.48 eV and the surfaces with twin saddle points were used for comparison to experiment. It is interesting that if ρ is made an increasing function of R (in accordance with the Heitler-London treatment of H_2), then the basin results from using too high a value at small R; ρ may be large (0.14 or even greater) at large R without making a basin. Such variable-ρ treatments are discussed later in this section.

The consequences of the predicted basin were discussed by Eyring (1932). First he noted that the predicted basin was so unstable that even if it could be collisionally stabilized it would survive only a few collisions. But the LEP method is approximate and the actual basin might have been deeper. Eyring considered a configuration point representing the instantaneous geometry of the H_3 system. He predicted that it moves very slowly through the first pass and into the shallow basin where it zigzags back and forth before it finds its way out through the second gap. The three atoms would then form a quasimolecule in the "sticky" collision. During this time, if the basin were deeper, the H_3 complex could be stabilized. To understand even the qualitative features of the dynamics it was necessary to have a more reliable calculation of the basic features of the surface. Thus attention was turned to these and we will consider what has been learned from ab initio calculations about the reaction-path part of the surface before returning to the history of the semiempirical work.

The ab initio variation method has also been used to calculate the H_3 energy. The first ab initio calculation on the H_3 potential-energy surface was performed by Coolidge and James (1934). They studied only one geometry (linear symmetric H_3 with $R_1 = 1.7a_0$). They used valence-bond theory with the two configurations mentioned above; for linear symmetric H_3 one may consider instead one symmetry-adapted configuration, which they did. They obtained an energy of -1.72 eV. (We give the energies of all ab initio calculations as binding energies, i.e., the zero of energy is three H atoms infinitely separated; but barriers and other positive energies referring to heights along the reaction path are given with respect to the energy of $H + H_2$.) They also considered several approximations to this result in order to study the justification for semiempirical valence-bond procedures. As mentioned above, they concluded that the approximations needed to obtain the LEP method are so severe that the justification for the latter must be wholly empirical. Their ab initio calculation was soon reproduced by Hirschfelder, Erying, and Rosen (1936), who also showed that at the same level of approximation a much lower energy can be obtained for linear symmetric H_3 by increasing R_1 to $2.0a_0$ where the energy is -3.14 eV. Using the experimental binding energy of H_2, this yields a rigorous upper bound on the classical barrier height of 1.61 eV. This was known to be much too high. It was recognized that a more reasonable approximation is to compare a calculated H_3 energy to an H_2 binding energy calculated at equivalent levels of approximation. Using a Heitler-London wave function for H_2 for comparison with Hirschfelder et al.'s valance-bond result, we obtain an estimate of the saddle-point height of 0.80 eV, still much too large. Hirschfelder et al. (1936) considered three improved approximations for linear symmetric H_3 in which they added ionic configurations with variable coefficients and/or optimized the orbital exponents. While these gave considerably lower energies for H_3, the estimated barrier height was not improved. These calculations are summarized in Table I (for comparison we have corrected the diatomic values Hirschfelder et al. quoted).

Subsequently, further calculations were performed at these same levels of approximation; they predicted that the saddle point is linear and symmetric. First, Hirschfelder et al. (1937) calculated energies for nonsymmetric linear geometries and found that the saddle point for the collinear reaction is symmetric. Stevenson and Hirschfelder (1937) then showed that the energy rose on bending. It is interesting to anticipate a result of the ab initio calculations to be discussed below, i.e., only one ab initio calculation (Conroy and Bruner, 1967) has ever predicted a saddle-point geometry other than linear symmetric. That calculation predicted a narrow shallow basin for linear symmetric geometries which would be far less significant than the broader deeper basins predicted by the early semiempirical calculations, and even that shallow basin disappeared in an improved calculation (Conroy and Bruner, 1967) by the same method. Thus Hirschfelder, Diamond, and Eyring's prediction of a symmetric saddle point has stood the test of time, and the latest calculations (Liu, 1973) have finally

TABLE I

Ab Initio Variational Upper Bound Calculations of the Binding Energy E and Zero-Point Energy $\frac{1}{2}h\nu_s$ for Harmonic Stretching of Linear Symmetric H_3 Along with Estimated Classical Barrier Heights E_b

Reference[a]	H_3			H_2			H_3	
	$R^{\ddagger b}$ (a_0)	E^c (eV)	$\frac{1}{2}h\nu_s$ (eV)	R^d (a_0)	$E^{c,d}$ (eV)	Reference	E_b^e (eV)	Description of H_3 calculation[f] (and one-electron basis[g])
Minimum basis set								
Coolidge and James, 1934	1.7	−1.72						VB: covalent only ($\zeta=1$)
Hirschfelder, Eyring, and Rosen, 1936	2.0	−2.33		1.64	−3.14	Coulson, 1937	0.80	VB: covalent only ($\zeta=1$)
	1.89	−2.46		1.41	−3.78	Wang, 1928	1.32	VB: covalent only (ζ opt)
	2.0	−2.65						VB: covalent+ionic: CCI ($\zeta=1$)
Hirschfelder et al., 1936, 1937	1.84	−2.94	0.20	1.42	−4.02	Weinbaum, 1933	1.08	VB: covalent+ionic: CCI (ζ opt)
Pearson, 1948	1.91	+1.27						MO (not antisymmetrized, ζ opt)
Walsh and Matsen, 1951	2.00	−1.67		1.61	−2.68	Coulson, 1937	1.01	MO ($\zeta=1$)
	1.82	−1.99		1.38	−3.47	Coulson, 1937	1.48	MO (ζ opt)
Griffing and Vanderslice, 1955	2.0	−1.17			−2.98	Griffing and Vanderslice, 1955	1.81	MO for H_3^- + Koopmans' theorem (ζ opt)
Ransil, 1957	1.95	−1.70		1.61	−2.68	Coulson, 1937	0.98	MO ($\zeta=1$)
	2.00	−1.78		1.38	−3.47	Coulson, 1937	1.69	MO (ζ opt)
		−2.63						MO: CCI (ζ opt$=1$)
Griffing et al., 1959	1.90	−2.06		1.61	−2.68	Coulson, 1937	0.62	MO: CI (ζ opt)
		−2.96						UHF: not spin eigenfunction (ζ not given)
Bradley, 1964	2.05	−2.10	0.120	1.61	−2.68	Coulson, 1937	0.56	UHF: not spin eigenfunction ($\zeta=1$)
	1.91	−2.44		1.38	−3.47	Coulson, 1937	1.03	UHF: not spin eigenfunction (ζ opt)
	1.91	−2.66						UHF: CI: not spin eigenfunction (ζ opt)
	1.91	−2.68		1.38	−3.47	Coulson, 1937	0.79	spin-projected UHF (ζ opt)
	1.91	−2.88						spin-projected UHF: CI (ζ opt)
Bowen and Linnett, 1966	1.89	−3.04		1.42	−4.02	Weinbaum, 1933	0.98	CCI (outer and middle orbital exponents separately opt)
Shavitt et al., 1968	1.88	−3.01		1.43	−4.02	Weinbaum, 1933	1.01	CI (ζ opt)
Gianinetti et al., 1969	1.91	−3.01		1.44	−4.02	Gianinetti et al., 1969	1.01	CI (ζ opt)
Blustin and Linnett, 1974	1.82	+4.31		1.47	+1.20	Frost, 1967	3.11	MO (GTF: one s shifted off nuclei)

Extended basis set
of hydrogen 1s orbitals

Boys et al., 1956; Boys and Shavitt, 1959	1.78	−3.48	0.120	1.42	−4.14	Boys and Shavitt, 1959	0.66	CCI (ETF: two s)
Shavitt et al., 1968	1.79	−3.55	0.123	1.41	−4.16	Shavitt et al., 1968	0.61	CCI (ETF: two s)
Gianinetti et al., 1969	1.83	−3.47		1.42	−4.16	Gianinetti et al., 1969	0.69	CI (ETF: two s)
	1.79	−2.41						MO (ETF: two s)
Bacskay and Linnett, 1972	1.79	−3.55			−4.16	Bacskay and Linnett, 1972a	0.61	CCI (ETF: two s)

Basis set including polarization
functions (or their equivalent)

Meador, 1958	1.84	−2.48						MO: CI (three hydrogenic $1s$ orbitals, two off nuclei; ζ opt)
Kimball and Trulio, 1958	1.93	−3.138						CCI (five equally spaced hydrogenic $1s$ orbitals; ζ opt)
Krauss, 1964	1.7	−2.530						MO (GTF: five s, one $p\sigma$)
Hoyland, 1964	1.78	−2.419						CI (six two-center elliptical orbitals with opt parameters)
Edmiston and Krauss, 1965	1.8	−3.891		1.4	−4.600	Edmiston and Krauss, 1965	0.709	PNO: CI (GTF: five s, one $p\sigma$, two $p\pi$)
Conroy and Bruner, 1967	1.75	−4.221						Conroy's method
Considine and Hayes, 1967	1.8	−3.050						CI: 52 terms (single-center expansion, 68 orbitals)
Hayes and Parr, 1967	1.8	−3.695						CI: 99 terms (single-center expansion, 85 orbitals)
Michels and Harris, 1968	1.76	−3.543		1.4	−4.191	Michels and Harris, 1968	0.648	CCI (ETF: one s, one $2p\sigma$)
Shavitt et al., 1968	1.764	−4.139	0.125	1.402	−4.615	Shavitt et al., 1968	0.476	CCI (ETF: two s, one $p\sigma$, one $p\pi$)
Schwartz and Schaad, 1968	1.725	−2.395						MO (twelve floating $1s$ Gaussians)
	1.7	−2.456						MO (eighteen floating $1s$ Gaussians)
Edmiston and Krauss, 1968	1.7	−2.530		1.4	−3.619	Edmiston and Krauss, 1968	1.090	MO (GTF: five s, one $p\sigma$, three $p\pi$)
	1.8	−4.064		1.4	−4.647	Edmiston and Krauss, 1968	0.583	PNO: CI (GTF as above)
Ladner and Goddard, 1969	1.765	−3.369		1.4	−4.123	Goddard and Ladner, 1969	0.754	SOGI (ETF: two s, one $p\sigma$)
Gianinetti et al., 1969	1.7965	−3.439		1.417	−4.193	Gianinetti et al., 1969	0.754	CI (ETF: one s, one $p\sigma$)
	1.795	−3.655		1.423	−4.329	Gianinetti et al., 1969	0.675	CI (ETF: two s, one $p\sigma$)
	1.792	−4.008		1.4165	−4.612	Gianinetti et al., 1969	0.604	CI (ETF: two s, one $p\sigma$, one $p\pi$)

147

Table I (continued)

Reference[a]	H₃ $R^{\ddagger b}$ (a_0)	H₃ E^c (eV)	H₃ $\frac{1}{2}h\nu_s$ (eV)	H₂ R^d (a_0)	H₂ $E^{c,d}$ (eV)	H₂ Reference	H₃ E_b^e (eV)	Description of H₃ calculation[f] (and one-electron basis[g])
Riera and Linnett, 1969	1.800	−3.774	0.122	1.416	−4.482	Riera and Linnett, 1969	0.707	CCI (ETF: one s, one pσ, one pπ)
Riera and Linnett, 1970	1.771	−3.913						CI (basis as above pllus five equally spaced 1s Gaussians)
Bacskay and Linnett, 1972	1.79	−2.516						MO (ETF: two s; plus two 1s Gaussians at bond midpoints)
	1.79	−3.786		1.4148	−4.326	Bacskay and Linnett, 1972a	0.540	CCI (basis as above)
	1.79	−4.128		1.4148	−4.594	Bacskay and Linnett, 1972a	0.466	CCI (basis as above plus twelve off-axial 1s Gaussians)
Liu, 1971	1.7, 1.75	−2.578		1.4	−3.635			MO (ETF: four s, three pσ, two dσ, one fσ, three pπ, three dπ, one fπ, three dδ, one fδ)
	1.76	−4.289		1.4	−4.725	Liu, 1971	0.436	PNO: CI (basis as above)
Liu, 1973	1.79	−2.578						MO (ETF: four s, three pσ, two dσ, three pπ, one dπ, one dδ)
	1.80	−3.397						Internal configuration MCSCF (basis as above)
	1.75	−4.296						CI in zero and first order subspaces (basis as above)
Exact H₂	1.757	−4.302	0.127	1.4014	−4.727	Liu, 1973	0.425	Near-complete CI (basis as above)
				1.4008	−4.748	Kolos and Wolniewicz, 1965		

[a] Reference for H₃ calculation. Hiershfelder et al., 1936 refers to Hirschfelder, Eyring, and Rosen, 1936.

[b] Nearest-neighbor distance for which the minimum energy was calculated for this method. In some cases the authors made an extensive search for the saddle-point geometry but in other cases only one or a few geometries were examined.

[c] Zero of energy is 3H; 1 eV = (1/27.2116)a.u.

[d] Internuclear distance and binding energy calculated for H₂ in a calculation comparable to the H₃ calculation (not always available).

[e] Difference of H₂ and H₃ binding energies.

[f] Abbreviations: VB, valence-bond method; MO, molecular orbital (i.e., Hartree-Fock-Roothaan) method; CCI, complete configuration interaction; UHF, unrestricted Hartree-Fock; CI, configuration interaction; PNO, pseudonatural orbitals; see, e.g., Krauss (1970) and Schaefer (1972). These methods and other ab initio methods mentioned in the text are explained in standard references; see, e.g., Krauss (1970) and Schaefer (1972).

[g] ζ, orbital exponent (in a_0^{-1}) of hydrogenic 1s orbital when all orbitals have same exponent (ζ = 1 for H); opt, optimized; ETF and GTF, number of one-electron exponential-type basis functions or Gaussian-type basis functions, of each symmetry centered on each nucleus. Optimization of orbital

148

established quite firmly that the saddle point on the linear surface is symmetric so that the "Lake Eyring" of the early semiempirical surfaces must be an artifact of the approximations involved.

After these calculations were completed there were no more ab initio calculations of the saddle-point properties until 1951. Since then, however, the interest has been very strong and the results of the calculations have slowly improved. In this process most H_3 calculations have been limited to the linear symmetric configuration in an attempt to calculate the energy of the saddle point. We discuss these ab initio attempts next and then we consider ab initio calculations of other features of the H_3 surface when all three atoms are close: nonsymmetric linear configurations, nonlinear configurations near the saddle point, and equilateral H_3. Then we consider the later semiempirical calculations of surface properties near the reaction path. Finally we consider all kinds of calculations of the very short-range repulsive forces and of the long-range forces and the region of the van der Waals' minimum.

The history of ab initio saddle-point calculations may conveniently be divided into three stages (or "ages" except that the chronology of the stages overlaps to some extent). The first stage involved using minimum-basis sets, that is, three $1s$ functions centered at the nuclei. The second stage involved using more than three spherical basis functions centered at the nuclei. And finally the calculations were performed using more general bases, especially extended basis sets of nuclear-centered functions including polarization functions (e.g., $2p\sigma$ functions in order to better represent the polarization of the charge distribution). The basis sets and results of all the calculations are summarized in Table I and we add only a few comments.

The calculation of accurate multicenter integrals was for decades a "bottleneck" of ab initio quantum calculations. Thus, for example, the accurate three-center integrals of Hirschfelder and co-workers (Hirschfelder et al., 1936; Hirschfelder and Weygandt, 1938) were used by several subsequent workers (even over 20 years later). Barker et al. (1954) and Barker and Eyring (1954) examined the use of Mulliken's (1949) method for approximating the three-center integrals in terms of two-center Coulombic-type integrals. They performed covalent valence-bond calculations with hydrogen-atom exponents. They found that the energy was raised from -2.33 eV to -2.00 eV. Using this same approximation scheme, Snow and Eyring (1957) found this was decreased to -2.62 eV when the outer orbital exponents and the middle one were optimized separately. Yasumori (1959), Oleari et al. (1961), and Harris et al. (1965) performed additional H_3 calculations using the Mulliken approximation.

Barker and Eyring (1957) also examined a "distance-normalized" approximation to the difficult integrals and obtained an energy of -3.17 eV for the covalent valence-bond wave function when the outer and middle orbital exponents were separately normalized. When this separate optimization of the exponents was finally carried out with correct integrals in a complete-configuration-interaction calculation, the energy lowering was less than 0.1 eV (Bowen and Linnett, 1966). The results using approximate integrals are not included in Table I because of the admittedly serious nature of the approximation. However, the reader should remember that the history of ab initio molecular calculations is filled with examples of calculations containing errors due to inadvertent use of incorrect integral values. No systematic attempt is made here to discuss the accuracy of the integrals involved in all the calculations. Presumably, however, the integrals were all evaluated accurately in calculations carried out within the last 10 years or so. Finally, we should emphasize that progress seen in the successive ab initio calculations reviewed in Table I and below is due not only to the cited efforts of the authors of these calculations but also in many cases to the quantum chemists, including many of these authors but also others, who devised improved methods for integral evaluation and for efficient utilization of computers for large-scale ab initio calculations. These parallel advances interacted strongly with advances in H_3 computations, but it is beyond the scope of the present chapter to discuss them in detail.

The next attempts to use accurate integrals were the ab initio minimum-basis-set molecular orbital calculations of Walsh and Matsen (1951) and Griffing and Vanderslice (1955) and the attempts by Ransil (1957), Meador (1958), and Griffing et al. (1959) to improve these results using configuration interaction based on molecular orbitals and in one case allowing the basis functions to be centered off the nuclei. The molecular orbital method leads to an energy about 0.5 eV higher than the valence-bond method, whereas the configuration-interaction calculations should be equivalent to Hirschfelder, Eyring, and Rosen's (1936) configuration-interaction calculations based on the valence-bond formalism. A complete-configuration-interaction (CCI) calculation is one in which the Hamiltonian is diagonalized in the space of all three-particle functions of the appropriate symmetry that can be formed from the chosen one-electron basis set. Such a result is exact within the restrictions imposed by the one-electron basis and yields the lowest energy that can be obtained with that basis. For three $1s$ functions a CCI consists in general of eight configurations, but for linear symmetric H_3 it may be reduced to four symmetry-adapted configurations. The calculation may be carried out equivalently in either the valence-bond or molecular orbital formalisms or without reference to either. Boys and Shavitt (1959) pointed out that the configuration-interaction wave functions of Walsh and Matsen (1951) and of Meador (1958) include only three of the four linearly independent configurations and that one of these is not a spin eigenfunction. The minimum-basis-set approach was later reexamined by Bradley (1964), Bowen and Linnett (1966), and others. Bowen and Linnett showed that a CCI calculation for a minimum basis set with separately optimized orbital exponents and the best value of R_1 yields an energy of -3.045 eV. This yields a rigorous upper bound of 1.333 eV on the classical barrier height and using the best possible minimum-basis-set calculation on H_2 for comparison yields an estimated barrier of 0.975 eV. Both values are too high to be useful. Rourke and Stewart (1968) showed that Bradley's (1964) minimum-basis-set wave function is not of sufficiently high quality for use with a local-energy method and that variation-method results are expected to be more accurate for this type of wave function.

Even before the configuration-interaction calculations discussed in the preceding paragraph it was very clear that one must go beyond a minimum basis set and Boys et al. (1956) had already done so. Their calculations (Boys et al., 1956; Boys and Shavitt, 1959), the first using an extended basis set, were CCI calculations for two $1s$ exponential-type functions on each nucleus. For linear symmetric geometries this involved 34 symmetry-adapted configurations. They obtained an energy of -3.48 eV and an estimated barrier of 0.66 eV, still much too large. Their result was affected slightly by the method used to calculate some of the difficult integrals. It is now known that the accurate CCI result for this basis with the best value of R_1 and optimization of two (but not four separate) values of the orbital exponents is -3.55 eV (see Table I).

Another extended-basis-set calculation was performed by Kimball and Trulio (1958). They carried out a complete configuration-interaction calculation in the space spanned by five $1s$ exponential functions equally spaced on a line. Their calculated binding energy (-3.138 eV) is of course also lower than any calculated with a minimum-basis-set.

Krauss (1964) carried out the first calculation including p functions. He used the molecular orbital (SCF) form of wave function and obtained an energy of -2.530 eV. However, a molecular orbital potential surface is not of useful accuracy for H_3. Edmiston and Krauss (1965) then used this basis to carry out their first configuration-interaction calculation. This yielded an energy 0.41 eV lower than obtained by Boys and Shavitt (1959), but the calculated barrier was not improved. This was the first calculation using present state-of-the-art techniques for potential-energy-surface calculations (they used the pseudonatural orbital method); thus it provides a convenient basis for comparison for later calculations. We will abbreviate it

EK1. In a similar but improved calculation, reported later (Edmiston and Krauss, 1968), they obtained an energy 0.17 eV lower than EK1, but by that time Shavitt et al. (1968; see also Karplus 1968) had already reported a result 0.25 eV lower than EK1. Shavitt et al. performed CCI calculations for a double-zeta-plus-polarization basis, which is better than the Gaussian bases of Edmiston and Krauss. This involved 200 symmetry-adapted configurations for linear symmetric geometries. Their calculated barrier was only 0.476 eV. Of special interest are the calculations of Conroy and Bruner (1965, 1967) using a method of Conroy (1964, 1970). They optimized their trial function by minimizing the energy variance functional rather than the usual energy functional. This permits extrapolation of the calculated results to obtain an approximation to the exact energy. This was combined with a new form of trial function explicitly incorporating interelectronic distances and a Monte Carlo-like or Diophantine method for the evaluation of the integrals. First (Conroy and Bruner, 1965) they reported a potential-energy surface for linear H_3 obtained using their extrapolation procedure. Then (Conroy and Bruner, 1967) they made an improved calculation for a greater variety of geometries and reported both upper bounds (see Table I) and extrapolated results (discussed below). A disadvantage of their calculations is the relatively large error in the numerical integration procedure, estimated at ± 0.03 to ± 0.05 eV. However, they also estimated this to be the overall accuracy of their extrapolated energy surface. Their upper bound energy was 0.33 eV better than the EK1 value.

After their second calculation Edmiston and Krauss concluded "further improvements will come in very small pieces as a result of more configurations, more pseudonatural orbitals, and more Gaussian basis functions." Here and in the discussion of Liu's results, "orbitals" means the linear combinations of basis functions used to construct configurations. Generalizing pseudonatural orbitals to whatever set of orbitals is used in the calculation and the last phrase to "more one-electron basis functions," we now recognize from current experience that this conclusion of Edmiston and Krauss is always true for the last few tenths of eV in large-scale configuration-interaction calculations of potential-energy surfaces.

Several workers (Michels and Harris, 1968; Shavitt et al., 1968; Linnett and Riera, 1969) suggested that, in particular, it would be necessary to add $d\sigma$ basis functions. This was done by Liu, who actually used one-electron bases many times larger than any previous one. In his preliminary report (Liu, 1971) he used the pseudonatural orbital method and obtained an energy 0.40 eV lower than EK1. His second set of calculations (Liu, 1973) differed mainly in the method used to select orbitals and configurations [for the basis set used in the second set of calculations a CCI would involve 14,949 symmetry-adapted configurations, compared to, for example, 35 for the basis of Harris and Michels (1968)]. This involved seven-configuration multiconfiguration self-consistent-field calculations and division of the configuration space into zeroeth-, first-, and second-order parts. The best energy obtained (Liu, 1973) was -4.302 eV, which is 0.41 eV lower than EK1. Comparison to the exact result for H$_2$ yields a rigorous upper bound to the barrier of only 0.446 eV, lower than all previous calculated values based on upper bound calculations even when those were obtained by subtracting comparable H$_2$ and H$_3$ calculations. Comparing Liu's H$_3$ result to his comparable one for H$_2$ yields a barrier of only 0.425 eV. Liu estimated the possible errors in his surface in two parts: error due to truncation of the one-electron basis (0.021 to 0.034 eV) and error due to not reaching the CCI limit for this basis set (0.001 eV). His calculated barrier would correspond to an H$_3$ energy of -4.323 eV and he estimated the actual result cannot be below -4.336 eV. Since he is extrapolating to the exact result from much closer than any previous worker his estimates are preferred over all others. This leads to $-4.323^{+0.021}_{-0.013}$ eV as the bounded energy of the H$_3$ saddle point, corresponding to a saddle-point height of $0.425^{+0.021}_{-0.013}$ eV. From Liu's arguments one could estimate with less certainty an even more closely bounded result. Thus if we assume that the residual error at the saddle point is at least as large as, but no more than 1.5 times, the residual error in H$_2$ at the same internuclear distance, then the estimated energy is -4.327 ± 0.005 eV, corresponding to a barrier of 0.420 ± 0.005 eV. For comparison with the kinetics literature, these two sets of bounds can be converted to $9.80^{+0.48}_{-0.30}$ and 9.69 ± 0.12 kcal/mole.

It is interesting to compare these final estimates from Liu's calculations to previous attempts to estimate the energy of the H$_3$ saddle point on the basis of ab initio calculations. Edmiston and Krauss (1965) improved their first H$_3$ calculation by including additional configurations in second order and obtained a binding energy of -3.983 eV. They also performed a comparable calculation on H$_2$. By assuming that the remaining error in the H$_3$ calculation relative to H$_2$ is proportional to the ratio of correlation energies, they estimated an H$_3$ binding energy of -4.23 eV. Conroy and Bruner (1967), in their improved calculation, obtained in extrapolated energy of -4.411 eV. Hayes and Parr (1967) estimated the errors due to angular and radial deficiencies in their 99-term wave function and predicted an exact energy (at $R_1 = 1.8a_0$) of -4.180 eV. Shavitt et al. (1968) assumed the residual error in their H$_3$ calculation was no more than 1.5 times the residual error in a comparable H$_2$ calculation, which implies that the energy of H$_3$ should be no more negative than -4.338 eV. Shavitt (1968) used this ab initio surface with one empirical parameter, a uniform scale factor for the energy profile along the reaction path, to compute transition-state theory rate constants for the isotopic H + H$_2$ reactions. The value of the scale factor that gave best agreement with the high-temperature experimental rates of Westenberg and deHaas was 0.89, implying an H_3 energy of -4.324 eV. After their second calculation, Edmiston and Krauss pointed out that their H_2 calculations at three internuclear distances accounted for 91% of the correlation energy. Thus they assumed that they were accounting for 91% of the correlation energy everywhere. This yielded an estimated binding energy of H_3 of -4.215 eV. Thus the estimates of Shavitt et al. (1968) and Shavitt (1968) agree remarkably well with those of Liu but none of the other estimates fall within even the wider of the bounds given above.

By considering the linear symmetric geometry we may also evaluate the force constant for symmetric stretching of the activated complex. From this one may calculate the zero-point energy $\frac{1}{2}h\nu_s$ of this normal mode in the harmonic approximation. There have been several such calculations and the results are given in Table I. [Harris et al. (1965), whose results are excluded from that table, obtained 0.131 eV.] This quantity, as well as the zero-point energy $\frac{1}{2}h\nu_b$ for a bending normal mode, is important for transition-state theory and the vibrationally adiabatic theory that are discussed elsewhere (Truhlar and Wyatt, 1976).

There has been relatively less attention devoted to geometries other than linear symmetric, but there have been several studies of nonsymmetric linear geometries (Hirschfelder et al., 1937; Boys et al., 1956; Boys and Shavitt, 1959; Conroy and Bruner, 1965, 1967; Edmiston and Krauss, 1968; Shavitt et al., 1968; Goddard and Ladner, 1969; Ladner and Goddard, 1969; Blustin and Linnett, 1975; Liu, 1973). Shavitt et al. (1968) and Liu (1973) have presented accurate analytic fits to their whole collinear surfaces. For use in one-dimensional tunneling corrections to transition-state theory it is of interest to calculate the second derivative at the barrier maximum and express it as the zero-point energy of the upside-down parabola with this force constant. This is called the imaginary zero-point energy of the asymmetric-stretch normal mode of the activated complex, and the few values calculated for it are as follows: $0.084i$ eV by Boys et al. (1956) and Boys and Shavitt (1959), $0.087i$ eV by Harris et al. (1965), $0.090i$ eV and $0.096i$ eV in double-zeta and double-zeta-plus-polarization bases by Shavitt et al. (1968), and $0.094i$ eV by Liu (1973) in his most accurate calculation.

Other important features of the nonsymmetric linear H_3 calculations are the position of the minimum energy path, the energy variation along and near this path, and the description of the bonding changes accompanying movement along this path. Harris et al. (1965), in CCI calculations employing a minimum basis set and the Mulliken approximation, mapped out the approximate position of the minimum energy path through the (R_1, R_2)

plane and the energy variation along it. Their results predicted that the curvature of the minimum energy path was confined to within about $0.4a_0$ of the saddle point and that the full width at half-maximum of the barrier along the curvilinear minimum energy path was about $2.0a_0$. Conroy and Bruner mapped out two complete potential-energy surfaces from their two sets of extrapolated results for the collinear calculation. In their first calculation (Conroy and Bruner, 1965) they found twin unsymmetrical saddle points with $(R_1, R_2) = (1.64a_0, 1.83a_0)$ and energies of -4.479 eV and a shallow symmetric depression with $R_1 = 1.76a_0$ and an energy 0.054 eV higher. These values appear by comparison to Liu's (1973) calculation to be too low by more than their estimated error of ± 0.05 eV in integral evaluation. They state that their second calculation (Conroy and Bruner, 1967) led to a linear symmetric saddle point but that calculation gave a barrier which is extremely flat near the top. It appears that their points near the top of the barrier are too widely spaced to judge whether a depression actually exists (Shavitt et al., 1968). Edmiston and Krauss (1968) and Shavitt et al. (1968) found the approximate position of the minimum-energy path through the (R_1, R_2) coordinate system by finding minima along cross sectional cuts almost perpendicular to the reaction path. Edmiston and Krauss did not examine points near enough to the barrier top to judge definitely whether a depression actually exists. The correct way to do this is to find the position of the minimum for linear symmetric geometries and then examine a small displacement along the asymmetric stretch normal mode to see if it is a saddle point or a local minimum. This has only been done by Harris et al. (1965), Shavitt et al. (1968), and Liu (1973). All these workers found that the saddle point was linear symmetric, and there is no longer any serious doubt that this is the case. Liu (1973) also used the method of steepest descents to start at the saddle point and find the accurate position of the whole minimum-energy path in the (R_1, R_2) coordinate system. (This differs from the minimum-energy path in a skewed-axis coordinate system.) The minimum-energy path determined in Liu's linear calculation approaches the equilibrium separation of H$_2$ much faster than the path of Shavitt et al. (1968). The potential energies along the five most accurate calculated minimum-energy paths at the places where they cross various $R_2 = $ constant lines are compared in Table II (these results were found by interpolation of the published data). In addition, we have followed the suggestion of Shavitt (1968) and in each case scaled the energy profile along the minimum-energy path by a constant determined to make the barrier correct. (For this purpose Liu's barrier is assumed correct.) These results are also shown. First we see that the barriers of Conroy and Bruner (1967) and Edmiston and Krauss (1968) are too flat at the top. We also see that the barrier of Conroy and Bruner is

TABLE II

Energies (in eV) Along Minimum-Energy Path in (R_1, R_2) Coordinates System for Given Values of R_2 (Where R_1 and R_2 Are the Interatomic Separations of the Two Nearest Neighbors)

Case	1	2	3	4	5	1	2	3	4
calculation $R_2(a_0)$	Conroy and Bruner, 1967[a]	Edmiston and Krauss, 1968[b]	Edmiston and Krauss, 1968[a]	Shavitt et al., 1968[b]	Liu, 1973[b]	Scaled	Scaled	Scaled	Scaled
Saddle point	0.34	0.58	0.53	0.48	0.42	0.42	0.42	0.42	0.42
1.9	0.33[c]	0.59[c]	0.55[c]	0.46	0.41	0.42[c]	0.43[c]	0.44[c]	0.41
2.2	0.28	0.54	0.53	0.40	0.34	0.36	0.39	0.43	0.35
2.6	0.16	0.38	0.37	0.28	0.23	0.20	0.28	0.30	0.25
3.0	0.05	0.23	0.22	0.17	0.14	0.06	0.17	0.18	0.16
3.4				0.10	0.07				0.09

[a] Extrapolated surface, relative to experimental energy of $H + H_2$.

[b] Upperbound surface, relative to calculated energy of $H + H_2$.

[c] These values are slightly uncertain because of the insufficient number of points calculated in this neighborhood.

too low at $R_2 \geqslant 3.0a_0$; that is, their interaction energy goes below 0.1 eV (and even below 0.0 eV) much too close to the saddle point. However, the scaled barrier of Shavitt et al. (1968) is in very good agreement with the barrier of Liu (1973).

Recently, Baskin et al. (1974), Blustin and Linnett (1974), and McCullough and Silver (1975) have pointed out that the correct minimum-energy path through the (R_1, R_2) plane cannot be obtained by considering the minima along cuts parallel to the axes. This is implicit in all the work discussed above and is also mentioned explicitly in an article discussed below (Truhlar and Kuppermann, 1971).

Weston (1959) and Shavitt (1968) pointed out that for tunneling corrections to transition-state theory one needs the minimum-energy path not in the (R_1, R_2) coordinate system but in normal-mode coordinate systems in which the axes are scaled and skewed so that the reduced mass is the same for motion in any direction. Of course, such a coordinate system depends on the ratios of the isotopic masses of the nuclei. Shavitt replotted the (R_1, R_2)-minimum-energy path in such a coordinate system for the equal-mass case. Choosing the scaling so the reduced mass is $\frac{2}{3}$ the mass of H showed that reaction-path curvature is non-negligible within $0.4a_0$ of the saddle point and the fullwidth at half-maximum of the barrier is about $1.8a_0$. Truhlar and Kuppermann (1971) determined the actual minimum energy path in this coordinate system for the surface of Shavitt et al. (1968). In addition, they developed an analytic approximation [based on the rotated-Morse-curve method of Wall and Porter (1962)] to the whole collinear surface following Shavitt's (1968) suggestions for scaling and found that the minimum-energy path for this surface in normal-coordinate space for the equal-mass case (Truhlar and Kuppermann, 1972). This surface is called the scaled SSMK or the Truhlar-Kuppermann surface.

Goddard and Ladner (1969, 1971; Ladner and Goddard, 1969; Goddard, 1972) provided a quantum-mechanical orbital picture of the reaction by using the best possible wave function for which the orbitals have an independent particle interpretation, that is, are eigenfunctions of one-electron Hamiltonians corresponding to the nuclear attraction and the nonlocal field of electrons in the other two orbitals. This was called the spin-coupling optimized group operator (SOGI) method. It provides a picture of the reaction in which the orbitals of the reactants gradually delocalize over all three centers, then relocalize to form product states. This orbital picture was used as a basis for a general theory (the orbital phase continuity principle) of orbital phase relationships accompanying bonding changes in reactions. It is a generalization of the valence-bond method and has certain conceptual advantages over the Woodward-Hoffman approach, which has been successfully applied to many molecule-molecule reactions. According to the latter, the H + H$_2$ reaction might be considered thermally forbidden [see, e.g., the orbital correlation diagram given by Hoffman (1968)], but the Woodward-Hoffman rules are actually not very useful for considering the reactivity of open-shell atoms, like H, with molecules. Of course, a quantitative understanding of the energetic changes along the reaction path requires a detailed consideration of the changes in the correlation energy (defined as the difference between the exact energy and that calculated from a molecular orbital or group-operator wave function) accompanying the reaction. This question, however, is closely tied to the question of completeness of the basis set since final convergence of the energy calculations requires simultaneous improvement of both. Several authors have discussed these problems (see, e.g., Edmiston and Krauss, 1965, 1968; Michels and Harris, 1968; Gianinetti et al., 1969; Linnett and Riera, 1969; Liu, 1971, 1973).

There has been very little work on nonlinear geometries. Stevenson and Hirschfelder (1937) evaluated $\frac{1}{2}h\nu_b$ (defined above) by taking derivatives of the energy expression with respect to bond angle. They obtained 0.059, 0.069, 0.052, and 0.067 eV in the four approximations considered by Hirschfelder et al. (1936) in the order they are listed in Table I. Thus their results were roughly the same as the value of 0.056 eV obtained by the semiempirical method of Eyring and Polanyi with $\rho = 0.20$. Boys et al. (1956) and Boys and Shavitt (1959) calculated a value of 0.059 eV; however, since this value is based on a nonlinear geometry with a 40° bond angle, the value is probably higher than should be obtained with their basis set (Johnston, 1966, p. 74). Harris et al. (1965) calculated 0.061 eV using the Mulliken approximation. The first reliable calculation was that of Shavitt et al. (1968), who obtained 0.060 eV in their double-zeta-plus-polarization basis. Liu and Siegbahn (private communication) have carried out some calculations for nonlinear geometries but they are not published. Their value for the harmonic bending zero-point energy is 0.057 eV. Except for a few calculations related to the equilateral triangle geomerty (see next paragraph), extensive treatments of the nonlinear geometries have been published only by Conroy and Bruner (1967) and Shavitt et al. (1968). Conroy and Bruner studied only isosceles triangle configurations with bond angles 120°, 90°, and 60°. Shavitt et al. studied systems with 150° and 120° bond angles. For each constant value of the bond angle, Shavitt et al. determined the minimum-energy paths in the (R_1, R_2) coordinate system and the associated energy profile and barrier. They obtained saddle-point geometries with nearest-neighbor distances of $1.78a_0$ and $1.81a_0$, respectively, and barrier heights of 0.55 and 0.79 eV. The scaling suggested by Shavitt (1968) would lower these values to 0.50 and 0.74 eV, respectively. For comparison, Conroy and Bruner obtained about 0.65 eV for the saddle-point height for a 120° bond angle. The changes of saddle-point heights with bond angle are in good agreement.

Siegbahn and Liu (private communication) made calculations for the following bond angles: 180°, 165°, 150°, 135°, 120°, 90°, and 60°. They used a basis of nine s, three p, and one d Gaussian-type functions (GTF) on each center. The nine s functions were contracted to linear-combination basis functions. This four s, three p, and one d hydrogen basis corresponds to 42 functions of a' symmetry and 18 of a'' symmetry. Using an approximate natural orbital transformation, this basis set was truncated to 30 a' functions and 13 a'' functions. A CCI calculation in this truncated basis involves 14,060 configurations and such a calculation was performed to give the final H_3 energy at each geometry. The GTF calculation yields $E_b = 0.429$ eV, compared to Liu's STO result of 0.424 eV. But the GTF collinear surface is parallel to the STO surface within 0.001 eV for $R_1 < R_2 < 4.0a_0$. For bond angles 150° and 120° the minimum-energy path of the new surface approaches the equilibrium separation of H_2 more rapidly than does the path of Shavitt et al. and this difference is exaggerated compared to the difference at 180°. It appears that the nonlinear portion of the surface of Shavitt et al. is not as accurate as the linear portion. Another indication of this is that their bending force constant is too large by about 15%.

There have been a few calculations on equilateral triangle H_3 (Hirschfelder, 1938; Conroy and Bruner, 1967; Porter et al., 1968; Blustin and Linnett, 1974). Hirschfelder's early calculations were at the minimum-basis set CCI level with unoptimized orbital exponents. He showed that the ground state becomes doubly degenerate and discussed this in terms of the Jahn-Teller effect. He and Coulson (1935) calculated that the energy was higher than the energy for three separated H atoms. Conroy and Bruner (1967), however, found a minimum energy for a distance of about $2.0a_0$ for equilateral H_3 and that the energy there was about 1.8 to 2.0 eV higher than linear geometries with the same nearest-neighbor distances or 2.0 to 2.3 eV higher than the linear saddle point. Porter et al. (1968) obtained qualitatively similar results and obtained $1.965a_0$, 2.2 eV, and 2.4 eV for these same quantities using CCI with the same size basis set as Shavitt et al. (1968) used for other geometries. (For nonsymmetric nonlinear geometries this involves 680 configurations.) Their major interest was the Jahn-Teller effect at this geometry and their work is discussed further in connection with excited states.

In relating the saddle-point properties of a potential-energy surface to experiments on reactive collisions, one must be careful to differentiate between classical barrier height E_b and transition-state theory activation energy at $0°K$ (which are properties of the surface) and phenomenological threshold energy E_{thr} and Arrhenius activation energy E_a (which are dynamical properties). The relationships of these quantities are discussed elsewhere (Menzinger and Wolfgang, 1969; LeRoy, 1969; Truhlar and Wyatt, 1976). An example of how these distinct concepts have been confused in the literature is given by the following quotation (Bacskay and Linnett 1972): "Experimental estimates of the activation energy, defined as the difference between H_3 and $H_2 + H$, range between 7 and 10 kcal/mole. The kinetic experiments of LeRoy et al. (1968) point to an activation energy of 9.2 kcal/mole whereas the more direct measurements of Kuppermann and White (1966) yield a value 7.6 ± 0.5 kcal/mole." This passage confuses E_b, E_a, and E_{thr}.

Now we return to the semiempirical valence-bond calculations. After the early work no new methods were developed until 1955 when Sato (1955, 1955a, 1955b) introduced a modified version of London's formula containing a new parameter k and a very approximate form of the H_2 triplet curve. He then obtained the Coulomb and exchange integrals by equating the Heitler-London expressions (including overlap) for the ground and triplet states of H_2 to these potential curves. For this purpose the constant k was treated as if it were the square of the orbital overlap integral although the real orbital overlap integral is not a constant but depends on internuclear distance. His modified London formula cannot be derived from valence-bond theory by letting k be the square of the overlap integral, and semiempirical values obtained for k are much smaller than the values of the square of the orbital overlap integral for the distances important along the reaction path. The parameter k must absorb all these inconsistencies plus the inconsistencies that remain from the LEP method (such as equating the Heitler-London expression to the accurate ground-state energy rather than the Heitler-London computed one). However, Sato's method, commonly abbreviated LEPS, had the advantage of leading to a symmetric saddle point for all values of k examined. Sato studied the barrier height as a function of k and suggested $k = 0.18$, which yields an E_b of 0.22 eV, as the "best" value. The Sato procedure was critically analyzed by Weston (1959; see also Weston, 1967). He found that $k = 0.1475$ led to a vibrationally adiabatic barrier height of 0.35 eV, which he deduced from experiments as the energy of activation. (The vibrationally adiabatic barrier height is the transition-state-theory activation energy at $0°K$ when the reaction coordinate is treated classically and the other degrees of freedom are treated quantum mechanically and it is given in the harmonic approximation by

$$E_0^{VAZC} = E_b + \tfrac{1}{2}hv_s + hv_b - \tfrac{1}{2}\hbar\omega_e$$

where the last term is the harmonic approximation to the zero-point energy of H_2.) Weston emphasized the empirical nature of the scheme. The saddle-point properties of the Sato and Weston surfaces for H_3 are given in Table III where they are compared with Liu's accurate values and other semiempirical surfaces to be discussed below.

TABLE III

Properties of Several Semiempirical and Analytic Surfaces with Linear Symmetric Saddle Points

Surface	$R^{\ddagger}$ (a_0)	E_b (eV)	$\tfrac{1}{2}hv_s$ (eV)	$\tfrac{1}{2}hv_b$ (eV)	$\tfrac{1}{2}hv_a$ (eV)
Sato ($k = 0.18$)	1.73	0.219	0.133	0.054	0.100i
Sato ($k = 0.1475$)	1.76	0.358	0.131	0.054	0.118i
Sato ($k = 0.144$)	1.75	0.380	0.132	0.055	0.128i
Cashion–Herschbach	1.82	0.468	0.133	0.051	0.152i
Porter–Karplus (No. 1)	1.70	0.373	0.136	0.059	0.143i
Porter–Karplus (No. 2)	1.70	0.398	0.135	0.061	0.137i
Porter–Karplus (No. 3)	1.70	0.413	0.135	0.062	0.130i
Pedersen–Porter (No. 6)	1.79	0.450	0.133	0.072	0.138i
Salomon (No. 4)	1.70	0.373	0.128	0.063	0.045i
Salomon (No. 6)	1.71	0.407	0.129	0.062	0.053i
Truhlar–Kuppermann	1.765	0.424	0.125	—	0.091i
Jones–Rosenfeld	1.74	0.425	0.130	0.070	0.092i
Malcome–Lawes	1.85	0.391	0.136	0.056	0.095i
Yates–Lester	1.74	0.425	0.128	0.067	0.104i
Accurate	1.76	0.425 or 0.420	0.127	0.056	0.094i

With the new constant (k) set equal to zero, the Sato method may be reinterpreted as an LEP calculation except with variable ρ (Weston, 1959). It was realized by the early workers that the assumption of constant ρ was not strictly valid. This was reemphasized by Hirschfelder (1941; see also Hirschfelder et al., 1954; Weston, 1959; Yasumori, 1959; Eyring and Eyring, 1963; and Cashion and Herschbach, 1964). The Coulomb fraction, as computed in the Heitler-London-Sugiura treatment of H$_2$, is near 0.14 at large distance but drops sharply as the distance is decreased. Eyring and Polanyi (1931) performed a calculation using this variable ρ; they obtained a symmetrical basin 0.52 eV above the energy of H + H$_2$. However, Wall et al. (1958) calculated a variable-ρ surface that did not have a basin. It is often stated that the assumption of constant ρ is directly responsible for the existence of a spurious basin in the semiempirical surfaces; unfortunately, the real situation is more complicated. We have already mentioned that constant ρ may or may not lead to a basin depending upon its value and that the first variable-ρ treatments gave a basin. Eyring and Eyring (1963; see also Eyring, 1962) used a distance-dependent ρ equal to an adjustable constant times the Sugiura values. The constant was chosen to be 1.4 to obtain a symmetric basin 0.33 eV above the energy of H + H$_2$ and twin saddle points of height 0.41 eV located about $0.6a_0$ from the basin. This gives a very wide barrier to reaction; thus the energy is still 0.26 eV above H + H$_2$ at a distance $1.7a_0$ from the linear symmetric configuration (compare Table II). Eyring and Eyring claimed that previous transition-state theory calculations (e.g., Weston, 1959) had shown that the Sato barrier is too thin and predicts too much tunneling and that their new surface was in better accord with experiment than any that had been obtained so far. Now we know the true surface has a symmetric saddle point; we will return to the question of the width of the barrier.

A different way of using experimental or theoretical energy curves for H$_2$ to predict the surface for H$_3$ is diatomics-in-molecules theory, introduced by Ellison (1963). In this method, matrix elements of the Hamiltonian in a basis of valence-bond configurations are approximated by diatomic potential curves and atomic energies. Using a two-configuration basis and what they considered to be the most accurate available potential energies for ground-state and triplet H$_2$, Ellison et al. (1963) calculated a barrier of about 0.56 eV [later corrected to 0.57 eV (Ellison, 1964)]. Surprisingly, within their scheme, neglect of overlap changes the calculated energies by less than 4×10^{-5} eV. About the same time Cashion and Herschbach (1964) proposed a modification of the LEP method in which the Coulomb and exchange integrals were evaluated by equating the Heitler-London energy expressions for the ground and triplet states of H$_2$, neglecting orbital overlap, to what they believed were the most accurate available potential curves for H_2. They pointed out the sensitivity of their results to the value of the then poorly known triplet curve energy at the end-atom-to-end-atom distance of the H_3 saddle point. They obtained a barrier of 0.42 eV [later corrected to 0.47 eV (Cashion and Herschbach, 1964a)]. It was pointed out by Ellison (1964) that at least for the case of linear symmetric configurations the treatments are identical. Actually they are completely identical (Pickup, 1973). Then the difference in results is just the result of the different input potential curves.

Also at about this time, Porter and Karplus (1964) reexamined the whole semiempirical valence-bond scheme and attempted to remove not only the constant-ρ approximation (also removed by Cashion and Herschbach as discussed above) but also the other three most criticized aspects of the method: (i) neglect of overlap integrals, (ii) calculation of single-exchange integrals in H_3 as if they were the same as in H_2 at a given interatomic distance, (iii) neglect of multiple-exchange integrals. The orbital overlap integrals were calculated using a distance-dependent orbital exponent. Rather than use the most accurate available H_2 data they used reasonably accurate but simple fits to the H_2 potential curves. They obtained a barrier of 0.37 eV in their first calculation (No. 1). With a different choice of parameters (No. 2, explained below) they obtained 0.40 eV. If the latter calculation is repeated using the corrected potential curves of Cashion and Herschbach (1964, 1964a), the calculated barrier is lowered to 0.20 eV. If this calculation and the calculation of Cashion and Herschbach are repeated using the now accurately known potential curves (Kolos and Wolniewicz, 1965), the calculated barriers become 0.31 and 0.57 eV, respectively (unpublished calculations).

It is important to distinguish two different types of semiempiricism. One is to use H_2 potentials and valence theory to predict the properties of H_3. This was attempted using the London equation but the extreme assumptions involved prevented this approach from being very fruitful. Thus the parameters of the LEP and Sato methods were adjusted to give what was thought to be the correct barrier height for H_3 with the hope that other features of the surface would then be reasonable. The investigations discussed in the previous two paragraphs raised the hope that such empiricism might not be necessary if one used more accurate potential curves or in addition removed the three other most criticized assumptions of the LEP method. These hopes were vitiated by the extreme sensitivity of the calculated surfaces to the H_2 data and, in the Porter-Karplus method, to the two semiempirical parameters (δ and ε) they introduced into the three-center terms needed to remove approximations (ii) and (iii). This indicates that the three-center terms are very important, and that the method is still sensitive to its assumptions even when reasonable efforts are made to remove the most criticized ones. Thus later workers should have abandoned the first type of semiempiricism, that is, they should not have tried to use semiempirical valence-bond theory or diatomics in molecules to "predict" the surface without using kinetic data or accurately known surface features to calibrate the method. This predictive approach has been called the ab initio semiempirical approach. It was abandoned immediately by Porter and Karplus. Thus while the values of δ and ε on their surface No. 1 were obtained by considerations of computed theoretical integral values, they also presented three additional surfaces (Nos. 2 to 4) in which δ and ε were used to alter the form of the semiempirical surface. These surfaces provide an example of what we mean by the second type of semiempiricism. In particular, the values for surface No. 2, which has been used very often for dynamical studies, were obtained by adjusting the vibrationally adiabatic barrier height so it agreed approximately with the Arrhenius activation energy [which Weston (1959) had extracted from the experimental data] while maximizing the nearest-neighbor distance because the semiempirical value was less than the ab initio one of Boys and Shavitt (see Tables I and III). But their surface No. 3 agrees better with Liu's.

The Porter-Karplus method was later reexamined by Pedersen and Porter (1967). They modified the formalism to use H$_2^+$ potential curves and the Mulliken approximation for removing approximations (ii) and (iii). They also used different approximations to the orbital exponents. Their best surface is given in Table III. Jones and Rosenfeld (1973) and Kung and Anderson (1974) recalibrated the Porter-Karplus method. The former authors used accurate cubic spline fits to the accurate H$_2$ potential curves and adjusted δ and ε so that the saddle-point properties are as close as possible to the scaled values recommended by Shavitt (1968). Thus they are also close to those for the accurate surface (see Table III). The surface of Kung and Anderson also uses fairly accurate H$_2$ curves but the parameters were not readjusted and the surface is less accurate.

Salomon (1969) reexamined the question of the width of the barrier. He modified the Cashion-Herschbach procedure to include orbital overlap integrals. These integrals were calculated using a distance-dependent orbital exponent with one adjustable parameter that was varied to obtain a symmetric saddle point and a wide barrier. The two surfaces he judged most accurate on the basis of transition-state theory calculations including tunneling are given in Table III. It is now seen, by comparison of $\frac{1}{2}h\nu_a$ to the last row, that there was some truth in the claim of Eyring and Eyring (1963) and Salomon that all previous semiempirical surfaces had a barrier that was too thin. But Salomon's surfaces overcompensated and produced a barrier that was too thick.

Steiner et al. (1973) have recently extended the diatomics-in-molecules calculation to include ionic configurations. The method was found to be stable with respect to addition of ionic terms only if overlap was neglected; without this assumption they found instabilities due to near linear dependences of the three-electron basis functions. But for H_3 the effect of ionic terms was small and did not remove the sensitivity of the barrier to the triplet curve. More recently Tully and Truesdale (1976) presented a more stable and consistent way to include overlap in the diatomics-in-molecules method.

A simple type of semiempirical scheme decomposes the binding energy of H_3 into a sum of two bonding functions for nearest neighbors and a triplet repulsion term between end atoms. Lipponcott and Liefer (1958) applied this to H_3 and found a barrier of 0.29 eV with attractive basins on either side. Johnston and Parr (1963; see also Johnston 1960, 1966) developed a scheme of this type, called the bond-energy bond-order scheme, which can be used to calculate the minimum-energy path, the energy profile along it, and saddle-point properties but, without new assumptions, not the whole surface. In this method the formation of a new bond "pays for" the breaking of the old bond in such a way that the sum of the bond orders $(n_1 + n_2)$ is unity. Pauling's relation between bond length and bond order yields the reaction path, and the relation $E_{bond}(n) = E_{bond}(1)n^p$ plus a triplet potential curve yields the energy profile. With no adjustable parameters the original results for the H_3 saddle point were very good (Johnston and Parr, 1963: $E_b = 0.43$ eV and $R_1 = 1.74a_0$) and the minimum-energy path is also good (Truhlar, 1972). Zavitsas (1972) developed an alternative form of this theory for calculating barriers; he obtained $E_b = 0.51$ eV. All three calculations involved very approximate triplet curves: an exponential in the first and 0.5 and 0.9 times Sato's triplet in the latter two. If more up-to-date values were used for the H_2 triplet curve or if more up-to-date experimental results are used in the Johnston-Parr calculation to calibrate p, the results are not as good. These methods are just as sensitive to the triplet curve as is the diatomics-in-molecules scheme.

The semiempirical valence-bond model has been used so extensively for H_3 that semiempirical molecular orbital models have been somewhat neglected. An early application of the crude form of Hückel theory to H_3 was not successful since H_3 was predicted to be bound with respect to $H + H_2$ by 0.83 times the bond energy of H_2 (Van Vleck and Sherman, 1935; Pearson, 1948). However, Bradley (1966) applied molecular orbital theory more successfully. He neglected overlap integrals, employed Pariser-Parr-Pople-type approximations (zero differential overlap and inclusion of nearest-neighbor interactions only), and evaluated the non-neglected integrals in terms of H_2 and H_2^+ potential curves. He obtained a linear symmetric barrier with $R_1 = 1.84a_0$ and a vibrationally adiabatic barrier height of 0.56 eV. Gimarc (1970) and Malcome-Lawes (1975) have applied extended Hückel theory to H_3. The latter adjusted one parameter so that the transition-state theory rate coefficient with no tunneling correction agrees with the result of Schulz and LeRoy (1965) at 423°K, but his barrier is too low (see Table III).

The semiempirical surfaces are usually easily programmed so it is easy to examine properties other than saddle-point height. Porter and Karplus (1964; see also Karplus, 1970) presented contour maps of one of their H_3 surfaces and profiles of the energy along the minimum-energy path for several bond angles. Several other collinear contour maps have been published. Apparently, the only accurately calculated minimum-energy paths that have been published for semiempirical surfaces are those determined in the (R_1, R_2) plane by Silver (1972) for four such surfaces and by Jones and Rosenfeld and Malcome-Lawes for their respective surfaces. Earlier minimum-energy paths for three semiempirical surfaces reported by Shavitt et al. (1968; see also Karplus, 1970) showed kinks. These were apparently due simply to errors in their determinations. A few other energy profiles have also been published (Bradley, 1966; Shavitt et al., 1968; Karplus, 1970; Malcome-Lawes, 1975). Malcome-Lawes's energy profile agrees remarkably well with that for surface No. 2 of Porter and Karplus.

Four analytical surfaces have not yet been mentioned. Russell and Light (1971) fitted the unscaled Shavitt et al. surface in natural collision coordinates and Anderson (1973) fitted it to a form suggested by work on bound triatomics. Anderson's fit had a spurious well 0.09 eV deep, but is worth consideration for fitting if augmented by other terms. Yates and Lester (1974) made an empirical modification of the formula used by Porter and Karplus and adjusted the parameters to fit Liu's collinear surface. They showed that the energy profiles for bent geometries agreed well with the scaled results of Shavitt et al. But Schatz (1975) has obtained a better fit using noncollinear points obtained by private communication from Liu.

Many properties of the surfaces are brought out more clearly when they are replotted in scaled and skewed coordinates that diagonalize the kinetic energy with the same reduced mass in each direction (Glasstone et al., 1941; Shavitt, 1968), in three-dimensional persepctive (Parr and Truhlar, 1971; Truhlar and Kuppermann, 1972), in natural collision coordinates (Jackson and Wyatt, 1973), or using a mapping in which all arrangement channels are represented even-handedly (Kuppermann, 1975; Ling and Kuppermann, 1975; Kuppermann et al., 1976).

There have been a few calculations on the very repulsive part of the potential-energy surface in the region where the atoms are very close. This part of the surface is important for the interpretation of high-energy atom-molecule scattering experiments. The first calculations involved approximations to the three-center integrals. Margeneau (1944) considered a single valence-bond configuration with special approximations designed for small distances. For an isosceles triangle with two sides $1.13a_0$ and one side $0.89a_0$ he calculated an interaction energy of 26 eV. He also considered closer separations for which the calculated interaction energy was even higher. Bauer (1951) obtained a simple spherically symmetric model potential by an approximate first-order perturbation-theory calculation. His potential is repulsive for $R[H - H_2(r_e)]$ less than $1.3a_0$ but for larger distances contains an attractive well about 1 eV deep, where $R[H - H_2(r_e)]$ is the distance from an H to the center of mass of an H_2 at its equilibrium distance. This well is much too deep and much too close. Aroeste and Jameson (1959) showed that the second valence-bond configuration neglected by Margenau is important. They performed four-configuration valence-bond calculations for perpendicular geometries that gave lower energies than Margeneau's by about a factor of 3 for $0.5a_0 \leqslant R[H - H_2(r_e)]$ $\leqslant 0.9a_0$. Trivedi (1970) evaluated the three-center integrals accurately. He used the two-configuration covalent valence-bond method with a minimum basis set and separately optimized exponents. He evaluated the energy only for isosceles triangles and found an interaction energy even lower than Aroeste and Jameson's. His results are fitted by

$$V = 19.248 \text{ eV} \exp\left\{ - 1.804a_0^{-1}R\big[H - H_2(r_e)\big]\right\}$$

for $0.5a_0 \leqslant R[H - H_2(r_e)] \leqslant 1.75a_0$. There are no reliable experiments to which these three sets of calculations may be compared. Vanderslice and Mason (1960) considered distances a little larger. They used a method that is very much like the LEP method with $\rho = 0$, but they obtained the exchange integral by approximating the H_2 triplet curve as an exponential for $1.63a_0 \leqslant R[H - H_2(r_e)] \leqslant 4.01a_0$. They obtained $V = 61.5$ eV $\exp\{ - 1.562a_0^{-1}R[H - H_2(r_e)]\}$ for the spherical average of the interaction potential. Recently, two more accurate calculations have been performed. Patch (1973) performed CCI minimum basis set and floating orbital calculations for linear, scalene, and isosceles geometries for $R[H - H_2(r_e)]$ in the range 1.0 to $4.0a_0$. Norbeck and Certain (1975; see also Norbeck, Certain, and Tang, 1975) calculated the interaction potential for $2.5a_0 \leqslant R[H - H_2(r_e)] \leqslant 5.0a_0$ for collinear and perpendicular approach. They used a valence-bond formalism with an extended basis set and 100 symmetry-adapted configurations at each geometry. For the collinear geometry at $R[H - H_2(r_e)] = 3.0a_0$, they obtained an interaction energy of 0.461 eV whereas the more accurate calculation of Shavitt et al. (1968) yielded 0.402 eV. At larger distances their calculations appear to be more accurate; thus for $R[H - H_2(r_e)] = 4.0a_0$, their interaction energy is 0.128 eV compared to 0.113 eV obtained by Shavitt et al. (1968).

There has been no attempt to calculate the electronically nonadiabatic corrections (i.e., the Born-Oppenheimer breakdown terms) for the H$_3$ potential-energy surface, but these corrections are expected to be small at low energy. For example, they are small for the bound states of H$_2$ (Orlikowski and Wolniewicz, 1974).

B.   Ground Electronic State: Long Range

At large $H - H_2$ separations, the attractive induced dipole-induced dipole dispersion interaction in the ground electronic state has the form

$$E_6(R,\chi) = -\frac{C_6}{R^6}\left[1 + \Gamma P_2(\cos\chi)\right]$$

where R is the distance between the atom and the molecule center of mass and χ is the angle between R and the molecular axis ($\chi = 0$, π denote collinear configurations). In most studies, the H$_2$ internuclear separation is regarded as frozen at $1.4a_0$ [see, however, Langhoff et al. (1971) where vibrational averaging is considered]. The anisotropy coefficient Γ determines the energy difference between the linear ($\chi = 0$) and perpendicular ($\chi = \pi/2$) configurations:

$$E_6\left(R, \frac{\pi}{2}\right) - E_6(R,0) = \frac{3}{2}\frac{C_6\Gamma}{R^6}$$

which shows that the collinear geometry is preferred if Γ is positive. Also note that the angular average $\langle E_6(R,\chi)\rangle$ equals $-C_6 R^{-6}$. Calculations of C_6 and Γ are summarized in Table IV. Margenau (1944) used experimental oscillator strengths and associated transition energies to provide the first estimate of C_6 for the $H - H_2$ interaction. He also provided virtually the only calculation of the long-range induced dipole-induced quadrupole ($-C_8/R^8$) dispersion interaction [the theory is developed in Margenau (1938)]. Mason and Hirschfelder (1957) provided the first information about the anisotropy coefficient. They estimated Γ from $\alpha_\parallel$ and $\alpha_\perp$, the components of the static electric-dipole polarizability of H$_2$ along and perpendicular to the bond axis, respectively, by the following approximate formula: $\Gamma = (\alpha_\parallel - \alpha_\perp)/(\alpha_\parallel + 2\alpha_\perp)$ [Hirschfelder, Curtiss, and Bird (1964), pp. 969–970, 1089]. The available calculations are compared in Table IV; they all indicate that Γ is positive.

TABLE IV

Calculations of Coefficients in the Long-Range $H + H_2$ Interaction Potential Given Approximately by $- C_6 R^{-6}[1 + \Gamma P_2(\cos\theta)] - C_8 R^{-8}$

Reference	C_6 (a.u.)[a]	Γ	C_8 (a.u.)[a]	Method, comments
Margenau (1944)	8.40	—	148	From semiempirical H_2 oscillator strengths
Mason and Hirschfelder (1957)	8.40	0.117	—	C_6 from Margenau; Γ from static polarizabilities for H_2
Dalgarno (1963)	9.20	—	—	From H_2 oscillator strengths forced to satisfy sum rules
Karplus and Kolker (1964)	9.91	0.154	—	From dynamic polarizabilities from uncoupled Hartree-Fock calculations
Dalgarno and Williams (1965)	9.24	—	—	From H_2 oscillator strengths that satisfy seven sum rules
Langhoff and Karplus (1970)	8.57	—	—	From bounds on C_6 established with Padé approximants
Victor and Dalgarno (1970)	8.92	0.104	—	From semiempirical dynamic polarizabilities of H_2
Langhoff, Gordon, and Karplus (1971)	8.57	0.099	—	From bounds on C_6 established with Gaussian quadratures or Padé approximants
	8.95	0.111	—	Same but corrected for H_2 vibration

[a] The atomic unit of energy is the hartree (1 hartree $= 27.2116$ eV $= 4.35981 \times 10^{-18}$ J); the atomic unit of length is the bohr (1 bohr $= 1a_0 = 0.529177 \times 10^{-10}$ m).

In the studies of Dalgarno (1963) and Dalgarno and Williams (1965), oscillator strengths from the ground state of H_2 were modified and extended so that sum rules were satisfied exactly. In the latter case, the oscillator strengths were required to satisfy seven sum rules. [See Langhoff et al. (1971) for other applications of sum rules.]

A different formulation of the dispersion interaction problem was provided by Casimir and Polder (1948). They showed that C_6 can be expressed as an integral over imaginary frequency of the product of the dynamic (frequency-dependent) polarizabilities of the interacting species. The first application to the $H - H_2$ interaction was provided by Karplus and Kolker (1964), who evaluated both C_6 and Γ. The Casimir-Polder formulation has also been applied to the $H - H_2$ interaction by Langhoff and Karplus (1970), Victor and Dalgarno (1970), and Langhoff et al. (1971). Langhoff and Karplus (1970) and Langhoff et al. (1971) have shown how to use bounds on the dynamic polarizabilities in the Casimir-Polder formulas to establish upper and lower bounds on the dispersion coefficients. The values of C_6 and Γ listed in Table IV are not in complete agreement, but the most recent calculations indicate a value of C_6 about 9 a.u. with Γ close to 0.1. There has been no recent work on the higher-order (R^{-8} etc.) interactions.

The shallow van der Waals well in the ground-state potential surface $V(r, R, \chi)$ is an important feature in the determination of cross sections for elastic and rotationally inelastic scattering (Sections III and IV.A) and in the transport properties of partly dissociated H$_2$ (Section IV.D). Margenau (1944) and Mason and Hirschfelder (1957) attempted to calculate the surface in the vicinity of the well by adding the first-order exchange forces (calculated by approximate valence-bond theory) to the second-order dispersion forces. They obtained well depths of 2 to 3 meV at $R = 6$ to $7a_0$. They obtained the result, still believed correct, that the well is deeper for a perpendicular geometry than for a collinear one. Conroy and Bruner (1965, 1967), Michels and Harris (1968), Shavitt et al. (1968), and Blustin and Linnett (1974) all briefly commented on the appearance of shallow wells at large R in their ab initio H$_3$ potential surfaces, but the basis sets employed were too small to accurately characterize the surface topology near the well minimum. However, the order of magnitude of the well depth can be estimated as 0.001 eV from the calculations of Shavitt et al. (1968). In contrast, the semiempirical Porter-Karplus (1964) surfaces and most other semiempirical surfaces are repulsive at large R. The characteristics of several semiempirical and ab initio surfaces have recently been compared by Norbeck, Certain, and Tang (1975; see also Norbeck and Certain, 1975) over the range $2a_0 \leqslant R \leqslant 5a_0$ (see Section IV.A). In this region the rotational anisotropy coefficient $V_2(R)$ in the expansion $V(R, \chi) = V_0(R) + V_2(R)P_2(\cos \chi) + \cdots$ (where $r = 1.4a_0$) changes rapidly. (Note that for $V_2 > 0$, the perpendicular approach of H to H$_2$ is favored). V_2 from Porter-Karplus surface No. 2 is negative at all R and becomes increasingly negative as R decreases, but V_2 from a new ab initio surface calculated by Norbeck et al. (1975) is positive near $R = 5a_0$, increases to a maximum near $3.5a_0$, and becomes negative for $R < 3.1a_0$.

Because of the difficulty of characterizing the van der Waals well and the long-range potential through ab initio calculations, a number of hybrid potentials have been proposed that smoothly join one set of potential expansion coefficients $\{V_0(R), V_2(R)\}$ at short range to another set at long range. For example, Dalgarno, Henry, and Roberts (1966) joined the Mason-Hirschfelder (1957) potential onto an $E_6(R, \chi)$ long-range attractive tail. Chu and Dalgarno (1975) joined the Porter-Karplus values of $V_0(R)$ and $V_2(R)$ onto the same long-range tail that Dalgarno et al. (1966) employed. (Rotationally inelastic-scattering calculations on both joined potentials are discussed in Section IV.A.) Takayanagi (1957; see also

Takayanagi and Nishimura, 1960) and Tang (1969) also constructed joined potentials. Tang's merges into Porter-Karplus surface No. 2 at small R. This potential was later used in the calculation of transport coefficients (see Section IV.D). In addition, Shui and Appleton (1971) constructed a spherically averaged potential $V_0(R)$ for use in trajectory studies of $H + H + H_2 \rightleftharpoons 2H_2$ recombination. The characteristics of a number of spherically averaged potentials near the well minimum are compared in Table V.

In order to fit the velocity dependence of the total scattering cross section of H on H_2, Stwalley et al. (1969) used several spherically symmetric potentials for which the reduced velocity parameter $v_0 = 2\varepsilon R_{min}/\hbar$, where ε is the well depth and R_{min} is the position of the minimum, was in the range 1.4 to 2.7 km/sec. The reduced velocity parameters for other proposed $V_0(R)$ potentials are listed in Table V. The potentials used by Stwalley et al. included those of Browning and Fox (1964) and Tang and Karplus (1968). To discuss their measurements of the $D + H_2$ total scattering cross section, Gengenbach et al. (1975) introduced a joined potential involving Born exponential repulsion at small R, cubic splines at intermediate R, and $-C_6/R^6 - C_8/R^8$ attraction at long range. Gengenbach et al. also used a number of older semiempirical $H - H_2$ spherically averaged potentials to predict $D + H_2$ total scattering cross sections. The Lennard-Jones potential of Clifton (1961), the exp-6 potential of Weissman and Mason (1962), the Dalgarno-Henry-Roberts (1966) potential, and the Tang (1969) potential all gave about the same disagreement (in a least-squares sense) with the experimental data. The Khouw et al. (1969) 12-6 potential with parameters fitted from diffusion data predicted cross sections in worse agreement with the scattering data than the previous four potentials. This clearly illustrated that bulk-transport measurements can be fitted by a variety of effective spherical potentials, some of which are not really very accurate.

C. Excited Electronic States

Van Volkenburgh et al. (1973) recently pointed out that, "Since H_3 is the simplest polyatomic molecule but one, several of its electronically excited states should be reasonably well known theoretically. This is by no means the case... ." Ab initio and semiempirical studies of H_3 excited states are summarized in Table VI, which shows that they have been limited to the lower surfaces at or near selected geometries (D_{3h}, C_{2v}, or $D_{\infty h}$) over small ranges of internuclear distances. In the first study of H_3 excited electronic states, Hirschfelder (1938) calculated a correlation diagram that illustrates how the energies of the $\tilde{X}^2\Sigma_u^+$, $^2\Sigma_g^+$, and $^4\Sigma_u^+$ states of symmetric linear H_3 change as the system is bent (with the constraint that $R_{ab} = R_{bc}$ for atoms

TABLE V
Van der Waals Well Characteristics in the H – H$_2$ Spherically Averaged Potential[a]

Reference	ε (meV)	R_0 (a_0)	R_{min} (a_0)	v_0 (km/sec)	C_6 (a.u.)	Form of potential,[b] comments
Clifton (1961)	2.78	5.20	5.84	2.61	8.04	12-6; fit to Margenau (1944) calculations
Wise (1961)	3.12	5.39	6.05	3.03	11.24	12-6; fit to Margenau (1944) calculations
Weissman and Mason (1962)	1.44	5.80	6.60	3.02	8.40	exp-6; fit to Margenau (1944) calculations
Browning and Fox (1964)	2.85	5.18	5.81	2.66	8.09	12-6; fit to Margenau's (1944) calculations; used to analyze viscosity data
Dalgarno, Henry, and Roberts (1966)	1.29	6.07	6.84	1.43	9.26	exp-6; fit to Mason-Hirschfelder (1957) potential at small R
Tang and Karplus (1968)	1.29	6.07	6.84	1.43	9.26	Fit to Porter-Karplus (1964) potential at small R and Dalgarno-Henry-Roberts (1966) potential at large R
Tang (1969)	0.46	6.31	7.41	0.55	9.26	exp-6; parameters adjusted to fit diffusion data
Sancier and Wise (1969)	1.44	6.48	7.37	1.71	16.36	12-6; parameters adjusted to fit diffusion data
Khouw, Morgan, and Schiff (1969)	7.32	4.39	4.73	5.56	7.70	Morse; used in trajectory studies of H+H+H$_2$ recombination
Shui and Appleton (1971)	3.27	5.54	6.43	3.38	—	12-6; parameters chosen to fit viscosity data
Cheng and Blackshear (1972)	2.79	4.84	5.43	2.43	5.27	exp. repulsion/spline/$(-C_6 R^{-6} - C_8 R^{-8})$; parameters chosen to fit D+H$_2$ total scattering cross section
Gengenbach, Hahn, and Toennies (1975)	2.34	5.94	6.71	2.52	8.80	

[a] ε is the well depth and R_{min} (or R_m) is the position of the minimum. At $R = R_0$, $V_0 = 0$. The reduced radial velocity is $v_0 = 2(\varepsilon R_{min})/\hbar$.

[b] exp-6 potential: $V_0(R) = \dfrac{\varepsilon}{1-(6/\alpha)}\left\{\left(\dfrac{6}{\alpha}\right)e^{\alpha(1-R/R_{min})} - \left(\dfrac{R_{min}}{R}\right)^6\right\}$; 12-6 potential: $V_0(R) = \varepsilon\left\{\left(\dfrac{R_{min}}{R}\right)^{12} - 2\left(\dfrac{R_{min}}{R}\right)^6\right\}$.

TABLE VI
Studies of H$_3$ Excited Electronic States

Reference	Method[a]/basis,[b] if any	Surfaces/geometries studied
Hirschfelder (1938)	VB/ETF: $1s$	D_{3h} for $2a_0 < R < 3a_0$: ^2E, $^4A_2$, $^2A_2$, $^2A_1$ $D_{\infty h} \to D_{3h} \to C_{2v}$ correlation diagram: $^2\Sigma_u^+$, $^2\Sigma_g^+ \to {}^2E$; $^4\Sigma_u^+ \to {}^4A_2$
Jameson and Aroeste (1960)	VB/ETF: $1s$ on each center of molecule; $2s$, $2p$ on atom	C_{2v} only: $^2A_1$, $^2B_2 \to H_2(X^1\Sigma_g^+) + H(n=2)$
Ellison, Huff, and Patel (1963)	DIM/$1s$	Table of energies for two lowest $^2\Sigma$ states, linear configurations only
Matsen (1964)	Spin-free quantum chemistry/discussion only	$D_{\infty h} \to D_{3h} \to C_{2v}$ correlation: $^2\Sigma_u^+ \to {}^2E \to {}^2A_1$; $^2\Sigma_g^+ \to {}^2E \to {}^2B_2$
Smirnov (1964)	VB/qualitative discussion only	Discussion of process $H_3(^2E;\ D_{3h}) \to \begin{cases} H + H_2(^1\Sigma_g^+) \\ H + H_2(^3\Sigma_g^+) \end{cases} \to H + H + H$
Porter, Stevens, and Karplus (1968)	Semiempirical VB[c] and ab initio/ETF: two s, one $p\sigma$, one $p\pi$	$D_{3h}(^2E)$ surfaces near intersection in normal-mode space $D_{\infty h} \to D_{3h} \to C_{2v}$ correlation for two lowest surfaces D_{3h} for $1a_0 < R < 3a_0$: X^2E, $^2A_1$, ^2E
Frenkel (1970)	MO and CI/GTF: four s, two $p\sigma$, two $p\pi$ on each center of molecule; four s on atom	Symmetric $D_{\infty h}$ for $2a_0 < R < 6a_0$: $^2\Sigma_u^+$, $^2\Sigma_g^+$, $^2\Sigma_g^+$, $^2\Pi_u^+$ $D_{\infty h} \to D_{3h} \to C_{2v}$ correlation for: $^2\Sigma_u^+$, $^2\Sigma_g^+ \to {}^2E \to {}^2A_1$, $^2B_2$ and $H_2(^1\Sigma_g^+) + H(2p)$ $^2\Pi_u \to {}^2A_1$
Tully (unpublished), quoted in Van Volkenburgh et al. (1973)	DIM/not given	Attractive long-range interaction for $H(n=2) + H_2(X^1\Sigma_g^+)$; $H_2(b^3\Sigma_u^+) + H$; $H_2^+(B^2\Sigma_u^+) + H$

[a] Abbreviations: VB, valence bond; DIM, diatomics-in-molecules; MO, molecular orbital; CI, configuration-interaction.
[b] On each center (except as indicated otherwise); ETF, exponential-type functions; GTF, Gaussian-type functions.
[c] Method of Porter and Karplus (1964).

labeled a-b-c). In this correlation, the $\tilde{X}\,^2\Sigma_u^+$ and $^2\Sigma_g^+$ states of the linear $D_{\infty h}$ geometry form the degenerate ^2E level of the equilateral triangle D_{3h} geometry while the $^4\Sigma_u^+$ excited state correlates with the $^4A_2$ state in D_{3h}. The variation of the energy of these states (and excited $^2A_2$ and $^2A_1$ states as well) was also studied as a function of internuclear distance for D_{3h} geometries. Correlation diagrams for the $D_{\infty h} \rightarrow C_{2v} \rightarrow D_{3h}$ deformation have also been discussed by Smirnov (1964), Matsen (1964), Porter, Stevens, and Karplus (1968), Frenkel (1970), and Van Volkenburgh et al. (1973). Only the two lowest states, which originate as $\tilde{X}\,^2\Sigma_u^+$ and $^2\Sigma_g^+$ in $D_{\infty h}$, were considered by Matsen (1964) and Porter et al. (1968).

The topology of the intersection between the two lowest electronic surfaces at the D_{3h} geometry depends upon what coordinates are used. For example, if the a-b-c angle is fixed at 60° and the two surfaces are plotted as functions of R_{ab} and R_{bc}, then intersection occurs along the diagonal "ridge" where $R_{ab} = R_{bc}$ (see Porter and Karplus, 1964 and Porter et al., 1968). However, if energy contours are plotted in the (q_1, q_2) normal-mode space for the equilateral triangular geometry (where q_1 and q_2 correspond to asymmetric stretch and bending deformations), then the lower and upper surfaces join at the vertices of two cones (the vertex for the lower cone points "up" at the descending cone of the upper surface). Pictures of both surfaces near the intersection were shown by Porter et al. (1964), while Matsen (1964) displayed only the lower cone. [A perspective picture of the intersection in natural collision coordinates was displayed by Jackson and Wyatt (1973).] The ridge or conical intersection between the two lowest H$_3$ surfaces in the D_{3h} geometry gives rise to Jahn-Teller instability (all nonlinear nuclear configurations for an electronically degenerate state of a polyatomic molecule are unstable), which is relieved as the linear geometry is approached and the degeneracy is removed.

In addition to the two lowest surfaces that intersect as the ^2E state with D_{3h} geometry, Frenkel (1970) calculated a correlation diagram for several states ($^2B_2$, $^2A_1$ for C_{2v} geometries) that correlate with $H_2(X\,^1\Sigma_g^+) + H\,(n = 2)$. It was found that several such states ($^2A_1$ and ^2E) possess deep minima near the D_{3h} geometry. Jameson and Aroeste (1960) also calculated a C_{2v} correlation diagram for states that dissociate to $H_2(X\,^1\Sigma_g^+) + H\,(n = 2)$. In unpublished studies, J. C. Tully (quoted in Van Volkenburgh et al., 1973) found that excited states correlating with $H\,(n = 2) + H_2(X\,^1\Sigma_g^+)$, $H + H_2(b\,^3\Sigma_g^+)$ or $H_2^+(B\,^2\Sigma_u^+) + H$ are attractive at long range. A number of features of the H_3 excited-state correlation diagram for doublet states have been discussed by Van Volkenburgh et al. (1973) (also see Section IV.C).

III. ELASTIC AND TOTAL SCATTERING CROSS SECTIONS

The final process competing with reaction, dissociation, and other inelastic scattering processes in $H + H_2$ collisions is elastic scattering. In fact, elastic scattering often dominates the integral total scattering cross section for $H + H_2$, where "integral" refers to the integration of the differential cross section over all scattering angles and "total" refers to a sum over all processes. Elastic scattering dominates when the large-impact-parameter small-angle scattering dominates the total scattering cross section and is itself mainly elastic scattering. Then measurements of the elastic scattering and the total scattering provide similar information and can be used to learn about the interaction potential. In addition, the differential elastic scattering cross section is needed to interpret other experiments such as the hot-atom experiments discussed elsewhere (Truhlar and Wyatt, 1976). For $H + H_2$ there have been four kinds of experiment on elastic or total scattering: (1) high-energy measurements of H-beam attenuation by H_2 gas, which yield the magnitude of the incomplete total scattering cross section $S(\theta_0, E_{rel})$ as a function of relative translational energy E_{rel} where "incomplete" means the integral over scattering angles θ excludes the region $\theta \leqslant \theta_0$; (2) low-energy measurements of H-beam attenuation by H_2 gas, which yield the relative magnitude of the integral total scattering cross section $S(E_{rel})$ as a function of E_{rel}; (3) a few low-energy measurements of the magnitude of $S(E_{rel})$ at selected E_{rel}; (4) low-energy measurements of the relative magnitude of the differential nonreactive scattering cross section $d\sigma_{nr}/d\theta (E_{rel})$ as a function of θ.

The first experiments were at high energy because it is easier to produce a beam of H atoms at very high energies. The beams involved in the high-energy experiments had laboratory energies of 196 to 7000 eV. Thus E_{rel} is 131 to 4667 eV. This is high enough (the de Broglie wavelength is of the order of $10^{-2}a_0$) and θ_0 should be large enough (0.1° to a few degrees in relative coordinates) for a classical description of the scattering and for the measurement to be sensitive to the repulsive region of the potential and not at all to the long-range van der Waals' attraction. But θ_0 is also small enough that the interaction energy during a collision that scatters at angle θ_0 is only a small fraction of E_{rel} and is of the order of magnitude of 1 eV.

The first high-energy experiments on $H + H_2$ were by Amdur and Pearlman (1940; Amdur, 1943, 1949) and were repeated by Amdur et al. (1950). More recently, high-energy experiments on $H + H_2$ were reported by Belyaev and Leonas (1967), whose results disagree with those of Amdur and co-workers. Using the same techniques, Belyaev and Leonas (1967) also reported results for other systems including $He + H_2$. A disadvantage of the small value of θ_0 used in these experiments is that the results are sensitive to the analysis of the data in terms of an effective angular aperture, which depends on beam size, length of the scattering chamber over which collisions occur, and detector aperture. Jordan and Amdur (1967) pointed out that the apparatuses used by Amdur and co-workers before the 1950s were primitive and that as effects such as beam-detector geometry and intensity distribution in the beam were better understood the apparatuses were improved and so was the accuracy of the experimental results. They also pointed out that it is necessary to measure the diameter of the intensity distribution of the beam because the results are very sensitive to the ratio of beam and detector radii when this is near unity. They corrected this error in their He-He experiments and resolved a longstanding discrepancy between theory and experiment for the He-He short-range interaction potential. It is interesting that a similar longstanding discrepancy existed between theory and the experiments of Amdur and co-workers for the short-range interaction in $H + H_2$ but it was overshadowed in the published discussions by the He-He controversy, presumably at least in part because the theoretical calculations were easier for He-He. Many suggestions, such as breakdown of the Born-Oppenheimer adiabaticity approximation, were considered and rejected in an attempt to resolve this discrepancy. Amdur and co-workers continued to perform new experiments and, when comparison was possible, all their results from 1967 and after were in reasonably good agreement with the results that have been reported by Leonas and co-workers. This includes studies on He-H_2 (Amdur and Smith, 1968). The interaction potential deduced from these experiments agrees with that obtained by Belyaev and Leonas within 55%. Thus is is reasonable for $H + H_2$ to reject the early results of Amdur and co-workers and to accept as essentially correct the experimental results of Belyaev and Leonas.

For most cases (usually atom-atom scattering) studied by Belyaev and Leonas, a log-log plot of $S(\theta_0, E_{rel})$ versus E_{rel} was linear. But for $H + H_2$ this was not the case. They attributed this anomalous energy dependence to the pronounced nonsphericity in the interaction potential and possibly to electronic nonadiabaticity at E_{rel} of 2.4 keV. Thus they analyzed only the lowest energy part of their curve in terms of an effective spherical interaction potential. For $1.89a_0 \leqslant R[\text{H-H}_2(R_e)] \leqslant 2.34a_0$, they obtained for this potential

$$V = 12.8 \text{ eV } a_0^{4.15} / R^{4.15}$$

For $R = 2.0a_0$ this is 0.72 eV.

Wartell and Cross (1971) have concluded that if electronic excitation and dissociation do not occur, high-energy scattering measurements may be interpreted in terms of a spherically symmetric potential and this potential is the average of the true potential over orientations evaluated at the equilibrium internuclear separation of the target. Wartell and Cross also added the proviso that the anisotropy and vibrational dependence of the potential must not be too high. Since $H + H_2$ is reactive, their conclusion is not necessarily applicable to $H + H_2$. For $H + H_2$, elastic scattering may not dominate rotationally and vibrationally inelastic scattering at a few hundred eV at the angles involved in the Amdur-type measurements. Nevertheless, it is interesting to attempt to compare the derived potential to theoretical calculations of the spherically averaged potential. The calculation of Mason and Vanderslice (1958) yields 2.70 eV for the spherical average of the potential at $2.0a_0$. The large difference from experiment is not surprising in view of the crudeness of the theoretical method (see above). The spherical average of Porter-Karplus surface No. 2 at $2.0a_0$ is 1.13 eV (Tang and Karplus, 1968), in better agreement with the experiment. The spherical averages of the potentials of Yates and Lester (discussed above) and Gengenbach et al. (discussed below) are more repulsive than the spherical average of Porter and Karplus's surface No. 2 (see comparison in Choi and Tang, 1975a).

Because of the Heisenberg uncertainty principle, the elastic differential cross section is flat for θ less than a small angle that is inversely proportional to velocity. Thus at low energies θ_0 can be made small compared to this angle and the complete integral total scattering cross section can be measured. The first such measurement was made by Harrison (1962). He used a thermal H beam and a mass spectrometer detector. The mean relative velocity of these experiments corresponds to E_{rel} in the range 0.20 to 0.29 eV. In this range Harrison found $S(E_{rel})$ constant within experimental error and equal to 196 a_0^2. Fluendy et al. (1967) used an H beam that was velocity selected in an inhomogeneous magnetic field and for detection they used a Pt bolometer. They measured only the relative cross section and found an $E_{rel}^{-0.09}$ dependence over the energy range 0.04 to 0.39 eV. Since $S(E_{rel})$ is essentially a measure of the range over which the interaction is strong enough that the classical deflection of the particle is greater than its quantal uncertainty, they and all subsequent workers have assumed that is is reasonable to interpret $S(E_{rel})$ in terms of an effective spherical interaction potential $V(R)$. The dependence measured for $S(E_{rel})$ is weaker than the $E_{rel}^{-0.20}$ expected for scattering from an R^{-6} potential, which indicates that the van der Waals' attraction is so weak that even at the lowest energy the scattering is caused mainly by the shorter-range repulsive forces. In this way one obtains a bound $\varepsilon R_m \lesssim 0.024$ eV a_0 where ε is the well depth and R_m the H to H$_2$ distance at the minimum of the well. Using a modified apparatus, Stwalley et al. (1969) remeasured the dependence of $S(E_{rel})$ on E_{rel} in the range 0.19 to 1.0 eV. They concluded the previous measurements for $E_{rel} < 0.08$ eV were unreliable. The data were not sufficient to determine a potential. But the two recent experiments described next, when combined, have been used to determine a $V(R)$ although its uniqueness and its precise meaning have not been established. Bauer et al. (1975) measured the E_{rel} dependence of $S(E_{rel})$ for the range 0.007 to 0.93 eV for H + D$_2$ and 0.007 to 0.63 eV for H + H$_2$. Their cross sections are in good agreement with those of Stwalley et al. (1969) in the region of overlap. By comparing to results calculated for various central potentials, they obtained εR_m equal to 0.016 eV a_0 from the H + D$_2$ data and 0.020 eV a_0 from the H + H$_2$ data. The difference is within the uncertainty of the determination. They compared this to a value of 0.016 eV a_0 obtained from semiempirical combining rules used for analyzing data on other systems. Gengenbach et al. (1975) measured the absolute value of $S(E_{rel})$ for D + H$_2$ at E_{rel} equals 0.31 eV and 0.48 eV with estimated accuracy of 1.5%. They obtained $142a_0^2$ at both energies. They used these results to normalize the data of Bauer et al. and used the normalized data to obtain a multiparameter $V(R)$. At $2.0a_0$, their potential is 2.18 eV but their experiment probably is not too sensitive to R values this small. Their potential has εR_m equal to 0.014 eV a_0 with $\varepsilon = 0.0023$ eV and $R_m = 5.93a_0$. They compare it to many of the available theoretical potentials for distances greater than about $2\frac{1}{2}a_0$. More recently, Toennies, Welz, and Wolf (unpublished, see Welz, 1976) have measured the integral cross section for H + H$_2$ down to E$_{rel}$ equals about 0.001 eV. Analysis of this experiment yields $\varepsilon = 0.0026$ eV and $R_m = 6.56$ a_0.

The differential cross section for nonreactive scattering has been measured by Fite and Brackman (1964, 1965) and Geddes et al. (1972) using thermal beams. The second measurement makes the first one obsolete. The differential cross section is forward peaked and drops by a factor of about 20 from 10° to 45° then flattens out. These measurements were not in absolute units and were therefore normalized to the measurement of Harrison (1962) but they have now been renormalized by Gengenbach et al. (1975). The differential cross section has not yet been measured with velocity-selected beams.

There have not been many calculations of the elastic scattering. We consider first the thermal and low-energy range. Almost all published calculations (Tang and Karplus, 1968; Stwalley et al., 1969; Gengenbach et al., 1975) are based on spherically symmetric approximations to the potential. One interesting result is that Gengenbach et al. (1975) showed that their multiparameter potential, obtained from analyzing integral total cross sections, yields a differential cross section for nonreactive scattering in good agreement with the experiment of Geddes et al. (1972). McCann and Flannery (1975, 1975a; see also Flannery and McCann, 1975) and Schatz and Kuppermann (1976) have peformed distinguishable-atom elastic scattering calculations that do include the anisotropy of the potential. McCann and Flannery used both a "multichannel semiclassical orbital treatment" and a multistate eikonal treatment for their calculations. The anisotropy seems to lower the pure elastic differential cross section at large angles. Schatz and Kuppermann calculated the differential and integral elastic scattering cross sections at $E_{rel} = 0.70$ eV using potential energy surface No. 2 of Porter and Karplus (1964). Their calculated integral total cross section $(221a_0^2)$ is considerably larger than experiment because of the inaccuracy of the potential surface so the results cannot be compared quantitatively to experiment. But they tested the assumption of a spherically symmetric potential by comparing their results to a calculation involving the spherical average of the potential. They found that the integral total cross section was affected only 0.1% by the anisotropy and that for center-of-mass scattering angles θ less than 30° the elastic differential and total differential cross sections were both essentially identical to that computed with the spherical approximation. Schatz and Kuppermann also found that the effect of particle indistinquishability on the elastic differential cross section was small (less than 10% for $30° < \theta < 90°$).

Ioup and Russek (1973) performed some calculations for energies in the 1 to 10 keV range. They are discussed in Section IV.A.

IV. INELASTIC SCATTERING CROSS SECTIONS AND TRANSPORT PROPERTIES

A. Rotational Energy Transfer

There are no experimental measurements of cross sections for rotationally inelastic nonreactive scattering of $H + H_2$, but there are several quantum-mechanical calculations for this scattering process in the ground electronic state. These calculations are described in this section. An earlier discussion was given by Takayanagi (1973).

For rotational excitation it is convenient to represent the interaction potential as an expansion in Legendre polynomials as

$$V(r, R, \chi) = \sum_{\lambda = 0, 2, \ldots} v_\lambda(r, R) P_\lambda(\cos \chi)$$

where r is the H$_2$ internuclear separation, $\mathbf{R}$ is a vector (of magnitude R) from the molecular center of mass to the atom, and χ is the angle between the internuclear axis and $\mathbf{R}$. The coefficients $v_\lambda(r, R)$ can be calculated by expanding one of the potential functions $V(r, R, \chi)$ discussed in Section II or they may be approximated directly. The interaction potential is thus specified by the set $\{v_\lambda(r, R), \lambda = 0, 2, \ldots\}$, and several such sets have been used for dynamical calculations. Cross sections for rotational inelastic transitions $(j \rightarrow j \pm 2, j \pm 4 \cdots)$ have been shown in these studies to depend primarily upon the first rotational anisotropy term, $v_2(r, R)$, and upon the spherical term $v_0(r, R)$, which may serve to "shield" $v_2(r, R)$ at certain values of R. Norbeck et al. (1975) have recently compared the $v_2(r_e, R)$ anisotropy coefficients on several H$_3$ surfaces; they also presented results of a new ab initio calculation of $v_2(r_e, R)$. In most scattering calculations the rigid-rotor approximation has been used; that is, r was set equal to r_e (the H$_2$ equilibrium distance) in both the $v_0(r, R)$ and $v_2(r, R)$ coefficients so they become functions $v_\lambda(R)$ of R alone. An examination of the breakdown of this approximation has been presented recently by Wolken et al. (1972). Before proceeding to results for H + H$_2$, we note that Secrest (1973) has presented an extensive review of the theory of inelastic energy transfer.

The first approximate calculations of cross sections for H + H$_2$ rotational inelastic processes were presented by Takayanagi (1957). He used the distorted-wave formalism to study the $j = 0 \rightarrow j' = 2$ process. An approximate fit [$v_0(R)$ approximated as a shallow Morse curve and $v_2(R)$ as an exponential repulsive term] to the early Margenau (1944) potential was used. In order to simplify the calculations, the modified-wave-number approximation was used; this involves replacing the transition probability for nonzero orbital angular momentum (l) at relative translational energy E_{rel} by the s wave ($l = 0$) probability evaluated at a modified (lower) energy $\overline{E}_{rel} = E_{rel} - l(l+1)h^2/2\mu R_0^2$, where R_0 is at (or near) the classical turning point in the relative motion. These calculations were then extended (within the modified-wave-number–distorted-wave framework) by Takayanagi and Nishimura (1960) to include transitions out of initial rotor states $j = 2$, 4, and 6. Thermal rate coefficients for temperatures up to 5000°K were then calculated.

Choi and Tang (1975) recently presented extensive calculations of H − H$_2$ rotational inelastic processes on three different model potential surfaces for translational energies in the 0.05 to 0.25 eV range. They compared their results with a number of earlier calculations, both approximate and accurate, that employed these potentials. We first briefly discuss the different potentials with the notation of Choi and Tang; then we compare the scattering results.

(a) DHR potential. Dalgarno, Henry, and Roberts (1966) defined a model $H + H_2$ potential that smoothly joined the short-range (exponential repulsion) Mason-Hirschfelder (1957) potential to a long-range R^{-6} attractive tail; the long-range component of the $v_0(R)$ term was based upon Dalgarno and Williams' (1965) study, and the long-range part of $v_2(R)$ was computed from static electric polarizabilities (Hirschfelder et al., 1954). In contrast to the potentials considered below, $v_2(R) > 0$ for all R in the DHR model; that is, the perpendicular approach of H to H_2 is favored over the collinear one for fixed r and R.

(b) PK potential. This is calculated from surface No. 2 of Porter and Karplus (1964). The resulting spherical potential $v_0(R)$ is considerably softer than the DHR spherical term. Also, the PK anisotropy term is negative at moderate to large R, meaning that the collinear approach of H to H_2 is preferred.

(c) TANG potential. Tang (1969) defined a potential [containing $v_0(R)$ and $v_2(R)$ terms] by smoothly joining the Porter-Karplus surface No. 2 at small R onto the DHR potential at large R. In this potential, $v_0(R)$ has a small van der Waals well at large R (unlike the spherical average of the PK surface). Also, $v_2(R)$ in the TANG potential contains a negative region at moderate R and a positive region for $R > 3.6a_0$. In the negative $v_2(R)$ region, the collinear geometry is preferred. Shavitt (private communication) has emphasized that a potential which favors the perpendicular over collinear approach for fixed r and R might still favor the collinear approach over a bent approach at fixed R_1 and R_2. Thus a detailed comparison is necessary to determine how calculations at fixed R_1 and R_2 compare with the Legendre expansions discussed here. However, Norbeck et al. (1975) have shown that the sign change of the TANG potential is qualitatively consistent with the diatomics-in-molecules potential of Steiner et al. (1973), with a potential obtained by combining Liu's (1973) collinear calculation and the spherically symmetric component determined by fitting beam measurements (Gengenbach et al. 1975), and with their own ab initio valence-bond calculations for two different basis set sizes.

In addition to the potentials discussed above, other Legendre-expanded potentials have been proposed for $H + H_2$ at various times. Thus, for example, Tang and Karplus (1968) considered not only the Legendre expansion for the $H + H_2$ potential with the H_2 distance fixed but also the spherically symmetric effective potential obtained under the assumption that the molecule adiabatically follows the incoming atom. They also considered a modified version of this adiabatic potential that includes a long-range attractive term. Of course, such spherically symmetric potentials do not lead to rotational excitation. Micha (1969) has used a different adiabatic approximation to obtain an effective potential that is nonspherical and complex (i.e., it has a negative imaginary component). Micha's starting point was a Cashion-Herschbach-like potential; his final potential has apparently not been used for scattering calculations. The TANG potential is not the only attempt to match the Porter-Karplus surface No. 2 at short range to a long-range attractive potential. An alternative version of such a potential has been presented by Wolken et al. (1972). The hybrid potential of Wolken et al. just consists of joining the PK values of $v_0(R)$ and $v_2(R)$ at small R to the DHR values at the positions [$R = 4.6a_0$ for $v_0(R)$ and $R = 7.0a_0$ for $v_2(R)$] where these curves cross. Thus the major difference between the hybrid potential and the PK potential is in $v_0(R)$. Although the resulting potential has cusps at the connection points, they do not directly affect the scattering. The hybrid potential of Wolken et al. will be abbreviated the WMK hybrid potential. Another attempt to make an analytic Legendre-expanded surface that is accurate at both small and large R was made by Ioup and Russek (1973); see below.

We first consider scattering calculations using the DHR potential. Dalgarno et al. (1966) used this potential in distorted-wave Born approximation (DWBA) calculations for the $j = 0 \rightarrow j' = 2$, $2 \rightarrow 4$, and $1 \rightarrow 3$ transitions in H + H$_2$ collisions, and in studies of the $0 \rightarrow 2$ transition in H + D$_2$. Differential cross sections, integral cross sections, and rate coefficients (up to 5000°K) were computed. In a later study, Allison and Dalgarno (1967) studied the $0 \rightarrow 2$ rotational transition in the close-coupling (CC) formulation. For translational energies up to 0.30 eV, the DWBA results were slightly larger than the CC cross sections. In their recent close-coupling studies, Choi and Tang (1975) obtained good agreement with the Allison-Dalgarno results for this potential.

We now turn to several calculations of rotational inelastic cross sections that employed the TANG potential. Tang (1969) first reported a series of DWBA calculations, then Hayes et al. (1971) reported CC results; they also compared their results with those of Tang. Integral cross sections for the $0 \rightarrow 2$ process agreed well with Tang's DWBA results; differential cross sections also agreed well, with strong forward peaking predicted in both calculations. Choi and Tang (1975) also obtained CC results for this potential; excellent agreement was obtained with the earlier results of Hayes et al.

Calculations of rotational inelastic processes on the PK potential include the CC results of Wolken et al. (1972), Choi and Tang (1975), and Schatz and Kuppermann (1976). Unlike all the other workers, Wolken et al. (1972) and Schatz and Kuppermann did not make the rigid rotor assumption. Integral cross sections for all three calculations are in good agreement, although many oscillations reported by Wolken et al. in the $0 \rightarrow 2$ differential cross section (particularly for scattering angles $\theta \lesssim 90°$) were not reproduced in the other calculations. Schatz (private communication) has compared the unpublished phases of his and Kuppermann's scattering matrix elements to the published ones of Wolken et al. There is good agreement for the detailed channel-to-channel $0\rightarrow2$ probabilities as a function of total angular momentum J and for $l=J\pm2$ there is good agreement for the phases. But for $l=J$ the phases differ by π. Thus it seems likely that these oscillations are spurious and are due to phase errors in the scattering matrix elements of Wolken et al. For both the PK and DHR potentials, the low-energy (relative translational energy $E_{rel}\lesssim0.5$ eV) $0\rightarrow2$ differential cross section is generally backpeaked with very smooth behavior in the backward direction (for $\theta\gtrsim90°$); the TANG potential predicts forward peaking at similar energies. In the calculations of Choi and Tang and of Wolken et al. the effect of higher anisotropy coefficients $v_\lambda(R)$ $(\lambda>2)$ was demonstrated to be very small, even for the $0\rightarrow4$ transition. In contrast to the other calculations mentioned so far, Schatz and Kuppermann (1976) included closed vibrational channels and allowed the anisotropy terms to have explicit r dependence (as well as R dependence) to produce vibrational coupling. However, it is interesting to note that the Wolken et al. calculations, which employed a single vibrational channel in their CC expansion, predicted $0\rightarrow2$ integral cross sections in good agreement with that of Schatz et al. Thus for low-energy integral cross sections for $0\rightarrow2$ rotational excitation on this surface, closed vibrational channels do not seem very significant (but this comment is not true for reactive scattering). Wolken et al. found that the rigid rotor approximation was unsatisfactory; it led to partial (fixed total angular momentum) integral cross sections as much as 25% less than the fully vibrationally averaged potential. The calculations by Schatz et al. allowed for simultaneous reactive and nonreactive $j\rightarrow j'$ excitation, so that antisymmetrized scattering amplitudes and cross sections were constructed to allow for interference between the direct and reactive amplitudes. The antisymmetrized $0\rightarrow2$ differential cross section has symmetry oscillations mostly confined to low angles, but the amplitude of the oscillations becomes larger as the relative translational energy increases from 0.5 to 0.7 eV. [Antisymmetrized cross sections had been computed previously by Saxon and Light (1972) and Shatz and Kuppermann (1976a) for coplanar $H+H_2$ scattering. In addition, antisymmetrization was explicitly included by Wolken and Karplus (1974) in their formulation of the H_3 reactive scattering problem. Their results for the $0\rightarrow2$ transition probability, when compared with purely nonreactive (Wolken et al., 1972), clearly show the effect of reaction at higher energies.]

Cross sections for $0\rightarrow2$ excitation differ greatly on these three potentials. Above threshold, the ordering is $\sigma_{02}(DHR)\gg\sigma_{02}(PK)>\sigma_{02}(TANG)$. Choi and Tang (1975) have qualitatively discussed why [in terms of the range and hardness of $v_0(R)$ and the sign and shape of $v_2(R)$] the three potentials are expected to produce such different integral and differential cross sections. Wolken et al. (1972) found that their hybrid potential led to significantly more rotational excitation than the unperturbed PK potential at $E = 0.1$ eV, although the results were similar at $E = 0.25$ eV. This also shows that $v_0(R)$ is important for the magnitude of σ_{02}.

Choi and Tang (1975) also compared DWBA and CC calculations on these surfaces; the DWBA results were generally quite favorable at low energies, and they were obtained with about 1% of the computer time required for full CC calculations.

As the total energy increases, the number of rotational channels required in a quantum close-coupling study increases very rapidly. For this reason, channel-decoupling methods have recently been employed to decrease the number of coupled equations to be solved. Chu and Dalgarno (1975) employed the effective-potential method (Rabitz, 1972; Zarur and Rabitz, 1973), which eliminates dependence of the potential on m_j quantum numbers for a space-fixed z-axis. They used the WMK hybrid potential. After testing the effective close-coupling method at a total energy E of 1.0 eV, they computed elastic and inelastic cross sections for total energies up to 1.5 eV. McGuire and Kruger (1975) used a different decoupling method, the body-fixed centrifugal decoupling approximation, which is based on neglect of coupling between different components of total angular momentum along the z-axis of a coordinate system that rotates with the three-body system (Pack, 1974; McGuire and Kouri, 1974). They used a spherical potential that was estimated in recent scattering experiments (Gengenbach et al., 1975) along with the $v_2(R)$ term of the PK potential. This will be called the MK potential. The 0→2 inelastic cross sections are lower than the results of Chu and Dalgarno (who employed a different potential). The 0→2 differential cross section gradually shifted to the forward direction as E increased from 0.5 to 1.5 eV. At each energy, it was observed that as the amount of rotational energy transferred in the $j = 0 \rightarrow j'$ collision increased, the differential cross section was more back-peaked. An information-theory analysis of the McGuire and Kruger 0→2 cross section for H + H₂ and D + H₂ has been presented (Levine et al., 1976).

Most recently, Choi and Tang (1976) performed CC and DWBA calculations at $E = 0.5$ eV for the WMK hybrid potential, for the MK potential, and for a potential (called potential I) that consists of the $v_0(R)$ of Gengenbach et al. (1975) and the $v_2(R)$ of Norbeck et al. (1975). For the WMK hybrid potential their CC results for the integral cross sections were in good agreement with effective close-coupling results of Chu and

Dalgarno for low j but there were very large errors at high j. DWBA results were slightly in error (by up to 26%) at low j but were more accurate for $\Delta j = 2$ transitions at high j. The effective close-coupling method did not predict the differential cross section very well but the DWBA did. The CC calculations for the MK potential were compared with the calculations of McGuire and Kruger (1975). The comparison showed that the body-fixed centrifugal decoupling approximation and the DWBA are much better than the effective close-coupling approximation for the 0→2 transition. However, the body-fixed centrifugal decoupling approximation also predicted the 0→4 and 0→6 integral within 24% but the DWBA is not useful for $\Delta j > 2$. Both approximations were fairly good for the differential cross section for $\theta > 10°$.

The integral cross sections obtained by Choi and Tang (1976) by converged CC calculations for potential I were over an order of magnitude smaller than those obtained for the other potentials for the 0→2 transition and were about 2-to-3 orders of magnitude smaller than those obtained with the other potentials for the 0→4 and 0→6 transitions. The differential cross sections for potential I, however, had very similar shape to those for the TANG potential. Hopefully, the question of which of these potentials, if any, is qualitatively correct will be answered when the calculations of Siegbahn and Liu become available.

The WMK hybrid potential was employed by McCann and Flannery (1975, 1975a; Flannery and McCann, 1975) in a "multistate semiclassical orbital treatment" (this is sometimes called the classical path method) of $H + H_2$ rotational inelasticity over the total energy range 0.5 to 1.5 eV. This time-dependent formulation treated translational motion classically, but allowed coupling between this motion and the quantum rotational states of H_2 through an "optical potential." Integral cross sections agreed well with the results of Chu and Dalgarno. The semiclassical differential cross sections averaged out quantum oscillations for $\theta > 90°$. McCann and Flannery (1975a) also considered the use of the Rabitz-Zarur effective potential formalism in conjunction with their multistate orbital treatment.

In an approximate study based upon the (high-energy) Born approximation, Ioup and Russek (1973) computed integral and differential cross sections for several nonreactive pure rotational and vibrational-rotational transitions for energies between 1 and 10 keV. They derived analytic formulas for differential and integral cross sections in terms of a parametrized potential. For $H + H_2$ calculations, they employed the ab initio surface of Shavitt et al. (1968) to adjust their parameters. In the Born-approximation calculations, the target molecule was treated as a harmonic oscillator, while relative atom-molecule motion was described by a plane wave; electronic nonadiabaticity, ionization, and molecular dissociation were neglected. The stress in this paper was on development of a fast approximate method that could eventually be used to extract potential surface information from experimental cross-section data in the keV range.

B. Vibrational Energy Transfer

Heidner and Kasper (1972) measured the rate coefficient at 299°K for

$$H + H_2(n_1 = 0) \rightarrow H_2(n_2 = 1) + H$$

where n_1 and n_2 are the initial and final vibrational quantum numbers. However, quasiclassical trajectory histogram calculations (Karplus and Wang, unpublished, quoted by Heidner and Kasper, 1972) indicate that the nonreactive contribution is negligible with respect to the reactive one. Thus this is not a measure of the rate coefficient for nonreactive vibrational energy transfer. In fact, there is no measurement available for this process and hardly any theory. The little theory that does exist is reviewed in this section.

Clark and Dickinson (1973) performed collinear calculations of vibrational excitation probabilities for the $H + H_2$ mass combination using an exponential repulsion interaction potential between the incoming H and the nearest H of H_2. Because the important rotation-vibration coupling is missing in such a calculation, it is difficult to assess its relevance to experiment. Quantum-mechanical (Truhlar and Kuppermann, 1972) and quasiclassical and semiclassical (Bowman and Kuppermann, 1973) calculations of collinear nonreactive vibrational excitation probabilities have also been carried out using the more accurate potential-energy surface of Truhlar and Kuppermann.

Smith and Wood (1973) reported both collinear and three-dimensional quasiclassical trajectory calculations in $H + H_2$ where H_2 is initially in its first or second excited vibrational state. They used a semiempirical potential-energy surface similar to the Weston (1959) surface. They used the histogram method to interpret these trajectories in terms of cross sections for energy transfer, but they included only those trajectories for which $R_1 < R_2$ or $R_3 < R_2$ at some point in the trajectory (where R_1 and R_3 are the interatomic distances that initially are infinite and R_2 is the other interatomic distance). This may be a serious approximation and it is not clear how much reliance can be placed on their results. It is also probable that even if their approximation were not made and a more accurate surface were used, the quasiclassical trajectory histogram method would not yield an accurate result for the vibrational excitation cross section for $H + H_2$ except at fairly high energies (at least a few eV).

Gengenbach et al. (1975) estimated a crude upper limit of 10^{-6} $\mathring{A}^2$ for the cross section for vibrational exciation at a relative translational energy of about 0.5 eV. This was based on an analogy to available data for $He + H_2$ and $Li^+ + H_2$.

If one excludes from consideration the out-of-data calculation by Bauer (1952), no three-dimensional quantum-mechanical calculations have yet been reported for nonreactive vibrational excitation cross sections in $H + H_2$ collisions at energies below 1 keV. The calculations of Ioup and Russek (1973) at energies 1 to 10 keV are discussed in the previous subsection. Schatz and Kuppermann (1975a) reported calculations on the potential energy surface No. 2 of Porter and Karplus (1964) of the reactive vibrationally inelastic probability for zero total angular momentum at $E_{rel} < 0.83$ eV. They also calculated (Schatz, private communication) the corresponding nonreactive probability and found (in contrast to the quasi-classical result) that both reactive and nonreactive vibrationally inelastic probabilities and cross sections are comparable in magnitude for this surface.

Shui (1973) made a classical trajectory calculation on the same Porter-Karplus surface of the one-way equilibrium transition-rate kernel of internal energy changes for $H + H_2$ collisions at 3000°K.

C.   Electronic Energy Transfer and Other Processes Involving Excited Electronic States

In this section, collisional processes involving electronically excited H atoms interacting with H_2 are reviewed. H atoms in the $2s$ state have an excitation energy of 10.2 eV and are metastable; in vacuum the population of this state decays by a two-photon process to $H(1s)$ with an extremely long natural lifetime of about $1/8$ sec. By contrast, $H(2p)$ has a radiative lifetime of 2×10^{-9} sec and decays via a one-photon process to $H(1s)$ with emission of Lyman-α (L_α) radiation. If $H(2s)$ is in the presence of an electric field (either external or from gas in the apparatus), $2s$-$2p$ mixing occurs with subsequent emission of L_α radiation. This mixing occurs quite readily because the energy separation of these states is less than 1 cm^{-1}. The following processes involving collisions of metastable $H(2s)$ with H_2 have been studied:

$$H(2s) + H_2 \xrightarrow{2s \to 2p} H(2p) + H_2 \to H(1s) + H_2 + L_\alpha$$

$$H(2s) + H_2 \xrightarrow{2s \to 1s} H(1s) + H_2$$

$$H(2s) + H_2 \xrightarrow{\substack{2s \text{ associative-} \\ \text{ionization}}} H_3^+ + e^-$$

The H-atom interconversion rate coefficient $k_{2s \to 2p}$ in the presence of H$_2$ was first studied by Fite et al. (1959), who produced metastable H by electron impact on H and measured L_α emission as a function of the concentration of the added quenching H$_2$ gas. The measured rate coefficient was converted to a cross section $\sigma_{2s \to 2p}$, presumably by the common approximation of dividing by the mean relative speed. For a mean "atom speed" of 8×10^5 cm/sec, the interconversion cross section was first reported as 70 Å^2; however, an erratum later indicated that the cross section should be increased by 50%. The interpretation of this experiment is complicated by the need to konw the polarization of the L_α emission; the value originally assumed has now been corrected but revised quenching rate coefficients were not presented (Fite et al., 1968; Ott et al., 1970). A recent re-evaluation (Czuchlewski and Ryan, unpublished, quoted by Van Volkenburgh et al., 1973) yields $\sigma_{2s \to 2p} = 120$ Å^2.

The associative ionization process was studied in a mass spectrometer by Chupka et al. (1968). This process is about 1 eV exothermic. *Assuming* that the excited H atoms they produced by photodissiciation were in the $2s$ state, they obtained an associative ionization cross section of about 1 Å^2, which decreased by a factor of about 1.8 as the kinetic energy of H($2s$) was increased from 0.07 ± 0.10 eV to 0.18 ± 0.16 eV. If their excited H atoms were in the $2p$ state, then the above analysis does not apply. Associative ionization was also observed by Comes and Wenning (1969). The reverse of the associative ionization process is electron recombination with H$_3^+$; it is discussed in Section V.

The cross section $\sigma_{2s \to 2p}$ for $2s \to 2p$ interconversion and its ratio to the cross section $\sigma_{2s \to nr}$ for nonradiative destruction of the $2s$ state can also be measured by producing H($2s$) by photodissociation in a bulb, determinimg the fraction of H atoms formed in the $2s$ state, and measuring the L_α flourescence intensity and its ratio to the photodissociation cross section. (Note: the nonradiative process is generally assumed to be the $2s \to 1s$ process.) Since photodissociation at a given wavelength produces excited atoms of known kinetic energy, these cross sections may be obtained as a function of relative kinetic energy. This technique was first applied by Comes and Wenning (1969, 1969a, 1970), who found $\sigma_{2s \to 2p} \cong 60$ to 70 Å^2 and $\sigma_{2s \to nr} \cong 50$ Å^2 at a mean relative speed of 3.5×10^5 cm/sec. At the same mean relative speed, deuterium substitution for H and H$_2$ increased $\sigma_{2s \to 2p}$ by a factor of 2. As the mean relative speed increased to 8×10^5 cm/sec, $\sigma_{2s \to 2p}/\sigma_{2s \to nr}$ was found to increase to 2 and the total destruction cross section $\sigma_{2s \to 2p} + \sigma_{2s \to nr}$ increased to over 150 Å^2. Mentall and Gentieu (1970) made similar measurements and they found that $\sigma_{2s \to 2p}/\sigma_{2s \to nr}$ increased from about $2\frac{1}{2}$ to 3 as the mean relative speed increased over the same interval.

Dose and Hett also used photodissociation to produce H(2s) and D(2s) of known velocity. They measured the sum of the elastic scattering cross section and the total cross section for destruction by measuring the exponential decrease of H(2s) intensity in a flight-timed beam as a function of molecular H_2 target gas thickness. For metastable H, their measured cross section over the speed interval 3.5×10^5 to 8×10^5 cm/sec decreased from about 120 Å^2 to 85 Å^2. As the mean relative speed further increased to 3×10^6 cm/sec, their cross section decreased to 55 to 60 Å^2. Very similar results were obtained for metastable D. Especially at high speeds, these results are obviously inconsistent with the bulb experiments discussed in the previous paragraph. Similar but unpublished beam experiments by Czuchlewski and Ryan are quoted by Van Volkenburgh et al. (1973).

In principle, one can also measure cross sections for collisions of H(2p). In fact, measurements of the quenching of L_α radiation might be interpreted this way. But in fact (as illustrated by the large interconversion cross sections discussed above), small perturbations may easily mix the 2s and 2p states and L_α-quenching measurements have also been interpreted as referring to some mixture of $n = 2$ states. The first such quenching measurements involving H_2 as collision partner were carried out by Wauchop et al. (1969). They measured quenching of L_α fluorescence in an optically thick discharge-flow system with added H_2. Their total rate coefficient for all processes that quench H(2p) was 2.4×10^{-12} cm^3/(molec ·sec), from which they calculated a cross section of about 0.03 Å^2. This value is much lower than subsequent measurements, possibly because of the difficulty of the analysis involving radiation trapping in their optically thick system.

The collisional deactivation of L_α fluorescence was next studied by Braun et al. (1970) in an optically thin discharge-flow system. From their measured rate coefficient at 300°K they calculated the quenching cross section to be about 84 Å^2. Then Van Volkenburgh et al. (1973) studied the quenching of H($n = 2$) and D($n = 2$) by H_2 and D_2; they measured rate coefficients in the range 1.9 to 2.5×10^{-9} cm^3/(molec·sec) at 295°K for the four isotopically different processes. Dividing these by the average relative thermal speed they obtained the quenching cross sections 84 Å^2 for H($n = 2$) by H_2 or D_2, 89 Å^2 for D($n = 2$) by D_2, and 91 Å^2 for D($n = 2$) by H_2. These quenching cross sections include all processes that deactivate H($n = 2$) and D($n = 2$) including associative ionization, reactive electronic energy exchange, nonreactive transfer of the excitation energy to electronic, vibrational, or rotational degrees of freedom or to relative translational energy, and dissociative deexcitation. Using dc ion-collection techniques, the cross section for associative ionization was measured as 1.11 Å^2

for the H($n=2$) + H$_2$ case with value from 0.82 to 1.06 Å^2 for the three isotopically substituted cases. In addition, by observing simultaneous L_α fluorescence from D($n=2$) and H($n=2$), the cross section for near-resonant reactive electronic energy transfer in the reaction H($n=2$) + D$_2$→HD + D($n=2$) was found to be 0.28 Å^2. Branching ratios in other channels were not measured. However, the diatomic electronic excitation channel has been observed by Chow and Smith (1970), who saw fluorescence of the Lyman bands ($B\,^1\Sigma_u^+ \rightarrow X\,^1\Sigma_g^+$) of H$_2$, which they interpreted as caused by the reaction

$$H(2p) + H_2\left(X\,^1\Sigma_g^+, v''\right) \rightarrow H(1s) + H_2\left(B\,^1\Sigma_u^+, v'\right)$$

where v'' and v' are vibrational quantum numbers. Energetic considerations showed that $v'' \geqslant 2$ was required for this process to occur.

It is unfortunate that some experimenters measured rate coefficients but presented only derived cross sections and confused relative speed and atomic speed in presenting their results. This leads to some confusion concerning their experimental results.

Two semiclassical calculations have been made for the $2s \rightarrow 2p$ interconversion cross section involving H$_2$ collision partner. Gersten (1969) computed elements of the S matrix, $S_{2sjm \rightarrow 2pj'm'}$ (where jm and $j'm'$ are the initial and final H$_2$ angular momentum and magnetic quantum numbers) from approximate time-dependent perturbation theory. An instantaneous atomic dipole-molecular quadrupole interaction was assumed, with classical relative translational (straight-line trajectory) and quantum or classical molecular rotational motion assumed. The calculated probability of quenching diverged at small impact parameter and so it had to be cut off at unity. Using classical rotational motion, averaging over the impact parameter, and assuming a relative speed of 8×10^5 cm/sec, a cross section of 76 Å^2 was obtained. Further assuming a large moment of inertia so the molecule does not rotate during the collision yielded a simpler formula that predicts $\alpha_{2s \rightarrow 2p} = 87$ Å^2 at the same relative speed and a velocity dependence of $v^{-2/3}$. In a similar treatment, Slocumb et al. (1971) also assumed a dipole-quadrupole interaction. They used the Born approximation with semiclassical approximations to evaluate S-matrix elements and transition probabilities and a cutoff at small impact parameters. Their final cross-section formula is similar to the large-moment-of-inertia formula of Gersten, differing only in a numerical coefficient that is smaller by a factor of $3^{-1/3}$; thus they predicted $\alpha_{2s \rightarrow 2p} = 63$ Å^2 at the same relative speed. Dose and Hett (1971) used the treatment of Gersten, taking "correct" account of rotational motion, to calculate the cross section for D($2s$) + H$_2$ →D($2p$) + H$_2$ over the relative-speed interval 3.5×10^5 to 8.0×10^6 cm/sec.

The calculated values agreed very well with their experiment over this whole range.

Frenkel (1970) discussed the interconversion, the quenching, and the associative ionization processes in terms of his calculated excited-state potential surfaces.

Less quantitative theoretical work has been done by others on the L_α quenching process. Comes and Wenning (1970) argued that their measured D/H isotope effect could be explained by the smaller rotational spacing of D_2 compared to H_2 if the D_2 and H_2 molecules are rotationally excited in collisions with the excited atoms. Braun et al. (1970) argued that the large quenching cross sections might be explained in terms of a virtual electron jump from $H(n=2)$, which has an ionization potential of only 3.4 eV, to H_2. Chow and Smith (1970) interpreted the electronic energy-transfer process they observed in terms of an avoided crossing of two potential surfaces and an H_2^- intermediate. They predicted that an isotopic experiment would reveal the process occurs with atom exchange. Slocumb et al. (1971) claimed the $2s \rightarrow 1s$ deexcitation cross section should be less than 1 Å^2, but this is contradicted by the experiments already discussed. Van Volkenburgh et al. (1973) discussed the quenching results in terms of electronic correlation diagrams linking asymptotic electronics states to equilateral triangle geometries of H_3 and H_3^+. They concluded that quenching occurs by the following mechanism:

$$H(n=2) + H_2\left(X\,^1\Sigma_g^+\right) \rightarrow H(1s) + H_2\left(b\,^3\Sigma_u^+\right)$$

where the $b\,^3\Sigma_u^+$ is strongly repulsive and hence dissociative and that there is little likelihood of

$$H(n=2) + H_2\left(X\,^1\Sigma_g^+\right) \rightarrow H(1s) + H_2\left(X\,^1\Sigma_g^+, \text{vibrationally excited}\right)$$

They also concluded that the lowest-energy $^2A''$ potential surface provides a likely route for reactive electronic energy exchange and they rationalized associative ionization in terms of vibronic coupling between a Rydberg state and the electronic continuum of the ion.

D. Transport Properties of Partly Dissociated H_2

The temperature dependence of viscosity and diffusion coefficients for H-H_2 mixtures has provided information about the effective spherically averaged (orientation-independent) interaction potential $V_0(R)$ between H and H_2. The first measurements on transport coefficients involving H-H_2 were Harteck's (1928) results on the viscosity of H-H_2 mixtures. The viscosity of the mixture η_{mix} was determined from the Poiseuille equation following measurement of the pressure drop for H-H$_2$ mixtures flowing through a capillary tube. The relative viscosity (η_{mix}/η_{H_2}) was reported at -80, 0, and $100°C$, but the lowest-temperature results may have been affected by ice formation on the inner walls of the U-tube viscometer. Later, Browning and Fox (1964) measured viscosities of H-H$_2$ mixtures over the same temperature range. Their viscosity data are in good agreement with those of Harteck, except at the lowest temperature. In order to analyze the data, Browning and Fox partitioned the total viscosity into contributions from H-H interactions (η_1), H$_2$-H$_2$ interactions (η_2), and H-H$_2$ interactions (η_{12}). (The term η_{12} has no direct physical significance, but it can be imagined as the viscosity of a hypothetical pure substance whose mass is twice the reduced mass of an H-H$_2$ pair.) Values of the multicomponent viscosity coefficient η_{12} were extracted from the data by assuming a theoretical value for η_1. Information about $V_0(R)$ can be extracted from η_{12} by fitting the parameters in an assumed form of $V_0(R)$ [the Lennard-Jones (12-6) or modified Buckingham (exp-6) forms have usually been used] so that the theoretical η_{12} values agree well with the experimental values. Collision theory provides the result that $\eta_{12} \approx T^{1/2}/\Omega_{12}^{(2,2)*}$, where T is the temperature and $\Omega_{12}^{(2,2)*}$ is a (reduced) collision integral, which can be evaluated from the assumed form of $V_0(R)$ (Hirschfelder, Curtiss, and Bird, 1954, pp. 523 to 528). The experimental η_{12} values were fitted by Browning and Fox with a 12-6 potential but a three-parameter exp-6 potential could also be fitted to the data. Parameters in spherically averaged potentials deduced from transport data are listed in Table V.

Cheng and Blackshear (1972) have recently remeasured the viscosity of H-H$_2$ mixtures between -72 and $100°C$. They extracted η_{12} from the data on η_{mix} and found that the temperature dependence was fitted by a 12-6 potential. The previously measured values of η_{mix} provided by Browning and Fox are somewhat lower the Cheng and Blackshear values; the discrepancy may be due to violation of the constant-flow-rate assumption (for the mixture versus pure H$_2$) used by Browning and Fox.

In addition to the viscosity measurements, several determinations of the multicomponent H-H$_2$ diffusion coefficient $D_{12}(T)$, where the temperature-dependence is explicitly indicated, have been reported. Wise (1961) measured the steady-state distribution of H atoms diffusing along a cylinder toward a catalytically active surface where recombination occurred. The steady-state distribution results from a compromise between diffusion away from the source and removal on the surface. [The diffusion equations for this experimental arrangement had been previously considered by Wise and Ablow (1958). In addition, Wise (1959) had used the steady-state method to measure the diffusion coefficient of H through Ar/H$_2$ mixtures.

See also Wise and Wood (1961).] These measurements provided *relative* diffusion coefficients $D_{12}(T_2)/D_{12}(T_1)$. From the collision-theory expression $D_{12}(T) \cong T^{1/2}/\Omega_{12}^{(1,1)*}$, where $\Omega_{12}^{(1,1)*}$ is another reduced collision integral [which can be computed from the deflection function for the H-H$_2$ spherically averaged potential (Hirschfelder, Curtiss, and Bird, 1954, pp. 523 to 528).] Wise fitted the temperature dependence of the experimental diffusion coefficient ratios with a 12-6 potential (see Table V). In a later study, Sancier and Wise (1969) measured the absolute value of $D_{12}(T)$ at one temperature, 298°K. Two types of measurements (of the relative atom concentration as a function of the distance from the source under steady-state conditions and of the relative atom concentration as a function of time just after the source, which produced steady-state conditions, was removed) were combined to produce $D_{12}(T)$. The previous relative diffusion coefficients of Wise (1961) were then converted to absolute values (over the range 298 to 719°K). Parameters in an exp-6 potential were then adjusted so that the collision-theory prediction for $D_{12}(T)$ agreed with the experimental temperature dependence.

A second direct determination of $D_{12}(T)$ was reported by Khouw et al. (1969). They studied the lower temperature range 202 to 364°K. They too measured the concentration of H atoms in a flowing H-H$_2$ mixture that was in contact with a catalytic sink. The H-atom concentration was determined by introducing small amounts of NO into the stream and measuring the HNO emission. A 12-6 potential, when used to compute the $\Omega_{12}^{(1,1)*}$ collision integrals, produced $D_{12}(T)$ values that reproduced the experimental temperature dependence. However, the authors were careful to point out that potentials involving other parameter sets also reproduced the experimental temperature dependence.

All of the $D_{12}(T)$ coefficients that have been directly measured or inferred from viscosity data are plotted in Fig. 1. Notice that the recent direct measurements of Khouw et al. (1969) are in good agreement with the results inferred by Cheng and Blackshear (1972) from viscosity data. An earlier discussion of the use of viscosity data to predict diffusion coefficients is given by Dalgarno and Henry (1964).

In analyzing the experimental studies discussed above, parameters in a model spherically symmetric potential wrere determined such that the temperature dependence of the experimental viscosity or diffusion coefficients, as calculated from the potential and the collision integrals, was reproduced. The properties of the various derived potentials are compared in Table V. From this table it is apparent that these potentials are not in close agreement, but as a group they do provide approximate limits on the effective spherically averaged H-H$_2$ interaction.

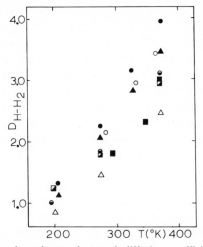

Fig. 1. Temperature dependence of mutual diffusion coefficient D_{H-H_2} (or D_{12}) in cm^2/sec at 1 atm from theory and experiment. △ Amdur (1936), from analysis of Harteck's (1928) viscosity data; ▨ Weissman and Mason (1962); computed from 12-6 potential fit to Margenau's (1944) calculations; ○ Browning and Fox (1964); computed from 12-6 potential fit to Margenau's (1944) calculations; ◓ Khouw, Morgan, and Schiff (1969), direct measurement; ■ Sancier and Wise (1969), direct measurement; ▲ Cheng and Blackshear (1972); computed from 12-6 potential inferred from viscosity data; ● Tang and Wei (1974); computed from Tang (1969) potential.

Other attempts have been made to extract potential parameters from the experimental transport data. The earliest was Amdur's (1936) reevaluation of Harteck's viscosity data. He assumed that H and H$_2$ are each "van der Waals gases" (hard elastic spheres with R^{-1} attraction), and then calculated D_{12} for the H-H$_2$ mixture. His results, shown in Fig. 1, are somewhat lower than the other D_{12} values. A second calculation of D_{12} from Harteck's data on η_{mix} was the study of Weissman and Mason (1962). They employed a theoretical calculation of η_1 from the best available H$_2$ potential curves, and also calculated the collision integral $\langle A_{12}^* \rangle$ (which is related to η_{12}) from 12-6 or exp-6 potentials fitted to Margenau's (1944) H$_3$ calculation. They then extracted D_{12} from Harteck's η_{mix} data. Their results, also shown in Fig. 1, are about 30% greater than Amdur's earlier estimates.

In a later theoretical study of the transport coefficients for H-H$_2$ mixtures, Clifton (1961) used a 12-6 potential to calculate high-temperature (1500 to 5000°K) values for η_{12} and D_{12} and for the coefficient of thermal conductivity (Clifton, 1962). The 12-6 potential used by Clifton was parameterized by fitting it to a weighted average of Margenau's (1944) early H_3 potential for parallel or perpendicular approach of H to H_2. Vanderslice et al. (1962) computed the transport properties (coefficients of viscosity, diffusion, thermal conductivity, and thermal diffusion) of dissociating H_2 as a function of the mole fraction in the mixture over the temperature range 1000 to 15,000°K. The H-H, H-H_2, and H_2-H_2 interactions were all considered, but ionization and electronic excitation were neglected. The spherically symmetric exponential repulsive potential of Vanderslice and Mason (1960) was used. At high temperatures, the collision integrals $\Omega_{12}^{(1,1)}$ and $\Omega_{12}^{(2,2)}$ for H-H_2 are smaller than the corresponding values for H-H, or H_2-H_2. The Vanderslice and Mason potential was also used by Estrup (1964) in a study of the cooling of hot T atoms in D_2.

Tang and Wei (1974) have recently computed D_{12} from what is probably a much more accurate spherically symmetric H-H_2 potential than was used in the previous studies. In earlier work, Tang (1969) smoothly joined the Porter-Karplus (1964) potential No. 2 to the Dalgarno-Henry-Roberts (1966) attractive long-range R^{-6} potential. (Rotationally-inelastic-scattering results obtained with this potential are discussed in Section IV.A.) Tang and Wei used a fit to this joined potential to evaluate the collision integrals that are required to compute η_{12} and D_{12} (see Fig. 1 for their D_{12} results). Their computed D_{12} values are somewhat higher than the experimental values of both Cheng and Blackshear (who extracted D_{12} from viscosity measurements) and of Khouw et al. (who directly determined D_{12}).

In an interesting study, Belov (1966) examined how the kinetic properties of dissociating H_2 are influenced by the $H + H_2$ exchange reaction at temperatures over 1000°K. At high temperature the differential reaction cross section for reactive scattering becomes forward peaked in scattering angle θ (see Truhlar and Wyatt, 1976). Belov therefore argued that the effective cross sections for diffusion and viscosity, which weight the differential cross section by $(1 - \cos^l\theta)$ (where l is 1 and 2 for diffusion or viscosity, respectively) before integration over $\sin\theta\,d\theta$, are much smaller than the diffusion and viscosity cross sections due to nonreactive scattering computed, for example, by Vanderslice et al. (1962). The reactive process would make a much greater contribution for a given integral cross section if the differential reaction cross section were isotropic. It is important to reexamine this question with more accurate estimates of the reactive differential cross sections.

Now that the potential energy surface for H_3 can be accurately calculated by ab initio electronic structure calculations and accurate calculations of rotational-vibrational transition probabilities and elastic scattering in atom-diatomic systems are possible, it is finally possible to test the assumptions of spherically symmetric potentials and no inelasticity that have been used to calculate transport coefficients in H-H$_2$ mixtures. In an early approximate quantum-mechanical treatment, Dalgarno and Henry (1964) estimated that rotationally inelastic processes contributed 7.5% of the H + D$_2$ diffusion cross section and 5.4% of the viscosity cross section at a relative translational energy of 0.0625 eV. More accurate calculations could now be made by integrating the more accurate differential cross sections now available (see Sections III and IV.A). It is now possible to compute accurate elastic scattering and rotational excitation differential cross sections, to integrate them to obtain transport cross sections, and to make a thermal average to obtain viscosities and diffusion coefficients as functions of temperature. This has not been done.

E. Other Processes and Applications

Rotational excitation of H$_2$ by collisions with H followed by emission of radiation has been considered as a possible mechanism for cooling of interstellar clouds in regions of neutral hydrogen, but the most recent work indicates it is not the most important mechanism for such cooling (Hollenbach et al., 1976). Takayanagi and Nishimura (1960) and Dalgarno et al. (1966) calculated rate coefficients for rotational excitation in H-H$_2$ collisions from their approximate quantum-mechanical cross sections. Their cross sections agree within a factor of 2.2, but it now appears that the DHR interaction potential used by Dalgarno et al. (1966) leads to rotational excitation cross sections that are much too large (see the work of Choi and Tang, 1975, discussed in Section IV.A). Hydrogen molecules formed on grains in interstellar space may leave with excess kinetic energy greater than the thermal average of the surrounding gas. To understand the loss of kinetic energy it is also necessary to know the differential cross sections for nonreactive collisions, especially elastic scattering (Chu and Dalgarno, 1975).

The rate of interconversion of o-H$_2$ and p-H$_2$ is very important in interstellar space because of the different radiative rates for these species (Takayanagi and Nishimura, 1960; Field, 1966) and observations of the ratios of o-H$_2$ to p-H$_2$ are important because their interpretation provides a clue as to the physical conditions in interstellar space. The mechanisms for this interconversion have been discussed most recently by Dalgarno et al. (1973), who conclude that the H + H$_2$ reaction, involving thermal or hot atoms, is probably not too important for this interconversion in interstellar space.

The pressure-induced vibrational absorption coefficient of H$_2$ is needed for various radiant heat-transfer calculations. At high temperatures, such as encountered in gaseous-core nuclear rockets and late-type stars, there is appreciable dissociation of H_2, and $H + H_2$ collisions become almost as important as $H_2 + H_2$ collisions. The only theoretical or experimental work on absorption due to $H + H_2$ collisions is the work of Patch (1973, 1974) in the region of the fundamental vibrational absorption. The collision-induced dipole moment was found by CCI calculations with a minimum-basis set for small H-to-H_2 separations of 1 to $4a_0$ and from the quadrupole-induced dipole in H at separations greater than $6a_0$. The ab initio calculations were performed with floating orbitals and with two sets of exponents for nuclear-centered orbitals. The latter choice with optimized exponents was judged best and yielded dipole moments in the range 0.0062 to 1.0 D. Unfortunately, the energies were not very accurate, so the induced absorption calculations were based on an interaction potential computed from the calculations of Mason and Hirschfelder (1957) for H-to-H_2 separations greater than $4.33a_0$ and from Porter-Karplus surface No. 2 for smaller separations. The $H + H_2$ collisions accounted for 32% of the absorption calculated, including both $H + H_2$ and $H_2 + H_2$ collisions at 5500 cm^{-1}, 3750°K, and 1 atm pressure.

The simplest inelastic event in $H + H_2$ collisions is change of the hyperfine state of the H atom. This process has been measured (Gordon et al. 1975), but since it gives information about the reaction probability it should have been included in our review of reactive collisions.

V. RECOMBINATION OF H_3^+ WITH ELECTRONS

Several beam studies aimed at producing H_3 by recombination of H_3^+ with electrons have been carried out.

The electron-ion recombination rate $\alpha(T)$ for e-H_3^+ collisions was measured by Leu, Biondi, and Johnson (1973) by using a microwave afterglow technique to measure the time rate of decay of electron density in a plasma containing H_3^+. The recombination rate coefficient decreased from 2.9×10^{-7} cm^3/(molec·sec) at 205°K to 2.0×10^{-7} cm^3/(molec·sec) at 450°K and varied as $T^{-1/2}$ over this temperature range. In an incline-beam experiment, Peart and Dolder (1974) measured cross sections σ for dissociative recombination of electrons with H_3^+ with negligible vibrational energy over the relative translational energy (E_{rel}) range 0.38 to 4.0 eV. The cross section varied from 23.8 Å^2 at 0.38 eV to 2.7 Å^2 at 4.0 eV with an $E_{rel}^{-0.87}$ energy dependence. Peart and Dodler then converted the recombination rate coefficients of Leu et al. to cross sections with the usual approximate relation $\alpha(T) = \langle v \rangle_T \sigma$, where $\langle v \rangle_T$ is the mean speed of the thermal beam (at temperature T). The cross sections obtained in this way from the Leu et al. data at low relative translational energies ($E_{rel} \lesssim 0.1$ eV) are about 200Å^2 and are in excellent agreement with cross sections extrapolated from the higher-energy beam measurements.

Caudano et al. (1975) used merged electron-ion beam techniques to measure the relative recombination cross sections for 0.05 eV $\leqslant E_{rel} \leqslant 4$ eV with better initial energy resolution than Peart and Dodler. Their results are consistent with Peart and Dodler's but show structure in the energy dependence that was attributed to "the resonant nature of the recombination process."

The first direct attempts to generate H$_3$ were the merging-beam experiments of Devienne (1967, 1968, 1968a) in which an H$_3^+$ beam was passed through H$_2$ to generate a neutral H$_3$ beam by charge exchange. The neutral beam was then merged with an He$^+$ beam to reform H$_3^+$, which was then detected. In a later double-chamber experiment, Devienne (1969) first passed an H$_3^+$ beam through D$_2$ in a charge-exchange chamber. The ions were deflected, and the neutral beam (presumed to be H$_3$) was passed into a collision chamber containing D$_2$. Mass-spectrometric analysis of ions formed in the collision chamber showed that H$_3^+$, H$_2^+$, and H$^+$ were all produced from ionization of the neutral beam that entered the second chamber. In a different set of experiments, Gray and Tomlinson (1969) passed an He/H$_2$ beam through an rf discharge. The resulting ion beam was directed through a charge-exchange region containing H$_2$ and D$_2$ and the ions were deflected from the beam. A fraction of the remaining neutral beam was reionized, but mass analysis did not show evidence for H$_3$ (or isotopic variants) with lifetimes the order of 10^{-8} sec (the transit time of the beam past electrostatic deflector plates that remove the ions). In more recent experiments, Barnett and Ray (1972) passed a beam of H$_3^+$ through H$_2$ to form H$_3$. The outer electron was stripped off in an intense electric field and the H$_3^+$ was detected. The H$_3$ "molecules" were thought to consist of a stable H$_3^+$ core and a highly excited Rydberg electron (principal quantum number $n \geqslant 11$). Additional experiments on the production and properties of excited H$_3$ would be very interesting.

VI. CONCLUDING REMARKS

The ground-state potential-energy surface of H$_3$ has now been calculated accurately for many geometries, especially in the vicinity of the barrier for reaction. The C_6 and Γ coefficients of the long-range H-H$_2$ interaction are also well known from theory (see Table IV), but the topology of the van der Waals well is not accurately known (see Table V). Recent scattering experiments (Gengenbach et al., 1975) should motivate more theoretical attention to this problem. The excited states are much more poorly understood, at least as far as quantitative results are concerned. The states correlating with $H(n=2) + H_2(X\,^1\Sigma_g^+)$ are particularly important in fluorescence-quenching experiments. The long-range interactions in excited states of H$_3$ may also be important for the interpretation of experiments. It is difficult to deduce an accurate interaction potential from experimental cross sections and rate coefficients for elastic scattering and the various energy-transfer processes. However, it is now possible to make realistic calculations for some of these processes and to use theory to understand the qualitative features of others. Nevertheless, the effect of the nonsphericity of the interaction potential on the elastic scattering and transport properties is not well studied, vibrational-rotational energy transfer in $H + H_2$ is poorly understood compared to the same processes in inert-gas collisions with homonuclear diatomics, electronic-energy transfer is poorly understood compared to analogous processes involving alkali atoms, the theory of associative ionization requires more work, and the transport data have not yet been interpreted using the most accurate available potentials. Experimental studies of rotational and vibrational energy transfer in $H + H_2$ collisions would be particularly valuable because reasonably accurate ab initio calculation of cross sections for simultaneous vibrational-rotational energy transfer in both reactive and nonreactive H-H_2 collisions should be possible in the near future. More detailed experimental studies of other energy-transfer processes and of associative ionization would also be valuable. In particular, the energy dependence of the branching ratios that determine the outcome in $H(n = 2) + H_2$ collisions needs further study.

In some respects, $H + H_2$ provides an important test case for studying inelastic processes in chemically reactive systems because the relatively high (compared to other atom-molecule atom-transfer reactions) energy barrier to reaction in either direction in the ground electronic state means that under many low-energy conditions reaction is negligible. However, the relevant interactions are still more representative of a reactive system than, for example, an inert-gas collision with a molecule. For this reason we hope the $H + H_2$ inelastic collision processes will be the subject of continued study in the near future.

ACKNOWLEDGEMENTS

The authors are grateful to Dr. S.-I. Chu, Dr. R. J. Cross, Jr., Dr. B. Liu, Dr. G. C. Shatz, and Dr. I. Shavitt for comments on various sections of the manuscript, to Dr. Liu for providing details of Per Siegbahn's and his unpublished calculations, and to Drs. Schatz, A. Kuppermann, and J. P. Toennies for also providing their results before publication. This manuscript was completed while one of the authors (D. G. T.) was a Visiting Fellow at the Joint Institute for Laboratory Astrophysics; he is grateful to the members of the JILA Scientific Reports Office (Leslie L. Haas, Gwendy L. Romey, and Lorraine H. Volsky) and to Olivia C. Briggs for their excellent typing and editorial and secretarial assistance during his stay and to all the members of JILA for their gracious hospitality.

References

Allison, A. C., and A. Dalgarno. 1967. *Proc. Phys. Soc. London,* **90,** 609.

Amdur, I., 1936. *J. Chem. Phys.,* **4,** 339.

Amdur, I., 1943. *J. Chem. Phys.,* **11,** 143.

Amdur, I., 1949. *J. Chem. Phys.,* **17,** 844.

Amdur, I., D. E. Davenport, and M. C. Kells. 1950. *J. Chem. Phys.,* **18,** 1676.

Amdur, I., and H. Pearlman. 1940. *J. Chem. Phys.,* **8,** 7.

Amdur, I., and A. L. Smith. 1968. *J. Chem. Phys.,* **48,** 568.

Anderson, A. B., 1973. *Chem. Phys. Lett.,* **18,** 303.

Aroeste, H., 1964. *Advan. Chem. Phys.,* **6,** 1.

Aroeste, H., and W. J. Jameson, Jr. 1959. *J. Chem. Phys.,* **30,** 372.

Bacskay, G. B., and J. W. Linnett. 1972. *Theor. Chim. Acta,* **26,** 15.

Bacskay, G. B., and J. W. Linnett. 1972a. *Theor. Chim. Acta,* **26,** 23.

Barker, R. S., and H. Eyring. 1954. *J. Chem. Phys.,* **22,** 1182.

Barker, R. S., H. Eyring, C. J. Thorne, and D. A. Baker. 1954. *J. Chem. Phys.,* **22,** 699.

Barnett, C. F., and J. A. Ray. 1972. *Phys. Rev. A,* **5,** 2120.

Baskin, C. P., C. F. Bender, C. W. Bauschlicher, Jr., and H. F. Schaefer III. 1974. *J. Am. Chem. Soc.,* **96,** 2709.

Bauer, E., 1951. *Phys. Rev.,* **84,** 315.

Bauer, E., 1952. *Phys. Rev.,* **85,** 277.

Bauer, W., R. W. Bickes, Jr., B. Lantzsch, J. P. Toennies, and K. Walaschewski. 1975. *Chem. Phys. Lett.,* **31,** 12.

Belov, V. A., 1966. *Teplofiz. Vysok. Temp.,* **4,** 625 [English transl.: *High Temp.,* **4,** 588 (1966)].

Belyaev, Yu. N., and V. B. Leonas. 1967. *Dokl. Akad. Nauk SSR,* **173,** 306 [English transl.: *Sov. Phys.—Dokl.,* **12,** 233 (1967)].

Blustin, P. H., and J. W. Linnett. 1974. *Chem. Soc. Faraday II,* **70,** 327.

Bowen, H. C., and J. W. Linnett. 1966. *Trans. Faraday Soc.,* **62,** 2953.

Bowman, J. M., and A. Kuppermann. 1973. *J. Chem. Phys.,* **59,** 6524.

Boys, S. F., G. B. Cook, C. M. Reeves, and I. Shavitt. 1956. *Nature,* **178,** 1207.

Boys, S. F., and I. Shavitt. 1959. Univ. of Wis. NRL Tech. Rep. WIS-AF-13.

Bradley, J. N., 1964. *Trans. Faraday Soc.,* **60,** 1353.

Bradley, J. N., 1966. *Chem. Commun.,* **1966,** 175.

Braun, W., C. Carlone, T. Carrington, G. Van Volkenburgh, and R. A. Young. 1970. *J. Chem. Phys.,* **53,** 4244.

Browning, R., and J. W. Fox. 1964. *Proc. Roy. Soc. London Ser. A,* **278,** 274.

Cashion, J. K., and D. R. Herschbach. 1964. *J. Chem. Phys.,* **40,** 2358.

Cashion, J. K., and D. R. Herschbach. 1964a. *J. Chem. Phys.,* **41,** 2199.

Casimir, H. B. G., and D. Polder. 1948. *Phys. Rev.,* **73,** 360.

Caudano, R., S. F. J. Wilk, and J. W. McGowan. 1975. ICPEAC, 9th, Abstracts Papers, p. 389.

Cheng, D. Y., and P. L. Blackshear. 1972. *J. Chem. Phys.,* **56,** 213.

Choi, B. H., and K. T. Tang. 1975. *J. Chem. Phys.,* **63,** 1783.

Choi, B. H., and K. T. Tang. 1975a. *J. Chem. Phys.,* **67,** 2854.

Choi, B. H., and K. T. Tang. 1976. *J. Chem. Phys.,* **64,** 942.

Chow, K. W., and A. L. Smith. 1970. *Chem. Phys. Lett.,* **7,** 127.

Chu, S. I., and A. Dalgarno. 1975. *Astrophys. J.,* **199,** 637.

Chupka, W. A., M. E. Russell, and K. Rafaey. 1968. *J. Chem. Phys.,* **48,** 1518.

Clark, A. P., and A. S. Dickinson. 1973. *J. Phys. B,* **6,** 164.

Clifton, D. G., 1961. *J. Chem. Phys.,* **35**, 1417.

Clifton, D. G., 1962. *J. Chem. Phys.,* **36**, 472.

Comes, F. J., and U. Wenning. 1969. *Z. Naturforsch.,* **24a**, 587.

Comes, F. J., and U. Wenning. 1969a. *Ber. Bunsenges. Phys. Chem.,* **73**, 901.

Comes, F. J., and U. Wenning. 1970. *Chem. Phys. Lett.,* **5**, 199.

Conroy, H., 1964. *J. Chem. Phys.,* **41**, 1327, 1331, 1336, 1341.

Conroy, H., 1970. In C. Schlier, Ed., *Molecular Beams and Reaction Kinetics,* Academic Press, New York, p. 349.

Conroy, H., and B. L. Bruner. 1965. *J. Chem. Phys.,* **42**, 4047.

Conroy, H., and B. L. Bruner. 1967. *J. Chem. Phys.,* **47**, 921.

Considine, J. P., and E. F. Hayes. 1967. *J. Chem. Phys.,* **46**, 1119.

Coolidge, A. S., and H. M. James. 1934. *J. Chem. Phys.,* **2**, 811.

Coulson, C. A., 1935. *Proc. Camb. Phil. Soc.,* **31**, 244.

Coulson, C. A., 1937. *Trans. Faraday Soc.,* **33**, 1479.

Dalgarno, A., 1963. *Rev. Mod. Phys.,* **35**, 522.

Dalgarno, A., J. H. Black, and J. C. Weisheit. 1973. *Astrophys. Lett.,* **14**, 77.

Dalgarno, A., and R. J. W. Henry. 1964. In M. R. C. McDowell, Ed., *Atomic Collision Processes,* North-Holland, Amsterdam, p. 914.

Dalgarno, A., R. J. W. Henry and C. S. Roberts. 1966. *Proc. Phys. Soc. London,* **88**, 611.

Dalgarno, A., and N. Lynn. 1965. *Proc. Roy. Soc. London Ser. A,* **69**, 821.

Dalgarno, A., and D. A. Williams. 1965. *Proc. Phys. Soc. London,* **85**, 679.

Devienne, F. M., 1967. *Compt. Rend.,* **264**, 1400.

Devienne, F. M., 1968. *Compt. Rend.,* **267**, 1279.

Devienne, F. M., 1968a. *Entropie,* **24**, 35.

Devienne, F. M., 1969. ICPEAC, 6th, Abstracts Papers, p. 789.

Dose, V., and W. Hett. 1971. *J. Phys. B,* **4**, L83.

Dose, V., and W. Hett, 1973. *J. Phys. B,* **7**, L454.

Edmiston, C., and M. Krauss. 1965. *J. Chem. Phys.,* **42**, 1119.

Edmiston, C., and M. Krauss. 1968. *J. Chem. Phys.,* **49**, 192.

Ellison, F. O., 1963. *J. Am. Chem. Soc.,* **85**, 3540.

Ellison, F. O., 1964. *J. Chem. Phys.,* **41**, 2198.

Ellison, F. O., N. T. Huff, and J. C. Patel. 1963. *J. Am. Chem. Soc.,* **85**, 3544.

Estrup, P. J., 1964. *J. Chem. Phys.,* **41**, 164.

Eyring, H., 1931. *J. Am. Chem. Soc.,* **53**, 2537.

Eyring, H., 1932. *Chem. Rev.,* **10**, 103.

Eyring, H., 1962. *Chem. Soc. Spec. Publ.,* **16**, 76.

Eyring, H., and E. M. Eyring, 1963. *Modern Chemical Kinetics,* Reinhold, New York.

Eyring, H., H. Gershinowitz, and C. E. Sun. 1935. *J. Chem. Phys.,* **3**, 786.

Eyring, H., and M. Polanyi. 1930. *Naturwiss.,* **18**, 914.

Eyring, H., and M. Polanyi. 1931. *Z. Physik. Chem.,* **B12**, 279.

Field, G. B., 1966. *Ann. Rev. Astron. Astrophys.,* **4**, 215.

Fite, W. L., and R. T. Brackmann. 1964. In M. R. C. McDowell, Ed., *Atomic Collision Processes,* North-Holland, Amsterdam, p. 955.

Fite, W. L., and R. T. Brackmann. 1965. *J. Chem. Phys.,* **42**, 4057.

Fite, W. L., R. T. Brackmann, D. G. Hummer, and R. F. Stebbings. 1959. *Phys. Rev.,* **116**, 363; erratum, **124**, 2051 (1961).

Fite, W. L., W. E. Kauppila, and W. R. Ott. 1968. *Phys. Rev. Lett.,* **20**, 409.

Flannery, M. R., and K. J. McCann. 1975. *J. Phys. B,* **8**, L387.

Fluendy, M. A. D., R. M. Martin, E. E. Muschlitz, Jr., and D. R. Herschbach. 1967. *J. Chem. Phys.,* **46**, 2172.

Frenkel, E., 1970. *Z. Naturforsch. A,* **25**, 1265.

Frost, A. A., 1967. *J. Chem. Phys.,* **47**, 3714.

Geddes, J., H. F. Krause, and W. L. Fite. 1972. *J. Chem. Phys.,* **56**, 3298; erratum, **59**, 566. (1973).

Gengenbach, R., Ch. Hahn, and J. P. Toennies. 1975. *J. Chem. Phys.,* **62**, 3620.

Gersten, J. I., 1969, *J. Chem. Phys.,* **51**, 637.

Gianinetti, A., G. F. Majorino, E. Rusconi, and M. Simonetta. 1969. *Int. J. Quantum Chem.,* **3**, 45.

Gimarc, B. M., 1970. *J. Chem. Phys.,* **53**, 1623.

Glasstone, S., K. J. Laidler, and H. Eyring. 1941. *The Theory of Rate Processes,* McGraw-Hill, New York.

Goodard, W. A., III. 1972. *J. Am. Chem. Soc.,* **94**, 793.

Goddard, W. A., III and R. C. Ladner. 1969. *Int. J. Quantum Chem., Symp.,* **3**, 63.

Goddard, W. A., III and R. C. Ladner. 1971. *J. Am. Chem. Soc.,* **93**, 6750.

Gordon, E. B., B. I. Ivanov, A. P. Perminov, E. S. Medvedev, A. N. Ponomarev, and V. L. Tal'roze. 1975. *Chem. Phys.,* **8**, 147.

Gray, J., and R. H. Tomlinson. 1969. *Chem. Phys. Lett.,* **4**, 251.

Griffing, V., J. L. Jackson, and B. J. Ransil, 1959. *J. Chem. Phys.,* **30**, 1066.

Griffing, V., and J. T. Vanderslice. 1955. *J. Chem. Phys.,* **23**, 1039.

Harris, R. E., D. A. Micha, and H. A. Pohl. 1965. *Arkiv Fysik,* **30**, 259.

Harrison, H., 1962. *J. Chem. Phys.,* **37**, 1164.

Harteck, P., 1928. *Z. physik. Chem.,* **A139**, 98.

Hayes, E. F., and R. G. Parr. 1967. *J. Chem. Phys.,* **47**, 3961.

Hayes, E. F., C. A. Wells, and D. J. Kouri. 1971. *Phys. Rev. A,* **4**, 1017.

Heidner, R. F., III and J. V. V. Kasper. 1972. *Chem. Phys. Lett.,* **15**, 179.

Hirschfelder, J. O., 1938. *J. Chem. Phys.,* **6**, 795.

Hirschfelder, J. O., 1941. *J. Chem. Phys.,* **9**, 645.

Hirschfelder, J. O., 1966. *J. Chem. Educ.,* **43**, 457.

Hirschfelder, J. O., C. F. Curtiss, and R. B. Bird. 1954. *Molecular Theory of Gases and Liquids,* Wiley, New York.

Hirschfelder, J. O., C. F. Curtiss, and R. B. Bird. 1964. *Molecular Theory of Gases and Liquids,* Wiley, New York, second printing.

Hirschfelder, J. O., H. Diamond, and H. Eyring. 1937. *J. Chem. Phys.,* **5**, 695.

Hirschfelder, J. O., H. Eyring, and N. Rosen. 1936. *J. Chem. Phys.,* **4**, 121.

Hirschfelder, J. O., H. Eyring, and B. Topley. 1936. *J. Chem. Phys.,* **4**, 170.

Hirschfelder, J. O., and C. N. Weygandt. 1938. *J. Chem. Phys.,* **6**, 806.

Hoffmann, R., 1968. *J. Chem. Phys.,* **49**, 3739.

Hollenbach, D., S. -I. Chu, and R. McCray. 1976. *Astrophys. J.,* **208**, 458.

Hoyland, J. R., 1964. *J. Chem. Phys.,* **41**, 1370.

Ioup, J. W., and A. Russek. 1973. *Phys. Rev. A,* **8**, 2848.

Jackson, J. L., and R. E. Wyatt. 1973. *Chem. Phys. Lett.,* **18**, 161.

Jameson, W. J., Jr. and H. Aroeste. 1960. *J. Chem. Phys.,* **32**, 374.

Johnston, H. S., 1960. *Advan. Chem. Phys.,* **3**, 131.

Johnston, H. S., 1966. *Gas Phase Reaction Rate Theory,* Ronald Press, New York.

Johnston, H. S., and C. A. Parr. 1963. *J. Am. Chem. Soc.,* **85**, 2544.

Jones, A., and J. L. J. Rosenfeld. 1973. *Proc. Roy. Soc. London Ser. A,* **333**, 419.

Jordan, J. E., and I. Amdur. 1967. *J. Chem. Phys.,* **46**, 105.

Karplus, M., 1968. *Discus. Faraday Soc.,* **44**, 168.

Karplus, M., 1970. In C. Schlier, Ed., *Molecular Beams and Reaction Kinetics,* Academic Press, New York, p. 320.

Karplus, M., and H. J. Kolker. 1964. *J. Chem. Phys.,* **41**, 3955.

Kassel, L. S., 1932. *Kinetics of Homogeneous Gas Reactions,* Chemical Catalog Co., New York.

Khouw, B., J. E. Morgan, and H. I. Schiff. 1969. *J. Chem. Phys.,* **50**, 66.

Kimball, G. E., and J. G. Trulio. 1958. *J. Chem. Phys.,* **28**, 493.

Kolos, W., and L. Wolniewicz. 1965. *J. Chem. Phys.,* **43**, 2429.

Krauss, M., 1964 *J. Res. Natl. Bur. Std.,* **68A**, 635.

Krauss, M., 1970. *Ann. Rev. Phys. Chem.,* **21**, 39.

Kung, R. T. V., and J. B. Anderson. 1974. *J. Chem. Phys.,* **60**, 3731.

Kuppermann, A., 1975. *Chem. Phys. Lett.,* **32**, 374.

Kuppermann, A., G. C. Schatz, and M. Baer. 1976. *J. Chem. Phys.* (to be published).

Kuppermann, A., and J. M. White. 1966. *J. Chem. Phys.,* **44**, 4352.

Ladner, R. C., and W. A. Goddard III. 1969. *J. Chem. Phys.,* **51**, 1073.

Laidler, K. J., and J. C. Polanyi, 1965. *Progr. Reaction Kinetics,* **3**, 1.

Langhoff, P. W., R. G. Gordon, and M. Karplus. 1971. *J. Chem. Phys.,* **55**, 2126.

Langhoff, P. W., and M. Karplus. 1970. *J. Chem. Phys.,* **53**, 233.

LeRoy, R. J., 1969. *J. Phys. Chem.,* **73**, 4338.

LeRoy, D. J., B. A. Ridley, and K. A. Quickert. 1968. *Discus. Faraday Soc.,* **44**, 92.

Leu, M. T., M. A. Biondi, and R. Johnson. 1973. *Phys. Rev. A,* **8**, 413.

Levine, R. D., R. B. Bernstein, P. Kahana, I. Procaccia, and E. T. Upchurch. 1976. *J. Chem. Phys.,* **64**, 796.

Ling, R. T., and A. Kuppermann. 1975. ICPEAC, 9th, Abstracts Papers, p. 353.

Linnett, J. W., and A. Riera. 1969. *Theor. Chim. Acta,* **15**, 196.

Lipponcott, E. R., and A. Liefer. 1958. *J. Chem. Phys.,* **28**, 769.

Liu, B., 1971. *Int. J. Quantum Chem. Symp.,* **5**, 123.

Liu, B., 1973. *J. Chem. Phys.,* **58**, 1925.

London, F., 1929. *Z. Elektrochem.,* **35**, 552.

Malcome-Lawes, D. J., 1975. *J. Chem. Soc. Faraday II,* **71**, 1183.

Margenau H., 1938. *J. Chem. Phys.,* **6**, 896.

Marganau, H., 1943. *Phys. Rev.,* **63**, 131.

Margenau, H., 1944. *Phys. Rev.,* **66**, 303.

Mason, E. A., and J. O. Hirschfelder. 1957. *J. Chem. Phys.,* **26**, 756.

Mason, E. A., and J. T. Vanderslice. 1958. *J. Chem. Phys.,* **28**, 1070.

Matsen, F. A., 1964. *J. Phys. Chem.,* **68**, 3282.

McCann, K. J., and M. R. Flannery. 1975. *Chem. Phys. Lett.,* **35**, 124.

McCann, K. J., and M. R. Flannery. 1975a. *J. Chem. Phys.,* **63**, 4695.

McCullough, E. A., Jr. and D. M. Silver. 1975. *J. Chem. Phys.,* **62**, 4050.

McGuire, P., and D. J. Kouri. 1974. *J. Chem. Phys.,* **60**, 2488.

McGuire, P., and H. Kruger. 1975. *J. Chem. Phys.,* **63**, 1090.

Meador, W. E., Jr. 1958. *J. Chem. Phys.,* **29**, 1339.

Meinke, C., and M. Reich. 1962. *Vak.-Tech.,* **11**, 86.

Mentall, J. E., and E. P. Gentieu. 1970. *J. Chem. Phys.,* **52**, 5641.

Menzinger, M., and R. Wolfgang. 1969. *Agnew. Chem. Int. Ed.,* **8**, 438.

Micha, D. A., 1969. *J. Chem. Phys.,* **50**, 722.

Michels, H. H., and F. E. Harris. 1968. *J. Chem. Phys.,* **48**, 2371.

Mulliken, R. S., 1949. *J. chim. Phys.,* **46**, 500.

Norbeck, J. M., and P. R. Certain. 1975. *J. Chem. Phys.,* **63**, 4127.

Norbeck, J. M., P. R. Certain, and K. T. Tang. 1975. *J. Chem. Phys.,* **63**, 590.

Oleari, L., S. Carra, and M. Simonetta. 1961. *Gazz. Chim. Ital.,* **91**, 1413.

Orlikowski, T., and L. Wolniewicz. 1974. *Chem Phys. Lett.,* **24**, 461.

Ott, W. R., W. E. Kauppila, and W. L. Fite. 1970. *Phys. Rev. A,* **1**, 1089.

Pack, R. T., 1974. *J. Chem. Phys.*, **60**, 633.
Parr, C. A., and D. G. Truhlar. 1971. *J. Phys. Chem.*, **75**, 1844.
Patch, R. W., 1973. *J. Chem. Phys.*, **59**, 6468.
Patch, R. W., 1974. *J. Quant. Spectrosc. Radiat. Transfer*, **14**, 101.
Pearson, R. G., 1948. *J. Chem. Phys.*, **16**, 502.
Peart, B., and K. T. Dolder. 1974. *J. Phys. B*, **7**, 1948.
Pedersen, L., and R. N. Porter. 1967. *Chem. Phys.*, **47**, 4751.
Pickup, B. T., 1973. *Proc. Roy. Soc. London Ser. A*, **333**, 69.
Porter, R. N., and M. Karplus. 1964. *J. Chem. Phys.*, **40**, 1105.
Porter, R. N., R. M. Stevens and M. Karplus. 1968. *J. Chem. Phys.*, **49**, 5163.
Rabitz, H., 1972. *J. Chem. Phys.*, **57**, 1718.
Ransil, B. J., 1957. *J. Chem. Phys.*, **26**, 971.
Riera, A., and J. W. Linnett. 1969. *Theor. Chim. Acta*, **15**, 181.
Riera, A., and J. W. Linnett. 1970. *Theor. Chim Acta*, **18**, 265.
Rourke, T. A., and E. T. Stewart, 1968. *Can. J. Phys.*, **46**, 1603.
Russell, J. D., and J. C. Light. 1971. *J. Chem. Phys.*, **54**, 4881.
Salomon, M., 1969. *J. Chem. Phys.*, **51**, 2406.
Sancier, K. M., and H. Wise. 1969. *J. Chem. Phys.*, **51**, 1434.
Sato, S., 1955. *Bull. Chem. Soc. Japan*, **28**, 450.
Sato, S., 1955a. *J. Chem. Phys.*, **23**, 592.
Sato, S., 1955b. *J. Chem. Phys.*, **23**, 2465.
Saxon, R. P., and J. C. Light. 1972. *J. Chem. Phys.*, **56**, 3885.
Schaefer, H. F., III. 1972. *The Electronic Structure of Atoms and Molecules*, Addison-Wesley, Reading, Massachusetts.
Schatz, G. C., 1975. Ph. D. thesis, California Institute of Technology, Pasadena.
Schatz, G. C., and A. Kuppermann, 1975a. *Phys. Rev. Lett.*, **35**, 1266.
Schatz, G. C., and A. Kuppermann. 1976. *J. Chem. Phys.*, **65**, 4668.
Schatz, G. C., and A. Kuppermann. 1976a. *J. Chem. Phys.*, **65**, 4624.
Schulz, W. R., and D. J. LeRoy. 1965. *J. Chem. Phys.*, **42**, 3869.
Schwartz, M. E., and L. J. Schaad. 1968. *J. Chem. Phys.*, **48**, 4709.
Secrest, D., 1973. *Ann. Rev. Phys. Chem.*, **24**, 379.
Shavitt, I., 1968. *J. Chem. Phys.*, **49**, 4048.
Shavitt, I., T. M. Stevens, F. L. Minn, and M. Karplus. 1968. *J. Chem. Phys.*, **48**, 2700; erratum, **49**, 4048 (1968).
Shui, V. H., 1973. *J. Chem. Phys.*, **58**, 4868.
Shui, V. H., and J. P. Appleton. 1971. *J. Chem. Phys.*, **55**, 3126.
Silver, D. M., 1972. *J. Chem. Phys.*, **57**, 586.
Slater, J. C., 1931. *Phys. Rev.*, **38**, 1105.
Slocumb, A., W. H. Miller, and H. F. Schaefer III. 1971. *J. Chem. Phys.*, **55**, 926.
Smirnov, B. M., 1964. *Zh. Eksp. Teor. Fiz.*, **46**, 578 [English transl.: *Sov. Phys.—JETP*, **19**, 394 1964].
Smith, I. W. M., and P. M. Wood. 1973. *Mol. Phys.*, **25**, 441.
Snow, R., and H. Eyring. 1957. *J. Phys. Chem.*, **61**, 1.
Steiner, E., P. R. Certain, and P. J. Kuntz. 1973. *J. Chem. Phys.*, **59**, 47.
Stevenson, D., and J. O. Hirschfelder, 1937. *J. Chem. Phys.*, **5**, 933.
Stwalley, W. C., A. Niehaus, and D. R. Herschbach. 1969, *J. Chem. Phys.*, **51**, 2287.
Sugiura, Y., 1927. *Z. Phys.*, **45**, 484.
Takayanagi, K., 1957. *Proc. Phys. Soc. London*, **A70**, 348.
Takayanagi, K., 1973. *Comments Atom. Mol. Phys.*, **4**, 59.
Takayanagi, K., and S. Nishimura. 1960. *Publ. Astron. Soc. Japan*, **12**, 77.

Tang, K. T., 1969. *Phys. Rev.,* **187,** 122.

Tang, K. T., and M. Karplus. 1968. *J. Chem. Phys.,* **49,** 1676.

Tang, K. T., and P. S. P. Wei. 1974. *J. Chem. Phys.,* **60,** 2454.

Trivedi, P. C., 1970. *Physica,* **48,** 486.

Truhlar, D. G., 1972. *J. Am. Chem. Soc.,* **94,** 7584.

Truhlar, D. G., and A. Kuppermann. 1971. *J. Chem. Soc.,* **93,** 1840.

Truhlar, D. G., and A. Kuppermann. 1972. *J. Chem. Phys.,* **56,** 2232.

Truhlar, D. G., and R. E. Wyatt. 1976. *Ann. Rev. Phys. Chem.,* **27,** 1.

Tully, J. C. and C. M. Truesdale. 1976. *J. Chem. Phys.,* **65,** 1002.

Vanderslice, J. T., and E. A. Mason. 1960. *J. Chem. Phys.,* **33,** 492.

Vanderslice, J. T., S. Weissman, E. A. Mason, and R. I. Fallon. 1962. *Phys. Fluids,* **5,** 155.

Van Vleck, J. H., and A. Sherman. 1935. *Rev. Mod. Phys.,* **7,** 167.

Van Volkenburgh, G., T. Carrington, and R. A. Young. 1973. *J. Chem. Phys.,***59,** 6035.

Victor, G. A., and A. Dalgarno. 1970. *J. Chem. Phys.,* **53,** 1316.

Wall, F. T., and R. N. Porter. 1962. *J. Chem. Phys.* **36,** 3256.

Wall, F. T., L. A. Hiller, Jr., and J. Mazur. 1958. *J. Chem. Phys.,* **28,** 255.

Walsh, M., and F. A. Matsen. 1951. *J. Chem. Phys.,* **19,** 526.

Wang, S. C., 1928. *Phys. Rev.,* **31,** 579.

Wartell, M. A., and R. J. Cross. 1971. *J. Chem. Phys.,,* **54,** 4519.

Wauchop, T. S., M. J. McEwan, and L. F. Phillips. 1969. *J. Chem. Phys.* **51,** 4227.

Weinbaum, S., 1933. *J. Chem. Phys.,* **1,** 593.

Weissmann, S., and E. A. Mason. 1962. *J. Chem. Phys.,* **36,** 794.

Welz, W., 1976. Ph.D. thesis, Bonn.

Weston, R. E., Jr. 1959. *J. Chem. Phys.,* **31,** 892.

Weston, R. E., Jr. 1967. *Science,* **168,** 332.

Wise, H., 1959. *J. Chem. Phys.,* **31,** 1414.

Wise, H., 1961. *J. Chem. Phys.,* **34,** 2139.

Wise, H., and C. M. Ablow. 1958. *J. Chem. Phys.,* **29,** 634.

Wise, H., and B. J. Wood. 1961. In L. Talbot, Ed., *Rarefied Gas Dynamics,* Academic Press, New York, p. 51.

Wolken, G., Jr. and M. Karplus. 1974. *J. Chem. Phys.,* **60,** 351.

Wolken, G., Jr., W. H. Miller, and M. Karplus. 1972. *J. Chem. Phys.,* **56,** 4930.

Yasumori, I., 1959. *Bull. Chem. Soc. Japan.* **32,** 1110.

Yates, A. C., and W. A. Lester. 1974. *Chem. Phys. Lett.,* **24,** 305.

Zarur, G., and H. Rabitz. 1973. *J. Chem. Phys.,* **59,** 943.

Zavitsas, A. A., 1972. *J. Am. Chem. Soc.,* **94,** 2779.

THEORETICAL ASPECTS OF IONIZATION POTENTIALS AND PHOTOELECTRON SPECTROSCOPY: A GREEN'S FUNCTION APPROACH

L. S. CEDERBAUM and W. DOMCKE

Institut für Theoretische Physik,
Physik-Department der Technischen Universität München,
8046 Garching, Germany

CONTENTS

I. INTRODUCTION

Within the last decade photoelectron (PE) spectroscopy has developed into an extremely useful experimental technique for studying the electronic structure of atoms and molecules in the gas phase and of condensed matter. The experimental aspects of both ultraviolet PE spectroscopy and x-ray PE spectroscopy are well described in several monographs and reviews.[1] The application of PE spectroscopy to molecules has emerged as particularly fruitful and has contributed significantly to our understanding of the chemical bonding in small and intermediate-size molecules. PE spectroscopy provides us above all with a direct method to measure the ionization potentials (IPs), that is, the binding energies, of the individual electrons of the system under consideration.

In order to interpret a PE spectrum and to exploit the information contained in it, a theoretical determination of the IPs is necessary. By definition, the IP can be obtained as the difference between the total energy of the ionic state and the total energy of the ground state. In this way IPs have been calculated by many workers, using total energies calculated in the HF approximation or within more sophisticated schemes, for example, configuration-interaction (CI) methods. It was soon realized, however, that these calculations suffer from the fact that the IPs—of the order of magnitude of an atomic unit or less for valence electrons—have to be determined as the difference of two large numbers of the order of magnitude of hundreds of atomic units. The initial- and final-state energies have therefore to be calculated to a very high degree of accuracy in order to obtain reliable values for the IPs. In particular, great care must be taken to ensure that the correlation in both the initial and the final state is treated in a consistent manner.

To overcome these difficulties, methods that allow the IPs to be calculated directly have been put forward by several authors. The need for subtraction of large numbers involved in the traditional approach can thus be overcome. These "direct" methods not only have computational advantages, but also supply us with a more intimate understanding of the ionization process. The initial- and final-state wave functions, which are calculated using the traditional approach, are difficult to interpret for many-electron systems. The direct approaches, on the other hand, do not involve the wave functions and lead to expressions for the IPs that may be interpreted in terms of physical processes. The quasiparticle picture, in particular, which is well established in other fields of physics, has been found to be a useful concept in atomic and molecular physics as well.

The same ideas can be applied when the vibrational structure in the PE spectrum is to be calculated. The traditional approach, requiring the calculation of the separate potential-energy surfaces of the initial and final states, is only applicable to the very smallest polyatomic molecules. Again, a direct approach can be formulated that is computationally much simpler and more suitable for interpretative purposes.

Much work has been done in the last few years on the direct determination of the IPs of atoms and molecules. Several authors have used Green's function techniques to calculate corrections to Koopmans' theorem.[2] A general survey of the Green's function method as applied to atoms and molecules has been given by Csanak, Taylor, and Yaris[3] and more specific aspects have been discussed by Linderberg and Öhrn.[4] An expansion of the self-energy part up to second order was first used by Reinhardt and Doll[5] to obtain IPs for atoms and by Cederbaum et al.[6] to calculate IPs of molecules. Later on, higher-order contributions to the self-energy part were considered (see Ref. 7). Schneider, Taylor, Yaris, and co-workers[8] applied the functional differentiation technique[9] to various problems associated with bound- and scattering-state properties of atoms. Recently, this method was used by Nerbrant[10] to calculate the IPs of molecules. Pickup and Goscinsci[11] developed a different approach, based on the superoperator formalism, and Purvis and Öhrn[12] made use of this formalism to obtain IPs of atoms and molecules.

Another direct approach closely related to the above Green's function methods, is due to Simons and co-workers.[13] They extended the equations-of-motion method developed by Rowe[14] to describe ionization. Other useful methods based on time independent perturbation theory have been developed and applied by Hubač and Kvasnička[15] and by Chong et al.[16]

We do not intend to enter into a detailed discussion and comparison of these different approaches here. Instead, our aim is to present a fully developed formalism that allows the interpretation and theoretical prediction of PE spectra, confining ourselves to a particular Green's function method. Some links with the other approaches are pointed out at appropriate places. In addition, some of the numerical results are compared with the results obtained by other groups.

It should be mentioned that Green's function methods, or, more generally, so-called many-body methods, can also be applied successfully to obtain other interesting properties of atoms and molecules. Kelly,[17] for example, has calculated correlation energies, oscillator strengths and photoabsorption cross sections of atoms using Brueckner's many-body perturbation theory. E. S. Chang et al.[18] and Cederbaum et al.[19] evaluated hyperfine coupling constants. Excitation energies and oscillator strengths were computed by McKoy and co-workers[20] using the equations-of-motion method, while Paldus and Čiček[21] studied excitation energies using the particle-hole propagator approach. Many-body perturbation methods were used by T. N. Chang et al.[22] to treat the double ionization of atoms.

In the following sections we present the Green's function theory, emphasizing the physical concepts behind the formalism. The connection between the PE spectrum and the Green's function is established in Section II. In Section III the self-energy part is introduced and its perturbation expansion discussed in detail for both closed- and open-shell systems. A more advanced formulation of the theory is given in Section IV. Section V presents a new approach to calculating the vibrational structure in the PE spectrum and can be read independently of the preceding sections. The theory outlined in this article has been tested numerically for a large number of examples. Several illustrative applications are collected in Section VI. Apart from Section IV, where some familiarity with many-body methods is assumed, the nonspecialist should be able to follow the text without recourse to other literature.

II. RELATIONSHIP BETWEEN GREEN'S FUNCTIONS AND PHOTOELECTRON SPECTRA

In a photoelectron experiment one measures the photocurrent as a function of the kinetic energy E_e of the ejected electrons. According to Einstein's relation this energy is equal to the difference between the photon energy $\hbar\omega_0$ and the binding energy I, which the measured electron has in the system

$$E_e = \hbar\omega_0 - I \tag{2.1}$$

The photocurrent is proportional to the intensity of the external field over many orders of magnitude of the latter. Therefore, the PE spectrum can be calculated with the well-known golden-rule formula to an extremely good approximation. The use of the golden-rule expression corresponds to the restriction to one-photon processes in the scattering formalism.[23] Thus the transition probability per unit time and per unit energy at the energy ω

is given by

$$P(\omega) = \frac{2\pi e^2}{m^2 c^2 \hbar} \sum_F \left| \langle F | \sum_n \mathbf{A}_n \cdot \mathbf{P}_n | \Psi_0^N \rangle \right|^2 \delta(\omega - E_e) \delta(E_F - E_0^N - \hbar \omega_0) \quad (2.2)$$

where $|F\rangle$ and $|\Psi_0^N\rangle$ are the final and initial states of the atom or molecule having N electrons, respectively. E_F and E_0^N are the corresponding energies and $\mathbf{A}_n \cdot \mathbf{P}_n$ is the scalar product of the external field at the nth electron and the momentum of this electron. Equation 2.2 incorporates the relation (2.1).

Defining annihilation and creation operators a_k and a_k^+ for an electron in a one-particle state $|\varphi_k\rangle$ we can express the sum over $\mathbf{A}_n \cdot \mathbf{P}_n$ in its second quantized form and obtain

$$P(\omega) = \frac{2\pi e^2}{m^2 c^2 \hbar} \sum_F \left| \langle F | \sum_{k,l} \tau_{kl} a_k^+ a_l | \Psi_0^N \rangle \right|^2 \delta(\omega - E_e) \delta(E_F - E_0^N - \hbar \omega_0)$$

$$= -\frac{2e^2}{m^2 c^2 \hbar} \sum_{k,l,m,n} \tau_{mn}^* \tau_{kl} \operatorname{Im}\{ iG_{lkmn}(-\hbar \omega_0 - i\eta)\} \delta(\omega - E_e) \quad (2.3)$$

where

$$\tau_{ij} = \langle \varphi_i | \mathbf{A} \cdot \mathbf{P} | \varphi_j \rangle$$

$G_{lkmn}(\omega)$ is the Fourier transform of the particle-hole component of the two-body Green's function[3,24,25] and η is a positive infinitesimal.

Starting from an appropriate basis of one-particle wave functions, which for definiteness we choose to be the Hartree-Fock (HF) orbitals, the τ_{ij} can be calculated as the matrix elements of $\mathbf{A} \cdot \mathbf{P}$ between the HF orbitals φ_i and φ_j. If we are able to calculate the two-body Green's functions, then (2.3) gives a complete description of the measured PE spectrum. The Green's functions contain all the internal properties of the molecule needed to describe the spectrum: electronic, vibrational and rotational interactions, spin-orbit coupling, and so on. The information about relative intensities of different bands in the spectrum as a function of the photon energy as well as the information about the angular distribution of the photoelectrons is contained in the matrix elements τ_{ij}. Thus the natural task of this work should be to evaluate the Green's functions $G_{lkmn}(\omega)$, since for a given one-particle basis set $\{\varphi_i\}$ the calculation of τ_{ij} is a straightforward problem. An *accurate* evaluation of these Green's functions is, however, a very difficult and costly problem. For some purposes, for example, the description of threshold ionization processes, many such functions might be needed for a reasonable calculation of $P(\omega)$. In addition, the continuum functions of the set $\{\varphi_i\}$ play an essential role in (2.3) and an accurate knowledge of them is especially important for describing threshold processes. For many atoms quite accurate one-particle continuum functions can be calculated,[17] but for molecules such a calculation runs into difficulties. Therefore, a simplified expression for $P(\omega)$ that still gives *exact* IPs and a fairly accurate description of the main features of the usual uv and x-ray PE spectra is most welcome. Such an expression is discussed in what follows.

Let us return to the first line in (2.3) and consider the final state $|F\rangle$ as an antisymmetrized product of a one-electron excited state $|e\rangle$ and a $(N-1)$-electron state $|\Psi_s^{N-1}\rangle$. Since the component of $|e\rangle$ in $|\Psi_0^N\rangle$ is almost exactly zero, if E_e is sufficiently large, a_k must annihilate the electron in $|e\rangle$ in the final state $|F\rangle$ in order to have a nonvanishing $P(\omega)$. One thus obtains

$$P(\omega)=\frac{2e^2}{m^2c^2\hbar}\sum_{e,n,l}\tau_{en}^*\tau_{el}\,\mathrm{Im}\{G_{ln}(\omega-\hbar\omega_0-i\eta)\}\delta(\omega-E_e)\qquad(2.4a)$$

where

$$G_{ln}(\omega-i\eta)=\sum_s\frac{\langle\Psi_0^N|a_n^+|\Psi_s^{N-1}\rangle\langle\Psi_s^{N-1}|a_l|\Psi_0^N\rangle}{\omega+E_s^{N-1}-E_0^N-i\eta}\qquad(2.4b)$$

is the Fourier transform of the advanced one-particle Green's function that is discussed in detail in the next section.

The assumption $|F\rangle=|e\rangle|\Psi_s^{N-1}\rangle$ is the only approximation used when going from (2.3) to our final equation, (2.4), and it deserves some comments. First, it is clear that the above ansatz implies that the ejected electron is not correlated with the electrons of the ion. This does not mean, however, that the state $|e\rangle$ describes a free electron; that is, $|e\rangle$ is, in general, not a plane wave. From (2.3) we can, in principle, construct the "optimum" one-electron states $|e\rangle$ to be used in (2.4). The ejected electron is then assumed to "feel" the other electrons, while its influence on these electrons is neglected. In the case $\hbar\omega_0\gg I$, the ejected electron is usually considered free and $|e\rangle$ taken as a plane wave of suitable symmetry. When $\hbar\omega_0-I\lesssim I$, then at least Coulomb-type waves describing the interaction of the electron with a static charge distribution of the ion should be used.[26] For ionization at the threshold ($\hbar\omega_0\approx I$), (2.3) involving the two-particle Green's function is the appropriate expression.

It is convenient to express $P(\omega)$ in terms of the eigenvalues $D_k(\omega)$ of $\mathbf{G}(\omega)$, which is the Green's function matrix with elements $G_{ij}(\omega)$:

$$P(\omega) = \frac{2e^2}{m^2c^2\hbar} \operatorname{Im}\left\{ \sum_e \operatorname{tr}\left[\tilde{\mathbf{D}}(\omega - \hbar\omega_0 - i\eta) \right] \delta(\omega - E_e) \right\} \qquad (2.5)$$

$$\tilde{D}_{ij}(\omega) = |\tau_{ei}(\omega)|^2 D_i(\omega)\delta_{ij}$$

$$\tau_{ei}(\omega) = \sum_j \tau_{ej} S_{ij}(\omega)$$

where $\mathbf{S}$ is the eigenvector matrix of $\mathbf{G}$. At this point it should be mentioned that

$$\operatorname{Im}\{ D_k(\omega - i\eta) \} = \pi \sum_s P_k(s)\delta(\omega + E_s^{N-1} - E_0^N) \qquad (2.6)$$

$$0 \leqslant P_k(s) \leqslant 1$$

The $P_k(s)$ are important physical quantities, which we call pole strengths.[27] Since the pole strengths are positive, it follows that $P(\omega) \geqslant 0$ for all ω, as must be true for a transition probability.

The state $|e\rangle$ and thus τ_{ei} obviously depend on the wave vector of the ejected electron. The absolute value of the wave vector is fixed because of $\delta(\omega - E_e)$ in (2.5), whereas its direction in space must be inserted into $|e\rangle$ according to the experimental situation. According to (2.1) E_e is equal to $\hbar\omega_0 - I_s$, where I_s simply stands for $E_s^{N-1} - E_0^N$, and $\tau_{ek}\delta(\omega - E_e)$ becomes $\tau_{e,k}\delta(\omega - E_e)$. Equation 2.5 is easily rewritten to give

$$P(\omega) = \frac{2\pi e^2}{m^2c^2\hbar} \sum_{s,k} |\tau_{e,k}(\hbar\omega_0 - I_s)|^2 P_k(s)\delta(\omega + I_s - \hbar\omega_0) \qquad (2.7)$$

To have a better understanding of $P(\omega)$, we discuss some of the consequences of (2.7). If the correlation of the electrons in the molecule is neglected, then the Green's function is equal to the so-called free Green's function $\mathbf{G}^0$ (see Section III.A). The latter being diagonal, one obtains for the transition probability $P^0(\omega)$ on the one-particle level

$$P^0(\omega) = \frac{4\pi e^2}{m^2c^2\hbar} \sum_{k \leqslant N/2} |\tau_{e_k k}|^2 \delta(\omega - \varepsilon_k - \hbar\omega_0) \qquad (2.8)$$

where ε_k denotes the kth orbital energy (for simplicity, only closed-shell molecules are considered). Thus at most $N/2$ bands will appear in the spectrum. This is in contradiction to experiment, which indicates that in the photoionization process there is a high probability for simultaneous excitation or ionization of a second electron in the same atom[28,1b] or molecule.[29,30] From (2.8) we see that such excitations accompanying the ionization, often referred to as shake-up (off) processes, can only be understood when the effects of electron correlation are considered.

This result can be expressed in a more pictorial form. Considering $G_{ln}(\omega - i\eta)$ as given in (2.4) and neglecting configuration interaction, it follows that an ionic state $|\Psi_s^{N-1}\rangle$ the electron configuration of which differs from that of the ground state of the reference molecule by the occupation of more than one orbital, should not be observed. However, if configuration interaction is taken into account, which means that the full $G_{ln}(\omega - i\eta)$ is considered, then this ionic state will "borrow" intensity from other states having suitable electronic configurations and the same symmetry, and will be observed.

Let us consider the case where the above ionic state $|\Psi_s^{N-1}\rangle$ borrows intensity from *only one* state $|\Psi_k^{N-1}\rangle$, which is characterized by an electron configuration differing from that of $|\Psi_0^N\rangle$ by the occupation of φ_k. This case is realized when the eigenvector matrix $\mathbf{S}$ of $\mathbf{G}$ is diagonal. Then the relative photocurrent $P(s/k)$ associated with the production of the states $|\Psi_s^{N-1}\rangle$ and $|\Psi_k^{N-1}\rangle$ is given by

$$P\left(\frac{s}{k}\right) = \frac{|\tau_{e_s k}|^2 P_k(s)}{|\tau_{e_k k}|^2 P_k(k)} \tag{2.9}$$

This relative photocurrent obviously depends on $\hbar\omega_0$, but in case $\hbar\omega_0 - I_k \gg |I_s - I_k|$ the dependence can be neglected and we obtain the simple result

$$P\left(\frac{s}{k}\right) = \frac{P_k(s)}{P_k(k)} \tag{2.10}$$

which implies that $P(s/k)$ depends only on the properties of $\mathbf{G}$. It should be mentioned that for most of the intense satellite lines $\mathbf{S}$ can be considered to be diagonal.[7d,31]

We have seen that the two-body Green's function or the one-body Green's function together with $|e\rangle$ determine the photoionization spectrum. Knowledge of $|e\rangle$ (and thus of the τ_{ei}'s) is important for obtaining the relative intensities of different bands at a given photon energy $\hbar\omega_0$ and for determining the angular distribution of the photoelectrons. There have been several attempts to calculate the intensities of the electronic bands and the angular distribution in the PE spectra of molecules, taking $|e\rangle$ as a plane wave.[12a, 32-34] In order to improve the results, orthogonalized plane waves[35] or Coulomb waves[36] have been considered. In the following, we do not pursue further the subject of the matrix elements τ_{ei}, since they depend heavily on the experimental conditions, for example, the energy of the incident light and the angle under which the photoelectrons are observed. The features in the PE spectrum that are independent of the experimental conditions and that represent the invariant information about the system under consideration are the ionization potentials and, as long as one is far from threshold, the vibrational structure of the bands. This invariant information is contained in the one-body Green's function G, to which the following sections are devoted.

III. ONE-BODY GREEN'S FUNCTION FOR CLOSED- AND OPEN-SHELL ATOMS AND MOLECULES

The greater part of the work done in the field of atomic and molecular Green's functions is restricted to closed-shell systems. In these cases the basic formalism is essentially identical to that developed in other fields and only the approximations introduced have a specific character. For open-shell atoms and molecules one can, in principle, choose a formalism that is simpler than those developed for open-shell problems in other fields, for example, in nuclear physics, by incorporating features characteristic for most atoms and molecules. In this section we present a Green's function approach to closed- *and* open-shell systems and follow a recent attempt[37] to establish a Green's function formalism for atomic and molecular systems with degenerate ground states. In this approach the Green's functions involved in the degenerate case can be evaluated by the same methods as used in the simpler case of closed-shell systems.

A. The Green's Function

For the sake of clarity the one-body Green's function for molecules with a nondegenerate ground state is discussed first. In the following we refer to the one-body Green's function simply as the Green's function and put $\hbar = 1$. The Green's function is defined as the N-electron ground-state average of a time-ordered product of an annihilation and a creation operator, a_k and a_l^+,[38]

$$G_{kl}(t,t') = -i\langle \Psi_0^N | T\{ a_k(t)a_l^+(t')\} | \Psi_0^N \rangle \tag{3.1}$$

where T is the Wick time-ordering operator. The creation and annihilation operators obey the usual relations

$$a_k(t) = e^{iHt} a_k e^{-iHt} \qquad a_k = a_k(0) \tag{3.2}$$

$$[a_k, a_l^+]_+ = \delta_{kl} \qquad [a_k, a_l]_+ = [a_k^+, a_l^+]_+ = 0$$

H is the full Hamiltonian of the system. If H is time independent, G_{kl} depends only on $t - t'$ and its Fourier transform is easily performed,

$$G_{kl}(\omega) = \int_{-\infty}^{\infty} G_{kl}(t - t') e^{i\omega(t - t')} d(t - t')$$

$$= \sum_s \frac{\langle \Psi_0^N | a_l^+ | \Psi_s^{N-1} \rangle \langle \Psi_s^{N-1} | a_k | \Psi_0^N \rangle}{\omega + I_s - i\eta}$$

$$+ \sum_r \frac{\langle \Psi_0^N | a_k | \Psi_r^{N+1} \rangle \langle \Psi_r^{N+1} | a_l^+ | \Psi_0^N \rangle}{\omega + A_r + i\eta}$$

$$I_s = E_s^{N-1} - E_0^N \qquad A_r = E_0^N - E_r^{N+1} \tag{3.3}$$

We immediately recognize that $\mathbf{G}(\omega - i\eta)$ in (2.4b) is the first sum of $\mathbf{G}(\omega)$ in the above equation. In addition, we see that $\mathbf{G}(\omega)$ also contains a retarded component $\mathbf{G}(\omega + i\eta)$, which has the electron affinities of the molecule as poles.

To proceed we decompose the Hamiltonian into two parts,

$$H = H_0 + H_I \tag{3.4}$$

and introduce a free Green's function $\mathbf{G}^0$, which is defined in complete analogy to $\mathbf{G}$ in (3.1), but with the unperturbed Hamiltonian H_0 instead of H:

$$G_{kl}^0(t) = -i \langle \Phi_0^N | T\{ a_k(t) a_l^+(0) \} | \Phi_0^N \rangle \tag{3.5}$$

$$a_k(t) = e^{iH_0 t} a_k e^{-iH_0 t}$$

where Φ_0^N is the ground-state eigenfunction of H_0. We now discuss briefly how $\mathbf{G}(t)$ can be expanded in terms of the interaction H_I. We first turn to the interaction representation,[26] where a wave function $\hat{\Psi}$ and an operator $\hat{A}$ are obtained from the Schrödinger wave function Ψ and operator A by $\hat{\Psi} = e^{iH_0 t} \Psi$ and $\hat{A} = e^{iH_0 t} A e^{-iH_0 t}$, respectively. From the Schrödinger equation it follows that

$$\hat{\Psi}(t) = S(t, t_0) \hat{\Psi}(t_0) \tag{3.6}$$

where $S(t,t_0)$ is given by the Dyson series[38]

$$S(t,t_0) = T \exp\left\{ -i \int_{t_0}^{t} \hat{H}_I(t') \, dt' \right\}$$

$$= 1 - i \int_{t_0}^{t} \hat{H}_I(t_1) \, dt_1 + (-i)^2 \int_{t_0}^{t} \hat{H}_I(t_1) \, dt_1 \int_{t_0}^{t_1} \hat{H}_I(t_2) \, dt_2 + \cdots \quad (3.7)$$

Making use of relation (3.6) and of the concept of the so-called "adiabatically turned-on interaction" one obtains the basic equation for the expansion of the Green's function,[38]

$$G_{kl}(t) = -i \frac{\langle \Phi_0^N | T\{ S(\infty, -\infty) \hat{a}_k(t) \hat{a}_l^+ \} | \Phi_0^N \rangle}{\langle \Phi_0^N | S(\infty, -\infty) | \Phi_0^N \rangle} \quad (3.8)$$

It should be noticed that only the unperturbed ground-state wave function appears in (3.8) and that the $\hat{a}_i(t)$ are the same operators as contained in the free Green's function. When $|\Phi_0^N\rangle$ can be written as $\Pi a_i^+ |0\rangle$, where $|0\rangle$ is the fermion vacuum, one can proceed via Wick's theorem[39] to obtain the well-known[24,25,38,40] diagrammatic expansion of G in terms of the interaction matrix elements and free propagators. The actual clue to (3.8) lies in this diagrammatic expansion of G combined with the linked-cluster theorem.[40]

For completeness, two important additional properties of G are mentioned. Since any one-particle operator O can be represented by $\Sigma O_{lk} a_l^+ a_k$, $O_{lk} = \langle \varphi_l | O | \varphi_k \rangle$, one readily obtains with the aid of the definitions in (3.1) to (3.3) the following expression for the ground-state expectation value of O:

$$\langle O \rangle = \frac{1}{2\pi i} \sum_{k,l} O_{lk} \oint G_{kl}(\omega) \, d\omega \quad (3.9)$$

A little more algebra is needed to derive an expression for the ground-state energy. When H_0 is a one-particle operator with matrix elements ε_{ij} and H_I is a two-particle operator, the result is[41]

$$E_0^N = \frac{1}{4\pi i} \oint \mathrm{tr}(\omega \mathbf{1} + \varepsilon) \mathbf{G}(\omega) \, d\omega \quad (3.10)$$

In what follows we discuss the open-shell case. Let us consider an atom or molecule with a degenerate ground state, where H is considered to commute with the spin operator $\mathbf{S}$. The case of spatial degeneracy is discussed later. The ground states are given as the $2S+1$ eigenstates

$|\Psi_0^N(S,M)\rangle$ of $\mathbf{S}^2$ and S_z. If there is no special preparation of the system, the photocurrent as a function of ω will, in analogy to (2.2), be described by

$$P(\omega) = (2S+1)^{-1} \sum_{M=-S}^{S} P^{(M)}(\omega) \qquad (3.11)$$

where $P^{(M)}(\omega)$ is simply given by (2.2) with $|\Psi_0^N(S,M)\rangle$ substituted for $|\Psi_0^N\rangle$. In complete analogy the appropriate Green's function is

$$\mathbf{G}(t) = (2S+1)^{-1} \sum_{M=-S}^{S} \mathbf{G}^{(M)}(t) \qquad (3.12)$$

$$G_{kl}^{(M)}(t) = -i\langle\Psi_0^N(S,M)|T\{a_k(t)a_l^+(0)\}|\Psi_0^N(S,M)\rangle$$

With this definition of $\mathbf{G}$ and with the analogous definition of the two-body Green's function, the relations derived in Section II, for example, (2.3) and (2.4), are still valid. It should be noted that in the case $S=0$, $\mathbf{G}$ in (3.12) reduces to the $\mathbf{G}$ defined in (3.1) for systems with a nondegenerate ground state. It is also straightforward to see that the expressions (3.9) and (3.10) for the ground-state expectation value of a one-particle operator and for the ground-state energy, respectively, are still valid for an open-shell system when the definition (3.12) is used.

In realistic problems the unperturbed Hamiltonian H_0 does not have a lower symmetry than H. Then usually there corresponds to each $|\Psi_0^N(S,M)\rangle$ an eigenstate $|\Phi_0^N(S,M)\rangle$ of H_0, $\mathbf{S}^2$ and S_z. These states are, of course, also degenerate. Turning on adiabatically the interaction H_I, the state $|\Phi_0^N(S,M)\rangle$ will develop into the state

$$\langle\Phi_0^N(S,M)|\Psi_0^N(S,M)\rangle^{-1}|\Psi_0^N(S,M)\rangle$$

and the Green's function $\mathbf{G}^{(M)}$ can be written in terms of the unperturbed state $|\Phi_0^N(S,M)\rangle$ in analogy to (3.8),

$$G_{kl}^{(M)}(t) = -i\frac{\langle\Phi_0^N(S,M)|T\{S(\infty,-\infty)\hat{a}_k(t)\hat{a}_l^+|\Phi_0^N(S,M)\rangle}{\langle\Phi_0^N(S,M)|S(\infty,-\infty)|\Phi_0^N(S,M)\rangle} \qquad (3.13)$$

We see again the formal equivalence between the closed- and open-shell cases.

Can we, however, apply Wick's theorem and the elegant diagrammatic methods that are available for the Green's function in the closed-shell case? In general, a state $|\Phi_0^N(S,M)\rangle$ is given by a linear combination of

Slater determinants, with each determinant containing different one-particle open-shell states. As a consequence, the one-particle states within the open-shell can no longer be separated by a canonical transformation into occupied and unoccupied states, and thus the diagrammatic approach via Wick's theorem breaks down.

The above formulas and discussions are not restricted to atoms and molecules, but apply to other fields as well, for example, to nuclei. To proceed, we now take into account specific features of atoms and molecules. For most atomic and molecular systems there are genuine electron configurations within the set of degenerate *unperturbed ground states*. As a consequence of Hund's rule, in particular, the states with maximum magnetic quantum numbers $M_L = \pm L$, $M_s = \pm S$ of a rotationally invariant Hamiltonian without spin-orbit coupling are given by *single* Slater determinants. It follows that the Green's functions $\mathbf{G}^{(S)}$ and $\mathbf{G}^{(-S)}$ can be, in complete analogy to $\mathbf{G}$ of closed-shell systems, evaluated by means of the well-known diagrammatic methods ($\mathbf{L}$ has been omitted for simplicity, but is considered below). What can be done, however, with the $\mathbf{G}^{(M)}$ with $M \neq S$, which also appear in (3.12)? Here we can gain assistance from results obtained by Cederbaum and Schirmer,[37] who showed by a tensorial analysis that $\mathbf{G}$ in (3.12) can be expressed in terms of $\mathbf{G}^{(S)}$ only:

$$\mathbf{G}_{\alpha\alpha} = \tfrac{1}{2}\left\{ \mathbf{G}^{(S)}_{\alpha\alpha} + \mathbf{G}^{(S)}_{\beta\beta} \right\} \tag{3.14}$$

$$\mathbf{G}_{\alpha\beta} = 0$$

where α, β denote the spin indices $m_s = \tfrac{1}{2}$ and $m_s = -\tfrac{1}{2}$, respectively, of the orbitals k and l involved in G_{kl}. We are thus able to treat open- and closed-shell atoms and molecules on the same level. In particular, the formalism derived for closed-shell systems is now directly applicable to open-shell systems as well. It has been shown by Yaris and Taylor[42] how by incorporating the above ideas a scattering theory for open-shell systems can be formulated within a Green's function approach.

If we are also interested in spatial degeneracy, we may write

$$\mathbf{G}(t) = (2L+1)^{-1}(2S+1)^{-1} \sum_{M_L = -L}^{L} \sum_{M_S = -S}^{S} \mathbf{G}^{(M_L, M_S)}(t) \tag{3.15}$$

The result analogous to (3.14) is now[37]

$$G_{nlm_l m_s, n'l'm_{l'} m_{s'}} = \frac{\delta_{ll'}\delta_{m_l m_{l'}}\delta_{m_s m_{s'}}}{2(2l+1)} \sum_{\mu_l \mu_s} G^{(L,S)}_{nl\mu_l \mu_s, n'l\mu_l \mu_s} \tag{3.16}$$

It is understood that the indices k and l in G_{kl} stand for a complete set of quantum numbers, for example, $(nlm_l m_s) \equiv k$ in (3.16). Instead of having to evaluate $2(2l+1)(2S+1)(2L+1)$ complicated individual Green's functions in (3.15), only $2(2l+1)$ Green's functions have to be considered, which are exactly those that can be evaluated by the diagrammatic analysis.

These considerations apply also to a set of ground states $\{|\Psi_0^N(\lambda, \mu_\lambda, S, M_s)\rangle\}$, characterized by any representation λ of the symmetry group of H. The corresponding Green's function is clearly

$$G(t) = r_\lambda^{-1}(2S+1)^{-1} \sum_{\mu_\lambda, M_s} G^{(\mu_\lambda, M_s)}(t)$$

$$G_{kl}^{(\mu_\lambda, M_s)}(t) = -i\langle\Psi_0^N(\lambda, \mu_\lambda, S, M_s)| T\{a_k(t)a_l^+\}|\Psi_0^N(\lambda, \mu_\lambda, S, M_s)\rangle \quad (3.17)$$

where r_λ is the dimension of λ. For a better understanding of the formalism, simple examples for the evaluation of G in (3.12) and (3.17) are discussed in Section IV.D.

Since the two-body Green's function appears in the most general golden-rule expression for $P(\omega)$, (2.3), we complete this section by defining this function for open-shell systems:

$$G_{lmkn} = \frac{1}{2S+1} \sum_{M=-S}^{S} G_{lmkn}^{(M)}$$

$$G_{lmkn}^{(M)} = -\langle\Psi_0^N(S,M)| T\{a_l(t_1)a_m(t_2)a_k^+(t_1')a_n^+(t_2')\}|\Psi_0^N(S,M)\rangle \quad (3.18)$$

The inclusion of L and/or λ is easily done. The equations analogous to the basic equation (3.14) are

$$G_{\alpha\alpha\alpha\alpha} = \tfrac{1}{6}\left[2G_{\alpha\alpha\alpha\alpha}^{(S)} + 2G_{\beta\beta\beta\beta}^{(S)} + G_{\alpha\beta\alpha\beta}^{(S)} + G_{\beta\alpha\beta\alpha}^{(S)} + G_{\alpha\beta\beta\alpha}^{(S)} + G_{\beta\alpha\alpha\beta}^{(S)} \right]$$

$$G_{\alpha\beta\alpha\beta} = \tfrac{1}{6}\left[G_{\alpha\alpha\alpha\alpha}^{(S)} + G_{\beta\beta\beta\beta}^{(S)} + 2G_{\alpha\beta\alpha\beta}^{(S)} + 2G_{\beta\alpha\beta\alpha}^{(S)} - G_{\alpha\beta\beta\alpha}^{(S)} - G_{\beta\alpha\alpha\beta}^{(S)} \right]$$

$$G_{\alpha\beta\beta\alpha} = G_{\alpha\alpha\alpha\alpha} - G_{\alpha\beta\alpha\beta} \qquad G_{\beta\beta\beta\beta} = G_{\alpha\alpha\alpha\alpha}$$

$$G_{\beta\alpha\beta\alpha} = G_{\alpha\beta\alpha\beta} \qquad G_{\beta\alpha\alpha\beta} = G_{\alpha\beta\beta\alpha} \qquad (3.19)$$

B. Dyson Equation and Self-Energy Part

1. General

To be specific the molecular second quantized Hamiltonian is first introduced. Apart from the pure electronic motion, this Hamiltonian may include, among others, terms describing the vibrational and rotational motions and their interaction with the electronic motion. The electron-vibration interaction, in particular, is important for the understanding of PE spectra. For the sake of simplicity we restrict ourselves for the moment to the discussion of the pure electronic problem of a many-electron molecule and return in Section V to the vibrational problem. In principle, terms from spin-orbit coupling down to hyperfine interaction can also be considered, but are not discussed here.

The electronic Hamiltonian reads

$$H = \sum_i \varepsilon_i a_i^+ a_i + \sum_{i,j} v_{ij} a_i^+ a_j + \tfrac{1}{2} \sum_{i,j,k,l} V_{ijkl} a_i^+ a_j^+ a_l a_k \qquad (3.20)$$

where ε_i, v_{ij} and V_{ijkl} are the matrix elements of H_0, a one-particle potential v, and of the Coulomb interaction between the electrons: $V_{ijkl} = \langle \varphi_i(\mathbf{r})\varphi_j(\mathbf{r}') \| \mathbf{r} - \mathbf{r}'|^{-1} | \varphi_k(\mathbf{r})\varphi_l(\mathbf{r}') \rangle$, respectively. $\{\varphi_i\}$ is any orthogonal basis of spin orbitals that diagonalizes H_0. The potential v can, in principle, include an external potential.

For the special case where we choose H_0 to be the so-called unrestricted HF operator,[43] the ε_i become the HF orbital energies and, if no external potential is included, we have

$$v_{ij} = - \sum_k (V_{ikjk} - V_{ikkj})n_k \qquad (3.21)$$

with

$$n_k = \begin{cases} 1 & \text{if } k \in \mathcal{F} \\ 0 & \text{if } k \notin \mathcal{F} \end{cases} \qquad \bar{n}_k = 1 - n_k \qquad (3.22)$$

$\mathcal{F}$ denotes the set of orbitals occupied in the HF ground state. If, on the other hand, we choose H_0 to be the so-called restricted HF (RHF) operator,[43-45] we have to distinguish between the open-shell and the closed-shell cases. In the case where we deal with a closed-shell molecule, v_{ij} in (3.21) is still correct. For an open-shell molecule the choice of v_{ij} is somewhat ambiguous and depends on the special method chosen.[45] We return to this point in Section IV.D. In the general case it is convenient to define the quantity

$$s_{ij} = v_{ij} + \sum_k (V_{ikjk} - V_{ikkj})n_k \qquad (3.23)$$

which vanishes when v_{ij} is given by (3.21).

Knowing H, one could, in principle, evaluate $\mathbf{G}$ by expanding it in terms of the matrix elements V_{ijkl} and v_{ij}. Such an expansion is, however, known to be unsuitable for most problems. A more adequate way to evaluate $\mathbf{G}$ is to start from the well-established Dyson equation[46] for the case of a nondegenerate ground state,

$$\mathbf{G}(\omega) = \mathbf{G}^0(\omega) + \mathbf{G}^0(\omega)\mathbf{\Sigma}(\omega)\mathbf{G}(\omega) \tag{3.24}$$

which relates the Green's function to the so-called self-energy part $\mathbf{\Sigma}$. $\mathbf{G}$ can now be evaluated by approximating $\mathbf{\Sigma}$ and solving the Dyson equation. Finding an accurate approximation to $\mathbf{\Sigma}$ is a difficult task that is discussed later.

Having defined the Hamiltonian, an explicit expression for the free Green's function in (3.5) can be given:

$$G_{kl}^0(t) = ie^{-i\varepsilon_k t}\delta_{kl} \begin{cases} -1, & t>0, \quad k \notin \mathscr{F} \\ 1, & t \leqslant 0, \quad k \in \mathscr{F} \end{cases} \tag{3.25a}$$

or in ω-space,

$$G_{kl}^0(\omega) = \frac{\delta_{kl}}{\omega - \varepsilon_k - ai\eta} \qquad a = \begin{cases} 1, & k \in \mathscr{F} \\ -1, & k \notin \mathscr{F} \end{cases} \tag{3.25b}$$

There is, of course, also a Dyson equation in the case of a degenerate ground state. From the discussion in Section III.A, especially that concerning (3.8) and (3.13), is is clear that $\mathbf{G}^{(S)}$ is the function for which the Dyson equation is valid in the open-shell case:

$$\mathbf{G}^{(S)} = {}^0\mathbf{G}^{(S)} + {}^0\mathbf{G}^{(S)}\mathbf{\Sigma}^{(S)}\mathbf{G}^{(S)} \tag{3.26}$$

where ${}^0\mathbf{G}^{(S)}$ is the free Green's function corresponding to $\mathbf{G}^{(S)}$. $\mathbf{G}_{\alpha\beta}$ vanishes, and therefore the Dyson equation for $\mathbf{G}^{(S)}$ splits into two Dyson equations, one for $\mathbf{G}_{\alpha\alpha}^{(S)}$ and one for $\mathbf{G}_{\beta\beta}^{(S)}$, which are identical for $S=0$. Again, since $\mathbf{G}$ in (3.24) is just $\mathbf{G}^{(S)}$ for $S=0$, we see the complete analogy between the open- and closed-shell formalisms. Moreover, there is actually only one formalism for both cases.

In the following we discuss the general properties of $\mathbf{\Sigma}^{(S)}$ that are essential for a better understanding of the formalism and its relation to the measured spectra. For simplicity, we drop the spin superscript S. From the spectral representation of $\mathbf{G}$ as given in (3.3) and from the Dyson equation, which is the definition of $\mathbf{\Sigma}$, the self-energy part can be shown to be of the form[31]

$$\Sigma_{kl}(\omega) = \Sigma_{kl}(\infty) + \sum_p \frac{f_{klp}}{\omega - N_p + i\eta} + \sum_q \frac{g_{klq}}{\omega - M_q - i\eta} \tag{3.27}$$

$$f_{kkp}, g_{kkq} \geqslant 0$$

where the two sets of poles $\{N_p\}$ and $\{M_q\}$ are separated by $\omega = \bar{\omega}_0$. It is shown below that $\Sigma(\infty)$ is given by

$$\Sigma_{kl}(\infty) = v_{kl} + \frac{1}{2\pi i} \sum_{p,q} (V_{kplq} - V_{kpql}) \oint G_{qp}(\omega) d\omega \qquad (3.28a)$$

which can be rewritten with the aid of the Dyson equation

$$\Sigma_{kl}(\infty) = s_{kl} + \frac{1}{2\pi i} \sum_{pp'q} (V_{kplq} - V_{kpql}) \oint G_{qq}^0(\omega) \Sigma_{qp'}(\omega) G_{p'p}(\omega) d\omega \qquad (3.28b)$$

Thus, if the ω-dependent part of $\Sigma(\omega)$ is known, then $\Sigma(\infty)$ can be obtained self-consistently from (3.28) and (3.26). For convenience we denote the ω-dependent part of the self-energy part by $M(\omega)$.

We may rewrite the Dyson equation to give

$$G^{-1} = \omega 1 - \varepsilon - \Sigma \qquad (3.29)$$

where 1 and ε are the unit matrix and the diagonal matrix of one-particle energies ε_i, respectively. To find the poles of G and the pole strengths $P_k(s)$, the zeros and the derivatives of the eigenvalues of G^{-1} have to be determined. Denoting the kth eigenvalue of $\varepsilon + \Sigma$ by $[\varepsilon + \Sigma]_k$, one obtains for the pole $\omega = -I_s$ of $D_k(\omega)$ and for the pole strength the simple relations

$$-I_s = [\varepsilon + \Sigma(-I_s)]_k \qquad (3.30)$$

$$P_k(s) = \left\{ 1 - \frac{\partial}{\partial \omega} [\varepsilon + \Sigma(-I_s)]_k \right\}^{-1}$$

For a more detailed discussion we assume, without loss of generality, G to be diagonal and introduce the notation $\Sigma_{+1}, \Sigma_{+2}, \ldots$ for the poles of the first sum in (3.27) in the order of increasing energy and $\Sigma_{-1}, \Sigma_{-2}, \ldots$ for the poles of the second sum in the order of decreasing energy. Any interval between two successive poles of Σ is specified by a subscript l as shown in Fig. 3.1, where a schematic plot of a diagonal term Σ_{kk} as a function of ω is given. The poles of G_{kk} are found as the energy values of the intersection points of the straight line $y = \omega - \varepsilon_k$ with $\Sigma_{kk}(\omega)$. The bands appearing in the photoelectron spectrum correspond to intersection points in the intervals characterized by $l \leqslant 0$; the other intersection points give rise to the electron affinities of the system. From (3.30) we see that the pole strengths [and thus a measure for the relative photocurrent $P(s/k)$ as defined in Section II] can also be readily deduced from Fig. 3.1. The steeper the slope of $\Sigma_{kk}(\omega)$ at the intersection point, the smaller the intensity of the corresponding line in the spectrum.

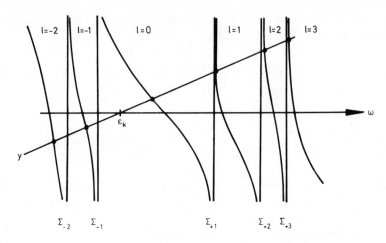

Fig. 3.1. A schematic plot of Σ_{kk} as a function of ω. The $-\omega$ values of the intersection points of the straight line $y = \omega - \varepsilon_k$ and $\Sigma_{kk}(\omega)$ are the ionization potentials and electron affinities.

For closed-shell systems the interval $l = 0$ is large compared to the other intervals and contains the orbital energies ε_i of the outer valence electrons. This can be easily seen from the perturbation expansion of $\Sigma(\omega)$, which is discussed below. Thus we see from Fig. 3.1 that $P_k(k)$ is large compared to $P_k(s)$, $s \neq k$, that is, ionization of an outer electron k "leads" to shake-up lines with only low intensity. The $P_k(k)$ calculated in Section VI for outer valence electrons usually have values of about 0.9, which means that only about $1 - P_k(k) \approx 10\%$ of its intensity has been borrowed by other states. The $P_k(k)$ of inner-shell electrons may have values of about 0.8 or even less.

It is instructive to consider a self-energy part with only one pole: $\Sigma_{kk} = Q^2(\omega - \mu)^{-1}$. With the Dyson equation we find that two IPs, I_k and I_μ, exist. For $|\varepsilon_k - \mu| \gg |Q|$, they and the corresponding pole strengths are given by

$$I_k \approx -\varepsilon_k - Q^2(\varepsilon_k - \mu)^{-1} \qquad P_k(k) \approx 1 - Q^2(\varepsilon_k - \mu)^{-2} \qquad (3.31)$$

$$I_\mu \approx -\mu + Q^2(\varepsilon_k - \mu)^{-1} \qquad P_k(\mu) \approx Q^2(\varepsilon_k - \mu)^{-2}$$

The pole strength corresponding to I_k is nearly unity, while the other one is small. Thus, if electron correlation is not taken into account, that is, $Q \to 0$, then $-I_k$ becomes the one-particle energy ε_k and I_μ is an ionization potential with probability zero.

It is clear from Fig. 3.1 that there is exactly one IP between two successive poles of Σ_{kk}. This can easily be understood from (3.27) for $k = l$ and can be generalized as follows:[31] there is exactly one pole of an eigenvalue D_k of G situated between two successive poles of the corresponding eigenvalue $[\varepsilon + \Sigma]_k$. Having derived an approximation to Σ we also have obtained, without calculating G, approximate IPs. Since the intervals $l \neq 0$ are often small, energies of processes corresponding to excitations accompanying the ionization can be obtained and assigned in this way.[47]

We have seen that if electron correlation is neglected, then the calculated spectrum will have lines at the one-particle energies ε_k. The lines observed in the experimental spectrum are usually assigned to the one-particle states specified by k. Behind this commonly used way of assigning photoelectron spectra, there is the assumption that the ionization process is well described by the "removal" of a quasiparticle in the state k, or, in other words, by the "creation" of a quasihole in the state k. The Green's function concept is ideally suited to studying the level of accuracy of this assumption. Our aim is to find out whether the corresponding eigenvalue of the Green's function in the range of the solution $\omega \approx - I_k$ has a form similar to the free function $G_{kk}^0(\omega)$. For this purpose we decompose the self-energy part into imaginary and real parts for real values of ω. Starting from the Dyson equation and expanding the real part of $[\varepsilon + \Sigma(\omega)]_k$ about $- I_k$ up to the linear term and Σ_k^I, which is the imaginary part of $[\varepsilon + \Sigma(\omega)]_k$, up to the constant term, one obtains for $\omega \approx - I_k$

$$D_k(\omega) = \frac{P_k(k)}{\omega + I_k - i\tau_k^{-1}} + f_k(\omega) \tag{3.32}$$

$$\tau_k^{-1} = P_k(k)\Sigma_k^I(-I_k)$$

where $f_k(\omega)$ is a correction function accounting for the higher orders of the expansion about $- I_k$. Hence, if $f_k(\omega)$ is negligible for $\omega \approx - I_k$ and $P_k(k) \approx 1$, then $D_k(\omega)$ has the form of a quasiparticle propagator for $\omega \approx - I_k$. The corresponding quasiparticle may have a finite lifetime τ_k. If a finite basis set of one-particle functions is used, τ_k is always infinite and τ_k^{-1} in (3.32) can be substituted by η. If an infinite basis set containing continuum functions is used, τ_k may become finite. Since the one-particle energies can only become continuous for unoccupied one-particle levels, the $\Sigma_k^I(-I_k)$ vanish when I_k is in the $l = 0$ interval of Fig. 3.1. This means that quasiparticles corresponding to the outer valence electrons have an infinite lifetime. Since for these electrons $P_k(k) \approx 0.9$, which implies with $\tau_k^{-1} = \eta$ and (3.30) that $f_k(\omega \approx -I_k)$ is negligible, the $D_k(\omega \approx -I_k)$ are proper quasiparticle propagators for outer valence ionizations.

2. Perturbation Expansion of the Self-Energy Part

It has already been mentioned in Section III.A that the Green's function can be expanded in terms of the interaction and that it is very convenient to describe this expansion in terms of diagrams. The diagrammatic expansion we use here is well known and can be found in many textbooks.[24,38] We shall not therefore go into the details of the method. Owing to their great importance, however, the rules for handling diagrams are summarized in what follows.

In a diagrammatic approach one starts by representing the diverse quantities and functions by pictorial symbols. The definitions we use are shown in Fig. 3.2. To obtain an idea of how equations can be represented by diagrams, the Dyson equation is shown in Fig. 3.3. The diagrams representing the nth order of the expansion of G are obtained by drawing all topologically nonequivalent linked diagrams having n $V_{ij[kl]}$ points and $(2n+1)$ G^0 lines. The elements of one kind can be connected only with elements of the other kind. Two diagrams are topologically equivalent if it is possible to transform them into one another by twisting or pulling the diagrams without breaking any G^0 line. Such diagrams are often referred to as Hugenholtz[48] or Abrikosov[38] diagrams. If we draw the diagrams with the wiggly lines representing V_{ijkl} instead of the $V_{ij[kl]}$ points, the diagrams are called Feynman diagrams. It is clear that each Abrikosov diagram contains several different Feynman diagrams, leading to fewer diagrams in each order of the Abrikosov expansion. An example is given in Fig. 3.4. In case $s_{ij} \neq 0$, there are n_1 elements s_{ij} and n_2 $V_{ij[kl]}$ points with $n_1 + n_2 = n$.

Let us return to the Dyson equation in Fig. 3.3. From the diagrammatic expansion of G we immediately see that the self-energy part can also be expanded diagrammatically. In Fig. 3.5 we show some terms of the expansion of G and with the aid of Fig. 3.3 one easily recognizes the corresponding diagrams of Σ. These diagrams are also shown in Fig. 3.5. Inserting one diagram of Σ into the Dyson equation implies an infinite number of diagrams in the expansion of G. The rules to draw the Σ diagrams are as easy as for the G diagrams. Instead of connecting n $V_{ij[kl]}$ points with $(2n+1)$ G^0 lines, one has to use only $(2n-1)$ G^0 lines and to take care that diagrams which split into two diagrams by removing a single G^0 line do not belong to Σ according to the Dyson equation.

Subsequently we discuss the rules to evaluate a given diagram of the self-energy part. There are two different sets of rules: one for $\Sigma(t, t')$ and

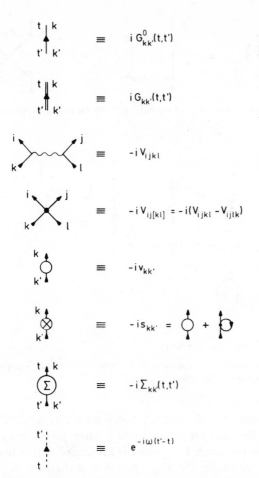

$$\equiv \ i\,G^0_{kk'}(t,t')$$

$$\equiv \ i\,G_{kk'}(t,t')$$

$$\equiv \ -i\,V_{ijkl}$$

$$\equiv \ -i\,V_{ij[kl]} = -i(V_{ijkl} - V_{ijlk})$$

$$\equiv \ -i\,v_{kk'}$$

$$\equiv \ -i\,s_{kk'} =$$

$$\equiv \ -i\,\Sigma_{kk'}(t,t')$$

$$\equiv \ e^{-i\omega(t'-t)}$$

Fig. 3.2. The definition of the symbols used in the diagrammatic expansion of the Green's function and the self-energy part.

Fig. 3.3. The Dyson equation.

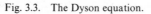

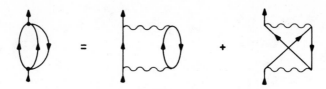

Fig. 3.4.   An Abrikosov diagram of second order and the Feynman diagrams contained in it.

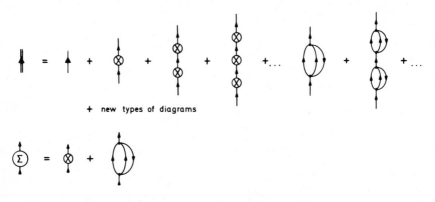

Fig. 3.5.   Some diagrams in the perturbation expansion of the Green's function and the corresponding diagrams of the self-energy part.

one for its Fourier transform $\Sigma(\omega)$. Since we rarely need Σ in time-space, only the rules for $\Sigma(\omega)$ are given. Each G^0 line points from a time t' to a time t, and therefore each $V_{ij[kl]}$ point is associated with a fixed time on the time axis. By permuting the $V_{ij[kl]}$ points in a diagram of nth order, one obtains $n!$ time-ordered diagrams. In Fig. 3.6 the two time-ordered diagrams of the diagram of second order shown in Fig. 3.4 are given. The rules to evaluate a given time-ordered diagram of $\Sigma_{kk'}(t,t')$ in ω-space are:

1.   Join the external indices k and k' with a $e^{-i\omega(t'-t)}$ line.

2.   Draw $n-1$ horizontal lines $-\cdot-\cdot-\cdot-$ between successive pairs of $V_{ij[kl]}$ points. Each of these horizontal lines i is associated with a contribution A_i to the diagram.

3.   Each G^0 line and $e^{-i\omega(t'-t)}$ line cut by a horizontal line i supplies an additive contribution to $1/A_i$, namely $+\omega$ $(-\omega)$ when the $e^{-i\omega(t'-t)}$ line points downward (upward), $+\varepsilon_j$ $(-\varepsilon_j)$ when the G^0_{jj} line points downward (upward). A G^0 line pointing downward is called a hole line; otherwise, a particle line. A hole line corresponds to $j \in \mathfrak{F}$, a particle line to $j \notin \mathfrak{F}$.

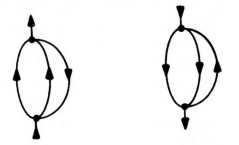

Fig. 3.6.   Time-ordered diagrams of second order.

4.   Multiply the interactions $V_{ij[kl]}$, the contributions of the horizontal lines and a factor $(-1)^{\Sigma_h + \Sigma_l}$, where Σ_h is the number of hole lines and Σ_l is the number of loops; then sum over the internal indices.

5.   Multiply the above contribution by 2^{-q}, where q is the number of permutations of two G^0 lines in the diagram leaving the diagram unchanged.

6.   The above rules 1 to 4 are also valid for a time-ordered Feynman diagram; just replace $V_{ij[kl]}$ points by V_{ijkl} wiggles.

7.   The sign of the $V_{ij[kl]}$ points and the number of loops is not uniquely determined in an Abrikosov diagram. To obtain the proper sign of the diagram compare with a Feynman diagram contained in it.

As an example, the evaluation of a time-ordered Abrikosov diagram of third order is illustrated in Fig. 3.7. The result is

$$\frac{1}{4} \sum_{\substack{s,u,l \in \mathcal{F} \\ m,n \notin \mathcal{F}}} \frac{V_{kl[su]} V_{su[mn]} V_{mn[k'l]}}{(\varepsilon_s + \varepsilon_u - \varepsilon_m - \varepsilon_n)(\omega + \varepsilon_l - \varepsilon_m - \varepsilon_n)} \qquad (3.33a)$$

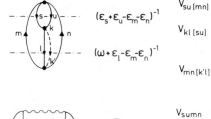

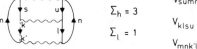

Fig. 3.7.   An example for the evaluation of a diagram (for more detail see text).

All orbitals occuring in (3.33a) are spin orbitals, that is, $m \equiv (m, \sigma_m)$. In case of a *closed-shell* system the orbitals defined by (m, σ_α) and (m, σ_β) have the same spatial function and a given Feynman diagram can easily be simplified by the following rule:

8. Multiply the contribution of a Feynman diagram by 2^{Σ_I} and replace spin orbitals by spatial orbitals.

As an example, the time-ordered Abrikosov diagram in Fig. 3.7 is evaluated by including rule 8 and taking account of the symmetry properties of V_{ijkl} $(= V_{jikl} = V_{kjil} = V_{ilkj} = V_{klij} = V_{jkli} = V_{lkji} = V_{lijk})$:

$$\sum \frac{(2V_{klsu} - V_{klus})V_{sumn}V_{mnk'l}}{(\varepsilon_s + \varepsilon_u - \varepsilon_m - \varepsilon_n)(\omega + \varepsilon_l - \varepsilon_m - \varepsilon_n)} \tag{3.33b}$$

where s, u, l are now doubly occupied orbitals and m, n empty orbitals.

In Section III.B.1 a general expression for the ω-independent component $\Sigma(\infty)$ of the self-energy part has been given without proof. With the aid of diagrams, (3.28) can easily be derived. First we must note that $\Sigma(\infty)$ is exactly the sum of the diagrams in the expansion of $\Sigma(\omega)$ that do not depend on ω. We call such diagrams constant diagrams. Examples for such diagrams are the diagrams 3 in Fig. 3.8 and 11 to 13 in Fig. 3.9. These diagrams are characterized by the fact that both external indices k and k' of any diagram of $\Sigma_{kk'}(\infty)$ are, apart from an infinitesimal, on the same time level. Obviously, these indices belong to the same $V_{kj[k'l]}$ point in each diagram. The other two indices j and l of this $V_{kj[k'l]}$ point must be joined together by a "dressed" line that contains all possible diagrams starting and ending with a G^0 line. This "dressed" line must be a G line. Writing down (3.28) in its diagrammatic version, Fig. 3.10, is consistent with the above inspection.

Let us now restrict ourselves to the closed-shell case and discuss the first few orders in the expansion of Σ. For simplicity, we choose the RHF

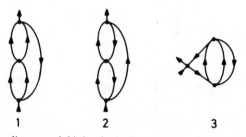

1   2   3

Fig. 3.8. The three diagrams of third order in the expansion of the self-energy part ($s_{ij} = 0$).

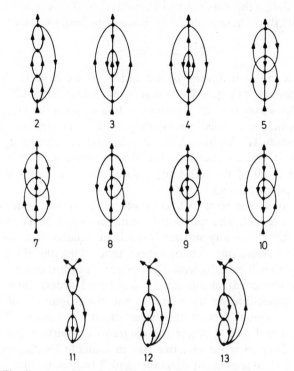

Fig. 3.9. The diagrams of fourth order in the expansion of the self-energy part ($s_{ij} = 0$).

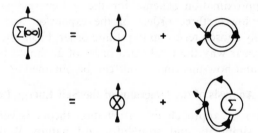

Fig. 3.10. A formula for the sum of the constant diagrams in the expansion of the self-energy part.

potential to define the unperturbed Hamiltonian, that is, $s_{ij} = 0$ as discussed in Section III.B.1. The expansion of Σ is symbolically written

$$\Sigma = \Sigma^{(1)} + \Sigma^{(2)} + \cdots \tag{3.34}$$

where $\Sigma^{(n)}$ is the contribution of nth order. (If we have to deal with an open-shell system with spin S, we denote the nth order of $\Sigma^{(S)}$ by $^{(n)}\Sigma^{(S)}$). From Fig. 3.5 we see that $\Sigma^{(1)}$ vanishes. It follows from the Dyson equation that $G^{(1)}$, which is defined analogously to $\Sigma^{(1)}$, is identical to the free Green's function G^0. Writing $I_k = -\varepsilon_k + \Delta_k$ and calculating Δ_k up to first order in the interaction between the HF particles, we obtain $I_k^{(1)} = -\varepsilon_k$, which is the result of the well-known Koopman's theorem.[2] We therefore call Δ_k Koopmans' defect.

The first correction term to I_k comes from the second-order diagram of Σ shown in Fig. 3.4. The numerical results presented in Section VI show clearly that $\Sigma^{(2)}$ in no way suffices to evaluate accurate IPs via the Dyson equation. Therefore, one cannot expect that taking the third order into account as well will be sufficient. The diagrams of third order are shown in Fig. 3.8. A numerical evaluation of additional orders turns out to be completely impossible, as the numbers and the magnitude of the expressions rises enormously with the order of the expansion. This is best demonstrated in Fig. 3.9, where the diagrams of fourth order are drawn. One should keep in mind that one has to evaluate 13 diagrams, each of them having 4! time-ordered diagrams and 7 indices to sum over. If one counts the diverse time-ordered Feynman diagrams, one obtains 4, 84, and 3120 terms in second, third, and fourth order, respectively. We therefore look for an approximation scheme for the self-energy part that takes account of the lowest three orders of the expansion and, in addition, contains carefully chosen terms up to infinite order. From the next sections it will become clear why the total third order of Σ should be included in the final result and how this final result can be obtained.

C. Analysis of the Diagrams of the Self-Energy Part

The diagrammatic approach to perturbation theory is very convenient because of its simplicity and straightforward nature. With some care, diagrams can be interpreted in physical terms and intuition may help in preselecting important diagrams or classes of diagrams. We have seen in Section III.B.2. that already the fourth order of the expansion contains too many terms to render a numerical evaluation possible. Moreover, even if we were able to calculate the fourth order, how could we be sure that convergence had been obtained? To make further progress, one has to work with physical and mathematical intuition to find out which terms of

Σ are important. As a first step toward the solution of the problem, we briefly investigate the results in other fields of physics with respect to their applicability in atomic and molecular physics.

1.   A Short Comparison with Electron Gas and Nuclear Matter

It is a well-established result that the sum of the ring diagrams (RPA) is an accurate approximation to the self-energy part of a high-density electron gas.[49] The RPA diagrams are shown in Fig. 3.11a together with the corresponding Abrikosov diagrams, Fig. 3.11b, which of course include all exchange diagrams not included in the RPA. We call the latter set of diagrams generalized RPA (GRPA). It is common in the treatment of metals to introduce a parameter r_s describing the volume per electron in an electron gas. With r_1 denoting the Bohr radius, the volume per electron in an homogeneous electron gas is

$$\Omega_e = \frac{4\pi}{3}(r_s r_1)^3 \tag{3.35}$$

One speaks of a high-density electron gas when $r_s \ll 1$.

In principle, one could treat atoms and molecules as an electron gas. We therefore have to investigate the parameters r_s for atoms and molecules. It is simpler to discuss atoms. Let R_n be the radius of the nth Bohr orbit, N the number of electrons, and Z_{eff} the screened nuclear charge[50]; then from

$$R_n \approx n^2 r_1 Z_{\text{eff}}^{-1} \tag{3.36}$$

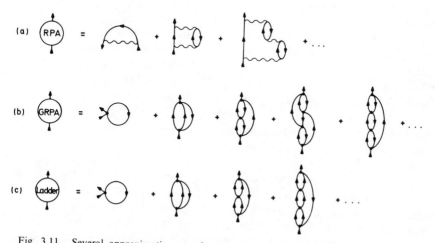

Fig. 3.11.   Several approximations to the self-energy part. (a) The random phase approximation (RPA). (b) The generalized RPA (GRPA). (c) The ladder approximation.

it follows that

$$r_s \approx n^2 Z_{eff}^{-1} N^{-1/3} \tag{3.37}$$

n in (3.37) refers to the outermost electron. In Table I the r_s for some atoms are listed. With the exception of He, all of these r_s are larger than 1, especially those of the alkali atoms, for which (3.37) is most suitable. Therefore, the RPA cannot be expected to work well for atoms. If we divide the electrons into shells, the r_s for each of these might be considerably smaller than the r_s in Table I. This suggests that one might investigate atoms in terms of a *nonuniform* high-density electron gas. Brueckner[51] tried to evaluate correlation energies of atoms within the framework of a high-density nonuniform electron-gas model. He arrived at the important conclusion that it makes no sense to treat atoms in such a model since the density gradient does not converge.

TABLE I

r_s for Some Atoms. The Effective Nuclear Charges Z_{eff} Are from Ref. 50.

Atom	N	n	Z_{eff}	r_s
He	2	1	1.35	0.6
Li	3	2	1.25	2.2
N	7	2	2.07	1.0
P	15	3	2.64	1.4
Ca	20	4	2.68	2.2
Kr	36	4	4.06	1.2

In the theory of nuclear matter it is common to introduce a parameter similar to r_s. With d denoting the effective range of the interaction and r_0 standing for the average distance between the interacting particles, one speaks of low-density, if $d/r_0 \ll 1$. Galitskii[52] showed that each hole line in a diagram is associated with a factor d/r_0. In the low-density case, the sum of diagrams having the lowest possible number of hole lines clearly dominates. Since each diagram must have at least one hole line, the so-called ladder diagrams, shown in Fig. 3.11c, are the most important ones.

For atoms and molecules d/r_0 is far from being small. Indeed, the ladder diagrams with only one hole line are smaller than the same diagrams with the reversed time order, that is, the diagrams with maximum number of hole lines. This will be seen in the next sections. In what follows we mean by ladder diagrams the Abrikosov diagrams and not the time-ordered version discussed above.

All these considerations lead to the result that atoms and molecules are located in an intermediate region, which is far from both the high- and low-density extremes. Therefore, one cannot expect that just a few diagrams dominate in each order. This renders a reasonable evaluation of the self-energy part much more difficult. We shall see below that both the ladder and the GRPA diagrams make considerable contributions, but often tend to compensate each other. This remarkable behavior helps in the search for an adequate approximation to Σ.

2. *Relations Between Diagrams of Third Order*

In the present section we restrict ourselves to closed-shell systems and choose the HF operator as the unperturbed Hamiltonian. With this choice of the Hamiltonian, Σ vanishes in first order and the only second-order term is represented by the diagram shown in Fig. 3.4. Both the GRPA and the ladder approximation contain this diagram. In third order Σ has three diagrams (Fig. 3.8), one of which is a constant diagram. This diagram can be evaluated with the relation shown in Fig. 3.10 by choosing Σ and G on the right-hand side to be $\Sigma^{(2)}$ and G^0, respectively. It is essential to note that the ladder series contains one of the remaining diagrams (1 in Fig. 3.8), which we refer to as diagram C, whereas the GRPA series contains the other diagram, referred to as D. Thus the whole ω-dependent self-energy part up to third order is included in a properly chosen combination of both series. Since we are, in the case of atoms and molecules, far from being near to one of the two extremes represented by the ladder and GRPA series, it is important to investigate and to compare the two diagrams of third order. Such an investigation might give us an idea how to select the important terms in the expansion of Σ. For completeness we also discuss the constant diagram of third order, which we denote by A.

In order to study the diagrams in some detail, we split each diagram into its six time orders. The resulting 18 time-ordered diagrams are shown in Fig. 3.12, where the nomenclature of Ref. 27 is used. For the sake of simplicity only the diagonal elements of $\Sigma_{kk'}$ are considered, that is, $k = k'$. It is easy to prove the following relations:

$$A3 = A4 \qquad A5 = A6$$

$$X2 = X3 \qquad X4 = X5 \qquad X = C, D \tag{3.38}$$

There remain 12 different time-ordered diagrams. It is difficult to establish additional exact relations, if indeed it is possible at all. We are also faced with the problem that the C and D diagrams depend on ω, and it is difficult to prove relations that are valid on the whole ω scale. The

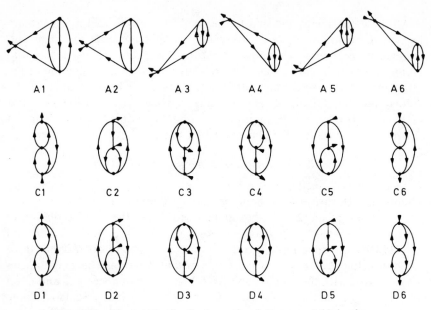

Fig. 3.12.   The notation for the time-ordered diagrams of third order.

following discussion is therefore restricted to $\omega \approx \varepsilon_k$, where $\varepsilon_k \in (\Sigma_{-1}, \Sigma_{+1})$, that is, to the description of the outer valence ionizations and of the corresponding electron affinities. This does not mean that the relations obtained below are not valid for other ω ranges.

By inspection of the expressions given in the Appendix, one obtains

$$A1 < 0 \qquad A2 > 0$$
$$D1, C6 < 0 \qquad D6, C1 > 0 \tag{3.39}$$

In addition, it can be shown that there exist pairs of diagrams that nearly compensate each other. Although the compensation is not complete, it is essential to consider each pair as one quantity in the perturbation expansion. Such diagrams are referred to as antigraphs.[27] In third order there exist six different pairs of diagrams, and five of these constitute antigraph pairs. It is important to notice that, apart from the constant diagram A, each antigraph pair contains one time-ordered diagram of the ladder series and one of the GRPA series.

To prove the antigraph relations we collect for each diagram those terms that have the greatest contribution to the diagram.[53] As the most simple example we discuss the pair $(A1, A2)$ explicitly. With the rules of Section

III.B.2 to evaluate diagrams for a closed-shell system it is easily found that

$$A1_k = - \sum \frac{(2V_{krkt} - V_{kkrt})(2V_{tsab} - V_{tsba})V_{abrs}}{(\varepsilon_s + \varepsilon_t - \varepsilon_a - \varepsilon_b)(\varepsilon_s + \varepsilon_r - \varepsilon_a - \varepsilon_b)} \qquad (3.40)$$

$$A2_k = \sum \frac{(2V_{kcka} - V_{kkca})(2V_{tscb} - V_{tsbc})V_{abts}}{(\varepsilon_s + \varepsilon_t - \varepsilon_a - \varepsilon_b)(\varepsilon_s + \varepsilon_t - \varepsilon_b - \varepsilon_c)}$$

where $s,r,t,\ldots$ denote doubly occupied orbitals, $a,b,c,\ldots$ unoccupied orbitals, and $i,j,k,\ldots$ stand for both. Collecting only those terms that lead to the main contributions to $A1_k$ and $A2_k$, one obtains the following equations:

$$A1_k + A2_k = - \sum_{k \in \mathcal{F}} \frac{(2V_{tsab} - V_{tsba})V_{abts}}{(\varepsilon_s + \varepsilon_t - \varepsilon_a - \varepsilon_b)^2} \left[V_{ktkt}(2 - \delta_{kt}) - 2V_{kaka} \right] \quad (3.41)$$

$$A1_k + A2_k = \sum_{k \notin \mathcal{F}} \frac{(2V_{tsab} - V_{tsba})V_{abts}}{(\varepsilon_s + \varepsilon_t - \varepsilon_a - \varepsilon_b)^2} \left[V_{kaka}(2 - \delta_{ka}) - 2V_{ktkt} \right]$$

For both occupied and unoccupied k, the antigraph relation $A1_k \approx - A2_k$ follows with the realistic assumption

$$\overline{V}_{ktkt}(2 - \delta_{kt}) \approx 2\overline{V}_{kaka}$$

$$\overline{V}_{kaka}(2 - \delta_{ka}) \approx 2\overline{V}_{ktkt}$$

where $\overline{V}$ denotes the average value over all possible t and a according to (3.41).

Analogously, one obtains additional antigraph relations. The complete set of relations is

$$A1_k \approx - A2_k \qquad A3_k \approx - A5_k$$

$$C1_k \approx - D1_k \qquad C6_k \approx - D6_k \qquad (3.42)$$

$$C2_k \approx - D2_k \qquad k \in \mathcal{F}$$

$$C4_k \approx - D4_k \qquad k \notin \mathcal{F}$$

The antigraph relation for the pair $(A3, A5)$ is less founded. For these diagrams it is hard to collect those terms leading to the main contributions. However, if $|A3|$ and $|A5|$ are considerable, they are antigraphs to each other because of $\overline{V}_{abas} \approx \overline{V}_{rbrs}$.

To illustrate the validity of the antigraph relations, the different contributions of the diagrams of third order are listed in Table II for F_2 and Ne. It can be seen that the contributions of the pair $(C4, D4)$, which is not a pair of antigraphs for $k \in \mathscr{T}$, are considerable. We suppose, however, that this large contribution is due to the high symmetry of small molecules. For larger molecules of low symmetry we have found the contribution of $(C4, D4)$ to be smaller.

To gain some insight into the origin of the antigraph relations, we consider a simple two-orbital model. In this model only one doubly occupied orbital s and one unoccupied orbital b contribute to the self-energy part. To make the model even simpler we take all six possible V_{ijkl} as equal. One then finds that the pairs $(A1, A2)$, $(A3, A5)$, $(C1, D1)$, $(C2, D2)$, $(C6, D6)$, and even $(C4, D4)$ are pairs of antigraphs for Σ_{ss} and for Σ_{bb}. Breaking the symmetry of this two-level model by introducing additional one-particle levels mostly affects the relation between $C4$ and $D4$ of Σ_{ss}. A two-level model is rather unrealistic, but may be approximately applicable to H_2 and Li^- (only for the outermost orbital). The contributions of the diagrams of third order for these systems have been calculated[53] by taking into account 12 orbitals and are listed in Table III. The various diagrams have only small contributions. Their sum is -0.044 eV for $\Sigma_{2s, 2s}$ of Li^- and -0.087 eV for $\Sigma_{1\sigma_g, 1\sigma_g}$ of H_2. Adding, however, the absolute values of these diagrams, one obtains 2.796 eV for Li^- and 2.941 eV for the hydrogen molecule.

Let us summarize these results. The contributions of some of the time-ordered diagrams of third order have been found to be large and are often comparable in magnitude to the diagram of second order. If we assume the convergence of the perturbation expansion, some of the diagrams should compensate each other at least partly. This is confirmed by the relations in (3.42). The a priori assumption that the contributions of the individual diagrams decrease with increasing order of the expansion cannot be maintained. Hence, it is useless to consider as many diagrams of a certain order as possible without closer examination. As each antigraph pair contains a ladder and a GRPA diagram, both the ladder and the GRPA series must be taken into account to an equal extent. We return to this point in Section IV, where our final approximation scheme is derived.

3.   Decomposition into Reorganization and Correlation Effects

To calculate atomic and molecular IPs via the Green's function approach, one starts from a given basis set $\{\varphi_i\}$ of spin orbitals i. It is convenient to choose $\{\varphi_i\}$ to be the one-particle orbitals of the neutral molecule. There is, of course, also the possibility to calculate the IPs by

TABLE II

The Contributions of the Diagrams of Third Order for F_2 and Ne. Each Diagram Has Been Evaluated at $\omega = \varepsilon + \Sigma^{(2)}(\omega)$. For F_2 the First Five Orbitals Are Occupied and for Ne the First Two. All Energies in eV.

System	Orb.	A1	A2	2A3	2A5	C1	D1	2C2	2D2	2C4	2D4	C6	D6
F_2	$2\sigma_g$	−2.30	2.30	0.97	−1.37	0.05	−0.06	0.16	−0.14	−0.04	−2.16	−8.04	8.52
	$2\sigma_u$	−2.25	2.20	0.94	−1.37	0.05	−0.06	0.16	−0.13	0.13	−2.60	−5.62	7.06
	$1\pi_u$	−2.19	2.21	0.84	−1.15	0.09	−0.10	0.27	−0.22	−0.28	−2.30	−3.15	3.02
	$3\sigma_g$	−1.77	2.04	0.85	−1.32	0.79	−0.86	1.42	−1.56	−0.26	−0.33	−1.01	0.89
	$1\pi_g$	−2.20	2.25	0.88	−1.21	0.15	−0.15	0.41	−0.30	−0.21	−1.84	−2.34	2.15
	$3\sigma_u$	−2.21	1.65	0.97	−1.46	0.69	−0.98	0.26	0.39	−1.42	1.68	−0.75	0.77
	$4\sigma_u$	−1.71	1.50	0.46	−0.75	1.40	−1.89	0.13	0.98	−0.25	0.27	−0.10	0.09
	$2\pi_g$	−1.73	1.61	0.42	−0.58	1.31	−1.27	0.18	0.79	−0.15	0.15	−0.12	0.09
	$4\sigma_g$	−1.77	1.63	0.46	−0.77	3.18	−4.83	0.14	1.44	−0.11	0.12	−0.06	0.05
Ne	$2s$	−1.04	0.99	1.40	−2.00	0.08	−0.08	0.20	−0.18	−0.14	−1.34	−2.30	1.48
	$2p$	−0.98	0.97	1.31	−1.88	0.15	−0.15	0.36	−0.28	−0.20	−1.10	−2.00	1.31
	$3p$	−0.62	0.53	0.30	−0.46	0.12	−0.14	0.07	0.10	−0.14	0.14	−0.11	0.08
	$3s$	−0.81	0.66	0.61	−0.95	0.33	−0.43	0.08	0.37	−0.12	0.14	−0.08	0.07

TABLE III

The Contributions of the Diagrams of Third Order for Li^- and H_2 in eV. Each Diagram Has Been Evaluated at $\omega = \varepsilon + \Sigma^{(2)}(\omega)$. It Should Be Noted that the Two Level Model (see text) is not Applicable to the $1s$ Orbital of Li^-.

System	Orb.	A1	A2	2A3	2A5	C1	D1	2C2	2D2	2C4	2D4	C6	D6
Li^-	$1s$	−0.54	0.57	0.47	−0.30	0.07	−0.12	0.18	−0.24	−0.04	−0.30	−2.32	3.44
	$2s$	−0.18	0.30	0.08	−0.05	0.21	−0.27	0.31	−0.55	−0.08	0.05	−0.30	0.42
H_2	$1\sigma_g$	−0.15	0.26	0.08	−0.05	0.22	−0.26	0.31	−0.52	−0.08	0.05	−0.45	0.51
	$1\sigma_u$	−0.14	0.13	0.03	−0.02	0.07	−0.12	0.00	0.00	−0.05	0.07	−0.02	0.04

237

separately determining the molecular ground-state energy and the energies of the diverse ionic states. If both the ground-state energy and the ionic-state energy are calculated on the HF level, one speaks of the ΔSCF procedure. In this case one takes account of reorganization (relaxation) effects. After the electron k has been ejected, the remaining $N-1$ electrons reorganize in order to minimize the energy. To obtain accurate IPs one must also include the electron correlation in both the ground and the ionic states. Thus there are corrections to an IP obtained via Koopmans' theorem because of the effects of reorganization and of changes of correlation. A decomposition of the corrections obtained in the Green's function approach into reorganization and correlation contributions is therefore of interest.

The contribution of the second-order self-energy part to Koopmans' defect is now considered. We follow the method of Pickup and Goscinski.[11] For simplicity, only terms up to $|V_{ij[mn]}|^2$ are retained. The IP I_k of electron k is given by

$$-I_k = \varepsilon_k + \Sigma_{kk}^{(2)}(\varepsilon_k) + O(V^3)$$

$$= \varepsilon_k + \frac{1}{2} \sum_{j,m,n} \frac{|V_{kj[mn]}|^2}{\varepsilon_k + \varepsilon_j - \varepsilon_m - \varepsilon_n} (\bar{n}_j n_m n_n + n_j \bar{n}_m \bar{n}_n) + O(V^3) \qquad (3.43)$$

where n_i and $\bar{n}_i$ are the occupation numbers defined in (3.22). Introducing orbital energies $\tilde{\varepsilon}_i$ and matrix elements $\tilde{V}_{ij[mn]}$ for the ionic state in question, the total HF energies for molecule and ion become

$$E_0^N(\text{HF}) = \sum_i \varepsilon_i n_i - \frac{1}{2} \sum_{ij} V_{ij[ij]} n_i n_j \qquad (3.44)$$

$$E_k^{N-1}(\text{HF}) = \sum_{i \neq k} \tilde{\varepsilon}_i n_i - \frac{1}{2} \sum_{ij \neq k} \tilde{V}_{ij[ij]} n_i n_j$$

Expressing $\tilde{\varepsilon}_i$ and $\tilde{V}_{ij[mn]}$ in terms of the ε_a and $V_{ab[cd]}$ via Rayleigh–Schrödinger perturbation theory, one obtains

$$-I_k(\Delta\text{SCF}) = E_0^N(\text{HF}) - E_k^{N-1}(\text{HF})$$

$$= \varepsilon_k + \sum_{j,n} \frac{|V_{kj[kn]}|^2}{\varepsilon_j - \varepsilon_n} \bar{n}_j n_n + O(V^3) \qquad (3.45)$$

Equation 3.43 can be rewritten to exhibit this reorganization term explicitly:

$$-I_k = \varepsilon_k + \sum_{j,n} \frac{|V_{kj[kn]}|^2}{\varepsilon_j - \varepsilon_n} \bar{n}_j n_n$$

$$+ \frac{1}{2} \sum_{j,m,n} \frac{|V_{kj[mn]}|^2}{\varepsilon_k + \varepsilon_j - \varepsilon_m - \varepsilon_n} n_j \bar{n}_m \bar{n}_n + \frac{1}{2} \sum_{m,n \neq k,j} \frac{|V_{kj[mn]}|^2}{\varepsilon_k + \varepsilon_j - \varepsilon_m - \varepsilon_n} \bar{n}_j n_m n_n + O(V^3)$$

$$(3.46)$$

It is also possible to interprete the last two terms in (3.46). To this end one may use conventional Rayleigh–Schrödinger perturbation theory to derive the second-order expression for the correlation energies:

$$E_0^N = E_0^N (\text{HF}) + \frac{1}{4} \sum_{ij,mn} \frac{|V_{ij[mn]}|^2}{\varepsilon_i + \varepsilon_j - \varepsilon_m - \varepsilon_n} n_i n_j \bar{n}_n \bar{n}_m \qquad (3.47)$$

$$E_k^{N-1} = E_k^{N-1} (\text{HF}) + \frac{1}{4} \sum_{\substack{ij \neq k \\ m,n}} \frac{|V_{ij[mn]}|^2}{\varepsilon_i + \varepsilon_j - \varepsilon_m - \varepsilon_n} n_i n_j \bar{n}_n \bar{n}_m$$

$$+ \frac{1}{2} \sum_{i,j \neq k,m} \frac{|V_{km[ij]}|^2}{\varepsilon_i + \varepsilon_j - \varepsilon_m - \varepsilon_k} n_i n_j \bar{n}_m$$

Determining $I_k = E_k^{N-1} - E_0^N$ recovers (3.46). Because of the removal of spin orbital k, a part of the pair correlation energies[54-56] in the ground state disappears, whereas the remaining pair correlation energies change because of reorganization. The corresponding contributions to Koopmans' defect in second order are given by the third and fourth terms in (3.46).

It is well known from a variety of calculations[57] that the ΔSCF proce-dure leads to rather accurate IPs of core electrons. This could be explained by Hedin and Johansson[58] and is easily understood from (3.46). The denominators of the terms in (3.46) that account for changes of correlation energies contain the core orbital energy ε_k, which is not the case for the reorganization term. When, in addition, the matrix elements appearing in the diverse terms are compared, it becomes clear that the reorganization term in (3.46) dominates the other terms. Since the inclusion of reorganiza-tion effects lowers the energy of the ion, it is not surprising to find the corresponding term in (3.46) to be positive definite.

The situation is much more complicated when the balance between reorganization and correlation effects is to be discussed for valence electron ionization. Here, both contributions are often of the same order of magnitude and the sign of the latter contribution may vary from molecule to molecule and even from orbital to orbital. An illustrative example is the nitrogen molecule. Koopmans' theorem predicts the sequence of IPs[59]: $1\pi_u, 3\sigma_g, 2\sigma_u$, in contradiction to the experimental sequence[1c]: $3\sigma_g, 1\pi_u, 2\sigma_u$. When reorganization effects are included, the sequence is still wrong[59]: $1\pi_u, 3\sigma_g, 2\sigma_u$. The reason is that for the $3\sigma_g$ and $2\sigma_u$ orbitals reorganization contributions have the same sign and about the same magnitude as the correlation contributions. For the $1\pi_u$ orbital, on the other hand, both contributions tend to cancel each other almost completely. This explains the good value obtained by applying Koopmans' theorem to the $1\pi_u$ orbital and the poor values obtained for the σ-type orbitals. In general, ΔSCF calculations[60] do not provide an improvement over Koopmans' theorem for valence electrons.

D. What Happens to the Green's Function When the Ground-State is Uncorrelated?

Should one be interested in the ionization potential of an atom or molecule with two electrons, one could equally look for the electron affinity of the ion. Since this ion has only one electron, its wave function is an eigenfunction of H_0 as well as of H. In such a case we may ask whether the Green's function formalism can be simplified. On the other hand, we may also pose this question when the ground-state wave function Ψ_0^N is not exactly an eigenfunction of H_0. In this case we could perform the calculation as if Ψ_0^N were an eigenfunction of H_0. The theoretical result of such a calculation is important, even when the ground-state eigenfunction of H is far from being an eigenstate of H_0. Apart from having an additional way of interpreting certain terms in the perturbation expansion, we can examine given approximation schemes with respect to their inclusion of the ground-state correlation energy.

For definiteness, we start from the Hamiltonian (3.20). Then

$$|\Psi_0^N\rangle = |\Phi_0^N\rangle = \prod_{i=1}^{N} a_i^+|0\rangle$$

$$(H - H_0)|\Phi_0^N\rangle = (E_0^N - E_0^{(0)N})|\Phi_0^N\rangle \tag{3.48}$$

where $E_0^{(0)N}$ is the sum of the one-particle energies ε_i with $i \in \mathcal{F}$ and

$$E_0^N = \sum_{i \in \mathcal{F}} \varepsilon_i + \frac{1}{2}\sum_{i,j \in \mathcal{F}} V_{ij[ij]} + \sum_{i \in \mathcal{F}} v_{ii} \tag{3.49}$$

In the case in which the ε_i are the HF orbital energies of a closed-shell system, E_0^N simply reduces to the HF ground-state energy.

Returning to the definition of the Green's function in (3.1) and making use of (3.48) we obtain

$$G_{kl}(t,t') = \begin{cases} -ie^{iE_0^N(t-t')}\langle\Phi_0^N|a_k e^{-iH(t-t')}a_l^+|\Phi_0^N\rangle & t > t' \text{ and } k,l \notin \mathcal{F} \\ ie^{-iE_0^N(t-t')}\langle\Phi_0^N|a_l^+ e^{iH(t-t')}a_k|\Phi_0^N\rangle & t \leqslant t' \text{ and } k,l \in \mathcal{F} \end{cases}$$

$$(3.50)$$

The Green's function matrix $\mathbf{G}$ splits into two blocks, one, $\mathbf{G}^l$, for orbitals that are occupied in the ground-state and one, $\mathbf{G}^u$, for unoccupied orbitals. Comparing with (3.3), it is clear that $G_{kl}(\omega)$ is analytic in the lower half-plane of complex ω for $k,l \in \mathcal{F}$ and analytic in the upper half-plane for $k,l \notin \mathcal{F}$. One may therefore start from two separate Dyson equations,

$$\mathbf{G}^u = \mathbf{G}^0 + \mathbf{G}^0\mathbf{\Sigma}^u\mathbf{G}^u$$

$$\mathbf{G}^l = \mathbf{G}^0 + \mathbf{G}^0\mathbf{\Sigma}^l\mathbf{G}^l$$

$$(3.51)$$

where $\mathbf{\Sigma}^u$ and $\mathbf{\Sigma}^l$ are analytic in the upper and lower half-planes, respectively.

We have seen that (3.51) follows from (3.48), but it is also possible to show that (3.51) implies a vanishing ground-state correlation energy. To prove the latter statement we start from the expression (3.10) for the ground-state energy, remembering that ε in (3.10) must be substituted by $\varepsilon + v$. With the aid of (3.51) one immediately obtains

$$E_0^N = \frac{1}{4\pi i}\oint^{\mathfrak{R}}\mathrm{tr}(\omega\mathbf{1} + \varepsilon + v)\mathbf{G}^l(\omega)\,d\omega$$

$$= \sum_{i \in \mathcal{F}}\left(\varepsilon_i + \tfrac{1}{2}v_{ii}\right) + \frac{1}{4\pi i}\oint^{\mathfrak{R}}\mathrm{tr}(\omega\mathbf{1} + \varepsilon + v)\mathbf{G}^0\mathbf{\Sigma}^l\mathbf{G}^l\,d\omega \qquad (3.52)$$

To evaluate the last term in the latter equation we decompose $\mathbf{\Sigma}^l$ according to (3.27) into a constant and a ω-dependent part and make further use of (3.51). The result is very simple:

$$E_0^N = \sum_{i \in \mathcal{F}}\left(\varepsilon_i + \tfrac{1}{2}v_{ii} + \tfrac{1}{2}\Sigma_{ii}(\infty)\right) \qquad (3.53)$$

To complete the proof we take advantage of (3.28b) and obtain

$$E_0^N = \sum_{i \in \mathcal{F}}\varepsilon_i + \sum_{i \in \mathcal{F}}v_{ii} + \tfrac{1}{2}\sum_{i,j \in \mathcal{F}}V_{ij[ij]} \qquad (3.54)$$

thus recovering (3.49). In the following we start from (3.51) as our definition of an uncorrelated ground state.

To proceed further we rewrite $a_i(t)$ to give $S^{-1}(t,0)\hat{a}_i(t)S(t,0)$, where $S(t,t_0)$ is the operator defined in (3.6) and $\hat{a}_i(t)$ is the destruction operator in the interaction representation. In complete analogy to (3.8) we now obtain

$$G_{kl}(t) = -i\frac{\langle\Phi_0^N|T\{\bar{S}(\infty,-\infty)\hat{a}_k(t)\hat{a}_l^+\}|\Phi_0^N\rangle}{\langle\Phi_0^N|\bar{S}(\infty,-\infty)|\Phi_0^N\rangle} \tag{3.55a}$$

$$\bar{S}(\infty,-\infty) = \sum_{n=0}^{\infty}\frac{(-i)^n}{n!}\int_{-\infty}^{\infty}\cdots\int_{-\infty}^{\infty}T\{\hat{\bar{H}}_I(t_1)\cdots\hat{\bar{H}}_I(t_n)\}\,dt_1\cdots dt_n$$

where, according to (3.51), $\hat{\bar{H}}_I(t_i)$ is not simply H_I in the interaction representation as in (3.8), but

$$\hat{\bar{H}}_I(t_i) = \hat{H}_I(t_i)\begin{cases} \theta(t-t_i)\theta(t_i) & \text{for } G^u \\ \theta(t_i-t)\theta(-t_i) & \text{for } G^l \end{cases} \tag{3.55b}$$

According to the linked cluster theorem, we again have to calculate only the connected diagrams appearing in the numerator in (3.55a):

$$G_{kl}(t) = -i\langle\Phi_0^N|T\{\bar{S}(\infty,-\infty)\hat{a}_k(t)a_l^+\}|\Phi_0^N\rangle_c \tag{3.56}$$

The main difference between the diagrammatic approach discussed in the preceding sections and the one implied by (3.56) is in the definition of $\bar{S}$. It is easy to see that *only time-ordered diagrams with $V_{ij[kl]}$ points lying between t and 0 have to be drawn: if $t < 0$ they contribute only to $k, l \in \mathfrak{F}$, otherwise to $k, l \notin \mathfrak{F}$. The same rules are also valid for the self-energy part $\Sigma_{kl}(t)$.* In Fig. 3.13 some diagrams are shown that contribute to Σ and some that have a zero contribution.

As a first consequence of the above rules only the constant diagram of *first* order contributes to $\Sigma(\infty)$. In addition, we see that among the time-ordered diagrams of third order only $C1$, $D1$, $C6$, and $D6$ contribute to Σ. It should be mentioned that the diagrams $D4$ and $D5$, which are often found to be essential for calculating IPs, vanish when the ground-state wave function is uncorrelated. When these diagrams and/or the constant diagrams are omitted in a given approximation to Σ, it is clear that a considerable amount of ground-state correlation has been neglected.

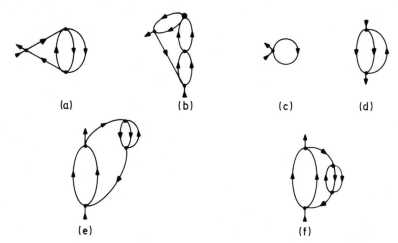

(a)   (b)   (c)   (d)

(e)   (f)

Fig. 3.13.   Time-ordered diagrams (a) to (f). When the ground state is uncorrelated, the diagrams (a), (b), and (e) do not contribute to Σ, (c) contributes to both Σ^u and Σ^l, while (d) contributes only to Σ^l and (f) only to Σ^u.

IV.   A STRUCTURE-CONSERVING RENORMALIZATION

In the preceding sections we have gained information about the self-energy part and its essential role in the Green's function formalism. In the present section this information is used to derive an adequate and compact approximation to the self-energy part. This approximation scheme is appropriate to describe both closed- and open-shell atoms and molecules.

The exact self-energy part has a constant term and an energy-dependent term which we denote by $\mathbf{M}(\omega)$:

$$\Sigma(\omega) = \Sigma(\infty) + \mathbf{M}(\omega) \tag{4.1}$$

We have seen that $\Sigma(\infty)$ can be obtained from $\mathbf{M}(\omega)$ via the self-consistent equation (3.28). Therefore only $\mathbf{M}(\omega)$ is investigated in what follows. The straightforward perturbation expansion of the self-energy part in terms of the matrix elements V_{ijkl} and v_{ij} is certainly limited to low orders, because of the immensely increasing effort involved in calculating higher orders. The lowest-order energy-dependent term is the second-order term $\mathbf{M}^{(2)}(\omega)$ of $\mathbf{M}(\omega)$:

$$M_{pq}^{(2)}(\omega) = \frac{1}{2} \sum_{j,k,l} V_{pj[kl]} V_{qj[kl]} \left[\frac{n_j \bar{n}_k \bar{n}_l}{\omega + \varepsilon_j - \varepsilon_k - \varepsilon_l + i\eta} + \frac{\bar{n}_j n_k n_l}{\omega + \varepsilon_j - \varepsilon_k - \varepsilon_l - i\eta} \right]$$

$$\tag{4.2}$$

It is important to note that this expansion up to second order has the analytical structure of the exact $M(\omega)$ as shown in (3.27). Unfortunately, $M^{(2)}$ has, in general, proved to provide only a poor approximation.[27] The extension to the third or a higher finite order will destroy the simple spectral structure of the second-order expression, since quadratic poles are added already by $M^{(3)}(\omega)$. It is clear that an expansion up to a finite order will completely fail to describe $M(\omega)$ in the neighborhood of its poles, whereas it might be useful if a region far from the poles is only of interest (see Fig. 3.1). The latter is the case for valence ionizations of closed-shell systems, fairly well described by a quasiparticle picture. Here, only one essential pole of G is situated between two largely separated poles of $M(\omega)$. An example of a process beyond the simple quasiparticle picture is the ionization of a core electron accompanied by secondary excitations. In this case several poles of G share in the total intensity, all lying relatively near to poles of $M(\omega)$. For open-shell systems a quasiparticle picture does not apply at all. Because of the multiplet splitting, the removal or addition of an electron will generally yield two or more energetically close final states represented by poles of G with comparable strengths. Evidently, the calculation of these poles—each being enclosed by two successive poles of M—via the Dyson equation requires a correct description of the pole region of $M(\omega)$.

Thus, in order to treat the closed- and the open-shell case, we have to look for a structure-conserving approximation, that is, an approximation for $M(\omega)$ exhibiting the spectral form reflected by the exact self-energy part. Clearly, such an approximation can only be obtained by performing some kind of infinite partial summation. Well-known examples of partial summations of diagrams are the summation of all ring-diagrams and of all ladder-diagrams already discussed in Section III.C.1. Unfortunately, neither the RPA nor the ladder self-energy parts are satisfactory approximations for finite electronic systems. A simple addition of the RPA and ladder series indeed contains all third-order terms of $M(\omega)$ that have been found to be essential, but leads to negative intensities.[31]

A. A Dyson-Like Equation for the Kernel of the Self-Energy Part

All diagrams contributing to $M_{pq}(\omega)$ start with a $V_{qj[kl]}$ point at the external index q and end with a $V_{pj'[k'l']}$ point at the external index p. It is therefore convenient to write $M_{pq}(\omega)$ in the form (for simplicity only real φ_s are considered)

$$M_{pq}(\omega) = \tfrac{1}{4} \sum_{\substack{j,k,l \\ j',k',l'}} V_{qj[kl]} \Gamma_{jkl,j'k'l'}(\omega) V_{pj'[k'l']} \tag{4.3a}$$

thus defining the ω-dependent kernel of the self-energy part. The $\Gamma_{jkl,j'k'l'}(\omega)$ are independent of p and q and can be shown to be antisymmetric in k and l and in k' and l', that is,

$$\Gamma_{jkl,j'k'l'} = -\Gamma_{jlk,j'k'l'} = -\Gamma_{jkl,j'l'k'} = \Gamma_{jlk,j'l'k'}$$

It follows that

$$M_{pq}(\omega) = \sum_{\substack{j,k,l \\ j',k',l'}} V_{qjkl}\Gamma_{jkl,j'k'l'}(\omega)V_{pj'k'l'} \tag{4.3b}$$

We may also introduce a matrix Γ with elements

$$(\Gamma)_{\lambda,\lambda'} = \Gamma_{\lambda\lambda'} = \Gamma_{jkl,j'k'l'}$$

where λ stands from now on for a three-index set. By analogy the vector $\mathbf{V}_p$ has elements $(\mathbf{V}_p)_\lambda = V_{p\lambda} = V_{pjkl}$ and (4.3b) is rewritten to give

$$M_{pq}(\omega) = \mathbf{V}_q^T\Gamma(\omega)\mathbf{V}_p \tag{4.4}$$

From the definition of Γ it is clear that it also has a spectral representation of the kind of (3.27) and is subject to an expansion in powers of $V_{ij[kl]}$ and v_{ij}, where, however, the nth-order term $\Gamma^{(n)}$ corresponds to the $(n+2)$th-order term $\mathbf{M}^{(n+2)}$. The zeroth-order term is directly obtained by comparison with (4.2), giving

$$\Gamma^{(0)}_{jkl,j'k'l'}(\omega) = \delta_{jj'}\left(\delta_{kk'}\delta_{ll'} - \delta_{kl'}\delta_{lk'}\right)\left[\frac{n_j\bar{n}_k\bar{n}_l}{\omega + \varepsilon_j - \varepsilon_k - \varepsilon_l + i\eta} + \frac{\bar{n}_j n_k n_l}{\omega + \varepsilon_j - \varepsilon_k - \varepsilon_l - i\eta}\right]$$

$$\tag{4.5}$$

For the following discussion we may omit the infinitesimal quantity η. The restriction on the indices in (4.5), explicitly given by $n_j\bar{n}_k\bar{n}_l$ and $\bar{n}_j n_k n_l$, is not maintained in the general expression for Γ in (4.3). We therefore introduce the diagonal matrix $(\omega\mathbf{1} - \mathbf{K})$,

$$(\omega\mathbf{1} - \mathbf{K})_{jkl,j'k'l'} = \delta_{jj'}\delta_{kk'}\delta_{ll'}\left(\omega + \varepsilon_j - \varepsilon_k - \varepsilon_l\right) \tag{4.6}$$

containing all possible expressions $\omega + \varepsilon_j - \varepsilon_k - \varepsilon_l$ with no restriction on the indices. Defining further a constant matrix γ^0 by

$$\gamma^0_{jkl,j'k'l'} = \delta_{jj'}\left[\delta_{kk'}\delta_{ll'} - \delta_{kl'}\delta_{lk'}\right] \tag{4.7}$$

the zeroth-order contribution $\Gamma^{(0)}$ can be rewritten in the following way:

$$\Gamma^{(0)}(\omega) = \mathbf{P}^0(\omega\mathbf{1} - \mathbf{K})^{-1}\gamma^0 \tag{4.8}$$

where the matrix $\mathbf{P}^0$,

$$P^0_{jkl,j'k'l'} = \delta_{jj'}\delta_{kk'}\delta_{ll'}\left[n_j\bar{n}_k\bar{n}_l + \bar{n}_j n_k n_l \right] \tag{4.9}$$

is a projector eliminating the unphysical poles contained in $(\omega\mathbf{1} - \mathbf{K})^{-1}$.

In analogy to the way the self-energy part of $\mathbf{G}$ has been introduced, we introduce a new quantity $\mathbf{C}(\omega)$ by writing a Dyson-like equation for Γ,[61]

$$\Gamma(\omega) = (\omega\mathbf{1} - \mathbf{K})^{-1}\gamma + (\omega\mathbf{1} - \mathbf{K})^{-1}\mathbf{C}(\omega)\Gamma(\omega) \tag{4.10}$$

where γ is a constant matrix. Here, the matrix $(\omega\mathbf{1} - \mathbf{K})^{-1}$ fulfills the role the free Green's function $\mathbf{G}^0$ fulfills in the Dyson equation. To be more flexible we do not fix $(\omega\mathbf{1} - \mathbf{K})^{-1}\gamma$ as equal to $\Gamma^{(0)}$ in (4.8), that is, γ need not be equal to $\mathbf{P}^0\gamma^0$.

From (4.10), which is the definition of the meaningful quantity $\mathbf{C}(\omega)$, we conclude that $\mathbf{C}(\omega)$ has properties similar to the self-energy part,

$$C_{\lambda\lambda'}(\omega) = C_{\lambda\lambda'}(\infty) + \sum_\nu \frac{a_\nu^{\lambda\lambda'}}{\omega - b_\nu} \tag{4.11}$$

To proceed further we remember that we have extended our space characterized by the projector $\mathbf{P}^0$ to the complete space spanned by three one-particle indices. This space fully describes the general set of operators which are constructed as all possible products of three fermion operators. If we restrict ourselves to a configurational space, where all the ionic states with $(N-1)$ and $(N+1)$ electrons can be obtained by applying all possible linear combinations of these operators to the ground-state wave function, we arrive[31] at the simplified recursion relation

$$\Gamma(\omega) = (\omega\mathbf{1} - \mathbf{K})^{-1}\gamma + (\omega\mathbf{1} - \mathbf{K})^{-1}\mathbf{C}\Gamma(\omega) \tag{4.12}$$

where $\mathbf{C}$ is now a constant matrix. The resulting poles of $\Gamma(\omega)$ in this equation,

$$\Gamma(\omega) = (\omega\mathbf{1} - \mathbf{K} - \mathbf{C})^{-1}\gamma \tag{4.13}$$

must again be separated into physical and unphysical poles depending on whether they are related to physical or unphysical poles of $(\omega\mathbf{1} - \mathbf{K})^{-1}$. As in (4.8) we can define an operator P that projects out all the unphysical poles of $(\omega 1 - K - C)^{-1}$. Such a projector can, in principle, already be included in γ. Then γ has a very complicated structure implicitly depending on the choice of C. We prefer, therefore, to decompose γ into a product of an operator P and a matrix which, for simplicity, is again denoted by γ. One obtains

$$\Gamma(\omega) = P(\omega 1 - K - C)^{-1}\gamma \qquad (4.14)$$

which is evidently a generalization of $\Gamma^{(0)}(\omega)$ in (4.8). In the above equation the operator P is written in a matrix representation. To find P we consider the eigenvector matrix X diagonalizing the matrix $(K + C)$,

$$(\omega 1 - K - C)^{-1} = X(\omega 1 - D)^{-1}X^{-1}$$

leading to

$$P = XP^0X^{-1} \qquad (4.15)$$

For a more detailed discussion of the projection process we refer to Section IV.B.

Until now we have not specified the matrices γ and C except that they are ω independent. Obviously, the zeroth-order contribution of $\Gamma(\omega)$ is $\Gamma^{(0)}(\omega)$ in (4.8) and one may also consider γ^0 as the zeroth-order expression of γ. It should be pointed out that γ^0 in (4.8) can be multiplied by any diagonal matrix $Y = P^0 + Z$ with $P^0Z = 0$ without changing $\Gamma^{(0)}(\omega)$. This is reason enough to consider γ to be different from γ^0. With $\Gamma^{(0)}(\omega)$ one obtains $M^{(2)}(\omega)$ via (4.4). An important point is that the *whole* third-order contribution $M^{(3)}(\omega)$ is determined by (4.4) when C is expanded up to first order. By comparing the first-order terms arising from the diagrams of $M^{(3)}$ and from the expansion defined by the recursion relation (4.12), one obtains[31]

$$C_{jkl,j'k'l'} = \tfrac{1}{2}\left[A_{jkl,j'k'l'} - A_{jkl,j'l'k'} \right]$$

$$A_{jkl,j'k'l'} = \delta_{jj'}\tfrac{1}{2}\left(\bar{n}_k\bar{n}_l - n_kn_l \right)V_{kl[k'l']}$$

$$+ \delta_{kk'}\left(n_l - n_j \right)V_{lj'[l'j]} + \delta_{ll'}\left(n_k - n_j \right)V_{kj'[k'j]} \qquad (4.16a)$$

An alternative choice of the matrix C is discussed elsewhere.[66] For the open-shell case $(s_{ij} \neq 0)$ we have to add to A the quantities

$$S_{jkl,j'k'l'} = \delta_{jj'}\delta_{kk'}s_{ll'} + \delta_{jj'}\delta_{ll'}s_{kk'} - \delta_{kk'}\delta_{ll'}s_{jj'} \qquad (4.16b)$$

For completeness we show in Fig. 4.1 the diagrams of third order of the expansion of M that have not been shown in Fig. 3.8 because of the special choice of $s_{ij} = 0$ made there.

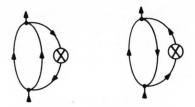

Fig. 4.1.   When $s_{ij} \neq 0$, these energy-dependent diagrams of third order have to be added to the diagrams of Fig. 3.8.

It is not useful to determine $\mathbf{P}\gamma$ from a comparison with diagrams of $\mathbf{M}$. For a given matrix $\mathbf{C}$, the exact projector $\mathbf{P}$, (4.15), has to be considered in order to project out the unphysical poles correctly. We shall see later how the matrix γ is obtained.

Finally two remarks should be made. First, we may restrict the index space to $k' > l'$ and $k > l$:

$$M_{pq}(\omega) = \sum_{\substack{j,k > l \\ j',k' > l'}} V_{qj[kl]} \Gamma_{jkl,j'k'l'} V_{pj'[k'l']} \qquad (4.17a)$$

In this index space Γ is given by

$$\Gamma(\omega) = \mathbf{P}(\omega\mathbf{1} - \mathbf{K} - 2\mathbf{C})^{-1}\gamma \qquad (4.17b)$$

and $\gamma^0 = \mathbf{1}$. Second, we may use the relation $A_{jkl,j'k'l'} = A_{jlk,j'l'k'}$ and obtain in the index space restricted to $k \neq l, k' \neq l'$

$$\Gamma(\omega) = \mathbf{P}(\omega\mathbf{1} - \mathbf{K} - \mathbf{A})^{-1}\gamma \qquad (4.18)$$

Therefore, one can choose the matrix $\mathbf{C}$ in (4.16) to be equal to the matrix $\mathbf{A}$ as was done previously.[31]

B.   Comparison with RPA Methods

Before we investigate further the above approximation scheme in the light of practical applications, we discuss its derivation from a completely different point of view. This stresses its central role and, in addition, clarifies the question as to which diagrams of $\mathbf{M}$ are actually included when the kernel is calculated via (4.14) with $\mathbf{C}$ of (4.16). To avoid too many complicated equations we make extensive use of diagrammatic techniques. In Section III.C.1 we discussed the RPA and the ladder approximation and their vital importance for the extreme cases of the high- and the low-density limit, respectively. The corresponding diagram series have been shown in Fig. 3.11. To obtain these self-energy parts one can also start from the particle-hole (ph) and the particle-particle (pp) response functions[62] on the RPA level, multiply them with a free Green's function and interaction matrix elements, and eliminate one of the two energy variables by integration. The procedure is shown in Fig. 4.2, where the squares denote the corresponding response functions. Because of the above, the notation pp-RPA for the ladder diagrams and ph-RPA for the RPA (GRPA) diagrams of **M** is commonly used. Remark: Because a GRPA diagram of nth order contains 2^n Feynman diagrams for $n > 2$ and only 2 Feynman diagrams for $n = 2$, we are not able to include all the GRPA diagrams with the correct weights in a ph-response formalism. We shall see below how the GRPA series can be obtained from a response formalism.

The graphical equations in Fig. 4.2 are actually RPA versions of more general equations for the response functions shown in Fig. 4.3. The pp-response function is a function of the irreducible interaction of two particles[62] denoted by J. Analogously, the ph-response function is a function of the irreducible interaction of a particle and a hole[62] denoted by I. J and I contain all diagrams of the corresponding response functions, which cannot be decomposed by cutting two particle lines and a particle and a hole line, respectively. The approximation leading to the RPA version is uniquely defined: the G lines have to be replaced by G^0 lines and the irreducible interaction part by its first order value.

The *exact* self-energy part can be obtained from the exact two particle-one hole (2p1h = 2h1p = 2ph) response function.[63] The graphical equation in terms of the irreducible interaction part X is represented in Fig. 4.4. When the G lines are substituted by G^0 lines and X by its first-order term, one obtains the 2ph-RPA equation, which is also drawn in Fig. 4.4. The kernel Γ is determined from the 2ph-response function $\Pi(\omega_1, \omega_2, \omega_3)$ by integration

$$\Gamma_{jkl,j'k'l'}(\omega) = \frac{1}{(2\pi)^2} \int d\omega_1 \, d\omega_2 \Pi_{jkl,j'k'l'}(\omega_1, \omega_2, \omega_1 + \omega_2 - \omega) \quad (4.19)$$

The graphical equation for the 2ph-RPA response function represents an integral equation that we give explicitly:

$$\Pi_{jkl,j'k'l'}(\omega_1, \omega_2, \omega_3) = G_{kk}^0(\omega_1) G_{ll}^0(\omega_2) G_{jj}^0(\omega_3) \left[\delta_{kk'} \delta_{ll'} - \delta_{kl'} \delta_{lk'} \right] \delta_{jj'}$$

$$+ \frac{1}{2} \sum_{k'',l''} V_{kl[k''l'']} G_{kk}^0(\omega_1) G_{ll}^0(\omega_2) \frac{i}{2\pi} \int d\omega' \Pi_{jk''l'',j'k'l'}(\omega', \omega_1 + \omega_2 - \omega', \omega_3)$$

$$+ \sum_{j''l''} V_{jl[j''l'']} G_{ll}^0(\omega_2) G_{jj}^0(\omega_3) \frac{i}{2\pi} \int d\omega' \Pi_{j''kl'',j'k'l'}(\omega_1, \omega', \omega_3 - \omega_2 + \omega')$$

$$+ \sum_{j''k''} V_{jk[j''k'']} G_{kk}^0(\omega_1) G_{jj}^0(\omega_3) \frac{i}{2\pi} \int d\omega' \Pi_{j''k''l,j'k'l'}(\omega', \omega_2, \omega_3 - \omega_1 + \omega')$$

$$(4.20)$$

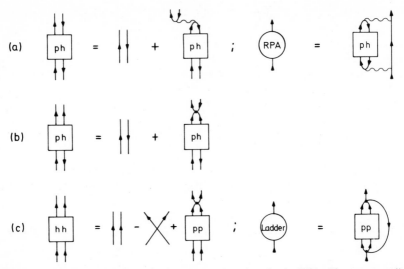

Fig. 4.2. Recursion relations for the response functions in the RPA. The corresponding self-energy parts are obtained according to the right-hand side of the figure. (*a*) *ph*-response function without exchange. (*b*) *ph*-response function with exchange. (*c*) *pp*-response function with exchange.

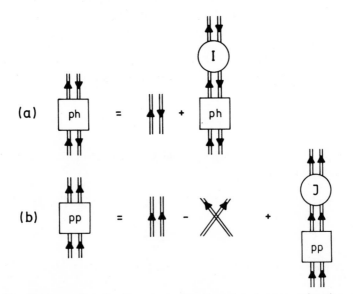

Fig. 4.3. Exact equations for the response functions. (*a*) *ph*-response function. I is the irreducible interaction of a particle and a hole. (*b*) *pp*-response function. J is the irreducible interaction of two particles.

250

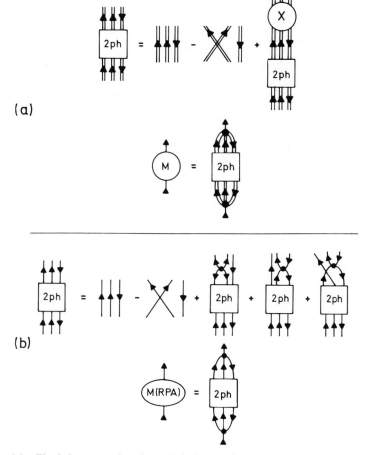

Fig. 4.4.   The 2ph-response function and the energy-dependent component of the self-energy part. (a) Exact equation and the exact self-energy part. (b) 2ph-RPA equation and the 2ph-RPA self-energy part.

This response function depends on three energy variables, all involved in the integral equation due to the three different types of integrals in (4.20). The analytic expression for Π is much more complicated than the expressions for the pp- and ph-RPA response functions. For the latter functions one of the two energy variables can easily be removed, leading to a straightforward solution. For completeness, the diagrams of the 2ph-RPA self-energy part are shown in Fig. 4.5.

As may be seen from Fig. 4.6, the pp-RPA and the ph-RPA (GRPA) summations are special cases of the full 2ph-RPA summation. The pp-RPA is obtained by omitting the last two terms in the graphical equation (Fig. 4.4) for the 2ph-RPA response. Analogously we introduce two pseudo-ph-RPA series by calculating $\mathbf{M}$ from the 2ph-RPA response, where the "ladder term" and one of the other homogenous terms in Fig. 4.4b are omitted. The corresponding integral equations can be solved exactly for these pseudo-ph-RPA problems as well as for the pp-RPA case. The solutions can be compared with the results of our approximation scheme with special choices of $\mathbf{C}$ and γ.

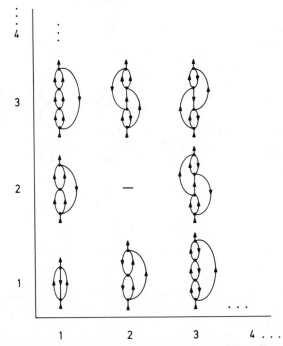

Fig. 4.5.　The Abrikosov diagrams contained in the 2ph-RPA self-energy part.

As an example, the calculation of the kernel Γ for the pp-RPA case is briefly sketched. One may start from (4.20) and omit the last two terms. We introduce a function Φ by $\Pi_{jkl,j'k'l'}(\omega_1,\omega_2,\omega_3) = (4\pi/i)G_{jj}^0(\omega_3)\Phi_{jkl,j'k'l'}(\omega_1 + \omega_2, \omega_1 - \omega_2)$. After integration over $(\omega_1 - \omega_2)$, Φ depends only on $(\omega_1 + \omega_2)$.

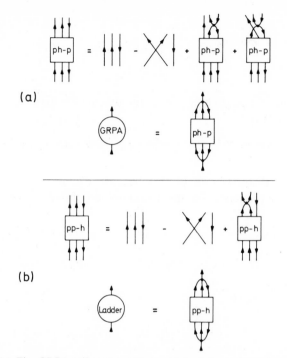

(a)

(b)

Fig. 4.6. (a) The GRPA self-energy part as obtained from the *ph* components of the 2*ph*-RPA response function. (b) The ladder self-energy part as obtained from the *pp* component of the 2*ph*-RPA response function.

One easily obtains the relations

$$\Gamma_{jkl,j'k'l'} = \frac{1}{2\pi i} \int d\bar{\omega} \, G_{jj}^0 \left(\bar{\omega} - \omega + \varepsilon_j\right) \Phi_{jkl,j'k'l'}(\bar{\omega})$$

$$\Phi_{jkl,j'k'l'}(\bar{\omega}) = \frac{\sigma \delta_{jj'}}{\bar{\omega} + \varepsilon_j - \varepsilon_k - \varepsilon_l + \sigma i\eta} \left\{ \left[\delta_{kk'}\delta_{ll'} - \delta_{kl'}\delta_{lk'} \right] \right.$$

$$\left. + \frac{1}{2} \sum_{k''l''} V_{kl[k''l'']}\Phi_{jk''l'',j'k'l'}(\bar{\omega}) \right\} \qquad (4.21)$$

$$\sigma = \bar{n}_k \bar{n}_l - n_k n_l$$

After solving for Φ and integrating over $\bar{\omega}$, it follows that

$$\Gamma = P(\omega 1 - K - C)^{-1}\gamma \qquad (4.22a)$$

where

$$C^{pp}_{\lambda\lambda'} \equiv C_{jkl,j'k'l'} = \delta_{jj'} \tfrac{1}{2} \left[\bar{n}_k \bar{n}_l - n_k n_l \right] V_{kl[k'l']}$$

$$\gamma = N\gamma^0 \tag{4.22b}$$

It is interesting to note that **N** is a simple diagonal matrix with elements 1, -1, or 0 according to

$$N^{pp}_{\lambda\lambda'} \equiv N_{jkl,j'k'l'} = \delta_{jj'}\delta_{kk'}\delta_{ll'} \left(\bar{n}_j - n_j \right)\left(\bar{n}_k \bar{n}_l - n_k n_l \right) \tag{4.22c}$$

The projector **P** appears in (4.22a) because of the integration over $\bar{\omega}$. Analogously, one obtains for one of the pseudo-ph-RPA cases

$$2C^{phI}_{\lambda\lambda'} \equiv 2C_{jkl,j'k'l'} = \delta_{kk'} \left(n_j - n_l \right) V_{lj'[l'j]} - (k' \leftrightarrow l') \tag{4.23a}$$

$$N^{phI}_{\lambda\lambda'} \equiv N_{jkl,j'k'l'} = \delta_{kk'}\delta_{ll'}\delta_{jj'} \left(n_k - \bar{n}_k \right)\left(n_l - n_j \right)$$

and for the other case

$$2C^{phII}_{\lambda\lambda'} \equiv 2C_{jkl,j'k'l'} = \delta_{ll'} \left(n_j - n_k \right) V_{kj'[k'j]} - (k' \leftrightarrow l') \tag{4.23b}$$

$$N^{phII}_{\lambda\lambda'} \equiv N_{jkl,j'k'l'} = \delta_{kk'}\delta_{ll'}\delta_{jj'} \left(n_l - \bar{n}_l \right)\left(n_k - n_j \right)$$

Owing to this correspondence to the results of the last section, we are led to the conclusion that the more general case described by (4.14) to (4.16) is algebraically equivalent to the 2ph-RPA summation shown in Fig. 4.4. The problem of obtaining γ for this general case is discussed in the next section.

One may gain further insight into the nature of the 2ph-RPA kernel by studying the equation of motion of the two-times six-point Green's function as done by Schuck et al.[64] After linearization of the equation of motion, a secular equation for the poles of this function is obtained with exactly the same matrix **C** as in (4.16a). The normalization condition on the amplitudes derived in Ref. 64 proves, however, to be a restriction to a very special case, which is discussed in the next section. The more general normalization condition that can be deduced with the equation-of-motion method depends on the ground-state wave function. Since the ground-state wave function consistent with the matrix **C** and implicitly contained in the six-point Green's function is unknown, one cannot use the formalism to determine the matrix γ.

C. Uncoupled 2ph-RPA and 2ph-TDA

1. Closed-Shell Case

To study the above formalism in the context of practical applications, it is convenient to decompose the basic matrix $\mathbf{C}$ into submatrices. An index $\lambda = (jkl)$ of this matrix belongs to one of the eight classes (hpp), (hhh), (pph), (php), (hph), (hhp), (ppp), (phh), where h (hole) stands for an occupied and p (particle) for an unoccupied one-particle state. Denoting these index classes briefly by the subscripts $1, 2, \ldots, 8$, one can build up the matrices $\mathbf{C}$ and $(\omega 1 - \mathbf{K} - \mathbf{C})$ from the set of submatrices $\mathbf{C}_{AB}$ and $(\omega 1 - \mathbf{K} - \mathbf{C})_{AB}, 1 \leqslant A, B \leqslant 8$. The matrix $(\omega 1 - \mathbf{K} - \mathbf{C})$ is represented in Fig. 4.7 for the closed-shell case $(s_{ij} = 0)$. The more general open-shell case is discussed later. We call the matrices $\{\mathbf{C}_{AB}\}, 4 \geqslant A, B \geqslant 1$, and $\{\mathbf{C}_{AB}\}, 8 \geqslant A, B \geqslant 5$, the upper block and lower block of $\mathbf{C}$, respectively. The remaining two blocks are referred to as the outer blocks. As shown in Fig. 4.7, many submatrices are zero. In the upper (lower) block only the first (last)

$j'k'l'$ \ jkl	1 hpp	2 hhh	3 pph	4 php	5 hph	6 hhp	7 ppp	8 phh
1 h p p	$\omega-K_{11}-C_{11}$	$-C_{12}$	$-C_{13}$	$-C_{14}$	$-C_{15}$	$-C_{16}$	$-C_{17}$	0
2 h h h	$+C_{12}^{T}$	$\omega-K_{22}-C_{22}$	0	0	$-C_{25}$	$-C_{25}$	0	0
3 p p h	$+C_{13}^{T}$	0	$\omega-K_{33}-C_{33}$	0	$-C_{35}$	0	$-C_{37}$	0
4 p h p	$+C_{14}^{T}$	0	0	$\omega-K_{44}-C_{44}$	0	$-C_{46}$	$-C_{47}$	0
5 h p h	0	$-C_{52}$	$-C_{53}$	0	$\omega-K_{55}-C_{55}$	0	0	$+C_{85}^{T}$
6 h h p	0	$-C_{62}$	0	$-C_{64}$	0	$\omega-K_{66}-C_{66}$	0	$+C_{86}^{T}$
7 p p p	0	0	$-C_{73}$	$-C_{74}$	0	0	$\omega-K_{77}-C_{77}$	$+C_{87}^{T}$
8 p h h	0	$-C_{82}$	$-C_{83}$	$-C_{84}$	$-C_{85}$	$-C_{86}$	$-C_{87}$	$\omega-K_{88}-C_{88}$

Fig. 4.7. The matrix $\omega 1 - \mathbf{K} - \mathbf{C}$ and its submatrices.

column and row and the diagonal differ from zero. In addition, we have the symmetry relations $C_{1A} = -C_{A1}^T$ for $A = 2, 3, 4$, $C_{8A} = -C_{A8}^T$ for $A = 5, 6, 7$ and $C_{AA} = C_{AA}^T$ for all A.

In the foregoing section we have seen that the matrix C can be decomposed into three terms that can be classified as a pp-RPA and two pseudo-ph-RPA contributions. It is important to see the submatrices C_{AB} in the light of this classification. Let us first discuss the upper block. In the submatrix C_{11} all three kinds of contributions occur. In C_{12}, C_{21}, and C_{22} we only have pp-RPA matrix elements, whereas C_{13}, C_{31}, C_{33} correspond to one of the pseudo-ph-RPA and C_{14}, C_{41}, C_{44} to the other ph-RPA. Analogous considerations apply to the lower block. Here the matrix C_{88} contains all three contributions, while C_{87}, C_{78}, C_{77} correspond to the pp-RPA and C_{86}, C_{68}, C_{66} and C_{85}, C_{58}, C_{55} to the ph-RPAs.

All matrix elements in the above submatrices of the upper and lower blocks are, according to (4.16a), simple sums of interaction matrix elements $V_{ij[kl]}$ with either four particle indices or four hole indices or two particle indices and two hole indices. If we now consider submatrices within the outer blocks, we find matrix elements that are not contained in any of the two-body response functions discussed above. All V_{ijkl} appearing here are characterized by either three particle indices and one hole index or three hole indices and one particle index. As can be seen from the diagrammatic expansion of the 2ph-RPA self-energy part in Fig. 4.5, these V_{ijkl} appear for the first time in fourth order. Here and in higher orders they *only* occur in certain time orderings of those diagrams that *mix* the pp- and the ph-contributions in the 2ph-RPA series. In Fig. 4.8 an example for such a time-ordered diagram is shown.

From Fig. 4.7 it is clear that the two outer blocks cause a severe asymmetry of the whole matrix C and thus of $(\omega 1 - K - C)$. Even if we could calculate the inverse of the latter matrix, is would still be impossible to construct the kernel $\Gamma(\omega)$, since we do not know the adequate inhomo-

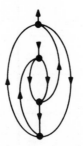

Fig. 4.8   A time-ordered diagram in the expansion of the self-energy part, which contains a V_{ijkl} matrix element with one particle index and three hole indices.

geneity γ. Because of this difficulty we have to leave open the question of how to perform the full summation and restrict ourselves to the case where the two outer blocks are neglected. Under these circumstances the upper and lower blocks in our matrix inversion problem decouple. We call this approach the "uncoupled" 2ph-RPA, whereas the full problem is referred to as the "coupled" 2ph-RPA. We shall see below that the uncoupled version is formally very similar to the ordinary RPA.[65]

From the diagrammatic point of view it is evident that the uncoupled 2ph-RPA leaves out some specific time orderings of the "mixing" diagrams, but still fully includes *all* third-order diagrams and *all* ladder and ph-RPA diagrams. In this sense the uncoupled 2ph-RPA is a justified approximation scheme of the kind we are looking for. Owing to the projection operator in (4.14), the physical poles develop from the C_{11} and C_{88} submatrices of the upper and lower blocks, respectively. Because $C_{A1} = 0$, $8 \geqslant A \geqslant 5$, and $C_{A8} = 0$, $4 \geqslant A \geqslant 1$, the influence of the other submatrices in the upper (lower) block on these physical poles is large compared to the influence of the submatrices in the outer blocks. A better understanding of the physics behind the uncoupled 2ph-RPA version may be gained from the discussions below and from Section IV.C.2.

In contrast to the coupled problem, we are able to solve the uncoupled problem exactly.[66] Here we make use of (4.22) and (4.23), which tell us that the kernels for both the pp- and the ph-RPA cases obey the general formula (4.14), which is also valid for the 2ph-RPA. Since the uncoupled 2ph-RPA contains all diagrams belonging to the pp-RPA and to the two pseudo-ph-RPA, one finds[31] (remembering that the diagram of second order must not be overcounted)

$$C^{2ph} = C^{pp} + C^{phI} + C^{phII} \tag{4.24a}$$

$$\gamma^{2ph} = N^{2ph}\gamma^0 = (N^{pp} + N^{phI} + N^{phII} - 2N^0)\gamma^0 \tag{4.24b}$$

where N^0 is a diagonal matrix with submatrices $N_{11}^0 = 1, N_{88}^0 = 1$ and zero elsewhere. Equations 4.24a and 4.24b are compatible when we restrict ourselves to the upper and lower blocks, that is, to the uncoupled 2ph-RPA. Equation 4.24a follows also from a comparison of (4.22) and (4.23) with (4.16). For completeness we give the submatrices of the diagonal matrix $N^{2ph}(=N)$: $N_{11} = 1, N_{22} = -1, N_{33} = -1, N_{44} = -1, N_{55} = -1, N_{66} = -1, N_{77} = -1$, and $N_{88} = 1$.

Let us start with the upper block. Apart from the projection operator and γ^0, we are left with the problem of evaluating the upper block of $(\omega 1 - K - C)^{-1}N$. Because $N = N^{-1}$, we may multiply $(\omega 1 - K - C)$ by N

and invert the result afterward, that is, we have to invert the *symmetric* matrix

$$U(\omega) =$$

$$\begin{bmatrix} \omega 1 - K_{11} - C_{11} & -C_{12} & -C_{13} & -C_{14} \\ -C_{12}^T & -(\omega 1 - K_{22} - C_{22}) & 0 & 0 \\ -C_{13}^T & 0 & -(\omega 1 - K_{33} - C_{33}) & 0 \\ -C_{14}^T & 0 & 0 & -(\omega 1 - K_{44} - C_{44}) \end{bmatrix}$$

$$(4.25)$$

This problem is similar to the problem in ordinary RPA where a matrix of the form

$$\begin{pmatrix} \omega 1 + A & B \\ B^T & -\omega 1 + A^T \end{pmatrix}$$

has to be inverted.[65] In order to calculate the lower block of $(\omega 1 - K - C)^{-1}N$ we analogously have to invert the symmetric RPA-like matrix

$$L(\omega) =$$

$$\begin{bmatrix} -(\omega 1 - K_{55} - C_{55}) & 0 & 0 & -C_{85}^T \\ 0 & -(\omega 1 - K_{66} - C_{66}) & 0 & -C_{86}^T \\ 0 & 0 & -(\omega 1 - K_{77} - C_{77}) & -C_{87}^T \\ -C_{85} & -C_{86} & -C_{87} & \omega 1 - K_{88} - C_{88} \end{bmatrix}$$

$$(4.26)$$

The solutions of both decoupled subproblems can be combined yielding

$$\Gamma(\omega) = XP^0(\omega 1 - D)^{-1}X^T \gamma^0 \qquad (4.27)$$

where D and X are the eigenvalue and eigenvector matrices of the uncoupled $(K + C)$ matrix, respectively, or equivalently of the pseudoeigenvalue problem

$$SX = NXD$$
$$XX^T = N$$

$$(4.28)$$

with

$$S = \begin{pmatrix} -U(0) & 0 \\ 0 & -L(0) \end{pmatrix}$$

Such a pseudoeigenvalue problem is well known from the ordinary RPA; being a standard procedure[65] it is not discussed here.

By definition, the projection operator P eliminates all unphysical poles. From (4.27) it is easily seen that the physical poles in the upper and lower blocks arise from $(\omega 1 - \mathbf{K} - \mathbf{C})_{11}$ and $(\omega 1 - \mathbf{K} - \mathbf{C})_{88}$, respectively. In the first case all physical poles possess a negative imaginary part $-i\eta$, whereas in the latter case all physical poles have a positive imaginary part $+i\eta$. Thus we may say that the upper block provides the "affinity-poles" ($\Sigma_\nu, \nu > 0$, in Fig. 3.1) of the self-energy part and the lower block the "ionization-poles" ($\Sigma_\nu, \nu < 0$).

Having completed the calculation of the kernel in the uncoupled 2ph-RPA, we discuss a simple approximation to it. From the preceding it is clear that the two submatrices $(\omega 1 - \mathbf{K} - \mathbf{C})_{11}$ and $(\omega 1 - \mathbf{K} - \mathbf{C})_{88}$ play a particular role in the 2ph-RPA. Confining ourselves to just these two submatrices, the only nonvanishing submatrices of the kernel Γ are[66]

$$\Gamma_{11}(\omega) = (\omega 1 - \mathbf{K} - \mathbf{C})_{11}^{-1}\gamma^0$$

$$\Gamma_{88}(\omega) = (\omega 1 - \mathbf{K} - \mathbf{C})_{88}^{-1}\gamma^0$$

(4.29)

It should be noted that the matrices $(\mathbf{K} + \mathbf{C})_{11}$ and $(\mathbf{K} + \mathbf{C})_{88}$ are symmetric. In the diagrammatic picture this means that of all 2ph-RPA diagrams of the self-energy part $\mathbf{M}(\omega)$, only two special time orderings are included. They may be called the "ascending" and "descending" time ordering. Each cut (for definition see Section III.B.2) between two subsequent interaction points intersects one hole line and two particle lines in the former case and one particle line and two hole lines in the latter case. These time-ordered diagrams can be summed directly, yielding Γ_{11} for the "ascending" time ordering and Γ_{88} for the "descending" time ordering. At this point it might be useful to consult Section III.D.

There is another derivation of the kernel in (4.29), giving direct insight into its physical meaning. In complete analogy to the well-known Tamm-Dancoff approximation[67,68] (TDA) for the particle-hole and particle-particle[69] cases, there is also a 2ph-TDA[70] approach to ionic wave functions. The exact $(N \pm 1)$-electron wave functions can be expanded in configurational space according to

$$|\Psi_s^{N-1}\rangle = \sum_i \alpha_i^s a_i |\Psi_0^N\rangle + \sum_{i,j,k} \alpha_{ijk}^s a_i^+ a_j a_k |\Psi_0^N\rangle + \cdots$$

$$|\Psi_s^{N+1}\rangle = \sum_i \beta_i^s a_i^+ |\Psi_0^N\rangle + \sum_{i,j,k} \beta_{ijk}^s a_i^+ a_j^+ a_k |\Psi_0^N\rangle + \cdots$$

(4.30)

In the 2ph-TDA the ground state $|\Psi_0^N\rangle$ is replaced by the unperturbed ground state $|\Phi_0^N\rangle$. Since one is only interested in such ionic states characterized by a set of three indices $s = \{ijk\}$, the α_i^s (β_i^s) and α_{ijklm}^s (β_{ijklm}^s) and higher coefficients are set to zero in the TDA. Considering the α_{ijk}^s and β_{ijk}^s as variational parameters, the energies $E_s^{N+1} - E_0^N(\text{HF})$ and $E_s^{N-1} - E_0^N(\text{HF})$ are obtained as roots of the secular equations for the matrices $(\mathbf{K} + \mathbf{C})_{11}$ and $(\mathbf{K} + \mathbf{C})_{88}$, respectively. The kernel in (4.29) is thus the 2ph-TDA kernel. Solving the Dyson equation with the 2ph-TDA self-energy part means that one takes account of the α_i^s (β_i^s) and of correlation in $|\Psi_0^N\rangle$. If the constant diagrams are calculated with (3.28) and Γ of (4.29), $\mathbf{\Sigma}(\infty)$ is correct through third order.

It is interesting to note that there is some correspondence between the method proposed by Simons and Smith[13a] and the present approach. It has been shown by Purvis and Öhrn[71] that the poles of the self-energy part used by Simons et al. are those obtained in the diagonal ($\mathbf{C}_{11}$ and $\mathbf{C}_{88}$ diagonal) 2ph-TDA and that if the procedure is taken to include all third-order diagrams of $\mathbf{M}(\omega)$, the poles become identical with those obtained in the 2ph-TDA.

Going beyond the TDA to the uncoupled and coupled RPA obviously means that some part of the ground-state correlation is taken into account. Thus there are two ground-state wave functions, $|\Psi_0^N\rangle_{\text{URPA}}$ and $|\Psi_0^N\rangle_{\text{CRPA}}$, consistent with the uncoupled and coupled RPA versions. Analogous to the TDA case, the $(N+1)$- and $(N-1)$-particle space is spanned by three-fermion operators acting on these N-particle ground states. From the discussion in the first part of this section it becomes clear that $|\Psi_0^N\rangle_{\text{CRPA}}$ contains more singly excited configurations than $|\Psi_0^N\rangle_{\text{URPA}}$. Since we are in the HF picture of a closed-shell molecule, the singly excited configurations usually are of minor importance. If we deal with an open-shell system and/or with a non-HF-picture, the same level of accuracy is achieved when the matrix elements s_{ij} are fully taken into account. This is discussed in the next section.

To conclude this section a very simple approximation to the 2ph-TDA is briefly discussed. If all off-diagonal elements of the matrix $\mathbf{C}$ in (4.16) are neglected, one obtains the so-called diagonal approximation,

$$M_{pq}(\omega) = \frac{1}{2} \sum \frac{V_{pj[kl]} V_{qj[kl]} (\bar{n}_j n_k n_l + n_j \bar{n}_k \bar{n}_l)}{\omega + \varepsilon_j - \varepsilon_k - \varepsilon_l + (\bar{n}_j n_k n_l - n_j \bar{n}_k \bar{n}_l) \Delta_{jkl}} \tag{4.31}$$

with shifts

$$\Delta_{jkl} = V_{kl[kl]} - V_{kj[kj]} - V_{lj[lj]}$$

This self-energy part can be derived by many other methods. The *self-consistent* form of this self-energy part in the one-pole approximation[27] has been used to calculate the shake-up lines in the PE spectrum of H_2O in the valence region[7d] and of C_3O_2 in the core region.[31]

2. Open-Shell Case

For an open-shell atom or molecule as well as for a closed-shell system where H_0 has not been chosen to be the HF operator, it is important to include the matrix elements s_{ij} defined in (3.23). According to (4.16) one must simply add the matrix S to the matrix A in order to obtain the correct matrix C. For clarity, we decompose S into its submatrices in a way analogous to that for C in the closed-shell case. This is shown in Fig. 4.9. In contrast to the full matrix C, the matrix S is symmetric. In the inner blocks the only nonvanishing submatrices of S are S_{AA}, $A = 1, 2, \ldots, 8$. It is thus immediately clear that in order to evaluate the 2ph-TDA and the uncoupled 2ph-RPA as defined above for the closed-shell case, one simply has to replace A_{BB} by $A_{BB} + S_{BB}$, leaving the corresponding inhomogenity γ unchanged.

Since S_{A1}, $S_{B8} \neq 0$ for $A = 5$ to 7, $B = 2$ to 4, the outer blocks of the matrix C are clearly more important than is the case for the outer blocks of the matrix C in the preceding section. Therefore, it is of advantage to

k' \ jkl	1 hpp	2 hhh	3 pph	4 php	5 hph	6 hhp	7 ppp	8 phh
1 hpp	S_{11}	0	0	0	S_{15}	S_{16}	S_{17}	0
2 hhh	0	S_{22}	0	0	S_{25}	S_{26}	0	S_{28}
3 pph	0	0	S_{33}	0	S_{35}	0	S_{37}	S_{38}
4 php	0	0	0	S_{44}	0	S_{46}	S_{47}	S_{48}
5 hph	S_{15}^T	S_{25}^T	S_{35}^T	0	S_{55}	0	0	0
6 hhp	S_{16}^T	S_{26}^T	0	S_{46}^T	0	S_{66}	0	0
7 ppp	S_{17}^T	0	S_{37}^T	S_{47}^T	0	0	S_{77}	0
8 phh	0	S_{28}^T	S_{38}^T	S_{48}^T	0	0	0	S_{88}

Fig. 4.9.   The matrix S and its submatrices.

eliminate **S** first and subsequently to solve the uncoupled 2ph-RPA. To this end we calculate the Green's function from the Dyson equation taking only **S** into account,

$$\tilde{G}_{kl} = G^0_{kl} + \sum_{p,q} G^0_{kp} s_{pq} \tilde{G}_{ql} \qquad (4.32)$$

This equation is symbolically shown in Fig. 4.10a. According to Fig. 4.10b we obtain the kernel $\Gamma(\omega)$ from

$$\Gamma_{jkl,j'k'l'}(\omega) = \frac{1}{(2\pi)^2} \int d\omega_1 d\omega_2 \tilde{G}_{jj'}(\omega_1 + \omega_2 - \omega)$$

$$\times \left[\tilde{G}_{kk'}(\omega_1) \tilde{G}_{ll'}(\omega_2) - \tilde{G}_{kl'}(\omega_1) \tilde{G}_{lk'}(\omega_2) \right] \qquad (4.33)$$

The integration can be performed to give[66]

$$\Gamma(\omega) = \mathbf{Y} \mathbf{P}^0 (\omega \mathbf{1} - \tilde{\mathbf{K}})^{-1} \mathbf{Y}^T \gamma^0 \qquad (4.34a)$$

where

$$Y_{jkl,j'k'l'} = y_{jj'} y_{kk'} y_{ll'} \qquad (4.34b)$$

$$\tilde{K}_{jkl,j'k'l'} = \delta_{jj'} \delta_{kk'} \delta_{ll'} \left(d_{ll'} + d_{kk'} + d_{jj'} \right)$$

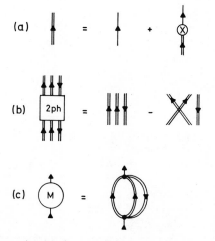

Fig. 4.10.   The Dyson equation (a), the exact 2ph-response function (b), and the corresponding self-energy part (c) for the case the matrix **A** in (4.18) is equal to **S**.

and y and d are the eigenvector and eigenvalue matrices of the matrix $\varepsilon + s$ ($s = \{s_{ij}\}$), respectively,

$$(\varepsilon + s)y = yd \qquad (4.34c)$$

$$yy^T = 1$$

Y and $\tilde{K}$, on the other hand, are the eigenvector and eigenvalue matrices of $K + S$,

$$(K + S)Y = Y\tilde{K} \qquad (4.34d)$$

$$YY^T = 1$$

Thus the analytical solution for $\Gamma(\omega)$, (4.34a), coincides with the algebraic solution of the coupled 2ph-RPA. The inhomogenity is clearly given by γ^0.

As the next step toward the solution of the problem, we replace the free Green's functions G^0 in Fig. 4.4b by the s-renormalized Green's functions $\tilde{G}$ defined in Fig. 4.10a. To obtain the corresponding uncoupled 2ph-RPA equations we repeat the procedure described in Sections IV.B and IV.C.1, that is we first solve the pp- and ph problems separately and then combine them to give the uncoupled 2ph-RPA. The final result is[66]

$$\Gamma(\omega) = YP(\omega 1 - \tilde{K} - \tilde{C})^{-1}N\gamma^0 Y^T \qquad (4.35)$$

where $\tilde{C}$ is the same matrix as C in (4.16a), but with the matrix elements $V_{ab[cd]}$ replaced by $\tilde{V}_{ab[cd]}$ defined by

$$\tilde{V}_{ab[cd]} = \sum_{i,j,k,l} y_{ia}y_{jb}y_{kc}y_{ld}V_{ij[kl]} \qquad (4.36)$$

In practice, this transformation can be confined to the open shell. In the uncoupled approximation the outer blocks of $\tilde{C}$ are set to zero.

To recover the form of the general equation (4.18) we start from (4.34d) and rewrite (4.35) to give

$$\Gamma(\omega) = P(\omega 1 - K - S - Y\tilde{C}Y^T)^{-1}\gamma \qquad (4.37)$$

$$\gamma = YNY^T\gamma^0.$$

Two different uncoupled 2ph-RPA approaches have been discussed above for the open-shell case. In the first approach one neglects the outer blocks of C, whereas in the second approach one solves the problem for s first and neglects the outer blocks of $\tilde{C}$. The latter approximation leads to a self-energy part that contains *all* diagrams up to third order. Neglecting the outer blocks **C** implies, on the other hand, that some time orders of the diagrams in Fig. 4.1 are not considered. The omitted time-ordered diagrams are characterized by matrix elements s_{ij} that couple an unoccupied orbital i to an occupied orbital j. In many cases these matrix elements are zero or play only a minor role. If all particle-hole elements s_{ij} vanish, both approaches are identical since $\mathbf{YNY}^T = \mathbf{N}$ in this case.

D. Simple Applications to Open-Shell Molecules

To become familiar with the concepts introduced in the preceding sections some examples are discussed. Our main interest is in demonstrating how the multiplet structure observed in PE spectra of open-shell molecules is correctly described by applying the structure-conserving renormalization procedure. For the sake of clarity we confine ourselves to the simplest symmetry-retaining procedure and to very restricted orbital subspaces. The procedure we choose is a simplified 2ph-TDA, where the 2ph-TDA kernel is used, but the ground-state wave function is considered to be uncorrelated *throughout* the calculation. According to Section III.D we then have to deal with two separate Dyson equations, one for the affinity poles and one for the ionization poles. Correspondingly, $^{(1)}\Sigma$ is the only constant term to be considered.

Introducing the relative occupation number $f = s/r$ for an open-shell of r spin orbitals occupied by s electrons, the free Green's function $^0\mathbf{G}$ becomes

$$^0G_{pq} = \delta_{pq} \left[\frac{\bar{n}_p}{\omega - \varepsilon_p - i\eta} + \frac{1 - \bar{n}_p}{\omega - \varepsilon_p + i\eta} \right] \tag{4.38}$$

where $\bar{n}_p = 1$, $\bar{n}_p = f$, or $\bar{n}_p = 0$ for a closed, open, or unoccupied shell p, respectively.

The ground state of the oxygen molecule is represented by the $^3\Sigma_g^-$ configuration

$$(1\sigma_g)^2(1\sigma_u)^2(2\sigma_g)^2(2\sigma_u)^2(3\sigma_g)^2(\pi_u)^4(\pi_g)^2$$

Since $S = 1$, we have three degenerate states $|\Psi_0(M)\rangle$, $M = 0, \pm 1$, which are associated with the three unperturbed ground states,

$$|\Phi_0(1)\rangle = |\pi_{x\alpha}\pi_{y\alpha}| \tag{4.39}$$

$$|\Phi_0(2)\rangle = \frac{1}{\sqrt{2}} \left[|\pi_{x\alpha}\pi_{y\beta}| + |\pi_{x\beta}\pi_{y\alpha}| \right]$$

$$|\Phi_0(3)\rangle = |\pi_{x\beta}\pi_{y\beta}|$$

where $|\pi_{x\nu}\pi_{y\nu'}|$ denotes a Slater determinant composed of all closed-shell single-particle states and the two states $\pi_{x\nu}$ and $\pi_{y\nu'}$ of the open π_g shell. To calculate the Green's function we start from

$$G(\omega) = \tfrac{1}{2}\left[\mathbf{G}_{\alpha\alpha}^{(S)} + \mathbf{G}_{\beta\beta}^{(S)} \right] \qquad S = 1 \tag{4.40}$$

and make use of the Dyson equations (3.26) and (3.51). The unperturbed Hamiltonian is chosen to be the RHF operator. Since the RHF operator is not defined uniquely[45], the matrix elements v_{ij} in (3.20) are left unspecified. We shall see below that the choice of the one-particle basis is irrelevant for the final results (multiplet splittings and intensities).

Let us first consider the ionization of an electron in any orbital of σ symmetry. To study the open-shell effects we restrict the calculation to an orbital subspace that includes the π_g orbital and the σ orbital in question. For this subspace the ground state configuration ($M = 1$) is shown in Fig. 4.11a.

The diagrams of first order are easily calculated:

$$^{(1)}\Sigma_{\sigma_\alpha\sigma_\alpha}^{(1)} = - V_{\sigma\pi\pi\sigma} + X_\sigma \tag{4.41}$$

$$^{(1)}\Sigma_{\sigma_\beta\sigma_\beta}^{(1)} = V_{\sigma\pi\pi\sigma} + X_\sigma$$

where $X_\sigma = v_{\sigma\sigma} + V_{\sigma\sigma\sigma\sigma} + 2V_{\sigma\pi\sigma\pi} - V_{\sigma\pi\pi\sigma}$ and the notation π is used instead of π_{gx} and π_{gy} whenever it is unambiguous. The resulting first-order Green's function is

$$^{(1)}G_{\sigma\sigma} = \frac{\tfrac{1}{2}}{\omega - \tilde{\varepsilon}_\sigma - V_{\sigma\pi\pi\sigma}} + \frac{\tfrac{1}{2}}{\omega - \tilde{\varepsilon}_\sigma + V_{\sigma\pi\pi\sigma}} \tag{4.42}$$

$$\tilde{\varepsilon}_\sigma = \varepsilon_\sigma + X_\sigma$$

The calculation of the kernel Γ with the 2ph-TDA equation (4.29) is simplified considerably as a result of symmetry. There is no contribution at

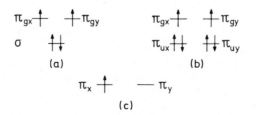

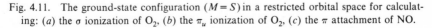

Fig. 4.11.   The ground-state configuration ($M = S$) in a restricted orbital space for calculating: (a) the σ ionization of O_2, (b) the π_u ionization of O_2, (c) the π attachment of NO.

all to $\Sigma^{(1)}_{\sigma_\beta\sigma_\beta}$ and Γ reduces to a one-dimensional submatrix in the case of $\Sigma^{(1)}_{\sigma_\alpha\sigma_\alpha}$. The results are

$$\Sigma^{(1)}_{\sigma_\alpha\sigma_\alpha} = -V_{\sigma\pi\pi\sigma} + X_\sigma + \frac{2V^2_{\sigma\pi\pi\sigma}}{\omega - \tilde{\varepsilon}_\sigma} \tag{4.43}$$

$$\Sigma^{(1)}_{\sigma_\beta\sigma_\beta} = V_{\sigma\pi\pi\sigma} + X_\sigma$$

The Dyson equations are easily solved to give

$$G^{(1)}_{\sigma_\alpha\sigma_\alpha} = \frac{\frac{1}{3}}{\omega - \tilde{\varepsilon}_\sigma - V_{\sigma\pi\pi\sigma}} + \frac{\frac{2}{3}}{\omega - \tilde{\varepsilon}_\sigma + 2V_{\sigma\pi\pi\sigma}} \tag{4.44}$$

$$G^{(1)}_{\sigma_\beta\sigma_\beta} = \frac{1}{\omega - \tilde{\varepsilon}_\sigma - V_{\sigma\pi\pi\sigma}}$$

and the Green's function (4.40) becomes

$$G_{\sigma\sigma} = \frac{\frac{2}{3}}{\omega - \tilde{\varepsilon}_\sigma - V_{\sigma\pi\pi\sigma}} + \frac{\frac{1}{3}}{\omega - \tilde{\varepsilon}_\sigma + 2V_{\sigma\pi\pi\sigma}} \tag{4.45}$$

The poles of $G_{\sigma\sigma}$ have to be interpreted as the negative IPs for the removal of an electron in the σ orbital of O_2. The ratio of the residues represents the intensity ratio of the corresponding lines in the PE spectrum. The lines correspond to the quartet and doublet states of the resulting ion. It should be mentioned that $G_{\sigma\sigma}$ in (4.45) is the *exact* Green's function in the subspace considered. This is because in this subspace the ground state is uncorrelated. A frozen orbital approach leads to the same result.[1b]

A less trivial problem is the ionization of an electron in the π_u orbital of O_2. Here the subspace is chosen to contain the π_u and π_g orbitals. The ground-state configuration for $M = 1$ is shown in Fig. 4.11b. Denoting the orbitals π_{ux}, π_{uy}, π_{gx}, and π_{gy} by 1, 1', 2, and 2', respectively, the self-energy parts in first order read

$$^{(1)}\Sigma^{(1)}_{1_\alpha 1_\alpha} = X_\pi \qquad ^{(1)}\Sigma^{(1)}_{1_\beta 1_\beta} = K_{12} + K_{12'} + X_\pi \tag{4.46}$$

where $X_\pi = v_{11} + J_{11} + 2J_{11'} + J_{12} + J_{12'} - K_{12} - K_{12'} - K_{11'}$ and $K_{ab} = V_{abba}$, $J_{ab} = V_{abab}$. With $\tilde{\varepsilon}_1 = \varepsilon_1 + X_\pi$, one obtains for the Green's function in first order

$$^{(1)}G_{11} = \frac{\frac{1}{2}}{\omega - \tilde{\varepsilon}_1} + \frac{\frac{1}{2}}{\omega - \tilde{\varepsilon}_1 - K_{12} - K_{12'}} \tag{4.47}$$

In the 2ph-TDA there is no additional contribution to $\Sigma^{(1)}_{1_\beta,1_\beta}$ whereas in the case of $\Sigma^{(1)}_{1_\alpha 1_\alpha}$ the matrix $\Gamma(\omega)$ has rank 4. One of the resulting four poles of the latter self-energy part has a vanishing residue, that is, Γ can be reduced to a matrix of rank 3. This pole leads to an $IP = \tilde{\varepsilon}_1 - V_{11'22'} + K_{12'}$, which corresponds to the $^2\Phi$ state of O_2^+. The vanishing residue implies that this IP is not observed in the PE spectrum.

In the next step the Dyson equations for $G^{(1)}_{1_\alpha 1_\alpha}$ and $G^{(1)}_{1_\beta 1_\beta}$ are solved. One of the four poles of the latter function is found to be equal to the pole of the first function. The results are

$$G^{(1)}_{1_\alpha 1_\alpha} = \frac{\frac{1}{3}}{\omega - \tilde{\varepsilon}_1 - K_{12} - K_{12'}} + \frac{2P_A}{\omega - \tilde{\varepsilon}_1 + a} + \frac{2P_B}{\omega - \tilde{\varepsilon}_1 + b} + \frac{2P_C}{\omega - \tilde{\varepsilon}_1 + c} \quad (4.48)$$

$$G^{(1)}_{1_\beta 1_\beta} = \frac{1}{\omega - \tilde{\varepsilon}_1 - K_{12} - K_{12'}}$$

The quantities a, b, c, P_A, P_B, and P_C can be obtained from the relation

$$\frac{P_A}{z+a} + \frac{P_B}{z+b} + \frac{P_C}{z+c} = \frac{\frac{1}{3}Q_2(z)}{Q_3(z)} \quad (4.49a)$$

where $Q_2(z)$ and $Q_3(z)$ are given by

$$Q_2(z) = (z-x)^2 + b_1(z-x) + b_0$$

$$Q_3(z) = (z-x)^3 + a_2(z-x)^2 + a_1(z-x) + a_0,$$

$$x = S - 2K_{22'} \qquad S = \frac{1}{2}(K_{12} + V_{11'22'})$$

$$b_0 = 2K_{12'}^2 - 8T^2 - 2SK_{12'} + 2SK_{22'} - 4K_{12'}K_{22'} + 4ST$$

$$b_1 = \frac{3}{2}S - 3K_{12'} + 2K_{22'} \qquad (4.49b)$$

$$a_0 = 8K_{12'}^2(K_{12'} - S - K_{22'}) - 12ST^2 - 8STK_{22'}$$

$$\quad - 8K_{12'}T^2 + 8K_{12'}K_{22'}^2 + 16T^2K_{22'} + 16STK_{12'} - 4SK_{22'}^2$$

$$a_1 = -4K_{12'}^2 - 8T^2 - 4K_{22'}^2 - 2SK_{12'} + 2SK_{22'} + 4K_{12'}K_{22'} + 4TS$$

$$a_2 = 3S - 2K_{12'} \qquad T = V_{121'2'}$$

The Green's function (4.40) reads

$$G_{11} = \frac{\frac{2}{3}}{\omega - \bar{\varepsilon}_1 - K_{12} - K_{12'}} + \frac{P_A}{\omega - \bar{\varepsilon}_1 + a} + \frac{P_B}{\omega - \bar{\varepsilon}_1 + b} + \frac{P_C}{\omega - \bar{\varepsilon}_1 + c}$$

$$P_A + P_B + P_C = \tfrac{1}{3} \tag{4.50}$$

According to this function four bands are to be expected in the PE spectrum owing to the ionization of a π_u electron. These bands correspond to a quartet state ($^4\Pi_u$) and to three doublet states ($A\,^2\Pi_u$, $B\,^2\Pi_u$, $C\,^2\Pi_u$) of O_2^+. The quartet-to-doublet intensity ratio is 2:1 and all three $^2\Pi_u$ bands have different intensities. To get an idea of the relative intensities of the $^2\Pi_u$ bands and of the multiplet splitting we use crudely estimated values for the integrals S, T, $K_{12'}$, $K_{22'}$ ($S = 0.1278$, $T = 0.0173$, $K_{12'} = 0.0138$, $K_{22'} = 0.0168$ a.u.). Choosing the energy of the $^4\Pi_u$ band as the origin of the energy scale, we obtain the energies 0.85, 3.13, and 10.70 eV for the three $^2\Pi_u$ bands. The intensity ratio $^4\Pi_u : A\,^2\Pi_u : B\,^2\Pi_u : C\,^2\Pi_u$ is found to be $2 : 0.44 : 0.01 : 0.55$.

The above results are in qualitative agreement with the experimental results of Edqvist et al.[72] and with the CI results of Dixon and Hull.[73] Edqvist et al. have observed a $^2\Pi_u$ band at an energy of 0.95 eV and a second $^2\Pi_u$ band at 7.90 eV relative to the $^4\Pi_u$ band. The remaining $^2\Pi_u$ band with its very low calculated intensity has not been observed. The experimental intensity ratios are $2 : 0.3 : \approx 0 : 0.75$. The quartet-to-doublet ratio is about the statistical ratio 2:1. The calculated quartet-to-doublet ratio is exactly 2:1 because the ground state wave function of O_2 has been taken to be uncorrelated.

Purvis and Öhrn[12b] have recently reported a Green's function calculation for O_2. The approach is a generalization of the method they use for closed-shell systems[12a] and is based on an unrestricted HF calculation. Thus their unperturbed Hamiltonian does not exhibit the symmetry properties of the system. As a consequence the method fails to describe the multiplet splitting properly. When the ionization of an electron in the π_u shell is calculated in the unrestricted HF approach, two IPs are obtained. Instead of obtaining the three $^2\Pi_u$ states discussed above, only one state, which might be considered as an average over the three actual states [cf. (4.47)], results. When the multiplet splitting is small, such a procedure is justified.

Since the ground state of O_2 is of $^3\Sigma_g^-$ symmetry, only spin degeneracy has to be taken into account when calculating the Green's function. To demonstrate how the formalism works in the case of both spin and spatial degeneracy, we choose the attachment of a π electron to the nitric oxide molecule as an example. The ground state of NO has the symmetry $^2\Pi$. The model ground-state configuration $(\Pi_x, M_s = \frac{1}{2})$ used here is shown in Fig. 4.11c.

The Green's function for the open π shell is

$$G_{\pi\pi} = \tfrac{1}{4}\left[G^{(1/2,x)}_{\pi_{x\alpha}\pi_{x\alpha}} + G^{(1/2,x)}_{\pi_{y\alpha}\pi_{y\alpha}} + G^{(1/2,x)}_{\pi_{x\beta}\pi_{x\beta}} + G^{(1/2,x)}_{\pi_{y\beta}\pi_{y\beta}} \right] \qquad (4.51)$$

The functions $G^{(1/2,x)}_{ab}$ are calculated in analogy to the above examples. The result is

$$G_{\pi\pi} = \frac{\tfrac{1}{4}}{\omega - \tilde{\varepsilon}_\pi - i\eta} + \frac{\tfrac{3}{8}}{\omega - \tilde{\varepsilon}_\pi - J_{\pi_x\pi_y} + K_{\pi_x\pi_y} + i\eta}$$

$$+ \frac{\tfrac{2}{8}}{\omega - \tilde{\varepsilon}_\pi - J_{\pi_x\pi_y} - K_{\pi_x\pi_y} + i\eta} + \frac{\tfrac{1}{8}}{\omega - \tilde{\varepsilon}_\pi - J_{\pi_x\pi_x} - K_{\pi_x\pi_y} + i\eta} \qquad (4.52)$$

This Green's function is easily interpreted. The first term describes the ionization of the π electron leading to a $^1\Sigma^+$ state of NO^+. The remaining terms in (4.52) describe different final states of NO^- obtained by adding a π electron to the ground state of NO. In particular, the second, third, and fourth terms are associated with the production of the $^3\Sigma^-$, $^1\triangle$, and $^1\Sigma^+$ states, respectively.

E. A Simple Geometric Approximation for the Valence Region

The assignment of the IPs of the outer-valence electrons of a molecule is essential for interpreting its photoelectron spectrum. IPs of inner-shell electrons are characteristic of the corresponding atom and differ usually only little for the same atom in different molecules. Therefore the assignment of inner-shell lines in a x-ray PE spectrum is straightforward, except for the chemical shifts observed for nonequivalent atoms. The outer-valence IPs, on the other hand, are characteristic of the molecule and provide direct information about the chemical bonding. Furthermore, many of the valence IPs fall into a narrow energy range, rendering their assignment a severe theoretical problem.

If we restrict ourselves to the evaluation of the Green's function in the outer-valence region of closed-shell atoms and molecules, we may take advantage of the fact that no poles of the self-energy part lie in the outer-valence region. Therefore, it is not necessary to calculate the poles of $\Sigma(\omega)$ accurately. The fact that $\Sigma(\omega)$ is a smooth function with a small slope $[P_i(i) \approx 0.9]$ in the region of the outer-valence IPs simplifies the calculation considerably.

From the discussion of the kernel we expect that a good approximation to the energy-dependent part $\mathbf{M}$ of Σ is

$$M_{pq}^{eff} = \sum_i \frac{V_i Z_i}{\omega - f_i - X_i} \tag{4.53}$$

where, for the sake of abbreviation, V_i stands for $\frac{1}{2} V_{pj[kl]} V_{qj[kl]}$, f_i stands for $\varepsilon_k + \varepsilon_l - \varepsilon_j$, and i should be understood as (j, k, l) with $k, l \in \mathcal{F}, j \notin \mathcal{F}$ and $k, l \notin \mathcal{F}, j \in \mathcal{F}$. Z_i and X_i are the change of the residues of $\mathbf{M}^{(2)}$ and the shifts of the poles of $\mathbf{M}^{(2)}$, respectively, both due to the renormalization $\boldsymbol{\Gamma}^{(0)} \to \boldsymbol{\Gamma}$ discussed in previous sections. Since the eigenvector matrix of $\mathbf{G}$ has been found to be nearly diagonal in the region of interest, we consider in the following only a diagonal $\mathbf{M}^{(n)}$ for n greater than 2.

It is convenient to write $Z_i = 1 - A_i$. Assuming the second-order term $M_{pp}^{(2)}$ to be the leading term in M_{pp}^{eff}, if follows that $|A_i| < 1$ and $|X_i/(\omega - f_i)| < 1$ for $\omega \approx \varepsilon_p$. Then we may expand M^{eff} according to

$$M^{eff} = \sum_i \frac{V_i}{\omega - f_i} \left[1 + \frac{X_i}{\omega - f_i} + \cdots \right] - \sum_i \frac{V_i A_i}{\omega - f_i} \left[1 + \frac{X_i}{\omega - f_i} + \cdots \right] \tag{4.54}$$

where the indices p and q have been suppressed. Now we may try to assign the different terms in (4.54) to particular types of diagrams. The $\sum_i V_i/(\omega - f_i)$ clearly is $M^{(2)}$. The time-ordered diagrams CI and DI, $I = 2$ to 5, obviously contribute to $- \sum_i V_i A_i/(\omega - f_i)$. The CI and DI, $I = 1, 6$, on the other hand, contribute mainly to $\sum_i V_i X_i/(\omega - f_i)^2$ and much less to the corresponding screened term $- \sum_i A_i V_i X_i/(\omega - f_i)^2$. Analogously, the diagrams of higher orders containing the poles of highest possible order contribute to the corresponding term in the first sum of (4.54) and have little influence on the screened term in the second sum of (4.54). This can be understood by investigating the diagonal approximation of the secular equation for $\boldsymbol{\Gamma}$.

Equation 4.54 is rewritten to give

$$M^{eff} = (1 - A(\omega)) \sum_i \frac{V_i}{\omega - f_i} \left[1 + \frac{X_i}{\omega - f_i} + \cdots \right] \tag{4.55}$$

where $A(\omega)$ is a smooth function for $\omega \approx \varepsilon_p$. Because of the small slope of M^{eff}, $A(\omega)$ is well approximated by $A = A(-I_p)$, where I_p is the IP corresponding to ε_p. From the above discussion we may estimate A in first order:

$$A = - \frac{\sum_{I=2}^{5} (CI + DI)}{M^{(2)}} \equiv F \tag{4.56}$$

A better approximation for A can be obtained from the following consider-ations. It is well known[49] that the RPA procedure "substitutes" the bare interaction by a screened one. Thus (4.56) suggests that A should be interpreted as a screening parameter. The first-order approximation $A = F$ means that the screening is done by the third-order time-ordered diagrams CI, DI, $I = 2$ to 5. To obtain the actual screening parameter we have to calculate these terms, but with the bare interaction substituted by the renormalized one. This again leads to a screening parameter, A', but now for the above terms CI, DI, $I = 2$ to 5. Since the screened interaction leading to A is the same as the one leading to A', we may take $A' \approx A$. One should keep in mind that this approximation is only used to estimate a part of the contribution of fourth and higher order of the self-energy part. It follows that

$$A = \frac{F}{1 + F} \tag{4.57}$$

Inserting this simple result into (4.55) leads to

$$M^{\text{eff}} = \sum_i \frac{V_i}{\omega - f_i} - \frac{F}{1 + F} \sum_i \frac{V_i}{\omega - f_i}$$

$$+ (1 + F)^{-1} \sum_i \frac{V_i X_i}{(\omega - f_i)^2} \left[1 + \frac{X_i}{\omega - f_i} + \cdots \right] \tag{4.58}$$

Now we may recall the definition of F and the correspondence between diagram types and terms in the expansion in (4.54) and obtain the final result

$$M^{\text{eff}} \approx M^{(2)} + (1 + F)^{-1} M^{(3)} \tag{4.59}$$

Let us discuss possible shortcomings of this approximation. In the step from (4.58) to (4.59) the terms with X_i^n, $n \geqslant 2$, have been neglected. This can be done to a good approximation if the pairs $(C1, D1)$ and $(C6, D6)$ are pairs of antigraphs as found in Section III.C.2. Since the antigraph relations are not exact relations, one might, in case they are not well satisfied, include terms of higher order in the X_i by the following simple procedure. The time-ordered diagrams that contribute to the diagonal approximation of Γ may also be approximated by a geometric series for $\omega \approx \varepsilon_p$. The geometric factor g is determined by $(C1 + D1 + C6 + D6)/M^{(2)}$ and the last sum in (4.58) is approximated by $(1 - g)^{-1}(C1 + D1 + C6 + D6)$. In practice, however, the antigraph relations for $(C1, D1)$ and $(C6, D6)$ have always been found to be well fulfilled and there is no need for the latter procedure. Another shortcoming of (4.59) is more serious, but can also be removed. In performing the expansion in (4.54) we have assumed $M^{(2)}$ to be the leading term in M^{eff} for $\omega \approx \varepsilon_p$. From numerical experience we deduce that this assumption is valid for nearly all cases. There are a few cases where the assumption is not valid for the following reasons. In principle, we may divide M^{eff} into two parts, $M^{\text{eff}}(1)$ and $M^{\text{eff}}(2)$, with poles having imaginary components $i\eta$ and $-i\eta$, respectively. This is also true for $M^{(2)}$, $M^{(2)}(1)$ and $M^{(2)}(2)$ being the second and the first sums in (4.2), respectively. It now is easy to see that $M^{(2)}(1)$ is positive, whereas $M^{(2)}(2)$ is negative for $\omega \approx \varepsilon_p$. Usually ε_p is nearer to poles of $M^{(2)}(1)$ than to poles of $M^{(2)}(2)$ and the residues of $M^{(2)}(1)$ are larger than those of $M^{(2)}(2)$. $M^{(2)}(1)$ therefore dominates over $M^{(2)}(2)$. When $M^{(2)}(1)$ and $M^{(2)}(2)$ nearly compensate each other, $M^{(2)}$ is no longer the dominating term in M^{eff}. In this case we may repeat the whole procedure leading to (4.59) for $M^{(2)}(1)$ and $M^{(2)}(2)$ separately and then combine both series to one. The result is identical with the one in (4.59), but with

$$F = \frac{G_1 M^{(3)}(1) + G_2 M^{(3)}(2)}{M^{(3)}} \tag{4.60a}$$

$$M^{(3)} = M^{(3)}(1) + M^{(3)}(2)$$

where $M^{(3)}(1)$ contains the time-ordered diagrams CI, DI, $I = 4$ to 6, and G_1, G_2 are the quantities analogous to F in (4.56):

$$G_1 = -\frac{\sum\limits_{I=4}^{5}(CI + DI)}{M^{(2)}(1)} \qquad G_2 = -\frac{\sum\limits_{I=2}^{3}(CI + DI)}{M^{(2)}(2)} \tag{4.60b}$$

To obtain the self-energy part we must add to $\mathbf{M}$ the contribution $\Sigma(\infty)$ of the constant diagrams. For this purpose one may start from (3.28a), which allows the self-consistent calculation of $\Sigma(\infty)$ for a given $\mathbf{M}(\omega)$. As the first step of the iteration we start with $\Sigma_{pq'}(\infty) = s_{pq'}$, which vanishes for a closed-shell system in the HF-representation. Substituting $\mathbf{G}^0$ for $\mathbf{G}$ and assuming the screening parameters to be, on the average, approximately equal to the screening parameter of the orbital of interest, one may write

$$\Sigma^{\text{eff}} \approx \Sigma^{(2)} + (1 + F)^{-1} \Sigma^{(3)} \tag{4.61}$$

For further improvements, which are only important when the antigraph relations are not accurately satisfied, one should start from the fundamental equation (3.28) using less crude approximations.

We have seen that it is possible to estimate the contributions of higher orders by calculating terms up to third order and taking into account analytical properties of $M(\omega)$. This conclusion is substantiated by the numerical results. The above scheme to approximate higher orders is very simple and requires a negligible amount of computational effort. It is, however, only applicable for outer valence ionizations, where the $-I_p$ are far from the poles of the self-energy part. We believe that considering the analytic properties of $M(\omega)$ in more detail will lead to improved approximation schemes of this kind.

For the reader interested in the calculation Σ^{eff}, the explicit expressions for the diagrams up to third order can be found in the Appendix.

V. THE VIBRATIONAL STRUCTURE IN MOLECULAR IONIZATION SPECTRA

The vibrational structure is a striking feature of molecular photoelectron spectra. In particular, the bands corresponding to the deeper valence electrons show pronounced vibrational effects. From the vibrational structure observed in the PE spectrum the bonding properties of the individual electrons can be inferred. The orbitals of diatomic molecules can thus simply be classified as bonding, antibonding, or nonbonding. The situation is less trivial for polyatomic molecules. Here the photoelectron spectrum reflects the bonding character of each particular electron with respect to the various normal coordinates of the molecule and contains, therefore, a considerable amount of information.

Unfortunately the vibrational structure in the PE spectrum becomes exceedingly complex for larger molecules as a result of the rapidly increasing number of normal coordinates. The interpretation of the vibrational structure (i.e., the assignment of fundamentals to the observed vibrational lines) is therefore often controversial, even for relatively small molecules. The analysis of the spectra is greatly facilitated if we can be guided by some theoretical prediction of the vibrational structure. For complex spectra, in particular, it is unnecessary to calculate accurately the large number of Franck-Condon factors that determine the intensities of the individual vibrational lines. Even a qualitative prediction of the shape of the spectral distribution suffices for interpretation purposes.

A theoretical determination of the vibrational structure could also provide a useful tool for assigning the ionization potentials. If two or more bands of the spectrum lie close together, the vertical IPs must be calculated to high accuracy in order to interpret the spectrum unambiguously. Such assignment problems may be tackled more easily by using typical (and theoretically predictable) features of the vibrational structure to identify the bands. Indeed, many assignments of molecular PE spectra are based largely on arguments concerning the vibrational structure of one or more bands. Theoretical calculations are thus desirable to put these arguments on a firmer basis.

It is well known that the vibrational structure in molecular electronic spectra can be calculated if the potential surfaces of the initial and final electronic state are known[74]. The spectroscopic data (i.e., equilibrium geometry, harmonic force constants, and possibly higher-order force constants) of the molecular ground state are well known in most cases, whereas the corresponding data for the ionic states are, in general, not experimentally available. In principle, the potential surfaces of the ionic states could be calculated by ab initio methods, but in practice this approach is too expensive, at least for polyatomic molecules. Furthermore, the Franck-Condon factors depend mainly on the difference between the potential functions of the initial and the final state.[75] Therefore both potential surfaces involved in the transition must be calculated to high accuracy in order to obtain good results.

To avoid these difficulties we extend the Green's function approach to include the vibrational effects. It is an important advantage of this approach that only the data for the initial electronic state (i.e., the electronic ground state of the parent molecule) are required. The Franck-Condon factors are expressed in terms of certain coupling constants, which can easily be calculated on the one-particle level. These coupling constants can be corrected (renormalized) to include many-body effects. Because of its simplicity, the method is applicable to fairly large polyatomic molecules, yielding reasonably accurate results with moderate computational expense.

A. The Hamiltonian

In this section the Hamiltonian including the vibrational degrees of freedom is expressed in the occupation number formalism. Neglecting spin-dependent interactions, the molecular Hamiltonian reads

$$H = T_N \left(\frac{\partial}{\partial \mathbf{X}} \right) + V_{NN} (\mathbf{X}) + H_{EN} (\mathbf{x}, \mathbf{X}) \tag{5.1}$$

where T_N denotes the nuclear kinetic energy, V_{NN} the electrostatic repulsion of the nuclei, and

$$H_{EN} = T_E \left(\frac{\partial}{\partial \mathbf{x}} \right) + V_{EN} (\mathbf{x}, \mathbf{X}) + V_{EE} (\mathbf{x}) \tag{5.2}$$

is the Hamiltonian of the electrons in the field of the nuclei. $\mathbf{x}$ represents the electron coordinates, $\mathbf{X}$ the set of $3N$ nuclear coordinates.

Since we are not interested in the translational and rotational motion of the molecule, we change to $3N$-6 ($3N$-5 for linear molecules) internal coordinates $\mathbf{R}$, fixing the six (five) coordinates of translation and rotation. When expressed in terms of the internal coordinates, the nuclear kinetic energy becomes in general a function of both $\mathbf{R}$ and $\partial / \partial \mathbf{R}$.[76]

Expanding the electronic field operator $\psi(\mathbf{x})$ in terms of HF single-particle orbitals,

$$\psi(\mathbf{x}) = \sum_i \varphi_i(\mathbf{x}, \mathbf{R}) a_i \tag{5.3}$$

H_{EN} takes the form (for a closed-shell system[77])

$$H_{EN}(\mathbf{x}, \mathbf{R}) = \sum_i \varepsilon_i(\mathbf{R}) a_i^+ a_i + \tfrac{1}{2} \sum_{ijkl} V_{ijkl}(\mathbf{R}) a_i^+ a_j^+ a_l a_k$$

$$- \sum_{ij} \sum_{k \in \mathcal{F}} \left[V_{ikjk}(\mathbf{R}) - V_{ikkj}(\mathbf{R}) \right] a_i^+ a_j \tag{5.4}$$

The HF orbitals $\varphi_i(\mathbf{x}, \mathbf{R})$ are obtained by diagonalizing the HF operator at each nuclear configuration $\mathbf{R}$. Although not indicated explicitly, the electronic creation and annihilation operators a_i^+, a_i depend on the nuclear coordinates $\mathbf{R}$.

After some algebra[75] (5.1) to (5.4) can be rewritten in the form

$$H = T_N\left(\mathbf{R}, \frac{\partial}{\partial \mathbf{R}}\right) + V_0(\mathbf{R}) + \sum_i \varepsilon_i(\mathbf{R})\left[a_i^+ a_i - n_i\right]$$

$$+ \tfrac{1}{2} \sum_{ijkl} V_{ijkl}(\mathbf{R})\left[\delta\sigma_1 a_i^+ a_j^+ a_l a_k\right.$$

$$\left. + \delta\sigma_2 a_l a_k a_i^+ a_j^+ + 2\delta\sigma_3 a_j^+ a_k a_l a_i^+ \right] \tag{5.5}$$

where $\delta\sigma_f = 1$ if $(ijkl) \in \sigma_f$ and $\delta\sigma_f = 0$ if $(ijkl) \notin \sigma_f$. The index set σ_1 means that at least φ_k and φ_l or φ_i and φ_j are unoccupied, σ_2 that at most one of the orbitals is unoccupied, and σ_3 that φ_k and φ_j or φ_l and φ_i are unoccupied. $V_0(\mathbf{R})$ in (5.5) is the potential energy for the nuclear motion in the electronic ground state within the HF approximation,

$$V_0(\mathbf{R}) = V_{NN}(\mathbf{R}) + \sum_{i \in \mathcal{F}} \varepsilon_i(\mathbf{R}) - \tfrac{1}{2} \sum_{ij \in \mathcal{F}} \left(V_{ijij}(\mathbf{R}) - V_{ijji}(\mathbf{R})\right) \tag{5.6}$$

$V_0(\mathbf{R})$ is expanded about the molecular equilibrium geometry, determined by

$$\left(\frac{\partial V_0}{\partial R_s}\right)_0 = 0 \qquad \text{for all } s$$

giving

$$V_0(\mathbf{R}) - V_0(0) = \tfrac{1}{2} \sum_{ss'} F_{ss'} R_s R_{s'} + \cdots \tag{5.7}$$

where $F_{ss'} = (\partial^2 V_0 / \partial R_s \partial R_{s'})_0$. The internal coordinates R_s are measured from the equilibrium geometry. The full expression for the nuclear kinetic energy T_N reads[76]

$$T_N\left(\mathbf{R}, \frac{\partial}{\partial \mathbf{R}}\right) = -\tfrac{1}{2} g^{1/4} \sum_{ss'} \frac{\partial}{\partial R_s} g^{-1/2} G_{ss'}(\mathbf{R}) \frac{\partial}{\partial R_{s'}} g^{1/4} \tag{5.8}$$

where $g = \det \mathbf{G}(\mathbf{R})$ and $\mathbf{G}(\mathbf{R})$ is the kinematic matrix. For small displacements from the equilibrium the above expression can be replaced by

$$T_N\left(\frac{\partial}{\partial \mathbf{R}}\right) = -\tfrac{1}{2} \sum_{ss'} G_{ss'}(0) \frac{\partial}{\partial R_s} \frac{\partial}{\partial R_s'} \tag{5.9}$$

where $\mathbf{G}(0)$ is the kinematic matrix evaluated at the equilibrium geometry.

The harmonic potential and the kinetic energy (5.9) can be diagonalized by introducing (dimensionless) normal coordinates[76]

$$\mathbf{Q} = \omega^{1/2} \mathbf{L}^{-1} \mathbf{R} \tag{5.10}$$

where

$$\mathbf{GFL} = \mathbf{L}\omega^2,$$
$$\mathbf{LL}^+ = \mathbf{G}$$

and ω denotes the diagonal matrix of ground-state vibrational frequencies ω_s. Introducing boson creation and annihilation operators b_s^+, b_s by

$$Q_s = \frac{1}{\sqrt{2}} (b_s + b_s^+) \tag{5.11}$$

$$\frac{\partial}{\partial Q_s} = \frac{1}{\sqrt{2}} (b_s - b_s^+)$$

the full Hamiltonian (5.5) takes the form

$$H = V_0(0) + \sum_{s=1}^{M} \omega_s\left(b_s^+ b_s + \tfrac{1}{2}\right) + \Delta T_N + \Delta V_0$$

$$+ \sum_i \varepsilon_i(\mathbf{Q})\left[a_i^+ a_i - n_i\right] + \sum_{ijkl} V_{ijkl}(\mathbf{Q})\left[\delta\sigma_l a_i^+ a_j^+ a_l a_k\right.$$

$$\left. + \delta\sigma_2 a_l a_k a_i^+ a_j^+ + 2\delta\sigma_3 a_j^+ a_k a_l a_i^+\right] \tag{5.12}$$

with

$$\Delta T_N = T_N \left(\mathbf{Q}, \frac{\partial}{\partial \mathbf{Q}} \right) + \frac{1}{2} \sum_s \omega_s \frac{\partial^2}{\partial Q_s^2} \tag{5.13}$$

$$\Delta V_0 = V_0(\mathbf{Q}) - \frac{1}{2} \sum_s \omega_s Q_s^2 - V_0(0) \tag{5.14}$$

ΔV_0 represents the deviation of the ground-state potential energy from the harmonic term. For vibrations of large amplitude the influence of ΔT_N and especially of ΔV_0 may become considerable. Both ΔT_N and ΔV_0 are referred to as anharmonic correction terms.

Expanding $\varepsilon_i(\mathbf{Q})$ and $V_{ijkl}(\mathbf{Q})$ in (5.12) in powers of $\mathbf{Q}$ we arrive at the Hamiltonian in its final form:

$$H = H_E + H_N + H_{EN}^{(1)} + H_{EN}^{(2)} \tag{5.15}$$

$$H_E = \sum_i \varepsilon_i(0) a_i^+ a_i + \frac{1}{2} \sum_{ijkl} V_{ijkl}(0) a_i^+ a_j^+ a_l a_k$$

$$- \sum_{ij} \sum_{k \in \mathscr{F}} \left[V_{ikjk}(0) - V_{ikkj}(0) \right] a_i^+ a_j \tag{5.16}$$

$$H_N = V_{NN}(0) + \sum_{s=1}^{M} \omega_s \left(b_s^+ b_s + \frac{1}{2} \right) + \Delta T_N + \Delta V_0 \tag{5.17}$$

$$H_{EN}^{(1)} = \sum_i \sum_{s=1}^{M} \frac{1}{\sqrt{2}} \left(\frac{\partial \varepsilon_i}{\partial Q_s} \right)_0 [a_i^+ a_i - n_i](b_s + b_s^+)$$

$$+ \sum_i \sum_{ss'=1}^{M} \frac{1}{4} \left(\frac{\partial^2 \varepsilon_i}{\partial Q_s \partial Q_{s'}} \right)_0 [a_i^+ a_i - n_i](b_s + b_s^+)(b_{s'} + b_{s'}^+) + \cdots \tag{5.18}$$

$$H_{EN}^{(2)} = \sum_{ijkl} \sum_{s=1}^{M} \frac{1}{2\sqrt{2}} \left(\frac{\partial V_{ijkl}}{\partial Q_s} \right)_0 \left[\delta\sigma_1 a_i^+ a_j^+ a_l a_k \right.$$

$$+ \delta\sigma_2 a_l a_k a_i^+ a_j^+ + 2\delta\sigma_3 a_j^+ a_k a_l a_i^+ \big](b_s + b_s^+)$$

$$+ \sum_{ijkl} \sum_{ss'=1}^{M} \frac{1}{8} \left(\frac{\partial^2 V_{ijkl}}{\partial Q_s \partial Q_{s'}} \right)_0 \left[\delta\sigma_1 a_i^+ a_j^+ a_l a_k + \delta\sigma_2 a_l a_k a_i^+ a_j^+ \right.$$

$$+ 2\delta\sigma_3 a_j^+ a_k a_l a_i^+ \big](b_s + b_s^+)(b_{s'} + b_{s'}^+) + \cdots \tag{5.19}$$

The anharmonic corrections ΔT_N and ΔV_0 may also be expanded in powers of $\mathbf{Q}$. The lowest-order term of ΔT_N is of the form

$$\sum_{ss's''} \alpha_{ss's''}(b_s - b_s^+)(b_{s'} + b_{s'}^+)(b_{s''} - b_{s''}^+)$$

whereas the expansion of ΔV_0 starts with a term of the form

$$\sum_{ss's''} \beta_{ss's''}(b_s + b_s^+)(b_{s'} + b_{s'}^+)(b_{s''} + b_{s''}^+)$$

It is easy to interpret the individual terms of the Hamiltonian (5.15). H_E describes the motion of the electrons for nuclei fixed at the equilibrium geometry. Calculating the electronic Green's function with this Hamiltonian as discussed at length in the preceding sections yields the vertical ionization potentials of the system. H_N describes the motion of the nuclei in the potential of the electronic ground state. ΔT_N and ΔV_0 represent deviations from the harmonic motion. In the formalism of second quantization these can be considered as boson-boson interactions. The remaining terms represent the coupling between the electronic and the nuclear motion. The coupling on the one-particle level is described by $H_{EN}^{(1)}$, (5.18), where only the linear and the quadratic coupling has been included explicitly. The expansion in (5.18) may be continued, however, to take into account anharmonic coupling terms if necessary. Finally $H_{EN}^{(2)}$ describes the modification of the interaction between electrons due to the coupling to the nuclear motion.

The Hamiltonian (5.15) is of interest since it describes a variety of physical effects. In fact, it contains too subtle and too many effects to allow a complete theoretical treatment. Fortunately, there exist some well-established approximations that allow the problem to be simplified considerably. The adiabatic approximation,[78] in particular, leads to a separation of the nuclear and the electronic motions. In the adiabatic approximation the electronic operators a_i and a_i^+ are considered to depend only "parametrically" on the nuclear coordinates $\mathbf{Q}$, that is, the nuclear kinetic energy operator is assumed to commute with a_i and a_i^+. The electron operators then commute with the boson operators b_s, b_s^+. The second well-known approximation is the harmonic approximation, which amounts to the neglect of ΔT_N and ΔV_0 and also of terms higher than second order in the expansion of $\varepsilon_i(\mathbf{Q})$ and $V_{ijkl}(\mathbf{Q})$. Finally, we may introduce the one-particle approximation, that is, neglect all terms containing the electron-electron interaction integrals V_{ijkl}.

In the next section the *exact* solution for the spectrum $P(\omega)$ within the adiabatic, harmonic, Franck-Condon, and one-particle approximations is given. The subsequent three sections are devoted to a discussion of these approximations. It is shown how many-body effects can be included within the Green's function approach and how nonadiabatic and anharmonic effects enter the formalism.

B.   Solution of the One-Body Problem

In Section II the transition probability $P(\omega)$ has been expressed in terms of the one-body Green's function. In this section we restrict ourselves to the one-particle approximation. Equation 2.4 then simplifies to (chosing ω_0 as the zero of the energy scale)

$$P(\omega) = 2 \sum_{ei} |\tau_{ei}|^2 \operatorname{Im} \left\{ G_{ii}^0 (\omega - i\eta) \right\} \delta (\omega - E_e) \tag{5.20}$$

Obviously, the vibrational structure in $P(\omega)$ can be calculated for each band i separately. Since the kinetic energy E_e of the outgoing electron has been assumed to be large compared to the vibrational frequencies ω_s, the matrix element τ_{ei} will be independent of E_e to a good approximation and we obtain for the ith band

$$P_i(\omega) = 2|\tau_{ei}|^2 \operatorname{Im} \left\{ G_{ii}^0 (\omega - i\eta) \right\} \tag{5.21}$$

Equation 5.21 can be rewritten in a slightly more general form[79] as

$$P_i(\omega) = \int dt\, e^{i\omega t} \langle \Phi_0^N | T_i^+ T_i(t) | \Phi_0^N \rangle \tag{5.22}$$

where $T_i(t) = e^{iHt} T_i e^{-iHt}$ and $T_i = \tau_{ei} a_i$ is the "transition operator" describing the ionization of an electron in orbital i. Equation 5.21 is recovered from (5.22) if the photoionization cross section τ_{ei} is assumed to be independent of the nuclear coordiantes $\mathbf{Q}$ and taken out of the matrix element in (5.22). This is the well-known Franck-Condon approximation. In principle, corrections to this approximation can be obtained by expanding τ_{ei} in powers of $\mathbf{Q}$.

In the adiabatic approximation the ground-state wave function $|\Phi_0^N\rangle$ is written as a product

$$|\Phi_0^N\rangle = |\varphi_0^N\rangle |0\rangle \tag{5.23}$$

where $|\varphi_0^N\rangle$ and $|0\rangle$ are the electronic and nuclear ground-state wave functions, respectively (we assume that the temperature is low enough so that no vibrations are excited in the parent molecule). In the one-particle approximation $|\varphi_0^N\rangle$ is the ground state for the HF operator $\sum_i \varepsilon_i(\mathbf{Q}) a_i^+ a_i$ and in the harmonic approximation $|0\rangle$ is the state containing no vibrational quanta, that is, $b_s|0\rangle = 0$ for all s. The Hamiltonian (5.15) takes in the harmonic and the one-particle approximation the simple form

$$H = V_{NN}(0) + \sum_i \varepsilon_i(0) a_i^+ a_i + \sum_s \omega_s \left(b_s^+ b_s + \tfrac{1}{2} \right)$$

$$+ \sum_i \sum_s \frac{1}{\sqrt{2}} \left(\frac{\partial \varepsilon_i}{\partial Q_s} \right)_0 \left[a_i^+ a_i - n_i \right] (b_s + b_s^+)$$

$$+ \sum_i \sum_{ss'} \frac{1}{4} \left(\frac{\partial^2 \varepsilon_i}{\partial Q_s \partial Q_{s'}} \right)_0 \left[a_i^+ a_i - n_i \right] (b_s + b_s^+)(b_{s'} + b_{s'}^+) \qquad (5.24)$$

Inserting the Hamiltonian (5.24) into (5.22) and assuming the a_i, a_i^+ to commute with the b_s, b_s^+ (adiabatic approximation) we obtain

$$P_i(\omega) = |\tau_{ei}|^2 \int dt \exp \left\{ i \left(\omega - \varepsilon_i(0) - \tfrac{1}{2} tr \omega \right) t \right\} \langle 0 | e^{i\tilde{H}t} | 0 \rangle \qquad (5.25)$$

with

$$\tilde{H} = \sum_s \omega_s \left(b_s^+ b_s + \tfrac{1}{2} \right) + \sum_s \kappa_s^i (b_s + b_s^+)$$

$$+ \sum_{ss'} \gamma_{ss'}^i (b_s + b_s^+)(b_{s'} + b_{s'}^+) \qquad (5.26)$$

and

$$\kappa_s^i = -\frac{1}{\sqrt{2}} \left(\frac{\partial \varepsilon_i}{\partial Q_s} \right)_0 \qquad (5.27)$$

$$\gamma_{ss'}^i = -\frac{1}{4} \left(\frac{\partial^2 \varepsilon_i}{\partial Q_s \partial Q_{s'}} \right)_0$$

κ_s^i and $\gamma_{ss'}^i$ are called linear and quadratic coupling constants, respectively.

To simplify the calculation of $P(\omega)$ we introduce the following matrix notation. $\boldsymbol{\kappa}$ denotes the M-dimensional vector of linear coupling constants κ_s^i, $\boldsymbol{\gamma}$ the M-dimensional matrix of quadratic coupling constants $\gamma_{ss'}^i$. In addition, we introduce the matrices and vectors

$$\mathbf{K} = \begin{pmatrix} \kappa \\ \kappa \end{pmatrix} \qquad \Gamma = \begin{pmatrix} \gamma & \gamma \\ \gamma & \gamma \end{pmatrix}$$

$$\Omega = \begin{pmatrix} \omega & 0 \\ 0 & \omega \end{pmatrix} \qquad \mathbf{B} = \begin{pmatrix} b_1 \\ \vdots \\ b_M \\ b_1^+ \\ \vdots \\ b_M^+ \end{pmatrix} \tag{5.28}$$

With this notation the Hamiltonian (5.26) reads

$$\tilde{H} = \tfrac{1}{2} \mathbf{B}^+ \Omega \mathbf{B} + \mathbf{B}^+ \mathbf{K} + \mathbf{B}^+ \Gamma \mathbf{B} \tag{5.29}$$

In order to diagonalize this Hamiltonian, new boson operators c_s, c_s^+ are introduced. Defining a column vector $\mathbf{C}$ in analogy to the column vector $\mathbf{B}$ in (5.28), we set

$$\mathbf{C} = \Lambda \mathbf{B} \qquad \Lambda = \begin{pmatrix} \lambda_1 & \lambda_2 \\ \lambda_2 & \lambda_1 \end{pmatrix} \tag{5.30}$$

where λ_1 and λ_2 are M-dimensional matrices with elements $\lambda_1^{ss'}$ and $\lambda_2^{ss'}$, respectively. The new boson operators are related to new normal coordinates $\hat{Q}_s$ according to

$$\hat{Q}_s = \frac{1}{\sqrt{2}} (c_s + c_s^+)$$

$$\frac{\partial}{\partial \hat{Q}_s} = \frac{1}{\sqrt{2}} (c_s - c_s^+) \tag{5.31}$$

Defining a transformation $\mathbf{J}$ that connects the new normal coordinates $\hat{\mathbf{Q}}$ with the ground-state normal coordinates $\mathbf{Q}$,

$$\hat{\mathbf{Q}} = \mathbf{J} \mathbf{Q} \tag{5.32}$$

we obtain from (5.30) and (5.31).

$$\lambda_1 + \lambda_2 = J \tag{5.33}$$

$$\lambda_1 - \lambda_2 = J^{-1+}$$

The requirement that the c's obey the commutation relations for bosons implies

$$\Lambda^{-1} = \begin{pmatrix} \lambda_1^+ & -\lambda_2^+ \\ -\lambda_2^+ & \lambda_1^+ \end{pmatrix} \tag{5.34}$$

Choosing now Λ to diagonalize the quadratic terms in $\tilde{H}$, that is, to give

$$\tilde{H} = \tfrac{1}{2} C^+ \hat{\Omega} C + C^+ \hat{K} \tag{5.35}$$

with $\hat{\Omega}$ and $\hat{K}$ defined analogously to Ω and K, we find

$$\Omega + 2\Gamma = \Lambda^+ \hat{\Omega} \Lambda$$

$$\hat{K} = \Lambda^{-1+} K \tag{5.36}$$

With (5.33) and (5.34) we rewrite (5.36) to give

$$\omega + 4\gamma = J^+ \hat{\omega} J \tag{5.37}$$

$$\hat{\omega} = J \omega J^+ \tag{5.38}$$

$$\hat{\kappa} = J^{-1+} \kappa \tag{5.39}$$

The general solution of (5.38) is

$$J = \hat{\omega}^{1/2} Z \omega^{-1/2} \tag{5.40}$$

with an orthogonal matrix Z. Inserting (5.40) into (5.37) we obtain the symmetric eigenvalue problem

$$\omega^{1/2} (\omega + 4\gamma) \omega^{1/2} Z^+ = Z^+ \hat{\omega}^2 \tag{5.41}$$

Solution of this secular equation yields the final-state vibrational frequencies $\hat{\omega}_s$ and, together with (5.40), the transformation matrix J between initial and final-state normal coordinates. It should be noted that, apart from the factors $\hat{\omega}^{1/2}$ and $\omega^{-1/2}$, which enter since we are working with dimensionless normal coordinates, J is an orthogonal transformation.

It remains to eliminate the linear term $C^+ \hat{K}$ from the Hamiltonian

(5.35). This is readily accomplished by the unitary transformation

$$U = \exp(-\mathbf{C}^+\mathbf{P}) \tag{5.42}$$

with

$$\mathbf{P} = \hat{\Omega}^{-1}\begin{pmatrix} \hat{\kappa} \\ -\hat{\kappa} \end{pmatrix} \tag{5.43}$$

The transformed Hamiltonian $U^+\tilde{H}U$ is diagonal and we obtain from (5.25) the following expression for the transition probability:

$$P_i(\omega) = |\tau_{ei}|^2 \sum_{n_1 \cdots n_M = 0}^{\infty} |\langle 0|U|n_1 \cdots n_M\rangle|^2 \, \delta\left(\omega - \varepsilon_i(0) + \Delta_i + \sum_s n_s \hat{\omega}_s\right)$$

$$\tag{5.44a}$$

with

$$\Delta_i = \tfrac{1}{2} tr(\hat{\omega} - \omega) - \hat{\kappa}^+ \hat{\omega}^{-1}\hat{\kappa} \tag{5.44b}$$

The physical meaning of the quantities in (5.44) is the following: $-\varepsilon_i(0)$ represents the vertical ionization energy, that is, the difference in energy between the ionic state and the molecular ground state, evaluated at the equilibrium geometry of the ground state. $-\varepsilon_i(0) + \Delta_i$ is the adiabatic ionization potential, that is, the difference in energy between the vibrational ground states of the ion and the molecule in its electronic ground state. The quantum numbers $n_1 \cdots n_M$ refer to vibrational quanta in the ionic state. Finally, the squared matrix elements $|\langle 0|U|n_1 \cdots n_M\rangle|^2$ are the well-known Franck-Condon (FC) factors determining the intensity distribution of the vibrational satellite lines.

To evaluate the FC factors we rewrite $\mathbf{C}^+\mathbf{P}$

$$\mathbf{C}^+\mathbf{P} = \sum_s \alpha_s b_s^+ + \sum_s \beta_s c_s \tag{5.45}$$

with

$$\alpha = -\lambda_1^{-1}\hat{\omega}^{-1}\hat{\kappa}$$
$$\beta = \hat{\omega}^{-1}\hat{\kappa} - \lambda_2\lambda_1^{-1}\hat{\omega}^{-1}\hat{\kappa} \tag{5.46}$$

where α and β are column vectors with elements α_s and β_s, respectively. Inserting (5.45) into (5.42), using the identity

$$e^{x+y} = e^x e^y e^{-\frac{1}{2}[x,y]}$$

([x,y] a c-number) and collecting all destruction operators c_s into y and all creation operators b_s^+ into x, we obtain [80]

$$\langle 0|U|n_1\cdots n_M\rangle = e^{\frac{1}{2}\beta^+\lambda_1\alpha} \sum_{m_1\cdots m_M=0}^{n_1\cdots n_M} \prod_{j=1}^{M} \left\{ \left[\frac{1}{m_j!}\binom{n_j}{m_j}\right]^{1/2} \beta_j^{m_j} \right\}$$

$$\times \langle 0|n_1-m_1\cdots n_M-m_M\rangle \qquad (5.47)$$

In (5.47) the overlap of the wavefunctions of two displaced M-dimensional harmonic oscillators with different frequencies has been expressed as a *finite* linear combination of the overlaps of the corresponding nondisplaced oscillators.

To calculate these latter overlaps we note that

$$\langle 0|b_s^+|n_1\cdots n_M\rangle = 0 \qquad \text{for } s=1,2,\ldots,M.$$

With (5.30) we obtain the recursion relation

$$\langle 0|n_1\cdots n_l+1\cdots n_M\rangle = \sum_{r=1}^{M} \left(\frac{n_r}{n_l+1}\right)^{1/2} f_{rl}\langle 0|n_1\cdots n_r-1\cdots n_M\rangle \qquad (5.48)$$

where

$$\mathbf{f} = \lambda_2\lambda_1^{-1}$$

Obviously $\langle 0|m_1\cdots m_M\rangle$ vanishes if $\sum_i m_i$ is an odd number. The remaining overlaps are proportional to the overlap of the ground states,[80]

$$\langle 0|0\cdots 0\rangle = [\det\lambda_1]^{-1/2}$$

These formulas complete the calculation of the transition probability $P(\omega)$.

The formalism simplifies considerably when the matrix γ is diagonal. The recursion relation (5.48) can then be evaluated explicitly and the FC factors expressed in terms of Hermite polynomials.[75] A particularly simple formula results when all quadratic coupling constants $\gamma_{ss'}^i$ vanish. Equation 5.44a then reduces to

$$P_i(\omega) = |\tau_{ei}|^2 \sum_{n_1\cdots n_M=0}^{\infty} \left\{ \prod_{s=1}^{M} \frac{a_s^{n_s}}{n_s!} e^{-a_s} \right\} \delta\left(\omega-\varepsilon_i(0)+\Delta_i+\sum_s n_s\omega_s\right) \qquad (5.49)$$

with

$$a_s = \left(\frac{\kappa_s^i}{\omega_s} \right)^2$$

and

$$\Delta_i = - \sum_s a_s \omega_s$$

In the one-dimensional case the line intensities thus follow simply a Poisson distribution $a^n / n! e^{-a}$.

The spectra of polyatomic molecules are usually rather complex and a large number of lines contributes to the vibrational structure. The observed PE spectra, on the other hand, provide in most cases only a poorly resolved picture of this complicated structure. For the interpretation of these spectra, it is unnecessary to determine all the FC factors: it is sufficient to calculate the shape of the spectral distribution. To this end a moment expansion of $P(\omega)$ is briefly discussed.

Setting, for convenience, the squared matrix element $|\tau_{ei}|^2$ equal to unity we have from (5.25)

$$P_i(\omega + \varepsilon_i(0)) = \int dt \, \langle 0| \exp\left\{ it \left(\omega - \tfrac{1}{2} tr\omega + \tilde{H} \right) \right\} |0\rangle \qquad (5.50)$$

It follows that the kth moment of $P_i(\omega)$, calculated relative to the vertical ionization energy $-\varepsilon_i(0)$,

$$M_k = \int d\omega \, \omega^k P_i(\omega + \varepsilon_i(0)) \qquad (5.51)$$

is given by

$$M_k = (-1)^k \langle 0| \left[\tilde{H} - \tfrac{1}{2} tr\omega \right]^k |0\rangle \qquad (5.52)$$

Using (5.44), then (5.51) and (5.52) may be rewritten as a sum rule for the FC factors

$$\sum_{n_1 \cdots n_M = 0}^{\infty} |\langle 0| U |n_1 \cdots n_M \rangle|^2 \left(\Delta_i + \sum_s n_s \hat{\omega}_s \right)^k = \langle 0| \left[\tilde{H} - \tfrac{1}{2} tr\omega \right]^k |0\rangle \qquad (5.53)$$

The moments (5.52) can be evaluated in a straightforward manner. It is clear that all the moments depend only on the coupling constants κ_s^i, $\gamma_{ss'}^i$ and the ground-state vibrational frequencies ω_s.

For $k = 0$, (5.52) and (5.53) just state the normalization of the spectrum.

For the first and second moment one obtains

$$M_1 = - tr\,\gamma \tag{5.54}$$

$$M_2 = \kappa^+\kappa + 2\,tr\,\gamma^2 + (tr\,\gamma)^2 \tag{5.55}$$

Equation 5.54 states that the "center of gravity" of the vibrational structure is shifted from the vertical ionization energy $-\varepsilon_i(0)$ by $-tr\,\gamma$. A measure of the width of the vibrational structure is given by

$$\left[M_2 - M_1^2 \right]^{1/2} = \left[\kappa^+\kappa + 2\,tr\,\gamma^2 \right]^{1/2} \tag{5.56}$$

Inspection of (5.50) shows that the low moments describe the short-time behavior of the vibrational motion in the ionic state. Looking for the short-time behavior of the system implies, on the other hand, a poor energy resolution. It is clear, therefore, that a low-resolution spectrum is well characterized by its first few moments. Equations 5.54 and 5.55 demonstrate that the center of gravity and the width of the vibrational spectrum can be determined without knowing anything about the details of the vibrational motion in the ionic state. For formulas that allow the approximation of a distribution in terms of its moments we refer the reader to the literature.[81]

C. Inclusion of Many-Body Effects

Having seen how the transition probability can be calculated in the adiabatic, harmonic, and one-particle approximations, we turn to a discussion of the effects neglected in the above calculation. In the present section we are concerned with the influence of the electron-electron interaction on the electron-vibration coupling.

We start again from (2.4), which relates $P(\omega)$ to the imaginary part of the Green's function. The presence of electronic shake-up lines in the spectrum $P(\omega)$ has been emphasized in Section II. When the Green's function is calculated starting from the Hamiltonian (5.15), which includes the vibrational degrees of freedom, the spectrum contains both electronic and vibrational shake-up lines.

The perturbation calculation of the Green's function G with the Hamiltonian (5.15) is now briefly discussed. The unperturbed Hamiltonian is chosen to be

$$H_0 = \sum_i \varepsilon_i(0)a_i^+ a_i + V_{NN}(0) + \sum_s \omega_s\left(b_s^+ b_s + \tfrac{1}{2}\right) \tag{5.57}$$

The electronic part of the above Hamiltonian is simply the HF operator for fixed nuclei. The zeroth-order electronic Green's function $\mathbf{G}^0$ is therefore the same as that used in the preceding sections. In addition, the boson propagator

$$iD_s(t) = e^{-i\omega_s t}\theta(t) + e^{i\omega_s t}\theta(-t) \tag{5.58}$$

is introduced. The electron-electron, electron-boson, and boson-boson interaction terms in the Hamiltonian (5.15) may now be treated by diagrammatic perturbation theory. A prescription to draw and to evaluate the diagrams for the self-energy part of the electronic Green's function (neglecting boson-boson interaction terms) has been given in Ref. 75. This diagrammatic approach offers the possibility of calculating $P(\omega)$ using only the HF data of the parent molecule. The drawback of the method is that, because of the large number of coupling terms, many different diagrams have to be evaluated, even in a low order of the perturbation theory. It should be noted that with this approach the equilibrium geometry and the force constants of the exact ground state are implicitly calculated from the corresponding quantities on the HF level.

Fortunately, it is unnecessary to deal with the full Hamiltonian (5.15) in order to include the influence of many-body effects on the vibrational structure. The adiabatic hypothesis tells us that we can also think in terms of well-defined potential-energy surfaces when the electrons interact with each other. It follows that the entire effect of the electron-electron interaction terms in (5.15) is to replace the bosons by "renormalized" bosons, $-\varepsilon_i(0)$ by the true vertical IP and the coupling constants by renormalized coupling constants. Thus all the results obtained in Section V.B can be taken over if the ω_s are identified with the molecular frequencies in the exact ground state and if $-\varepsilon_i(0)$, κ_s^i, and $\gamma_{ss'}^i$ are replaced by the corresponding renormalized quantities. Since the spectroscopic constants of the electronic ground state are well known experimentally for most molecules of interest, it is only necessary to calculate the renormalized coupling constants in order to obtain the vibrational structure.

The renormalized κ_s^i and $\gamma_{ss'}^i$ can be determined in the following way. We start from that part of the Hamiltonian that does not contain the nuclear repulsion and the nuclear kinetic energy:

$$H_{EN} = \sum_i \varepsilon_i(\mathbf{Q})a_i^+ a_i + \frac{1}{2}\sum_{ijkl} V_{ijkl}(\mathbf{Q})a_i^+ a_j^+ a_l a_k$$

$$- \sum_{ijk} \left[V_{ikjk}(\mathbf{Q}) - V_{ikkj}(\mathbf{Q}) \right] n_k a_i^+ a_j \tag{5.59}$$

$\mathbf{Q}$ now denotes the normal coordinates in the exact ground state. The

Green's function calculated with the Hamiltonian (5.59) along the lines discussed in Sections III and IV is a function of $\mathbf{Q}$. Denoting by $E_i(\mathbf{Q})$ the pole of the Green's function describing the ionization of an electron in orbital i, the vertical ionization potential I_i is given by $-E_i(0)$ and the renormalized coupling constants are

$$
\kappa_s^i = -\frac{1}{\sqrt{2}}\left(\frac{\partial E_i}{\partial Q_s}\right)_0 \tag{5.60}
$$

$$
\gamma_{ss'}^i = -\frac{1}{4}\left(\frac{\partial^2 E_i}{\partial Q_s \partial Q_{s'}}\right)_0
$$

The "free" coupling constants obtained by replacing $E_i(\mathbf{Q})$ by $\varepsilon_i(\mathbf{Q})$ in (5.60) will be denoted by $\kappa_s^{i(0)}$ and $\gamma_{ss'}^{i(0)}$. They are *not* identical with the κ_s^i and $\gamma_{ss'}^i$ in the preceding section, since the derivative is now taken with respect to the normal coordinates of the exact ground state.

It is an advantage of the renormalization procedure discussed above that it allows the calculation of the vertical IPs and of the vibrational structure of the bands at the same level of approximation. The approach is economic, taking account of the fact that the spectroscopic data for the electronic ground state are well known in most cases, whereas the corresponding data for the ionic states are generally not accessible experimentally. The renormalized coupling constants can be calculated with moderate numerical expense, rendering the method applicable to fairly large polyatomic molecules. The free coupling constants $\kappa_s^{i(0)}$ and $\gamma_{ss'}^{i(0)}$, in particular, can be obtained using standard MO-LCAO HF programs. It seems that even semiempirical CNDO schemes can be used to obtain the coupling constants.[82, 83]

The above discussion has been confined to the vibrational structure of the quasiparticle lines. It is clear, however, that the same arguments hold for any other pole of the Green's function. The vibrational structure of an electronic satellite line can be calculated in the same way as discussed above for the main lines. Each satellite line possesses its own vibrational coupling constants, which are of course different from the coupling constants of the corresponding quasiparticle line. One has to identify the E_i in (5.60) with that pole of $\mathbf{G}$ which corresponds to the satellite line in question.

It is interesting to note that the vibrational structure of a satellite line can be calculated also on the one-particle level. Considering a shake-up line involving the ionization of orbital i and simultaneous excitation of an electron from orbital k to orbital l, we may use (5.22) with a transition operator

$$T = \tau_{ikl} a_l^+ a_k a_i$$

In analogy to the derivation of (5.27) we obtain the coupling constants

$$\kappa_s^{(0)} = \frac{1}{\sqrt{2}} \frac{\partial}{\partial Q_s} (\varepsilon_l - \varepsilon_k - \varepsilon_i)_0$$

$$\gamma_{ss'}^{(0)} = \frac{1}{4} \frac{\partial^2}{\partial Q_s \partial Q_{s'}} (\varepsilon_l - \varepsilon_k - \varepsilon_i)_0 \tag{5.61}$$

In lowest order, therefore, any coupling constant of the shake-up line is the sum of the corresponding coupling constants of the three orbitals involved in the process. Since the occupied orbitals have frequently strong bonding and the lowest unoccupied orbitals antibonding character, rather large coupling constants are to be expected for many of the shake-up lines.

D. Nonadiabatic Effects

The adiabatic approximation is based on the fact that the energy difference between electronic states is large compared to the spacing of the rotational and vibrational excitations. A breakdown of the adiabatic approximation may occur, therefore, when two electronic states become degenerate or nearly degenerate. Most important is the case of the symmetry-induced degeneracy, leading to the well-known Jahn-Teller and (for linear molecules) Renner-Teller effects. Particularly the Jahn-Teller effect manifests itself in the PE spectrum. Molecules of high symmetry such as BH_3, CH_4, SF_6, or C_6H_6 possess spatially degenerate orbitals which, according to the Jahn-Teller theorem,[84] can couple linearly to vibrations of suitable symmetry. If the coupling is sufficiently strong, a very complex vibrational structure and a splitting of the band is observed.

We do not enter into a detailed discussion of the Jahn-Teller and Renner-Teller effects here, since comprehensive reviews on the subject exist.[85–87] We restrict ourselves to a brief discussion of the physical origin of the nonadiabatic coupling terms and point out how the formalism outlined in Section V.B has to be modified to include these effects.

In what follows, the ground-state wave function, which enters the calculation of the transition probability according to (5.22), is supposed to be nondegenerate. Indeed, if the ground state were electronically degenerate, the molecule would, according to the Jahn-Teller theorem, distort until the degeneracy were removed (an exception is provided by linear molecules, where the coupling vanishes in first order). Complications could arise if the Jahn-Teller coupling were weak, leaving a near degeneracy in the ground state. For the overwhelming majority of molecules, however, the electronic ground state is nondegenerate and well separated from the electronically excited states. Then the adiabatic product ansatz (5.23) for the ground-state wave function is well justified.

When the orbital i, from which ionization takes place, is spatially degenerate, the ion is formed in a degenerate electronic state. The breakdown of the adiabatic approximation expected in this case is reflected by the fact that some of the commutators of the boson operators b_s and b_s^+ with the electronic operators a_i and a_i^+, which have been neglected in Section V.B, become not only large but even divergent. Since the field operators (5.3) are independent of the normal coordinates, it follows that

$$a_i(\mathbf{Q}) = \sum_j \langle \varphi_i(\mathbf{Q})|\varphi_j(0)\rangle a_j(0) \tag{5.62}$$

With the definition (5.11) one obtains the commutator relations

$$\left[b_s^+, a_i(\mathbf{Q}) \right] = -\left[b_s, a_i(\mathbf{Q}) \right]$$

$$= \frac{1}{\sqrt{2}} \sum_j \langle \varphi_i(\mathbf{Q})|\frac{\partial}{\partial Q_s}|\varphi_j(\mathbf{Q})\rangle a_j(\mathbf{Q}) \tag{5.63}$$

In the adiabatic approximation the matrix elements $\langle \varphi_i(\mathbf{Q})|\partial/\partial\mathbf{Q}|\varphi_j(\mathbf{Q})\rangle$ and $\langle \varphi_i(\mathbf{Q})|\partial^2/\partial\mathbf{Q}^2|\varphi_j(\mathbf{Q})\rangle$ are neglected. If these matrix elements are singular, there is always the possibility of removing the divergency by a transformation to a suitable one-particle basis. The resulting one-particle Hamiltonian is, however, no longer diagonal, that is, it contains nondiagonal terms $a_i^+ a_j$.

The above-mentioned behavior of the electronic operators is best understood by considering a simple example, the equilateral triangular molecule belonging to the D_{3h} symmetry group.[85, 88] Denoting with r_1, r_2, and r_3 the three internuclear distances, the internal symmetry coordinates of the molecule in the D_{3h} conformation are

$$S_1 = 3^{-1/2}(r_1 + r_2 + r_3)$$

$$S_2 = 6^{-1/2}(2r_1 - r_2 - r_3)$$

$$S_3 = 2^{-1/2}(r_2 - r_3)$$

S_1 belongs to the A_1' representation, whereas S_2 and S_3 form a basis for the

E' representation. Apart from dimensional factors, the above symmetry coordinates are already the normal coordinates of the molecule in the D_{3h} conformation. It is the distortion of the molecule with respect to the normal coordinates of E' symmetry that removes the degeneracy of electronic orbitals of e' or e'' symmetry. A displacement along Q_2 leaves the molecule in C_{2v} symmetry, whereas a displacement along Q_3 retains only C_s symmmetry. Nevertheless, the splitting of the energy of the e orbitals is equal in both directions for small displacements. Introducing polar coordinates ρ, α in the Q_2, Q_3 plane,

$$Q_2 = \rho \cos \alpha$$

$$Q_3 = \rho \sin \alpha$$

the orbital energies are independent of α in first order in ρ,

$$\varepsilon(\rho, \alpha) = \varepsilon(0) \pm \sqrt{2}\, \kappa \rho \tag{5.64}$$

Adding $-\varepsilon(\rho, \alpha)$ to the ground-state potential energy $\frac{1}{2}\omega_2 \rho^2$, the well-known "Mexican hat" potential-energy surface[85-87] is obtained. In second order in ρ, ε depends also on α and a more complicated potential surface results.[85-87]

The form (5.64) of the orbital energies induces the following $\mathbf{Q}$ dependence of the single-particle wave functions φ_1, φ_2 for small displacements from the D_{3h} configuration:

$$\varphi_1(\alpha) = \cos(\alpha/2)\varphi_1(0) + \sin(\alpha/2)\varphi_2(0)$$
$$\varphi_2(\alpha) = -\sin(\alpha/2)\varphi_1(0) + \cos(\alpha/2)\varphi_2(0) \tag{5.65}$$

The orbitals thus depend in zeroth order in ρ on the azimuthal angle α. Equation 5.65 implies that φ_1 goes into φ_2 and φ_2 into $-\varphi_1$ when α varies from $0°$ to $180°$. Both φ_1 and φ_2 change sign when α varies from $0°$ to $360°$.[85, 89] Since

$$\frac{\partial}{\partial Q_2} = \cos \alpha \frac{\partial}{\partial \rho} - \frac{\sin \alpha}{\rho} \frac{\partial}{\partial \alpha}$$

$$\frac{\partial}{\partial Q_3} = \sin \alpha \frac{\partial}{\partial \rho} + \frac{\cos \alpha}{\rho} \frac{\partial}{\partial \alpha} \tag{5.66}$$

it is clear that the derivatives of φ_1 and φ_2 with respect to Q_2 or Q_3 diverge as $1/\rho$ for $\rho \to 0$. This singular behavior reflects the fact that even very small changes in the nuclear configuration require a considerable change of the electronic orbitals if ρ is sufficiently small. The physical concept of electrons following instantaneously the motion of the nuclei must therefore break down in the vicinity of $\rho = 0$, even for very small nuclear velocities.

Since the electronic operators a_1, a_2 transform as the wave functions φ_1, φ_2 in the ρ, α space, they exhibit the same singular behavior. In particular, the commutators of b_2 and b_3 with a_1 and a_2 are singular. To overcome this difficulty, new single-particle orbitals $\hat{\varphi}_1$, $\hat{\varphi}_2$ are introduced by the transformation

$$\hat{\varphi}_1 = \cos(\alpha/2)\varphi_1 - \sin(\alpha/2)\varphi_2$$
$$\hat{\varphi}_2 = \sin(\alpha/2)\varphi_1 + \cos(\alpha/2)\varphi_2$$

(5.67)

leaving the nondegenerate orbitals unchanged. The new wave functions $\hat{\varphi}_1$, $\hat{\varphi}_2$ depend no longer on α in zeroth order in ρ. Therefore their derivatives with respect to Q_2 and Q_3 do not exhibit the singularity discussed above. The matrix elements of $\partial/\partial Q_{2,3}$ and $\partial^2/\partial Q_{2,3}^2$ taken with the functions $\hat{\varphi}_{1,2}$ are of the same order of magnitude as in the nondegenerate case and can therefore be neglected to within the same degree of accuracy.

Expressing the one-particle part of the electronic Hamiltonian $\sum_i \varepsilon_i(\mathbf{Q})a_i^+ a_i$ in terms of the operators $\hat{a}_i$ and $\hat{a}_i^+$, which destroy or create particles in the new orbitals $\hat{\varphi}_i$, we obtain in first order in ρ the Hamiltonian

$$H = \omega(b_2^+ b_2 + b_3^+ b_3 + 1) + \varepsilon(0)(\hat{a}_1^+ \hat{a}_1 + \hat{a}_2^+ \hat{a}_2)$$
$$+ \kappa(\hat{a}_1^+ \hat{a}_1 - \hat{a}_2^+ \hat{a}_2)(b_2 + b_2^+) + \kappa(\hat{a}_1^+ \hat{a}_2 + \hat{a}_2^+ \hat{a}_1)(b_3 + b_3^+)$$

(5.68)

For simplicity, only the degenerate orbitals 1, 2 and the degenerate normal coordinates Q_2, Q_3 have been included. The transformation of the orbital basis has produced electron-boson coupling terms that are nondiagonal in the electronic coordinates. The original nonadiabatic coupling, resulting from singular matrix elements of the nuclear kinetic-energy operator, has thus been replaced by a nondiagonal electronic coupling. Model Hamiltonians of the form (5.68) have been discussed by several authors.[90, 91] The Hamiltonian (5.68) can be easily generalized to include quadratic coupling terms.

In most theoretical treatments of the Jahn-Teller problem, the coupling constant κ and possibly higher-order coupling constants have been considered as adjustable parameters and little attempt has been made towards an ab initio determination of the Jahn-Teller coupling constants. It is clear from (5.64) that κ can be obtained on the one-particle level as the derivative of the HF orbital energy with respect to ρ. Since the form (5.64) of the orbital energy follows from symmetry arguments only, the pole of the exact Green's function will exhibit the same ρ, α dependence. The renormalized coupling constant can therefore be obtained as the derivative of the pole of the Green's function as in the nondegenerate case. Quadratic and higher-order coupling constants can be obtained analogously. It should be noted that this determination of the coupling constants avoids the so-called "crude adiabatic approximation,"[85, 87] where the $\mathbf{Q}$ dependence of the electronic wave functions is neglected explicitly. It has been argued that this approximation is not capable of describing vibronic coupling effects properly.[92]

The calculation of the transition probability $P(\omega)$ has proved to be very difficult in the Jahn-Teller case. Even for the simplest model involving only one pair of degenerate vibrations and only linear coupling, no exact solution could be given. In principle, the problem can be solved numerically by diagonalizing the Hamiltonian in a basis of two-dimensional harmonic-oscillator wave functions.[91, 93] Very large matrices have to be diagonalized, however, when the coupling strength increases. Approximate solutions, based on perturbation theory or canonical transformation methods, have been derived by several authors.[90, 91, 94] The numerical calculations show that the vibrational spacing becomes irregular and that a splitting of the band occurs if the coupling is sufficiently strong.

E.   What Is the "Best" Harmonic Approximation?

It is well known from many calculations on diatomic molecules that the FC factors are not insensitive to the anharmonicity of the potential-energy curves. Rather accurate FC factors can be calculated for diatomic molecules using potential functions constructed by the Rydberg-Klein-Rees method from spectroscopic data[95] or by fitting Morse potentials to the spectroscopic data of the initial and final states.[96] The calculation of FC factors for multidimensional potential functions of polyatomic molecules, on the other hand, has so far been confined to the harmonic approximation,[97, 98] although the importance of anharmonic effects is known. It should be noted that this traditional harmonic approximation and the harmonic approximation as defined in Section V.A are not identical. As is shown below, a major part of anharmonic effects neglected in the traditional harmonic approach is taken into account in the present harmonic approximation.

The anharmonic terms in our basic Hamiltonian (5.15) are ΔV_0 and ΔT_N, representing corrections to the harmonic ground-state potential and kinetic energy, respectively, and higher-order coupling terms, that is, terms involving the third or higher derivatives of $\varepsilon_i(\mathbf{Q})$ or $V_{ijkl}(\mathbf{Q})$. Usually, the anharmonic coupling, that is, the change of the anharmonicity of the potential-energy surface due to ionization, is much smaller than the anharmonic corrections to the ground-state potential energy itself. In some cases, however, for example, when the ionic potential surface is strongly distorted by an avoided crossing, anharmonic coupling may become important as well.

The Hamiltonian (5.15) has been obtained by expanding all Q-dependent quantities (with the exception of the electronic operators a_i, a_i^+) about the equilibrium geometry of the electronic ground state. Calculating $P(\omega)$ with this Hamiltonian implies that both the electronic ground state (initial state) *and* the ionic state (final state) potential-energy surfaces are expanded about the ground-state equilibrium geometry. Correspondingly, the kinematic matrix $G(Q)$ is expanded about the initial-state equilibrium geometry both in the initial- and the final-state vibrational Hamiltonians. In the harmonic approximation, in particular, the potentials are replaced by their quadratic expansions about $Q = 0$, and $G(Q)$ is approximated by $G(0)$ both in the initial and the final state.

It must be emphasized, however, that the traditional approach to calculate FC factors within the harmonic approximation is to expand both potential surfaces up to second order about their *respective* minima. Correspondingly, the kinematic matrix $G(Q)$ is approximated by its value at the equilibrium geometry of the initial and final state, respectively.[98] The FC factors are then obtained as the overlap integrals of the harmonic-oscillator wave functions of the initial and final electronic states.

Since within this traditional approach the kinematic matrices for the initial and the final state differ in general, the nuclear kinetic energy is not the same in the initial and in the final state. Ionization of an electron leads (via the change in equilibrium geometry) to a change in the nuclear kinetic energy. It is interesting to note that the formalism of Section V.B can easily be extended to include this change in nuclear kinetic energy by introducing additional coupling terms that are quadratic in the momentum. It has been shown[80] that the FC factors can still be calculated exactly. The main consequence of this additional coupling is that the matrix $\hat{\omega}^{-1/2} J \omega^{1/2}$ is no longer orthogonal; the influence on the FC factors is rather small, however.[80]

In what follows it is shown that anharmonic corrections are of much less importance when both potential-energy surfaces and kinetic energies are expanded about the ground-state equilibrium geometry (in contrast to the traditional approach). It is for this reason that the applicability of the present formalism is not confined to potential functions that are nearly harmonic.

The drawback of the traditional approach is easily understood by considering Fig. 5.1, which shows schematically the initial- and final-state potential-energy surfaces. The initial-state vibrational wave function can be assumed to be well described within the harmonic approximation. In the ionic state, however, the nuclei perform large-amplitude vibrations if the coupling is strong. In particular, the minimum of the upper potential-energy surface may lie considerably outside the "Franck-Condon region," indicated by the shaded area in Fig. 5.1. As is shown below, the overall shape of the spectrum depends only on the behavior of the final-state energy surface *within* the Franck-Condon region. It is of advantage, therefore, to expand the final-state energy about a point within the FC region (i.e., the ground-state equilibrium configuration, point A in Fig. 5.1). An expansion about the final-state equilibrium configuration (point B in Fig. 5.1) will give a poor description of the final-state energy within the FC region if the coupling is strong. It is clear that an expansion about point B is of advantage if one is interested in an accurate description of the first few vibrational energy levels in the final state. In the strong-coupling case, however, these levels are excited with negligible intensity and do not contribute to the observed vibrational structure.

As discussed in Section V.B, the overall shape of the spectrum is determined by the low moments of $P(\omega)$. It is of interest, therefore, to

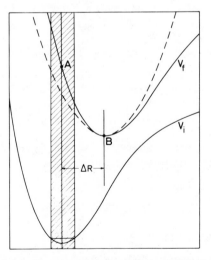

Fig. 5.1. A schematic one-dimensional drawing of the initial (i) and final (f) potential energy surfaces. The FC region is indicated by the shaded area. The harmonic expansion of V_f about its equilibrium geometry is represented by a broken line.

consider the influence of anharmonicity on the moments of $P(\omega)$. We have from (5.52)

$$M_k = (-1)^k \langle 0|\tilde{H}^k|0\rangle \tag{5.69}$$

with

$$\tilde{H} = \sum_s \omega_s b_s^+ b_s + \Delta T_N + \Delta V_0 + \Delta V \tag{5.70}$$

where ΔV represents the sum of all electron-vibration coupling terms

$$\Delta V = \Delta V_h + \Delta V_a$$

$$\Delta V_h = \sum_s \kappa_s (b_s + b_s^+) + \sum_{ss'} \gamma_{ss'} (b_s + b_s^+)(b_{s'} + b_{s'}^+)$$

Since the anharmonic coupling terms ΔV_a are usually much smaller than ΔT_N and ΔV_0, only the influence of the latter will be considered. Let us, for a first qualitative discussion, assume that the vibrational ground-state wave function is well described within the harmonic approximation. We expect that the first few excited vibrational states can still be approximated by harmonic-oscillator wave functions, but that the energy levels and wave functions of higher excited vibrational states will be influenced considerably by the anharmonic corrections ΔT_N and ΔV_0. Formally, this means that the matrix elements $\langle \mathbf{n}|\Delta T_N + \Delta V_0|0\rangle$ ($\mathbf{n}$ denotes the set of ground-state vibrational quantum numbers $n_1, n_2, \cdots n_M$) are negligible compared to matrix elements of the type $\langle \mathbf{m}|\Delta T_N + \Delta V_0|\mathbf{n}\rangle$ with sufficiently high quantum numbers m_s, n_s.

With the above assumptions we obtain for the first moment

$$M_1 = -\langle 0|\Delta V_h|0\rangle \tag{5.71}$$

The first moment depends therefore only on the electron-vibration coupling terms. The anharmonic corrections ΔT_N and ΔV_0 do not enter. The same holds for the second moment

$$M_2 = \langle 0|(\Delta V_h)^2|0\rangle \tag{5.72}$$

However, in the evaluation of the third moment

$$M_3 = -\langle 0|\tilde{H}^3|0\rangle$$

$$= -\sum_{\mathbf{nm}} \langle 0|\tilde{H}|\mathbf{m}\rangle\langle \mathbf{m}|\tilde{H}|\mathbf{n}\rangle\langle \mathbf{n}|\tilde{H}|0\rangle$$

terms of the form

$$\sum_{nm} \langle 0|\Delta V|m\rangle\langle m|\Delta T_N + \Delta V_0|n\rangle\langle n|\Delta V|0\rangle \tag{5.73}$$

appear. Analogous expressions enter the higher moments. Since in ΔV only terms linear and quadratic in $(b_s + b_s^+)$ are of importance, the anharmonic corrections to the third moment are still small if $\langle m|\Delta T_N + \Delta V_0|n\rangle \approx 0$ for $m_s, n_s = 0, 1, 2$. For higher moments, however, the anharmonic corrections become increasingly important. We conclude, therefore, that the gross features of the spectrum are rather insensitive to the anharmonic terms ΔT_N and ΔV_0, whereas the fine structure of the spectrum, which is determined by the higher moments, is influenced significantly by anharmonic corrections. The latter result is quite clear from inspection of Fig. 5.1: it is evident that the harmonic expansion of the final-state energy around point A cannot lead to the correct level spacing. The spacing of the vibrational levels calculated in this way will be larger than the true level spacing.

To compare these findings with the results of the traditional approach, the moments are recalculated in this approach. Denoting the true final-state potential energy by $\hat{V}(\mathbf{R})$, that is, $\hat{V}(\mathbf{R}) = V_0(\mathbf{R}) + \Delta V(\mathbf{R})$, the final-state vibrational Hamiltonian reads

$$\tilde{H} = \Delta + \sum_s \hat{\omega}_s c_s^+ c_s + \Delta \hat{T}_N + \Delta \hat{V} \tag{5.74}$$

where the $\hat{\omega}_s$ are the final state harmonic frequencies. $\Delta \hat{T}_N$ and $\Delta \hat{V}$ are defined by

$$\Delta \hat{T}_N = T_N\left(\mathbf{R}, \frac{\partial}{\partial \mathbf{R}}\right) + \frac{1}{2}\sum_{ss'} G_{ss'}(\hat{0}) \frac{\partial}{\partial R_s} \frac{\partial}{\partial R_{s'}} \tag{5.75}$$

$$\Delta \hat{V} = \hat{V}(\mathbf{R}) - \frac{1}{2}\sum_{ss'} \hat{F}_{ss'} R_s R_{s'} - \hat{V}(\hat{0}) \tag{5.76}$$

where $\hat{0}$ denotes the final-state equilibrium geometry and the $\hat{F}_{ss'}$ are the final-state harmonic force constants, $\hat{F}_{ss'} = (\partial^2 \hat{V} / \partial R_s \partial R_{s'})_0$. $\Delta \hat{T}_N$ and $\Delta \hat{V}$ represent the anharmonic corrections within the traditional approach [they differ, of course, from ΔT_N and ΔV_0 defined in (5.13) and (5.14)].

Neglecting, as above, anharmonic corrections to the vibrational ground state, we obtain for the first moment of $P(\omega)$

$$M_1 = -\langle 0|\tilde{H}|0\rangle = -\Delta - \sum_n |\langle 0|\hat{n}\rangle|^2 \left(\sum_s \hat{\omega}_s n_s\right)$$

$$- \sum_{nm} \langle 0|\hat{m}\rangle\langle \hat{m}|(\Delta \hat{T}_N + \Delta \hat{V})|\hat{n}\rangle\langle \hat{n}|0\rangle \tag{5.77}$$

The summations are over the eigenstates of the final-state harmonic Hamiltonian $\sum_s \hat{\omega}_s(c_s^+ c_s + \frac{1}{2})$. The last term in (5.77) gives the influence of anharmonicity effects on the first moment. When the coupling of the electron to one or several normal vibrations s is strong, the overlap matrix element $\langle 0|\hat{m}\rangle$ reaches its maximum for $m_s \gg 1$. Since anharmonic corrections are certainly not small for these highly excited vibrational states $|\hat{m}\rangle$, it is clear that calculating M_1 within the harmonic approximation [i.e., from the first two terms on the right-hand side of (5.77)] introduces a large error in the strong-coupling case. The same considerations apply also to the second and the higher moments. Therefore both the position (center of gravity) and the width of the spectrum may be affected by considerable errors when $P(\omega)$ is calculated within the harmonic approximation in its traditional form.

The above discussion was based on the argument that the vibrational ground-state wave function should be reasonably described within the harmonic approximation, whereas for the wave functions of highly excited vibrational states anharmonic corrections might be essential. A more rigorous treatment would be based on a systematic perturbation expansion with respect to the anharmonic terms, including corrections both to the final-state vibrational Hamiltonian and the initial-state vibrational wave function up to a given order. In order to further illustrate the results obtained above, we briefly discuss such a systematic expansion for a simple one-dimensional model. We represent the potential energy in the initial state by a third-order polynomial. The initial-state vibrational Hamiltonian then reads

$$\tilde{H}_i = \omega b^+ b + \eta(b + b^+)^3 \tag{5.78}$$

For stretching vibrations the coefficient η is always negative (because the potential energy tends to a dissociation limit as $Q \to \infty$) and usually $|\eta|/\omega \ll 1$. We further assume that the electron-vibration coupling involves only terms up to second order in Q. The final-state vibrational Hamiltonian then has the form

$$\tilde{H}_f = \omega b^+ b + \eta(b + b^+)^3 + \kappa(b + b^+) + \gamma(b + b^+)^2 \tag{5.79}$$

For the ground-state wave function of $\tilde{H}_i$ we obtain, in first order in η,

$$|\bar{0}\rangle = |0\rangle - \frac{3\eta}{\omega}|1\rangle - \frac{\frac{1}{3}\sqrt{6}\,\eta}{\omega}|3\rangle \tag{5.80}$$

It follows for the first moment that

$$M_1 = -\langle \bar{0}|\tilde{H}_f|\bar{0}\rangle$$

$$= -\gamma + \frac{6\eta\kappa}{\omega} \tag{5.81a}$$

The width of the spectrum is determined by

$$M_2 - M_1^2 = \kappa^2 + 2\gamma^2 - \frac{32\eta\kappa\gamma}{\omega} \tag{5.81b}$$

Equations 5.81a and 5.81b represent, up to terms linear in η, the exact position and width of the spectrum. Let us next compare these results with the expressions obtained in the harmonic approximation.

The harmonic approximation, as defined in Section V.A, trivially yields

$$M_1 = -\gamma \tag{5.82a}$$

$$M_2 - M_1^2 = \kappa^2 + 2\gamma^2 \tag{5.82b}$$

In order to apply the harmonic approximation in the traditional sense we first have to search for the minimum of the final-state potential energy. Then we have to expand the potential energy about the minimum geometry, neglecting terms of third and higher order in the displacement from the minimum geometry. Reexpressing the potential in terms of b and b^+ and neglecting terms of higher order in η we obtain

$$\tilde{H} = \omega b^+ b + \left[\kappa - \frac{12\eta\kappa^2}{(\omega+4\gamma)^2}\right](b+b^+) + \left[\gamma - \frac{6\eta\kappa}{(\omega+4\gamma)}\right](b+b^+)^2$$

$$- \frac{8\eta\kappa^3}{(\omega+4\gamma)^3} \tag{5.83}$$

The moments of this Hamiltonian are easily calculated with the harmonic vibrational ground-state wave function $|0\rangle$, giving

$$M_1 = -\gamma + \frac{6\eta\kappa}{(\omega+4\gamma)} + \frac{8\eta\kappa^3}{(\omega+4\gamma)^3} \tag{5.84a}$$

$$M_2 - M_1^2 = \kappa^2 + 2\gamma^2 - \frac{24\eta\kappa^3}{(\omega+4\gamma)^2} - \frac{24\eta\kappa\gamma}{(\omega+4\gamma)} \tag{5.84b}$$

To compare the results (5.82) and (5.84) with (5.81), we take account of the fact that usually $\gamma \ll \omega$. Introducing the abbreviation $(\kappa/\omega)^2 = \bar{n}$, the error in (5.82a) as compared to (5.81a) reads

$$6\eta \, \text{sign} \, (\kappa) \bar{n}^{1/2} \tag{5.85}$$

whereas the error in (5.84a) is

$$-8\eta \, \text{sign} \, (\kappa) \bar{n}^{3/2} \tag{5.86}$$

Apart from corrections due to quadratic coupling and anharmonicity effects, $\bar{n}$ is just the mean number of vibrational quanta excited in the final state [cf. (5.49)] and is thus a measure of the coupling strength. While the error (5.85) is proportional to the square root of $\bar{n}$, the error (5.86) increases with the third power of the square root of $\bar{n}$. In the case of strong coupling, therefore, the error inherent in the expression (5.84a) for the first moment will become very large compared to the error in (5.82a). Only for very weak coupling ($\bar{n} < 1$) is the expression (5.84a) superior. This is the expected result, since for $\bar{n} < 1$ most of the spectral intensity is concentrated in the zeroth vibrational line, and it is clear that the lowest vibrational state in the final state is best described by expanding the final-state energy about its own minimum. For $\bar{n} < 1$, however, the anharmonic effects are in any case small.

Similar results are obtained for the width of the spectrum. The error inherent in (5.82b) is

$$-32 \, \text{sign} \, (\kappa) \eta \gamma \bar{n}^{1/2} \tag{5.87}$$

whereas the expression (5.84b) is affected by the error

$$24 \, \text{sign} \, (\kappa) \eta \omega \bar{n}^{3/2} - 8 \, \text{sign} \, (\kappa) \eta \gamma \bar{n}^{1/2} \tag{5.88}$$

Since $\gamma \ll \omega$, the first term is the dominant term in (5.88) for all $\bar{n}$ of interest and clearly exceeds the error (5.87). It is clear from the calculation of the moments that the traditional harmonic approximation will provide a rather poor description of the shape of the vibrational structure when the coupling is strong.

It is interesting to note that the error in the first moment (5.82a) is due to the neglect of anharmonic corrections to the vibrational ground-state wave function $|0\rangle$, whereas the error in the expression (5.84a) results from the incorrect description of the final-state potential-energy curve within the Franck-Condon region. Likewise, the large defect in (5.84b) [i.e., the first term in (5.88)] is the consequence of a poor representation of the final-state potential energy within the FC region. It is this poor description of the final-state energy that limits the applicability of the traditional approach to the weak-coupling case.

The same arguments hold, in principle, for the anharmonic corrections to the kinetic energy, denoted by ΔT_N above. However, these corrections have usually much less influence on the shape of the spectrum.[80]

The objection might be raised that a third-order polynomial, as introduced in (5.78), is not a realistic representation of actual molecular potential-energy curves. We therefore conclude this section with a brief discussion of realistic one-dimensional potential-energy curves.

As is well known, the ground state and many of the excited and ionic state potential-energy curves of diatomic molecules are well represented by the Morse function,[74a]

$$V(R) = D\left(1 - e^{-\beta(R - R_0)}\right)^2 \qquad (5.89)$$

The advantage of this potential is that both the vibrational energy levels and wave functions can be determined analytically. The FC factors can then be obtained by numerical integration.[96] Approximate methods to calculate the overlap integrals have also been proposed.[99, 100]

It is interesting to compare the FC factors calculated in the harmonic approximation with the exact Morse function overlap integrals. To be sure that realistic parameters are used in the potential (5.89), we choose the parameters to represent the potential curves of two well-known states of the N_2 molecule, the $X\,^1\Sigma_g^+$ ground state and the $A\,^3\Sigma_u^+$ excited state. The $A\,^3\Sigma_u^+ - X\,^1\Sigma_g^+$ transition represents a strong coupling case: the intensity maximum lies at about the ninth vibrational line in the observed electron impact spectrum.[101] Correspondingly, the influence of anharmonicity on the FC factors is expected to be considerable for this example.

Figure 5.2 shows the results of the calculations. For clarity, only the FC factor envelope, that is, the smooth line connecting the tops of the individual vibrational lines, has been drawn. The envelopes have been normalized to equal area. In a way similar to the moment expansion discussed above, the FC factor envelope represents the overall shape of the spectrum. It should be noted that the form of the envelope depends not only on the FC factors, but also on the spacing of the vibrational lines.

Curve 1 in Fig. 5.2 represents the envelope of the numerically calculated Morse function overlap integrals, that is, the "exact" FC factor envelope. It is well known that the FC factors calculated from Morse potentials reproduce the experimentally observed intensity distribution quite well.[96] Curve 2 is obtained with the traditional harmonic approach, that is, by expanding both Morse potentials about their respective minima up to

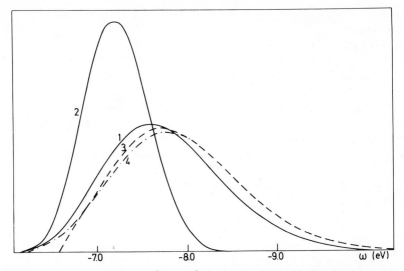

Fig. 5.2.   FC envelopes for the $A\,^3\Sigma_u^+ - X\,^1\Sigma_g^+$ transition in N_2. (1) Calculated from Morse potentials. (2) Calculated in the traditional harmonic approach. (3) Calculated with the present formalism using linear and quadratic coupling constants. (4) Obtained as curve 3, considering linear coupling only.

second order and calculating the overlap integrals of the harmonic-oscillator wave functions. Envelope 3 has been obtained with the formalism of Section V.B. Curve 4 is discussed below.

Figure 5.2 illustrates clearly the essential difference between the traditional harmonic approximation and the present approach. While curve 2 gives an acceptable description of the spectrum only in the very vicinity of the adiabatic excitation potential, curve 3 represents reasonably the overall shape of the spectrum. The features discussed above for the first and the second moment of the spectrum are readily identified in the figure. The center of gravity of the spectrum, calculated with the formalism of Section V.B, is slightly displaced to lower energy [cf (5.85)]. The traditional harmonic approach, on the other hand, gives the center of gravity shifted considerably to higher energy [cf (5.86)]. (Note that $\kappa < 0$ in the present case and that the energy variable ω is negative and increases from right to left in Fig. 5.2.) Likewise, the width of the spectrum, calculated in the present approach, is only slightly too large [cf (5.87)], whereas the traditional approach gives the width of the spectrum much too small [cf (5.88)].

The results obtained within the present approach can be further improved by taking account of some particular properties of potential-energy curves of diatomic molecules. To exhibit these properties, let the initial-state potential energy be represented by the Morse potential

$$\cdot V_0(R) = D_0 \left(1 - e^{-\beta_0(R-R_0)}\right)^2 \tag{5.90}$$

and the final-state energy by

$$V_1(R) = D_1 \left(1 - e^{-\beta_1(R-R_1)}\right)^2 \tag{5.91}$$

where R_0 and R_1 are the equilibrium distances in the initial and final states, respectively. Expanding $V_1(R)$ up to second order about R_0, we obtain

$$V_1(R) = D_1 (1 - Y)^2 + 2D_1 \beta_1 Y (1 - Y)(R - R_0)$$
$$+ D_1 \beta_1^2 Y (2Y - 1)(R - R_0)^2 \tag{5.92}$$

with the abbreviation

$$Y = e^{-\beta_1(R_0 - R_1)} \tag{5.93}$$

We now replace in the third term of (5.92) the second derivative of V_1,

$$V_1''(R_0) = 2D_1 \beta_1^2 Y (2Y - 1)$$

by the second derivative of V_0,

$$V_0''(R_0) = 2D_0 \beta_0^2$$

This replacement is identical with the neglect of the quadratic coupling constant γ, since

$$\gamma = (4\mu\omega_0)^{-1} (V_1''(R_0) - V_0''(R_0))$$

where μ denotes the reduced mass and ω_0 the initial-state vibrational frequency, given by

$$\mu\omega_0^2 = 2D_0 \beta_0^2 \tag{5.94}$$

The potential energy $\tilde{V}_1(R)$ thus obtained has its minimum at the internuclear distance

$$\tilde{R} = R_0 - \frac{D_1 \beta_1}{D_0 \beta_0^2} Y (Y - 1)$$

and the minimum value is

$$\tilde{V}_1(\tilde{R}) = D_1 (1 - Y)^2 \left(1 - \frac{\omega_1^2}{\omega_0^2} Y^2 \right) \tag{5.95}$$

where ω_1 is the final-state vibrational frequency, defined in analogy to (5.94).

The true final-state energy (5.91) has the minimum value $V_1(R_1) = 0$. The approximate final-state energy $\tilde{V}_1$ possesses the same minimum value, $\tilde{V}_1(\tilde{R}) = 0$, provided that

$$\frac{\omega_1^2}{\omega_0^2} Y^2 = 1$$

or, with (5.93),

$$\ln\left(\frac{\omega_1^2}{\omega_0^2} \right) = -2\beta_1 (R_1 - R_0)$$

Introducing the force constants k_0 and k_1 instead of ω_0^2 and ω_1^2, we arrive at

$$\ln\left(\frac{k_1}{k_0} \right) = -2\beta_1 (R_1 - R_0) \tag{5.96}$$

Equation 5.96 allows us to establish some interesting correlations with experimental data. First of all, it is an empirical rule that the parameter β is roughly the same for all states of the molecule that can be reasonably represented by a Morse function. This rule is easily verified for molecules like N_2 and CO, where the spectroscopic data of many ionic and excited states are available. Considering β_1 as a constant, (5.96) states that the logarithm of the force constant varies linearly with the equilibrium internuclear distance for all ionic or excited states of a given molecule. This statement can again be tested for molecules where sufficient spectroscopic data on ionic or excited states are available. We have found that (5.96) is fulfilled with considerable accuracy for all singly ionized and valence excited states of N_2 and CO. It is interesting to note that Anderson and Parr[102] have found, by considering the ground-state data of about 200 diatomic molecules, that $\ln k_0$ is linear in R_0 to a good approximation. This means that the ground-state force constant k_0 depends exponentially on the ground-state equilibrium internuclear distance R_0,

$$k_0 = Ce^{-\alpha R_0}$$

with universal constants C and α. It is not surprising that this rule holds to even higher accuracy when different electronic states of a given molecule are considered.

Having seen that (5.96) reflects a universal property of diatomic molecule potential curves, we can trace the argument leading to (5.96) back and conclude that setting the quadratic coupling constant γ zero will lead to an improved description of the final-state energy in the vicinity of its minimum. It is clear that an expansion of $V_1(R)$ about R_0 provides a good representation of $V_1(R)$ within the FC region, but not necessarily in the vicinity of its minimum. As a result, the adiabatic ionization potential obtained from (5.44a) may be affected by a considerable error when the coupling is strong. Equation 5.96 guarantees, however, that we arrive, independent of the coupling strength, at a fairly accurate value for the adibatic IP when we simply set the quadratic coupling constant γ to zero. Taking $\gamma = 0$ has little influence on the overall shape of the spectrum, but improves the description of the spectrum in the vicinity of the adiabatic IP.

Figure 5.2 provides a good illustration of these results: curve 4 shows the FC factor envelope, calculated with $\gamma = 0$. Compared to curve 3 the description of the spectrum near the adiabatic excitation potential has been greatly improved. Although the FC factors corresponding to curve 4 are obtained by an extremely simple formula [cf (5.49)], curve 4 provides the best approximation to the true FC factor envelope.

It is expected that the particular properties of diatomic molecule potential-energy curves discussed above are found, more generally, for all stretching normal coordinates of polyatomic molecules. It is known that the stretching potentials can be well approximated by Morse functions, with parameters quite similar to those of diatomic molecules.[103] Therefore the foregoing considerations should apply, at least in part, to the potential surfaces of polyatomic molecules as well.

VI.   APPLICATIONS TO ATOMS AND MOLECULES

A.   The $2p$ Ionization Potential of Neon

The theory discussed in the preceding sections does not distinguish between atoms and molecules. On the other hand, one may argue that the approximations introduced might be of different accuracy for each kind of system. Such arguments can be based on symmetry considerations and on the fact that occupied atomic valence orbitals are more localized in space than is the case for molecules. It is our opinion that these arguments do not apply as far as the accuracy of the basic approximations is concerned, but may prove to be correct when an incomplete basis set is used. The use of an incomplete basis set may influence the results for atoms and molecules in different ways.

The $2p$ IP of Ne has been calculated previously[7c] within a formalism basically equivalent to the present one. The basis set used does not contain polarization functions (d-type functions). Here we use[114] a more extensive basis set of Cartesian Gaussian functions: $11s/7p/2d$ contracted to $8s/6p/2d$. The exponential parameters for the s- and p-type functions and the contraction coefficients are taken from Huzinaga and Sakai[104], whereas the exponential parameters for the d-type functions have been chosen as 0.6 and 1.8. The total SCF energy calculated with this basis set is $E_{tot} = -128.5447$ a.u. For comparison, the best results given in the literature are listed in Table IV together with the energies of the occupied orbitals.

TABLE IV
Orbital Energies (in eV) and Total SCF Energies (in a.u.) for Neon

	Clementi[105]	Bagus Gilbert[106]	Froese-Fischer[107]	This work
ε_{1s}	-891.55	-891.54	-891.55	-891.70
ε_{2s}	-52.52	-52.51	-52.51	-52.50
ε_{2p}	-23.14	-23.13	-23.13	-23.12
E_{tot}	-128.5470	-128.5471	-128.5474	-128.5447

In the following we discuss the contributions of the various terms of the self-energy part. To calculate the self-energy part all occupied orbitals (apart from the $1s$ orbital) and the 23 unoccupied orbitals lowest in energy have been considered. At this stage, variation of the number of orbitals has been found to have but little influence on the final results. By applying Koopman's theorem one obtains 23.12 eV for the $2p$ IP of Ne as compared with the experimental value[108] of 21.60 eV for this IP. If the Koopmans' defect of 1.52 eV is decomposed into contributions due to reorganization and correlation effects, one obtains values of 3.24 and -1.72 eV, respectively.

In second order of the expansion of the self-energy part we obtain a $2p$ IP of 19.86 eV. Taking the self-energy part matrix to be diagonal, the result is 19.82 eV. The low value obtained in second order is not surprising when compared to the results obtained for small molecules.[7] The second-order IP evaluated previously[7c] without inclusion of d-type functions is 19.73 eV and differs only little from the present result.

Expanding the self-energy part up to third order yields an IP of 22.40 eV. The shift of -2.54 eV compares well with that obtained with the lower-grade HF calculation.[7c] The final IP is determined with the method described in Section IV.E. The resulting value is 21.56 eV and compares well with the experimental IP.

The above results, summarized in Table V, show a behavior which is also typical for small molecules. If the Koopmans' defect is large, the second order IP, $IP^{(2)}$, is usually too small when compared with the experimental value. The absolute error in second order is occasionally even larger than the absolute error on the HF level. The IP obtained in third order, $IP^{(3)}$, is then too large, but smaller than the value obtained via Koopmans' theorem. Remembering the latter energy to be the IP obtained in first-order perturbation theory, the expansion of the IP seems to be a series with alternating signs. For moderate Koopmans' defects the situation is not as unique.

TABLE V

Results for the $2p$ Ionization Potential of Neon (All Energies in eV). $IP^{(2)}$, $IP^{(3)}$, and $IP^{(R)}$ denote the IP calculated in second order, third order, and with the final formula, respectively. P is the pole strength.

Exp.	Koopmans	ΔSCF	$IP^{(2)}$	$IP^{(3)}$	$IP^{(R)}$	P
	23.12^b	19.84^c	19.86^b	22.40^b	21.56^b	0.95^b
21.60^a						
	23.13^d	19.89^d	20.76^d	—	20.71^d	—

[a] Experimental IP taken from Ref. 108.
[b] Present results.
[c] Reference 110.
[d] Green's function results from Ref. 109.

The pole strength $P_{2p}(2p)$ has been calculated to have a value of 0.95. Since this number is close to 1, no intense satellite lines should be observed in the low-energy region of the PE spectrum. Many satellite lines have been observed by Siegbahn et al.[1b] in the x-ray PE spectrum for energies above 870 eV. The pole strength $P_{1s}(1s)$ for the $1s$ orbital has indeed been found[7c] to be considerably smaller than $P_{2p}(2p)$ and $P_{2s}(2s)$.

Recently, Ribarsky[109] has calculated the IPs of Ne in a Green's function approach using approximations different from ours. He starts from a $6s/6p/2d$ basis set of 4 contracted and 30 uncontracted Gaussian orbitals. Twenty-one functions of this set were used in the self-energy calculation. His final and second-order results (see Table V) differ considerably from the values obtained here. It should be noticed that his renormalization procedure alters the second-order result only slightly.

The IPs of neon have also been computed by other methods. These methods are based on a subtraction of separately calculated total energies of atom and ion. Boys and Handy[110] approximate the atomic wave function as a product of a correlation function and a Slater determinant formed from space spin orbitals. The ground-state energy is then obtained as the solution of the so-called transcorrelated wave equation. They obtain 21.58 eV for the $2p$ IP of Ne.

Moser and Nesbet computed the lowest configurations of atoms by solving the Bethe-Goldstone equations. They have discussed two different approaches towards the solution of these equations and obtained 21.97 eV[111] and 21.25 eV[112] for the $2p$ IP of neon. Weiss[113] uses a symmetry-adapted variation of Nesbet's formulation of the Bethe-Goldstone scheme to calculate the first IP of Ne and obtains a value of 21.52 eV. The differences between these three closely related approaches are discussed in Ref. 112.

B. The Nitrogen Molecule

The nitrogen molecule represents a particularly interesting example, because the experimentally observed sequence of the IPs differs from the sequence obtained from applying Koopmans' theorem. Since the very extensive near HF-limit calculations of Cade et al.,[59] it was clear that this "breakdown" of Koopmans' theorem must be a correlation effect. For this reason the nitrogen molecule has been chosen by most authors engaged in the direct calculation of IPs of molecules as a touchstone of their methods.[7a, 12a, 13d, 16d, 10, 115] It is therefore worthwhile to discuss N_2 in some detail. The vibrational structure in the HeI PE spectrum of N_2 is considered as well. We shall see that interesting correlation effects are encountered in calculating the vibrational coupling constants. In addition, the vibrational structure of an intense satellite line showing up in the HeII spectrum is discussed.

The values of the three lowest IPs obtained by applying Koopmans' theorem are[59] 17.10 eV ($1\pi_u$), 17.36 eV ($3\sigma_g$), and 20.92 eV ($2\sigma_u$) compared to the experimental values[1c] 15.60 eV ($3\sigma_g$), 16.98 eV ($1\pi_u$), and 18.78 eV ($2\sigma_u$). Interestingly, the wrong sequence still remains when the IPs are obtained by a ΔSCF calculation. The ΔSCF values for the IPs are[59] 16.01 eV ($3\sigma_g$), 15.67 eV ($1\pi_u$), and 19.93 eV ($2\sigma_u$). The inclusion of reorganization effects is thus not sufficient to obtain the correct sequence of the IPs. The first Green's function calculation[7a] revealed that already the self-energy part in second order, when inserted into the Dyson equation, reproduces the correct ordering of the IPs. This can be understood from simple symmetry considerations.[6d] Third- and higher-order corrections have been found[7a] to be important, however, in order to obtain good agreement with experiment.

It has been pointed out by Purvis and Öhrn[12a] that not only higher-order contributions, but also the size of the basis set used in the calculations has a considerable influence on the results. Recently, a Green's function calculation on N_2 employing a more extensive basis set has been published.[115] The basis used consists of $11s/7p/1d$ functions on each of the N atoms, contracted to $5s/4p/1d$ leading to a SCF energy $E_{SCF} = -108.9750$ a.u. The near HF-limit value of Cade et al.[59] is $E_{SCF} = -108.9928$ a.u. All occupied and the 21 unoccupied orbitals lowest in energy were included in the calculation of the self-energy part. The enlargement of the basis set was found to improve the agreement of the calculated IPs with experiment.

In the meantime, a still more extensive calculation has been performed,[114] the results of which are reported here. The basis set employed is the Gaussian s-p basis described above, supplemented, however, with two sets of d-type functions with exponents 1.2 and 0.5. The HF orbital energies and the total SCF energy are given in Table VI. The orbital energies are seen to be practically identical with those of Cade et al.[59] The self-energy part is calculated up to third order and the interaction renormalized as described in Section IV.E. All occupied orbitals (with the exception of the core orbitals) and the 42 unoccupied orbitals lowest in energy are taken into account. At this stage variation of the number of unoccupied orbitals has been found to have very little influence on the final results. The results obtained in different orders as well as the final results are shown in Table VI. It should be mentioned that the $1\pi_u$ orbital of N_2 represents one of the rare cases where the second-order contribution is much smaller than the third-order contribution. Therefore (4.60) and (4.61) have to be used in the renormalization procedure.

TABLE VI

Total SCF Energies, Orbital Energies ε, Vertical Ionization Potentials, and Pole Strengths P for N_2. $IP^{(2)}$ = IP Calculated in Second Order; $IP^{(3)}$ = IP Calculated in Third Order; $IP^{(R)}$ = Final Result. Total Energies in a.u., other energies in eV.

Orbital	$-\varepsilon^a$	$-\varepsilon^b$	$IP^{(2)c}$	$IP^{(2)}$	$IP^{(3)}$	$IP^{(R)}$	P	exp. IP^d
$3\sigma_g$	17.28	17.28	14.87	14.87	15.89	15.45	0.91	15.60
$1\pi_u$	16.75	16.74	17.01	17.02	16.68	16.76	0.93	16.98
$2\sigma_u$	21.17	21.17	17.96	18.01	19.66	18.91	0.89	18.78

$$E_{SCF} = -108.9928^a \qquad\qquad E_{SCF} = -108.9841^b$$

[a] Reference 59.
[b] This work;
[c] obtained with the diagonal approximation for the second order self-energy part.
[d] From Ref. 1c.

The results shown in Table VI are satisfactory. The error for the $1\pi_u$ orbital is 0.22 eV, whereas for both σ orbitals the calculated IPs are within 0.15 eV of the experimental values. Indeed, the results for the $1\pi_u$ orbital are found to depend most sensitively on the basis set and the number of unoccupied orbitals, indicating that a further enlargement of the basis set is necessary to improve the result for this IP. The data in Table VI reveal the importance of third- and higher-order contributions to the IPs. This conclusion holds fairly independent of the quality of the basis set used. We would like to mention that N_2 is a particularly critical example and that for most systems, especially for larger and less symmetric molecules, basis set effects are less important.

Let us now compare the above results with those obtained with related methods. Purvis and Öhrn[12a] and Nerbrant[10] have used modified second-order self-energy parts to obtain the IPs of N_2 in the Green's function approach. Smith et al.[13d] have applied their equations-of-motion method and Chong et al.[16d] obtained the IPs starting from Rayleigh-Schrödinger perturbation theory. The total SCF energies, orbital energies, and final results reported by the various authors are listed in Table VII. From the quoted total SCF energies it is seen that basis sets of very different quality have been used. It is therefore not possible to infer the accuracy of the different methods from the data in Table VII. In principle, all direct approaches (as defined in the Introduction) calculate corrections to Koopmans' theorem and cannot be expected to correct for errors in the orbital energies. To eliminate the considerable differences in the orbital energies obtained by the different authors, it is useful to add their calculated Koopmans' defects to the near HF-limit orbital energies. The results thus obtained are shown in Table VII as well. It is seen that the basis set has to be enlarged until the orbital energies agree fairly well with the near HF-limit values, before definite conclusions on the accuracy of the method can be drawn.

Let us next consider the vibrational structure in the HeI PE spectrum of N_2. Vibrational coupling constants on the HF level have been obtained[75,116] from the data in the literature. Here, we report renormalized coupling constants for the N_2 molecule. To save computer time, a somewhat smaller basis set[115] than that employed in the calculation of the vertical IPs has been used. All occupied orbitals (with the exception of the core orbitals) and 34 unoccupied orbitals have been taken into account. The renormalized vibrational coupling constants are obtained as the normal coordinate derivatives of the poles of the Green's function, as described in Section V.C. Two calculations near the equilibrium geometry are sufficient to obtain the linear coupling constants κ. The free coupling constants obtained from the present HF data, those obtained from the HF

TABLE VII

A Comparison of Calculated Ionization Potentials of N_2. ε Denotes the HF Orbital Energy, IP(A) the Final Result for the Ionization Potential. IP(B) is Obtained by Adding the Calculated Koopmans' Defects to the Near HF-Limit Orbital Energies.[59] Total Energies and Internuclear Distances in a.u., Other Quantities in eV.

	Cade et al.[59]	Smith, Chen, and Simons[13d]			Chong, Herring, and McWilliams[16d]			Purvis and Öhrn[12a]			Nerbrant[10]			This work			Experiment[1c]
	$-\varepsilon$	$-\varepsilon$	IP(A)	IP(B)	$-\varepsilon$	IP(A)	IP(B)	$-\varepsilon$	IP(A)	IP(B)	$-\varepsilon$	IP(A)	IP(B)	$-\varepsilon$	IP(A)	IP(B)	IP
$3\sigma_g$	17.28	17.58	15.69	15.39	17.00	14.90	15.18	—	14.91	—	17.19	14.74	14.83	17.28	15.45	15.45	15.60
$1\pi_u$	16.75	17.76	17.03	16.02	17.04	16.67	16.38	—	17.23	—	16.76	16.86	16.85	16.74	16.76	16.77	16.98
$2\sigma_u$	21.17	21.75	18.63	18.05	21.02	18.39	18.54	—	17.55	—	21.05	17.90	18.02	21.17	18.91	18.91	18.78
E_{tot}	−108.9928	−108.8644			—			—			−108.956			−108.9841			
R	2.068	2.068									2.0694			2.0693			

data of Cade et al.,[59] and the renormalized coupling constants are listed in Table VIII, together with the "coupling parameters" $a = (\kappa/\omega)^2$. In addition, coupling constants and coupling parameters obtained from the ΔSCF data of Ref. 59 are given.

TABLE VIII

Free Coupling Constants $\kappa^{(0)}$, Renormalized Coupling Constants κ and Coupling Constants Including Reorganization Effects $\kappa^{\Delta SCF}$ (in eV) and Corresponding Coupling Parameters for N_2.

Orbital	$\kappa^{(0)a}$	$\kappa^{(0)b}$	κ	$\kappa^{\Delta SCF}$	$a^{(0)a}$	$a^{(0)b}$	a	$a^{\Delta SCF}$
$3\sigma_g$	-0.078	-0.078	-0.071	-0.084	0.072	0.072	0.058	0.082
$1\pi_u$	-0.382	-0.378	-0.324	-0.348	1.71	1.67	1.23	1.42
$2\sigma_u$	0.286	0.286	0.189	0.283	0.96	0.96	0.42	0.94

[a] Calculated with the near HF-limit data of Ref. 59.
[b] This work.

Figure 6.1 shows a comparison between the calculated and the experimental PE spectra. The vibrational structure is calculated with the simple formula (5.49). Since quadratic coupling constants and anharmonic effects are neglected explicitly, the frequency change due to ionization is not taken into account. For the $1\pi_u$ ionization, where a noticeable reduction of the vibrational frequency occurs, the calculated spacing of the vibrational lines is therefore too large. From the discussion of Section V.E it is expected, however, that the intensity distribution is well reproduced by (5.49). The lines in the calculated spectra are drawn as Gaussians of width 0.03 eV to simulate the limited resolution in the experimental spectrum. The total intensities of the bands, that is, the squared matrix elements $|\tau_{ei}|^2$ in (5.49), have not been calculated. In Fig. 6.1 the total intensities in the calculated spectra (a, b, c) have been chosen equal to the intensities in the experimental spectrum.

In Fig. 6.1a the PE spectrum calculated on the HF level is shown. The incorrect ordering of the IPs is evident. The calculated vibrational structure is in satisfactory agreement with experiment for both $3\sigma_g$ and $1\pi_u$ ionization. For $2\sigma_u$ ionization, however, a rather strong vibrational excitation is predicted on the HF level, in sharp contrast to the observed very weak vibrational excitation. This discrepancy has already been noted previously.[75,116] In Ref. 75 it has been shown by estimating second-order corrections that many-body effects are responsible for the inability of the free coupling constant $\kappa^{(0)}$ $(2\sigma_u)$ to reproduce the experimental spectrum. The renormalized coupling constants listed in Table VIII show indeed a

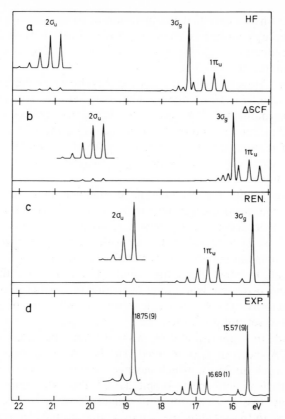

Fig. 6.1.   A comparison of the calculated and experimental PE spectrum of N_2. (*a*) The spectrum calculated on the HF level. (*b*) The spectrum calculated with ΔSCF data. (*c*) The spectrum calculated including many-body effects. (*d*) The experimental spectrum.[1a]

remarkable behavior. The coupling parameter for the $2\sigma_u$ orbital is strongly reduced when many-body effects are taken into account. The relative changes of the coupling parameters of the $3\sigma_g$ and $1\pi_u$ orbitals are much smaller. The spectrum in Fig. 6.1*c*, drawn with the vertical IPs of Table VIII and the renormalized coupling constants, agrees favorably with the experimental spectrum. The calculated vibrational coupling of the $2\sigma_u$ orbital is still too strong. Indeed, our calculations show that enlarging the basis set and increasing the number of unoccupied orbitals causes the $2\sigma_u$ coupling constant to decrease monotonically. Therefore, a still more extensive calculation seems necessary to obtain the full correction to the $2\sigma_u$ coupling constant.

It is interesting to note that the inclusion of reorganization effects does not alter the $2\sigma_u$ coupling constant at all (see Table VIII and Fig. 6.1b). The strong reduction of the coupling constant of the $2\sigma_u$ orbital is thus a pure correlation effect.

It has been argued[116] that the discrepancy between the $2\sigma_u$ bands in Figs. 6.1a and 6.1d is partly due to anharmonic effects. It is certainly true that the anharmonicity of the potential curves may influence the results when the vibrational structure is calculated using the free coupling constant, which is much too large. When a properly renormalized coupling constant is used, however, the coupling is small and anharmonic corrections play a minor role.

The above results for N_2 give us confidence that, in general, the influence of the electron-electron interaction on the vibrational coupling constants can be accounted for within reasonable amounts of computer time. Indeed, we have found in dealing with a lot of other molecules that correlation contributions such as for the $2\sigma_u$ coupling constant of N_2 are quite unusual. In most cases the coupling constants on the HF level reproduce the observed vibrational structure of valence electron bands quite satisfactorily. The N_2 molecule is thus a rather complicated system with respect to several properties.

To conclude this discussion of N_2, a satellite line showing up in the HeII and the ESCA spectrum is considered. The ESCA spectrum of N_2 exhibits a weak line at about 25 eV, which has been assigned to the $C^2\Sigma_u^+$ state of N_2^+ by comparison with spectroscopic data.[1b] This interpretation has been confirmed by two Green's function calculations.[12a, 117] The ionic state corresponding to the satellite line is represented by the electron configuration $(2\sigma_u)^2 (1\pi_u)^3 (3\sigma_g)^1 (1\pi_g)^1$, differing from the ground state of N_2^+ by the excitation of a $1\pi_u$ electron to the unoccupied $1\pi_g$ orbital. That the $C^2\Sigma_u^+$ state borrows intensity from the $B^2\Sigma_u^+$ state has also been shown by a configuration-interaction calculation.[118]

Recently, the 23 to 28-eV region of the PE spectrum of N_2 has been recorded with high resolution using HeII radiation.[119] The 25-eV satellite line is found to possess a pronounced vibrational structure with maximum intensity at the seventh or eighth vibrational line. The spectrum is shown in Fig. 6.2b.

The vibrational structure of the shake-up line is easily calculated on the one-particle level by going back to (5.61). In the present case, the linear coupling constant in zeroth order in the electron-electron interaction reads

$$\kappa^{(0)} = \frac{1}{\sqrt{2}} \frac{\partial}{\partial Q} \left(\varepsilon_{1\pi_g} - \varepsilon_{1\pi_u} - \varepsilon_{3\sigma_g} \right)_0 \tag{6.1}$$

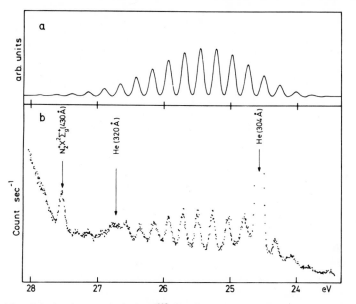

Fig. 6.2. Calculated (*a*) and observed[119] (*b*) vibrational structure of the 25-eV shake-up line in the PE spectrum of N_2. The vibrational lines in the calculated spectrum are drawn as Gaussians to simulate the limited resolution in the HeII spectrum. [Reproduced with permission from *J. Chem. Phys.*, **64**, 612 (1976).]

This expression has been evaluated,[120] leading to a vibrational coupling parameter $a^{(0)} = 9.5$. The vibrational structure calculated with (5.49) is shown in Fig. 6.2*a*. The observed extended vibrational structure is clearly reproduced by the calculation. Although the calculated vibrational coupling is somewhat stronger than observed experimentally, the agreement with experiment is very satisfactory in view of the simplicity of the calculation.

The unusual behavior of the coupling constant of the $2\sigma_u$ orbital discussed above is now easily understood. The presence of the satellite line of $2\sigma_u$ symmetry in the PE spectrum indicates, in agreement with the theoretical calculations,[12a, 117, 118] a relatively strong mixing of the $(2\sigma_u)^1$ $(1\pi_u)^4$ $(3\sigma_g)^2$ and $(2\sigma_u)^2$ $(1\pi_u)^3$ $(3\sigma_g)^1$ $(1\pi_g)^1$ configurations. The former configuration is more strongly bonding than the ground-state configuration of N_2, whereas the latter configuration is much less bonding than the ground-state configuration. Therefore, already a small admixture of the $(2\sigma_u)^2$ $(1\pi_u)^3$ $(3\sigma_g)^1$ $(1\pi_g)^1$ configuration to the $(2\sigma_u)^1$ $(1\pi_u)^4$ $(3\sigma_g)^2$ configuration will reduce its coupling constant κ substantially. This explains the extraordinarily large contribution of correlation effects to the coupling constant for $2\sigma_u$ ionization. On the other hand, an admixture of the $(2\sigma_u)^1$ $(1\pi_u)^4$ $(3\sigma_g)^2$ configuration to the $(2\sigma_u)^2$ $(1\pi_u)^3$ $(3\sigma_g)^1$ $(1\pi_g)^1$ configuration will be less efficient in reducing the coupling constant of the latter. The fact that the vibrational coupling observed for the satellite line is somewhat weaker than calculated in the one-particle approximation may be attributed to this configuration-mixing effect. The many-body effects can be quantitatively incorporated into the coupling constant κ of the satellite line by calculating κ as the derivative of the corresponding secondary pole[7d] of the one-particle Green's function.

C. A Theoretical PE Spectrum of a Polyatomic Molecule: Cyanogen

Cyanogen, C_2N_2, is in some respects similar to the nitrogen molecule. The HeI PE spectrum[1c, 121] exhibits four bands, which have been assigned[121] (in order of increasing energy) as $1\pi_g$, $5\sigma_g$, $4\sigma_u$, and $1\pi_u$. For C_2N_2 very near HF-limit data are available.[122] From these data and Koopmans' theorem the following sequence of the ionization potentials is obtained: $1\pi_g, 1\pi_u, 5\sigma_g, 4\sigma_u$. This ordering of the IPs is clearly in a serious contradiction to the experimental result. Since the discrepancy cannot be traced back to a deficiency in the HF calculation, many-body effects must be responsible for the interchange of the $1\pi_u$ orbital with both σ orbitals.

The vibrational structure in the HeI PE spectrum of C_2N_2 is well resolved and easy to interpret. While the σ bands exhibit only a very weak vibrational coupling, both the $1\pi_g$ and the $1\pi_u$ bands show considerable vibrational structure.[121] It is an interesting feature that only the high-frequency mode (of C-N stretching character) is excited in the $1\pi_g$ band, whereas in the fourth band the low-frequency mode (of C-C stretching character) dominates. C_2N_2 is therefore a suitable example for checking quantitatively how the method described in Section V works for a polyatomic molecule.

The calculation of the vertical IPs is discussed first. To exhibit the sensitivity of the results to the quality of the basis, two basis sets have been employed.[123] The first one (basis A) is of double-zeta quality consisting of $9s/5p$ Cartesian Gaussian functions on each atom contracted to $4s/2p$. The second basis set (basis B) includes the first one and has a set of d-type functions added with exponent $\alpha = 0.80$ on the C atom and $\alpha = 0.85$ on the N atom. In Table IX the orbital energies and the total SCF energies are listed for both basis sets and are compared with the near HF-limit data of McLean and Yoshimine.[122]

The self-energy part has been calculated and the Dyson equation solved as described for N_2 above. In the calculation employing basis A, all the

TABLE IX

Comparison of Orbital Energies (ε) and Total SCF Energies for C_2N_2.

ε in eV, E_{tot} in a.u.

Orbital	$-\varepsilon^a$	$-\varepsilon$ (basis A)	$-\varepsilon$ (basis B)
$1\pi_g$	13.62	13.75	13.60
$1\pi_u$	16.42	16.73	16.42
$5\sigma_g$	17.09	16.79	16.93
$4\sigma_u$	17.52	17.15	17.35
E_{tot}	-184.6568	-184.4553	-184.5703

^{a} From Ref. 122.

occupied orbitals with the exception of the C and N core orbitals and the 19 unoccupied orbitals lowest in energy have been taken into account. With basis B, the same occupied orbitals and the 29 unoccupied orbitals lowest in energy have been considered. At these stages variation of the number of unoccupied orbitals showed only little influence on the final results.

The results obtained with basis A and basis B are shown in Table X. The behavior of the IPs (obtained with basis B) in the different orders of the perturbation theory is illustrated in Fig. 6.3. The ordering of the IPs calculated in first order (Koopmans' theorem) is incorrect, as already mentioned above. In contrast to N_2, a wrong sequence is obtained in second and third order as well. It is seen that both σ orbitals are strongly shifted to give the sequence $5\sigma_g$, $1\pi_g$, $4\sigma_u$, and $1\pi_u$ in second order and $1\pi_g, 5\sigma_g, 1\pi_u, 4\sigma_u$ in third order. However, in second order the $5\sigma_g$ and $1\pi_g$ IPs are nearly degenerate and in third order the $4\sigma_u$ and $1\pi_u$ are very close

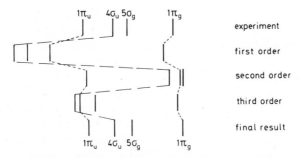

Fig. 6.3. A comparison of the IPs of cyanogen calculated in different orders of the perturbation expansion of the self-energy part. For clarity, the $5\sigma_g$ results are not connected by a broken line as the other results are. All calculations are performed with basis B. [Reproduced with permission from *Chem. Phys.*, **10**, 459 (1975).]

TABLE X

IPs of C_2N_2 Calculated with Basis A and Basis B. $IP^{(i)}, i = 1, 2, 3$, Stands for the IP Calculated in First-, Second-, and Third-Order Perturbation Theory, Respectively. $IP^{(R)}$ Represents the Final Result. The P Are the Pole Strengths Corresponding to the $IP^{(R)}$. All Energies in eV.

	Orbital	$IP^{(1)}$	$IP^{(2)}$	$IP^{(3)}$	$IP^{(R)}$	P	Exp. vertical IP^a	IP^b
Basis A	$1\pi_g$	13.75	13.34	13.53	13.41	0.91	13.36	13.28
	$1\pi_u$	16.73	15.72	16.05	15.89	0.90	15.6	15.49
	$5\sigma_g$	16.79	12.70	14.62	13.94	0.89	14.49	14.24
	$4\sigma_u$	17.15	13.06	15.05	14.31	0.88	14.86	14.68
Basis B	$1\pi_g$	13.60	13.17	13.24	13.20	0.91	13.36	13.22
	$1\pi_u$	16.42	15.50	15.66	15.56	0.90	15.6	15.56
	$5\sigma_g$	16.93	13.09	15.30	14.40	0.89	14.49	14.56
	$4\sigma_u$	17.35	13.44	15.79	14.80	0.88	14.86	14.97

[a] From Ref. 1c.

[b] These values are obtained by adding the final Koopmans' defects to the near HF-limit orbital energies of Ref. 122.

together. It might be the case that further enlarging the basis set changes the relative positions of these orbitals. The renormalization causes the σ IPs to shift only moderately compared to the above shifts and yields the correct ordering as well as good numerical values for the IPs.

The final results obtained with basis A are in qualitative agreement with those obtained with the more extensive basis set B. Quantitatively, the σ IPs obtained with basis A are somewhat lower than the corresponding experimental IPs. Adding the calculated Koopmans' defects to the accurate orbital energies of McLean and Yoshimine,[122] the values in the last column of Table X are obtained. It is seen that the results obtained with basis A are now of an accuracy that is comparable to that obtained with basis B.

To evaluate the vibrational coupling constants we must first construct the normal coordinates of C_2N_2 in its ground state. As a four-atomic linear molecule, C_2N_2 has seven normal coordinates, three of them representing bond stretching and four angle deformations. For the latter four and for the antisymmetric bond stretching coordinate the pole E_i of the Green's function is a symmetric function. Therefore the corresponding linear coupling constants κ^i are equal to zero. In the harmonic and Franck–Condon approximation these five modes can only appear in combinations of an even total number of quanta. Since the corresponding FC factors are usually very small, we do not consider these modes here.

We are thus left with two normal coordinates. It is convenient to express the normal coordinates in terms of internal symmetry coordinates. Denoting the change of distance between the two carbon atoms by R, the change of the one C–N bond length by r_1 and of the second bond length by r_2, the following symmetry coordinates S_1 and S_2 represent the symmetric C-N stretching and C-C stretching coordinates, respectively:

$$S_1 = \frac{1}{\sqrt{2}}(r_1 + r_2) \qquad S_2 = R \qquad (6.2)$$

In these coordinates, the force field of C_2N_2 is known to be[124] $F_{11} = 17.66$, $F_{22} = 7.08$, and $F_{12} = 0.82$ mdyn/Å. With well-known methods[76] one obtains the normal coordinates in terms of S_1 and S_2. The vibrational coupling constants κ_1^i and κ_2^i are easily expressed in terms of the derivatives of the pole E_i with respect to the internal symmetry coordinates S_1 and S_2. To save computer time the smaller basis set A has been employed in the calculation of the vibrational coupling constants.

The free coupling constants $\kappa_1^{(0)}, \kappa_2^{(0)}$ and the renormalized coupling constants κ_1, κ_2 are listed in Table XI together with the corresponding coupling parameters $a_1^{(0)}, a_2^{(0)}$, and a_1, a_2. Equation 5.49 is employed to calculate the vibrational structure. In Fig. 6.4 the calculated spectrum is compared with the experimental PE spectrum of C_2N_2.[1c] Figure 6.4a shows the spectrum calculated on the HF level. The spectrum of Fig. 6.4b has been obtained with the final vertical IPs and renormalized vibrational coupling constants. The lines in Figs. 6.4a and 6.4b are drawn as Lorentzians with a half-width corresponding to the experimental resolution in the HeI spectrum. Besides the electronic transition matrix elements $|\tau_{ei}|^2$, the only experimental data used to obtain the theoretical spectra is the well-known force field of C_2N_2 in its ground state.

The inability of the HF approximation to predict the correct ordering of the bands in the PE spectrum is obvious from Fig. 6.4a. However, the vibrational structure calculated on the HF level agrees in all essential

TABLE XI

The Free Coupling Constants $\kappa_s^{(0)}$ and Vibrational Coupling Parameters $a_s^{(0)}$ Compared to the Renormalized Coupling Constants κ_s and Vibrational Coupling Parameters a_s. $\kappa_s^{(0)}$ and κ_s in eV.

Orbital	$\kappa_1^{(0)}$	κ_1	$a_1^{(0)}$	a_1	$\kappa_2^{(0)}$	κ_2	$a_2^{(0)}$	a_2
$1\pi_g$	0.262	0.225	0.805	0.594	0.008	0.020	0.006	0.033
$1\pi_u$	0.074	0.117	0.064	0.161	−0.149	−0.108	1.879	0.990
$5\sigma_g$	−0.056	−0.034	0.037	0.014	−0.008	0.013	0.006	0.015
$4\sigma_u$	−0.080	−0.053	0.074	0.034	0.028	0.042	0.068	0.147

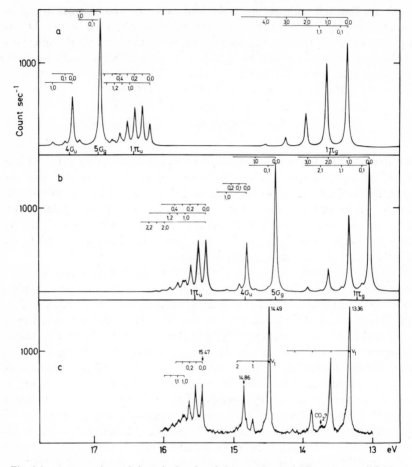

Fig. 6.4. A comparison of the calculated and the experimental PE spectrum of C_2N_2. (a) The spectrum calculated on the HF level. (b) The spectrum calculated including many-body effects. (c) The experimental spectrum. The calculated vertical IPs are indicated in the figure.

features with the observed vibrational structure. In agreement with experiment, both σ orbitals emerge as strictly nonbonding. The fact that the high-frequency mode ν_1 is excited in the $1\pi_g$ band and the low-frequency mode ν_2 is the dominant mode in the $1\pi_u$ band is clearly reproduced, although for both π bands the calculated coupling is too strong. These results are very interesting from the practical point of view. Whereas the vertical IPs calculated in the HF approximation are evidently of little use in assigning the PE spectrum of C_2N_2, the calculated vibrational structure allows the identification of the bands (the two σ bands, showing virtually identical vibrational structure, can of course not be distinguished in this way).

Considering Fig. 6.4b, the improvement obtained by the inclusion of many-body corrections is evident. All four bands are now within 0.2 eV of their experimental positions. The vibrational structure of the two π bands is in quantitative agreement with experiment. The fact that the vibrational coupling is very weak for both σ bands is reproduced by the calculation. For the $5\sigma_g$ band, however, the parameter a_1 is too small. We think that this deviation is a basis-set effect.

In dealing with a polyatomic molecule the computational advantages of the present approach to the calculation of the vibrational structure become obvious. If we were to expand the final-state potential surface about its minimum, as is usually done, we would first have to search for this minimum. This is a labourious task, even when only two normal coordinates are involved as in the case of C_2N_2. All second derivatives at this minimum geometry had to be calculated before the FC factors could be obtained with the usual formulas. This work had to be done for each ionic state separately. In the present approach, on the other hand, only normal coordinate derivatives at the ground-state equilibrium geometry are needed. In particular, the coupling constants for all the ionic states are obtained simultaneously. The calculations are further simplified by the fact that for practical purposes usually only the linear coupling constants are needed. The good agreement with experiment found above for C_2N_2 fully justifies the neglect of the quadratic coupling constants and of anharmonicity effects, confirming the conclusions drawn in Section V.E. It is for these reasons that the present approach allows the calculation of the vibrational structure for polyatomic molecules within reasonable amounts of computer time.

D. Clarifying Assignment Problems: Formaldehyde

The HeI PE spectrum of formaldehyde, H_2CO, has been extensively studied in the past.[1c, 125–127] The spectrum exhibits four bands, indicating that four IPs are lower than 20 eV. There have been many discussions in the literature about the assignment of the third and the fourth IP. According to the original proposal of Baker et al.[126] the third IP corresponds to the $1b_2$ orbital and the fourth IP to the $5a_1$ orbital. Their argument was based on the analysis of the vibrational structure of both bands. Brundle et al.,[127] on the other hand, favored the reverse assignment on grounds of an ab initio SCF calculation and application of Koopmans' theorem. Recently, a SCF-X_α scattered wave calculation (employing overlapping atomic spheres) has been performed[128] and the ordering originally proposed by Baker et al.[126] has been obtained. Previous Green's function calculations[6a, 7b] as well as a Rayleigh–Schrödinger perturbation calculation[16b] predicted the ordering given by Koopmans' theorem to be the correct one. To clarify this problem definitely, the IPs have been recalculated with the Green's function method employing an extensive basis set.[129] The results of this calculation are discussed below.

The interpretation of the vibrational structure in the PE spectrum of formaldehyde has proved to be difficult as well, owing to the accidental degeneracy of vibrational frequencies in some of the ionic states. Especially for the second band of H_2CO and D_2CO no assignment of the normal vibrations involved could be given,[1c, 125–127] although the vibrational structure is well resolved and apparently simple. In view of these difficulties, a theoretical calculation of the vibrational structure in the PE spectrum is also of interest.

The calculation is completely analogous to the one described in the preceding section and need only be described briefly. Two basis sets are employed. The first one (basis A) is of double-zeta quality and consists of $9s/4p$ Cartesian Gaussian functions on C and O and four s-type functions on H, contracted to $4s/2p$ on C and O and two s-type functions on H. The second set (basis B) includes a set of d-type functions on C and O. In Table XII the total SCF energies and orbital energies obtained with basis A and B are listed and compared with the results of Garrison et al.,[130] who to the authors' knowledge have reported the lowest total energy.

TABLE XII
Total SCF Energies (in a.u.) and Orbital Energies (in eV) for
Formaldehyde

Orbital	$-\varepsilon^a$	$-\varepsilon^b$	$-\varepsilon^c$
$2b_2$	12.14	12.08	12.03
$1b_1$	14.70	14.63	14.60
$5a_1$	17.57	17.76	17.77
$1b_2$	19.19	18.87	18.82
E_{tot}	-113.7891	-113.9012	-113.9149

[a] This work, basis A.
[b] This work, basis B.
[c] From Ref. 130.

In calculating the vibrational structure, only the three totally symmetric normal vibrations ν_1, ν_2, and ν_3 are considered. The coupling to the non-totally symmetric vibrations occurs only through the quadratic coupling constants $\gamma_{ss'}$ and is expected to be very weak. The appropriate internal symmetry coordinates are

$$S_1 = \frac{1}{\sqrt{2}}(r_1 + r_2) \qquad S_2 = R \qquad S_3 = \sqrt{\tfrac{3}{2}}\, r_0 \Phi \qquad (6.3)$$

where r_1 and r_2 denote the change of the two C—H distances, R denotes the change of the C—O bond length, and Φ the deviation of the H—C—H bond angle from its equilibrium value. The force field of Shimanouchi and Suzuki[131] is used to construct the normal coordinates of H_2CO and D_2CO in their electronic ground states.

The final results for the ionization potentials are listed in Table XIII. The values reported recently by Chong et al.[16b] are included for comparison. Whereas the IPs calculated via Koopmans' theorem differ by as much as 2 eV from the experimental ones, the final results obtained with basis A are within 0.4 eV of the experimental values. The results obtained with the more extensive basis B are within 0.2 eV of the experimental IPs. Adding the Koopmans' defects calculated with basis A to the orbital energies of the more accurate SCF calculation of Ref. 130, the results are again within 0.2 eV of the experimental values, indicating that the smaller basis set A is already quite adequate for the calculation of the many-body corrections. The ordering of the IPs is clearly seen to be independent of basis-set effects. We conclude that the third band has to be assigned as $5a_1$ and the fourth band as $1b_2$. Koopmans' theorem thus correctly predicts the ordering of the IPs in the case of formaldehyde. The ordering deduced from the overlapping spheres SCF-X_α calculation[128] is incorrect.

TABLE XIII
Vertical IPs of Formaldehyde (in eV)

Orbital	IP^a	IP^b	IP^c	Exp. IP^d	IP^e
$2b_2$	10.81	10.84	10.70	10.9	11.15
$1b_1$	14.62	14.29	14.52	14.5	14.73
$5a_1$	16.20	16.36	16.16	16.2	16.19
$1b_2$	17.36	17.13	16.99	$\approx$17.0	17.59

^{a} Final results obtained with basis A.
^{b} Final results obtained with basis B.
^{c} Values obtained by adding the final Koopmans' defects calculated with basis A to the orbital energies of Ref. 130.
^{d} Estimated center of gravity of the corresponding band in the PE spectrum of H_2CO.[1c]
^{e} Final results of Chong et al.[16b]

The free vibrational coupling parameters $a_s^{(0)}$ and the renormalized coupling parameters a_s, $s = 1, 2, 3$ (calculated with basis A) are listed in Table XIV. To have a simple way of comparing the results with experiment, we show the calculated vibrational structure together with the corresponding band in the experimental spectrum[1c] in Figs. 6.5 to 6.8. For the second, third, and fourth bands the individual vibrational lines in the calculated spectra are drawn as Lorentzians with a half-width of 0.03 eV, representing the experimental resolution. The first band of H_2CO and D_2CO has been recorded on an expanded energy scale by Turner et al.[1c] and in that case the line shape in the PE spectrum resembles more closely a Gaussian. For the first band, therefore, the lines in the calculated spectra are drawn as Gaussians. The IPs calculated with basis A and the renormalized coupling parameters a_s are used in drawing the theoretical spectra. Formaldehyde and formaldehyde-d_2 represent complicated examples, since the structure of several bands is the result of a superposition of two degenerate or nearly degenerate normal modes. It is therefore essential to employ correct ionic-state vibrational frequencies in drawing the spectra. Having calculated only linear vibrational coupling constants, we are not in a position to predict the frequency change due to ionization. Therefore, the experimental ionic frequencies are taken to draw the calculated vibrational structure. Although the ionic vibrational frequencies are not unambiguously known for most of the ionic states of H_2CO and D_2CO,[1c] it is found that the ionic vibrational frequencies can easily be determined from the structure observed in the PE spectrum with the help of the calculated FC factors. (It is an essential advantage of the approach outlined in Section V that the FC factors can be calculated *independently* of the ionic frequen-

TABLE XIV

Vibrational Coupling Parameters for Valence Ionization of Formaldehyde and Deuteroformaldehyde. The $a_s^{(0)}$ Are Calculated on the HF Level, the a_s with Inclusion of Many-Body Effects.

		$a_1^{(0)}$	$a_2^{(0)}$	$a_3^{(0)}$	a_1	a_2	a_3
$2b_2$	H_2CO	0.217	0.375	0.087	0.033	0.129	0.126
	D_2CO	0.456	0.229	0.028	0.092	0.122	0.090
$1b_1$	H_2CO	0.128	3.555	0.310	0.036	2.792	0.270
	D_2CO	0.778	3.125	0.008	0.414	2.611	0.011
$5a_1$	H_2CO	0.003	2.176	0.147	0.004	1.156	0.301
	D_2CO	0.168	1.660	0.731	0.098	0.765	0.855
$1b_2$	H_2CO	0.886	0.201	1.439	0.757	0.195	1.533
	D_2CO	0.765	0.995	1.420	0.635	0.950	1.534

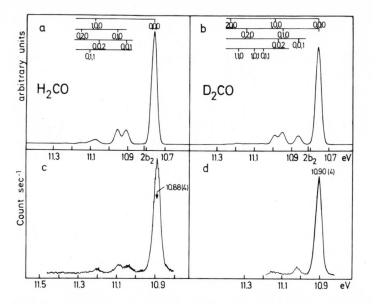

Fig. 6.5.   The calculated (*a*), (*b*) and observed[1c] (*c*), (*d*) first band in the spectrum of H_2CO and D_2CO. The vertical IPs and coupling parameters are calculated with basis *A*. [Reproduced with permission from *J. Chem. Phys.*, **64**, 612 (1976).]

cies.) An exception is the fourth band. The low intensity, the very complex vibrational structure and the servere overlapping with the third band prevent the assignment of vibrational frequencies. However, the vibrational structure calculated for this band is very complex and does not depend critically on the frequencies chosen in drawing the spectra.

The results obtained for the individual bands are now briefly discussed. The first band in the PE spectrum (see Fig. 6.5) is seen to correspond to a nonbonding electron. The calculated vibrational structure is in satisfactory agreement with experiment, reproducing correctly the fact that all three normal modes are excited with comparable intensity.

The second band in the experimental spectrum corresponds to the ionization of the $1b_1$ CO-π-bonding electron and exhibits the expected strong excitation of the C—O stretching mode ν_2 (see Fig. 6.6). The higher-resolution spectra reveal, however, that the vibrational structure consists of a series of doublets for both H_2CO and D_2CO, and thus at least one other vibrational mode must be excited. The assignment of this mode has caused considerable difficulties.[1c, 125–127] According to Turner et al.,[1c] the structure is due to the excitation of ν_2 together with one quantum of the C—H stretching mode ν_1. It is then difficult to explain, however, the

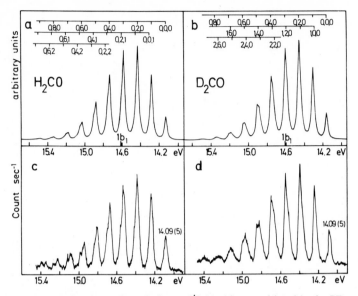

Fig. 6.6. The calculated (*a*), (*b*) and observed[1c] (*c*), (*d*) second band in the PE spectrum of H_2CO and D_2CO. The vertical IPs and coupling parameters are calculated with basis *A*. [Reproduced with permission from *J. Chem. Phys.*, **64**, 612 (1976).]

apparent insensitivity of both frequencies to deuteration. One would expect ν_1 to be considerably reduced for D_2CO, and thus the accidental near degeneracy of both frequencies in H_2CO should be removed for D_2CO.

This problem is resolved by the calculation. We find that for H_2CO the vibrational structure is due to the excitation of ν_2 and ν_3, the coupling of ν_1 being negligibly small. For D_2CO, on the other hand, the coupling of ν_3 is very small and the structure is due to the excitation of ν_1 and ν_2. Thus the vibrational coupling is completely different for the two isotopic species. It should be noted that this result is obtained already on the HF level (see Table XIV). As shown in Fig. 6.6, the calculated vibrational structure is in nearly quantitative agreement with experiment for both H_2CO and D_2CO.

The third band in the experimental spectrum of H_2CO (see Fig. 6.7) consists of a single series of narrow lines. From a comparison with the third band of D_2CO, which shows a considerably more complex structure due to strong excitation of both ν_2 and ν_3, Turner et al.[1c] concluded that the simplicity of the third band of H_2CO was due to an equality of the frequencies ν_2 and ν_3 in this ionic state. This interpretation of the vibrational structure is confirmed by the present calculation. As shown in Table XIV we find strong excitation of ν_2 accompanied by weak excitation of ν_3

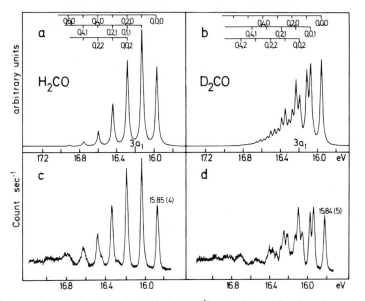

Fig. 6.7.   The calculated (*a*), (*b*) and observed[1c] (*c*), (*d*) third band in the spectrum of H_2CO and D_2CO. The vertical IPs and coupling parameters are calculated with basis *A*. The diffuse structure at higher energies in parts (*c*) and (*d*) belongs to the fourth band. [Reproduced with permission from *J. Chem. Phys.*, **64**, 612 (1976).]

in the case of H_2CO. For D_2CO, on the other hand, both ν_2 and ν_3 are calculated to couple strongly, the coupling of ν_3 being even stronger than that of ν_2. The coupling parameter of the C—H stretching vibration ν_1 is very small in both cases. Though the calculated vibrational coupling is somewhat too small, the typical structure of the third band of both H_2CO and D_2CO is clearly reproduced (see Fig. 6.7).

The fourth band appears as a complex and diffuse structure in the experimental spectra of H_2CO and D_2CO. From the calculated coupling parameters we see that all three normal vibrations are strongly excited in the case of D_2CO, whereas for H_2CO only ν_1 and ν_3 couple strongly. In both cases a very complex vibrational structure results. In drawing the calculated spectra, the following ionic vibrational frequencies have been used: $\hat{\omega}_1 = 1400$ cm^{-1}, $\hat{\omega}_2 = 1500$ cm^{-1}, $\hat{\omega}_3 = 1050$ cm^{-1} for H_2CO, and $\hat{\omega}_1 = 990$ cm^{-1}, $\hat{\omega}_2 = 1300$ cm^{-1}, $\hat{\omega}_3 = 800$ cm^{-1} for D_2CO. As shown in Fig. 6.8, the spectral shape calculated for D_2CO is in good qualitative agreement with the broad and structureless hump observed in the PE spectrum. It is clear from Fig. 6.8 that strong excitation of all three normal vibrations

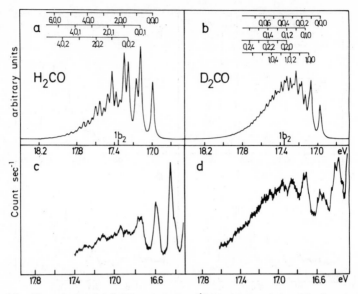

Fig. 6.8.   The calculated (*a*), (*b*) and observed[1c] (*c*), (*d*) fourth band in the spectrum of H_2CO and D_2CO. The vertical IPs and coupling parameters are calculated with basis *A*. Note that the intense lines at lower energies in parts (*c*) and (*d*) belong to the third band. The fourth band is the diffuse structure centered at about 17.0 eV. Note also that the experimental spectrum of H_2CO (*c*) has been recorded with less amplification than that of D_2CO (*d*).

leads to a vibrational structure of such a complexity that it cannot be resolved with present spectrometers.

It can be seen from Table XIV that the free coupling parameters $a_s^{(0)}$ reproduce correctly all the trends discussed above. The quantitative agreement with experiment is definitely better, however, when the renormalized coupling constants a_s are used.

From the above discussion of the vibrational structure it is obvious that the third band has to be assigned as $5a_1$ and the fourth band as $1b_2$. The calculation predicts the excitation of ν_2 and ν_3 in the case of $5a_1$ ionization, whereas ν_1 and ν_3 and—in the case of D_2CO—also ν_2 are calculated to be strongly excited for $1b_2$ ionization. These features are easily identified in the experimental spectra: the third band in the PE spectrum of D_2CO shows clearly the strong excitation of both ν_2 and ν_3 with no evidence for the excitation of ν_1. The fourth band in the D_2CO spectrum, on the other hand, exhibits the very complex vibrational structure that is to be expected when three normal modes are strongly excited. The calculation of the vibrational structure together with the calculation of the vertical IPs leaves, therefore, no doubt as to the assignment of the ionization potentials of formaldehyde.

The calculations on H_2CO and D_2CO reveal a surprising sensitivity of the vibrational coupling parameters a_s to deuteration for most of the bands. In the case of the second band, for instance, ν_2 and ν_3 are excited for H_2CO, but only ν_1 and ν_2 are present in the spectrum of D_2CO. A pronounced isotopic effect is also found for the third band: whereas ν_3 couples only weakly for H_2CO, it is the mode that dominates in the spectrum of D_2CO. These results demonstrate that electronic bonding properties derived from so-called overlap populations are of a limited value for interpreting the vibrational structure in the PE spectra of poly-atomic molecules. It is the bonding strength of a particular electron with respect to a particular normal coordinate, which is reflected in the PE spectrum, and not the bonding strength with respect to any two nuclei of the molecule. Therefore the normal coordinate derivatives of the orbital energy, or, in a more sophisticated treatment, of the pole of the Green's function, are the appropriate quantities to discuss the vibrational structure of a band in the PE spectrum. Isotopic effects on the vibrational coupling strengths are then automatically taken into account in the construction of the ground-state normal coordinates.

E.   Application to a Larger Molecule: Pyridine

The Green's function method has been applied in the preceding sections to the neon atom and to the molecules N_2, C_2N_2, and H_2CO. In these cases we can eliminate to a large extent the uncertainties due to basis-set effects. It is, of course, of interest to investigate whether the method is also useful for studying larger molecules, where errors due to basis-set effects cannot be excluded. The PE spectra of molecules with about 10 or more atoms are usually very difficult to interpret, since it is often the case that many broad bands overlap. Without further knowledge, even the question of how many electronic bands are involved must be left unanswered. A powerful method to calculate IPs is, therefore, of great importance.

Von Niessen et al.[132] have applied the present Green's function method to several molecules with considerably more atoms and/or more electrons than the molecules discussed in the previous sections (e.g., C_6H_6, SF_6). Here we choose to discuss the pyridine molecule (C_5H_5N), which is a prominent exponent of hetereocyclic aromatic molecules and plays an important role in biochemistry. The PE spectrum of pyridine has been extensively investigated both experimentally and theoretically. The large number of investigations has led to several controversies in the experimen-tal as well as in the theoretical work. These controversies have been discussed in detail in Ref. 132b. Here we discuss the calculation of the IPs and compare them with the experimental values without making a comparison with previous work.

To perform the SCF calculation a basis set of double-zeta quality has been used. This basis set consists of Cartesian Gaussian functions: $9s/5p$ functions on each C and N atom and four s-type functions on each H atom, contracted to $4s/2p$ and two s-type functions, respectively. The exponential parameters and contraction coefficients are taken from Huzinaga.[133] The total SCF energy obtained with this basis set is $E_{tot} = -246.5491$ a.u. and is slightly lower than the best E_{tot} obtained previously: $-246.417,$[134] $-246.327,$[135] -245.622[136] a.u.

The IPs have been calculated by taking into account 15 occupied orbitals (the core orbitals hardly contribute) and the 17 unoccupied orbitals lowest in energy. The results obtained in the various steps of the perturbation expansion are listed in Table XV together with the experimental vertical IPs of Åsbrink as cited by Almlöf et al.[134] The three outermost valence IPs are the more interesting ones. Here the HF calculation predicts the ordering $1a_2, 2b_1, 7a_1$, whereas the orderings $7a_1, 1a_2, 2b_1$ and $1a_2, 7a_1, 2b_1$ are obtained in second and third order, respectively. The final ordering obtained by applying the renormalization procedure is $1a_2, 7a_1, 2b_1$, where, in agreement with experiment, the $1a_2$ and $7a_1$ IPs are very close together. The relative ordering of these two IPs may change owing to basis set effects. This is of little relevance, however, since the corresponding bands cannot be separated in the experimental spectrum. The situation is illustrated in Fig. 6.9, where the experimental spectrum[1c] is shown together with the calculated IPs.

TABLE XV
Results for the Valence IPs of Pyridine (All Energies in eV)

orbital	$-\varepsilon$	$IP^{(2)}$	$IP^{(3)}$	$IP^{(R)}$	P	IP^a
$1a_2$	9.81	9.30	9.55	9.57	0.92	9.7
$7a_1$	11.29	8.27	9.89	9.66	0.90	9.7
$2b_1$	10.65	9.82	10.23	10.24	0.91	10.5
$5b_2$	14.25	11.92	13.02	12.87	0.93	12.5
$1b_1$	14.98	12.89	13.50	13.43	0.82	13.2
$6a_1$	15.89	13.11	14.34	14.18	0.89	13.7
$4b_2$	16.61	14.16	15.18	15.11	0.90	14.4
$5a_1$	17.84	15.40	16.38	16.32	0.88	15.7
$3b_2$	18.26	14.98	16.49	16.33	0.88	15.7
$4a_1$	19.93	16.97	18.08	18.00	0.85	17.1

^{a}Experimental IPs of Åsbrink as cited in Ref. 134.

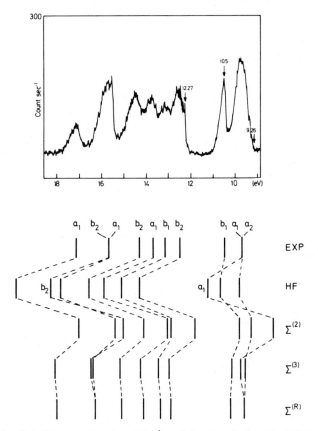

Fig. 6.9. The HeI PE spectrum of pyridine[1c] and the IPs calculated in different orders of the perturbation theory. [Reproduced with permission from *Chem. Phys.*, **10**, 345 (1975).]

From the figure we see that there is another interesting grouping of orbitals, namely the $5a_1$ and $3b_2$ orbitals. On the HF level the corresponding IPs differ by 0.4 eV. This difference is retained in second order, but with a reversed ordering. In third order and also after applying the renormalization procedure both IPs have, in agreement with experiment, approximately the same value. The overall agreement of the calculated IPs with the experiment is satisfactory.

When the IPs of a molecule of the size of pyridine are to be calculated within a reasonable amount of computer time, two major difficulties arise. First, one is restricted to the use of basis sets of rather limited quality. Second, the number of orbitals that can be taken into account may not exhaust the basis. The double-zeta basis set used here is considered to be accurate enough to solve most assignment problems and to obtain IPs of moderate accuracy. If enough orbitals are considered in the calculation, the Koopmans' defects calculated for outer valence orbitals are assumed to be quite accurate and the main error comes from the orbital energies. This error cannot be estimated. The second difficulty mentioned above is less severe. It has been found in all cases that the core orbitals contribute very little for the valence IPs. The inclusion of many virtual orbitals is usually more important for inner valence orbitals than for the outer valence orbitals. In addition, we can test the effect of exhausting the basis by varying the number of virtual orbitals. It is of advantage that the perturbation series usually has alternating signs. As a consequence, the error that arises due to the neglect of higher virtual orbitals is considerably smaller for the *final* result than for the contributions of the *individual* orders of the perturbation expansion.

F.   Application to Molecules Containing Second-Row Atoms: H_2S

In the preceding sections molecules composed of first-row atoms have been discussed. Inclusion of d-type functions in the HF basis sets has been found to be important for some of these molecules. The influence of d- and even of f-type functions on the calculated IPs of molecules containing second row atoms is expected to be even more essential. Von Niessen et al.[132] have applied the present Green's function theory to molecules with a second-row atom (e.g., phosphoridine, C_5H_5P) and compared the results with those obtained for similar molecules containing first-row atoms only (e.g., pyridine, C_5H_5N). The basis sets used were of double-zeta quality. Some of the calculated outer valence IPs have been found to be of lower accuracy for the molecules containing a second-row atom. For the time being the calculations are too costly for these large molecules to allow an extensive investigation of basis set effects. Therefore, a small molecule, H_2S, has been chosen to study basis-set effects.[137]

Let us start with a double-zeta basis set (basis A) consisting of $12s/9p$ functions on the S atom and four s functions on each of the H atoms, contracted to $6s/4p$ and two s functions, respectively.[133, 139] The HF ground-state energy obtained is -398.4979 a.u. The final values of the IPs are 9.91, 12.76, and 15.72 eV for the $2b_1$, $5a_1$, and $2b_2$ orbitals, respectively. For comparison, the experimental values are 10.48, 13.4, 15.5 eV. The calculated IPs are of an accuracy comparable to that obtained by Chong et al.[16d] starting from a "$1\frac{1}{2}$-zeta" basis. The $2b_1$ and $5a_1$ IPs are seen to be too low.

As a next step, three s-type and three p-type functions are added to basis A, improving the total SCF energy considerably. The outer valence orbital energies and the calculated IPs change, however, only very slightly.

Adding instead a d-type function on the S atom with an exponential parameter $\alpha = 0.5$ does not improve the calculated IPs either. This has been found to be due to the unfavorable choice of the paramenter α. Adding d-type functions with smaller exponential parameters to basis A, the results shown in Table XVI are obtained.[137] One can see that inclusion of two d-type functions on the S atom (and of a p-type function on each H atom) changes the HF orbital energies by about 0.2 eV and the calculated IPs by up to 0.57 eV. These IPs are already in good agreement with the experiment. When a further d-type function is added, the change in the IPs is less than 0.1 eV.

TABLE XVI

Orbital Energies, Ionization Potentials, and Total SCF Energies of H_2S Calculated with Different Basis Sets. Experimental Values for the IPs Are Given in the Last Column. Total Energies Are in a.u. and All Other Energies in eV.

n.Orbital	$-\varepsilon^a$	IP^a	$-\varepsilon^b$	IP^b	$-\varepsilon^c$	IP^c	$-\varepsilon^d$	IP^d	IP^e
$2b_1$	10.66	9.91	10.48	10.24	10.48	10.24	10.51	10.38	10.48
$5a_1$	13.43	12.76	13.63	13.25	13.65	13.32	13.62	13.36	13.4
$2b_2$	16.24	15.72	16.12	15.54	16.12	15.57	16.09	15.58	15.5
E_{tot}^{SCF}	-398.4979^a		-398.5455^b		-398.5492^c		-398.5467^d		

$^a 12s/9p + 4s$ contracted to $6s/4p + 2s$.
$^b 12s/9p/2d + 4s/1p$ contracted to $6s/4p/2d + 2s/1p$.
$^c 12s/9p/3d + 4s/1p$ contracted to $6s/4p/3d + 2s/1p$.
$^d 12s/9p/2d/1f + 4s/1p$ contracted to $6s/4p/2d/1f + 2s/1p$.
^{e}Centroids estimated from the PE spectrum in Ref. 1c.

Several additional basis sets have investigated for H_2S. One of them $(12s/9p/3d + 6s/2p$ contracted to $8s/6p/3d + 3s/2p)$ leads to a total SCF energy of -398.7083 a.u. compared to -398.6862 a.u. obtained by Rothenberg et al.[140] which is the lowest HF energy we could find in the literature. From these calculations we conclude that less strongly contracting the basis functions improves the final results for the IPs insignificantly.

To study the influence of f-type functions on the calculated IPs, two d-type functions and one f-type function on the S atom (and one p-type function on each H atom) have been added to basis A. The results, listed in Table XVI show a slight improvement due to f-type functions. All three IP's are now within 0.1 eV of the experimental values.

We may conclude that a well-balanced basis set is required for obtaining accurate IPs for molecules containing second-row atoms. It seems that the total HF energy is not a unique criterion for selecting the appropriate basis set.

G. The Vibrational Structure in Inner-Shell Ionization Spectra

The ESCA spectra of some light molecules show core electron lines that are unusually broad.[1b, 141] It has been attempted to explain this effect by lifetime broadening,[142] but it became clear later that vibrational excitation must be responsible for the observed broading.[141, 143] Phonon broadening of core lines has also been found in the XPS spectra of solids.[144] Recently, Gelius et al.[145] succeeded in resolving the vibrational structure of the C1s line in methane. It is interesting to see whether a simple ab initio calculation can explain the unexpectedly strong vibrational coupling of the C1s electron.

It is sufficient to consider only the totally symmetric C-H stretching vibration ν_1, because only this vibration can be coupled linearly to the $1a_1$(C1s) and $2a_1$(C2s) electrons. The triply degenerate $1t_2$ orbital is not considered here, since the corresponding band in the PE spectrum exhibits a pronounced Jahn-Teller splitting involving the triply degenerate deformation mode ν_4.

Extensive HF calculations on CH_4 and its positive ions have been performed by Clementi and Popkie.[146] From their data the free coupling parameters $a_1^{(0)}(1s)$ and $a_1^{(0)}(2s)$ are easily determined to be 1.58 and 1.09 respectively (see Table XVII). For comparison, "experimental" values for $a_1(1s)$ and $a_1(2s)$, determined from the intensity ratio I_1/I_0 of the first two lines in the ESCA or the HeII spectrum, are also given in Table XVII. It is most surprising to see that C1s electron emerges as strongly antibonding on the HF level. Figure 6.10a shows the vibrational structure of the C1s line calculated with $a_1^{(0)}(1s)$ and convoluted with a Gaussian to account for the limited experimental resolution. The experimental C1s ESCA spectrum[145] together with a least-squares fit to three Gaussians is shown in Fig. 6.10c. Obviously, the one-particle approximation is not adequate to describe the vibrational coupling of the C1s electron in CH_4.

To analyze the influence of many-body effects, renormalized coupling constants have been calculated for CH_1.[147] It is well known from numeri-

TABLE XVII

Calculated and Experimental Coupling Parameters for the
C1s and C2s Orbitals of Methane

Orbital	$a_1^{(0)}$	a_1	$a_1^{\Delta SCF}$	a^{exp}
$1a_1$(C1s)	1.58	0.47	0.66	0.54[a]
$2a_1$(C2s)	1.09	1.13	1.14	1.67[b]

[a]From the ESCA spectrum of Ref. 145.
[b]From the HeII spectrum of Ref. 149.

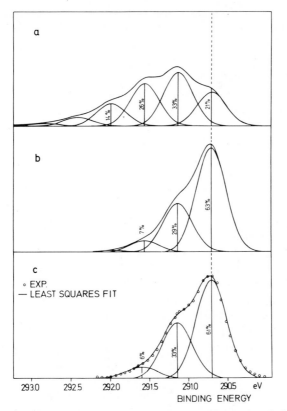

Fig. 6.10. The calculated and experimental vibrational structure of the $C1s$ line in methane. (*a*) The spectrum calculated with the free coupling parameter. (*b*) The spectrum calculated with a renormalized coupling parameter. (*c*) The $C1s$ ESCA spectrum together with a least-squares fit to three Gaussians.[145] The spectra are normalized to equal intensity. [Reproduced with permission from *Chem. Phys. Lett.*, **31**, 582 (1975).]

cal calculations that the second-order self-energy part reproduces the IPs of *core* electrons quite well.[10, 12, 147] It is reasonable, therefore, to renormalize the coupling constants in second order of the perturbation theory. The resulting coupling parameters are given in the second column of Table XVII. The coupling parameter of the $C1s$ orbital is seen to be drastically reduced from its free value, whereas the coupling parameter of the $C2s$ electron remains nearly unchanged. In Fig. 6.10*b* the resulting vibrational structure of the $C1s$ line is shown. It is seen to be in excellent agreement with the observed spectrum.

As is well known (see Section III.C.3), the ΔSCF procedure is a rather accurate scheme for calculating the IPs of core electrons. It is reasonable to expect that a ΔSCF calculation predicts also the vibrational coupling constants of core orbitals correctly. The ΔSCF values for the coupling parameters $a_1(1s)$ and $a_1(2s)$ have been determined from the data of Clementi and Popkie[146] and are given in the third column of Table XVII. These $a_1^{\Delta SCF}$ are seen to be in good agreement with the a_1 calculated in second-order perturbation theory. It is now clear that the strong reduction of the C1s coupling parameter from its free value is a *reorganization* effect. It seems, moreover, that the ΔSCF scheme is appropriate for calculating reliable vibrational coupling constants for *core* electrons.

FC factors for the ionization of CH_4 have also been determined by Meyer[148] from extensive configuration-interaction calculations on both the ground state and the various ionic states. His FC factors are in good agreement with the present results.

The C1s orbital in carbon monoxide has been found[147] to exhibit a behavior quite similar to the CH_4 1s orbital discussed above. The vibrational coupling on the one-particle level is rather strong, but is drastically reduced in second order of the perturbation theory. The coupling parameter of the O1s electron, on the other hand, is vanishingly small in the one-particle approximation.

To gain an understanding of the influence of reorganization effects on the coupling constants we consider (3.45) together with the definition (5.60) of the renormalized coupling constants. Keeping in mind that the derivatives of the V_{ijkl} are usually small and that the largest derivatives of the orbital energies are positive for occupied valence orbitals (bonding character) and usually negative for unoccupied valence orbitals (antibonding character), one finds a negative reorganization contribution to the coupling constant of core orbitals. This means that the inclusion of reorganization effects makes the core electrons more bonding (or less antibonding).

The strong vibrational coupling of the C1s electron in CH_4 and CO gives us reason to expect similar coupling strengths in other molecules also and for the core electrons of other first-row atoms. This is in accordance with the observed line broadening in the ESCA spectra of various molecules.[141, 143] Since the lifetime broadening of the core levels of first-row atoms in molecules is small compared to typical vibrational spacings, the vibrational fine structure of core ionization lines can be resolved in principle. Hopefully, further improvement of the resolution in ESCA will render the observation of the vibrational fine structure of core electron lines possible in the near future. The analysis of the vibrational structure of core lines could provide us with valuable information on the chemical bonding in molecules, supplementing the information contained in the chemical shifts usually measured with ESCA. As shown for CH_4 above, the vibrational structure of core electron lines can be theoretically predicted with the help of standard ΔSCF calculations.

1. APPENDIX

In this Appendix we present explicit expressions for the diagrams of the self-energy part up to third order. In the following expressions the summation over spin indices has already been carried out. The indices $i, j, k \ldots$ stand for *doubly occupied* orbitals and $a,b,c,\ldots$ for unoccupied orbitals.

$$\Sigma_{pq}^{(2)}(\omega) = \sum \frac{(2V_{paij} - V_{paji})V_{qaij}}{\omega + \varepsilon_a - \varepsilon_i - \varepsilon_j} + \sum \frac{(2V_{piab} - V_{piba})V_{qiab}}{\omega + \varepsilon_i - \varepsilon_a - \varepsilon_b}$$

$$\Sigma_{pq}^{(3)}(\omega) = \sum_{I=1}^{6} (AI + CI + DI)$$

$$A1 = -\sum \frac{(2V_{pkqj} - V_{pkjq})(2V_{jiab} - V_{jiba})V_{abki}}{(\varepsilon_j + \varepsilon_i - \varepsilon_a - \varepsilon_b)(\varepsilon_k + \varepsilon_i - \varepsilon_a - \varepsilon_b)}$$

$$A2 = \sum \frac{(2V_{pcqb} - V_{pcbq})(2V_{jiab} - V_{jiba})V_{jica}}{(\varepsilon_j + \varepsilon_i - \varepsilon_a - \varepsilon_b)(\varepsilon_j + \varepsilon_i - \varepsilon_a - \varepsilon_c)}$$

$$A3 = \sum \frac{(2V_{pcqj} - V_{pcjq})(2V_{jiab} - V_{jiba})V_{abci}}{(\varepsilon_j + \varepsilon_i - \varepsilon_a - \varepsilon_b)(\varepsilon_j - \varepsilon_c)}$$

$$A4 = \sum \frac{(2V_{pjqc} - V_{pjcq})(2V_{jiab} - V_{jiba})V_{abci}}{(\varepsilon_j + \varepsilon_i - \varepsilon_a - \varepsilon_b)(\varepsilon_j - \varepsilon_c)}$$

$$A5 = -\sum \frac{(2V_{pbqk} - V_{pbkq})(2V_{jiab} - V_{jiba})V_{ijka}}{(\varepsilon_j + \varepsilon_i - \varepsilon_a - \varepsilon_b)(\varepsilon_k - \varepsilon_b)}$$

$$A6 = -\sum \frac{(2V_{pkqb} - V_{pkbq})(2V_{jiab} - V_{jiba})V_{ijka}}{(\varepsilon_j + \varepsilon_i - \varepsilon_a - \varepsilon_b)(\varepsilon_k - \varepsilon_b)}$$

$$C1 = \sum \frac{(2V_{piab} - V_{piba})V_{abcd}V_{qicd}}{(\omega + \varepsilon_i - \varepsilon_a - \varepsilon_b)(\omega + \varepsilon_i - \varepsilon_c - \varepsilon_d)}$$

$$C2 = \sum \frac{(2V_{piab} - V_{piba})V_{abjk}V_{qijk}}{(\omega + \varepsilon_i - \varepsilon_a - \varepsilon_b)(\varepsilon_j + \varepsilon_k - \varepsilon_a - \varepsilon_b)}$$

$$C3 = \sum \frac{(2V_{pijk} - V_{pikj})V_{abjk}V_{qiab}}{(\omega + \varepsilon_i - \varepsilon_a - \varepsilon_b)(\varepsilon_j + \varepsilon_k - \varepsilon_a - \varepsilon_b)}$$

$$C4 = \sum \frac{(2V_{paij} - V_{paji})V_{ijbc}V_{qabc}}{(\omega + \varepsilon_a - \varepsilon_i - \varepsilon_j)(\varepsilon_i + \varepsilon_j - \varepsilon_b - \varepsilon_c)}$$

$$C5 = \sum \frac{(2V_{pabc} - V_{pacb})V_{ijbc}V_{qaij}}{(\omega + \varepsilon_a - \varepsilon_i - \varepsilon_j)(\varepsilon_i + \varepsilon_j - \varepsilon_b - \varepsilon_c)}$$

$$C6 = -\sum \frac{(2V_{pakl} - V_{palk})V_{klij}V_{qaij}}{(\omega + \varepsilon_a - \varepsilon_i - \varepsilon_j)(\omega + \varepsilon_a - \varepsilon_k - \varepsilon_l)}$$

$$D1 = \sum \left\{ \frac{V_{piab}\left[V_{ajic}(V_{qjcb} - 2V_{qjbc}) + V_{ajci}(V_{qjbc} - 2V_{qjcb}) \right]}{(\omega + \varepsilon_i - \varepsilon_a - \varepsilon_b)(\omega + \varepsilon_j - \varepsilon_b - \varepsilon_c)} \right.$$

$$\left. + \frac{V_{piba}\left[V_{ajic}(4V_{qjbc} - 2V_{qjcb}) + V_{ajci}(V_{qjcb} - 2V_{qjbc}) \right]}{(\omega + \varepsilon_i - \varepsilon_a - \varepsilon_b)(\omega + \varepsilon_j - \varepsilon_b - \varepsilon_c)} \right\}$$

$$D2 = \sum \left\{ \frac{V_{pica}\left[V_{abij}(4V_{qbcj} - 2V_{qbjc}) + V_{abji}(V_{qbjc} - 2V_{qbcj}) \right]}{(\omega + \varepsilon_i - \varepsilon_a - \varepsilon_c)(\varepsilon_i + \varepsilon_j - \varepsilon_a - \varepsilon_b)} \right.$$

$$\left. + \frac{V_{piac}\left[V_{abij}(V_{qbjc} - 2V_{qbcj}) + V_{abji}(V_{qbcj} - 2V_{qbjc}) \right]}{(\omega + \varepsilon_i - \varepsilon_a - \varepsilon_c)(\varepsilon_i + \varepsilon_j - \varepsilon_a - \varepsilon_b)} \right\}$$

$$D3 = \sum \left\{ \frac{V_{pcja}\left[V_{jicb}(V_{qiba} - 2V_{qiab}) + V_{jibc}(V_{qiab} - 2V_{qiba}) \right]}{(\omega + \varepsilon_i - \varepsilon_a - \varepsilon_b)(\varepsilon_j + \varepsilon_i - \varepsilon_b - \varepsilon_c)} \right.$$

$$\left. + \frac{V_{pcaj}\left[V_{jicb}(4V_{qiab} - 2V_{qiba}) + V_{jibc}(V_{qiba} - 2V_{qiab}) \right]}{(\omega + \varepsilon_i - \varepsilon_a - \varepsilon_b)(\varepsilon_j + \varepsilon_i - \varepsilon_b - \varepsilon_c)} \right\}$$

$$D4 = \sum \left\{ \frac{V_{pakj}\left[V_{jiab}(4V_{qikb} - 2V_{qibk}) + V_{jiba}(V_{qibk} - 2V_{qikb}) \right]}{(\omega + \varepsilon_a - \varepsilon_j - \varepsilon_k)(\varepsilon_i + \varepsilon_j - \varepsilon_a - \varepsilon_b)} \right.$$

$$\left. + \frac{V_{pqjk}\left[V_{jiab}(V_{qibk} - 2V_{qikb}) + V_{jiba}(V_{qikb} - 2V_{qibk}) \right]}{(\omega + \varepsilon_a - \varepsilon_j - \varepsilon_k)(\varepsilon_i + \varepsilon_j - \varepsilon_a - \varepsilon_b)} \right\}$$

$$D5 = \sum \left\{ \frac{V_{pibk}\left[V_{jiab}\left(V_{qajk} - 2V_{qakj} \right) + V_{jiba}\left(V_{qakj} - 2V_{qajk} \right) \right]}{(\omega + \varepsilon_a - \varepsilon_j - \varepsilon_k)(\varepsilon_i + \varepsilon_j - \varepsilon_a - \varepsilon_b)} \right.$$

$$\left. + \frac{V_{pikb}\left[V_{jiab}\left(4V_{qakj} - 2V_{qajk} \right) + V_{jiba}\left(V_{qajk} - 2V_{qakj} \right) \right]}{(\omega + \varepsilon_a - \varepsilon_j - \varepsilon_k)(\varepsilon_i + \varepsilon_j - \varepsilon_a - \varepsilon_b)} \right\}$$

$$D6 = -\sum \left\{ \frac{V_{paki}\left[V_{ibaj}\left(4V_{qbkj} - 2V_{qbjk} \right) + V_{ibja}\left(V_{qbjk} - 2V_{qbkj} \right) \right]}{(\omega + \varepsilon_a - \varepsilon_i - \varepsilon_k)(\omega + \varepsilon_b - \varepsilon_j - \varepsilon_k)} \right.$$

$$\left. + \frac{V_{paik}\left[V_{ibaj}\left(V_{qbjk} - 2V_{qbkj} \right) + V_{ibja}\left(V_{qbkj} - 2V_{qbjk} \right) \right]}{(\omega + \varepsilon_a - \varepsilon_i - \varepsilon_k)(\omega + \varepsilon_b - \varepsilon_j - \varepsilon_k)} \right\}$$

Acknowledgments

It is a pleasure to acknowledge many fruitful discussions with W. Brenig over the last years. A great part of the results presented here is due to collaboration with W. von Niessen and J. Schirmer, to whom we wish to express our sincere thanks. We further thank K. Schön-hammer, M. Majster, H. S. Taylor, R. Yaris, and A. Bradshaw for many useful discussions. The authors are indebted to G. Diercksen and W. Kraemer for the permission to use their SCF programs.

References

1. a. K. Siegbahn, C. Nordling, A. Fahlman, R. Nordberg, K. Hamrin, J. Hedman, G. Johansson, T. Bergmark, S. E. Karlson, I. Lindgren, and B. Lindberg, *ESCA—Atomic, Molecular and Solid State Structure Studied by Means of Electron Spectroscopy*, North-Holland, Amsterdam, 1967.
    b. K. Siegbahn, C. Nordling, G. Johansson, J. Hedman, P. F. Heden, K. Harmin, U. Gelius, T. Bergmark, L. O. Werme, R. Manne, and Y. Baer, *ESCA—Applied to Free Molecules*, North-Holland, Amsterdam, 1969.
    c. D. W. Turner, C. Baker, A. D. Baker, and C. R. Brundle, *Molecular Photoelectron Spectroscopy*, Wiley, New York, 1970.
    d. J. H. D. Eland, *Photoelectron Spectroscopy*, Butterworth, London, 1974.
    e. W. C. Price, *Advan. Atomic Mol. Phys.*, **10**, 131 (1974).
2. T. Koopmans, *Phys. (Utr.)*, **1**, 104 (1933).
3. Gy Csanak, H. S. Taylor, and R. Yaris, *Advan. Atomic Mol. Phys.*, **7**, 287 (1971).
4. J. Linderberg and Y. Öhrn, *Propagators in Quantum Chemistry*, Academic Press, New York, 1973.
5. W. P. Reinhardt and J. D. Doll, *J. Chem. Phys.*, **50**, 2767 (1969); J. D. Doll and W. P. Reinhardt, *J. Chem. Phys.*, **57**, 1169 (1972).
6. a. L. S. Cederbaum, G. Hohlneicher, and S. Peyerimhoff, *Chem. Phys. Lett.*, **11**, 421 (1971).
    b. F. Ecker and G. Hohlneicher, *Theoret. Chim. Acta*, **25**, 289 (1972).
    c. G. Hohlneicher, F. Ecker, and L. S. Cederbaum, in D. A. Shirley, Ed., *Electron Spectroscopy*, North-Holland, Amsterdam, 1972, p. 647.

d. L. S. Cederbaum, *Chem. Phys. Lett.*, **25**, 562 (1974).

7. a. L. S. Cederbaum, G. Hohlneicher, and W. von Niessen, *Chem. Phys. Lett.*, **18**, 503 (1973).

b. L. S. Cederbaum, G. Hohlneicher, and W. von Niessen, *Mol. Phys.*, **26**, 1405 (1973).

c. L. S. Cederbaum and W. von Niessen, *Chem. Phys. Lett.*, **24**, 263 (1974).

d. L. S. Cederbaum, *Mol. Phys.*, **28**, 479 (1974).

8. a. B. Schneider, H. S. Taylor, and R. Yaris, *Phys. Rev. A*, **1**, 855 (1970).

b. B. Schneider, *Phys. Rev. A*, **2**, 1873 (1970); **7**, 557 (1973).

c. B. S. Yarlagadda, Gy Csanak, H. S. Taylor, B. Schneider, and R. Yaris, *Phys. Rev. A*, **7**, 146 (1973).

d. B. Schneider, B. S. Yarlagadda, H. S. Taylor, and R. Yaris, *Chem. Phys. Lett.*, **22**, 381 (1973).

e. Gy Csanak, H. S. Taylor, and D. N. Tripathy, *J. Phys. B: Atom. Molec. Phys.*, **6**, 2040 (1973).

f. L. D. Thomas, Gy Csanak, H. S. Taylor, and B. S. Yarlagadda, *J. Phys. B: Atom. Molec. Phys.*, **7**, 1719 (1974).

9. P. C. Martin and J. S. Schwinger, *Phys. Rev.*, **115**, 1342 (1959).

10. P. -O. Nerbrant, *Int. J. Quant. Chem.*, **9**, 901 (1975).

11. B. T. Pickup and O. Goscinski, *Mol. Phys.* **26**, 1013 (1973).

12. a. G. Purvis and Y. Öhrn, *J. Chem. Phys.*, **60**, 4063 (1974).

b. G. Purvis and Y. Öhrn, *J. Chem. Phys.*, **62**, 2045 (1975).

c. G. Purvis and Y. Öhrn, *Chem. Phys. Lett.*, **33**, 396 (1975).

13. a. J. Simons and W. D. Smith, *J. Chem. Phys.*, **58**, 4899 (1973)

b. J. Simons, *Chem. Phys. Lett.*, **25**, 122 (1974).

c. T. T. Chen, W. D. Smith, and J. Simons, *J. Chem. Phys.* **61**, 2670 (1974).

d. W. D. Smith, T. T. Chen, and J. Simons, *Chem. Phys. Lett.*, **26**, 296 (1974).

e. K. Griffing and J. Simons, *J. Chem. Phys.*, **62**, 535 (1975).

14. D. J. Rowe, *Rev. Mod. Phys.*, **40**, 153 (1968); *Phys. Rev.*, **175**, 1283 (1968).

15. a. I. Hubač, V. Kvasnička, and A. Holubec, *Chem. Phys. Lett.*, **23**, 381 (1973).

b. V. Kvasnička and I. Hubač, *J. Chem. Phys.*, **60**, 4483 (1974).

c. S. Biscupič, L. Valko, and V. Kvasnička, *Theoret. Chim. Acta*, **38**, 149 (1975).

16. a. D. P. Chong, F. G. Herring, and D. McWilliams, *J. Chem. Phys.*, **61**, 78 (1974).

b. D. P. Chong, F. G. Herring, and D. McWilliams, *J. Chem. Phys.*, **61**, 958 (1974).

c. D. P. Chong, F. G. Herring, and D. McWilliams, *Chem. Phys. Lett.*, **25**, 568 (1974).

d. D. P. Chong, F. G. Herring, and D. McWilliams, *J. Chem. Phys.*, **61**, 3567 (1974).

e. D. P. Chong, F. G. Herring, and D. McWilliams, *J. Electr. Spectr.*, **7**, 445 (1975).

17. H. P. Kelly, *Advan. Chem. Phys.*, **14**, 129 (1969); *Phys. Rev.*, **136**, B896 (1964); J.-J. Chang and H. P. Kelly, *Phys. Rev. A*, **12**, 92 (1975).

18. E. S. Chang, R. T. Pu, and T. P. Das, *Phys. Rev.*, **174**, 1 (1968).

19. L. S. Cederbaum, F. E. P. Matschke, and W. von Niessen, *Phys. Rev. A*, **12**, 6 (1975).

20. T. Shibuya and V. McKoy, *Phys. Rev. A*, **2**, 2208 (1970); J. Rose, T. Shibuya, and V. McKoy, *J. Chem. Phys.*, **58**, 74, 500 (1973).

21. J. Čížek and J. Paldus, *Int. J. Quant. Chem.*, **6**, 435 (1970); J. Paldus and J. Čížek, *J. Chem. Phys.*, **60**, 149 (1974).

22. T. N. Chang, T. Ishihara, and R. T. Poe, *Phys. Rev. Lett.*, **27**, 838 (1971); T. N. Chang and R. T. Poe, *Phys. Rev. A*, **12**, 1432 (1975).

23. G. D. Mahan, *Phys. Rev. B*, **2**, 4334 (1970).

24. D. J. Thouless, *The Quantum Mechanics of Many-Body Systems*, Academic Press, New York, 1961.

25. A. B. Migdal, *Theory of Finite Fermi Systems and Application to Atomic Nuclei*, Wiley, New York, 1967.
26. K. Gottfried, *Quantum Mechanics*, Benjamin, New York, 1966.
27. L. S. Cederbaum, *Theor. Chim. Acta*, **31**, 239 (1973).
28. M. O. Krause, M. L. Vestal, W. H. Johnston, and T. A. Carlson, *Phys. Rev.*, **133**, A385 (1964).
29. U. Gelius, E. Basilier, S. Svensson, T. Bergmark, and K. Siegbahn, *J. Elect. Spectr.*, **2**, 405 (1974); A. W. Potts and T. A. Williams, *J. Elect. Spectr.*, **3**, 3 (1974).
30. T. A. Carlson, M. O. Krause, and W. E. Moddemann, *J. de Phys. (Paris)*, **C4**, 76 (1971).
31. L. S. Cederbaum, *J. Chem. Phys.*, **62**, 2160 (1975).
32. U. Gelius, in D. A. Shirley, Ed., *Electron Spectroscopy*, North-Holland, Amsterdam, 1972, p. 311.
33. I. G. Kaplan and A. P. Markin, *Opt. Spectrosc.*, **24**, 475 (1968).
34. A. Schweig and W. Thiel, *J. Elect. Spectrosc.*, **3**, 27 (1974).
35. F. O. Ellison, *J. Chem. Phys.*, **61**, 507 (1974); J. W. Rabalais, T. P. Debies, J. L. Berkosky, J. J. Huang, and F. O. Ellison, *J. Chem. Phys.*, **61**, 516, 529 (1974).
36. H. C. Tuckwell, *J. Phys. B: Atom. Molec. Phys.*, **3**, 293 (1970); S. Iwata and S. Nagakura, *Mol. Phys.*, **27**, 425 (1974); B. Ritchie, *J. Chem. Phys.* **61**, 3279 (1974).
37. L. S. Cederbaum and J. Schirmer, *Z. Physik*, **271**, 221 (1974).
38. A. Abrikosov, L. Gorkov, and J. Dzyaloshinskii, *Quantum Field Theoretical Methods in Statistical Physics*, Pergamon Press, Oxford, 1965.
39. G. C. Wick, *Phys. Rev.*, **80**, 268 (1950).
40. J. Goldstone, *Proc. Roy. Soc. (London)*, **A239**, 267 (1957).
41. V. M. Galitskii and A. B. Migdal, *Sov. Phys.—JETP*, **34**, 96 (1958).
42. R. Yaris and H. S. Taylor, *Phys. Rev. A*, **12**, 1751 (1975).
43. P. O. Löwdin, *Advan. Chem. Phys.*, **14**, 283 (1969).
44. C. C. J. Roothaan, *Rev. Mod. Phys.*, **23**, 69 (1951).
45. C. C. J. Roothaan, *Rev. Mod. Phys.*, **32**, 179 (1960).
46. F. J. Dyson, *Phys. Rev.*, **75**, 486 (1949).
47. This is only true when a finite basis set of one-particle functions is used. If an infinite basis set is used, Σ might have a cut instead of discrete poles.
48. L. van Hove, N. Hugenholtz, and L. Howland, *Quantum Theory of Many-Particle Systems*, Benjamin, New York, 1961.
49. D. Pines, *The Many-Body Problem*, Benjamin, New York, 1961.
50. W. Finkelnburg, *Einführung in die Atomphysik*, Springer, Berlin, 1967.
51. K. A. Brueckner, *Advan. Chem. Phys.*, **14**, 215 (1969).
52. V. Galitskii, *Soviet Phys. —JETP*, **7**, 104 (1958).
53. L. S. Cederbaum, *J. Phys. B: Atom. Molec. Phys.*, **8**, 290 (1975).
54. K. F. Freed, *Phys. Rev.*, **173**, 1 (1968).
55. O. Sinanoğlu, *Advan. Chem Phys.*, **14**, 237 (1969).
56. R. K. Nesbet, *Advan. Chem. Phys.*, **14**, 1 (1969).
57. See, for example, P. S. Bagus, *Phys. Rev.*, **139**, A619 (1965); C. M. Moser, R. K. Nesbet, and G. Verhaegen, *Chem. Phys. Lett.*, **12**, 230 (1971); L. J. Aarous, M. F. Guest, M. B. Hall, and I. H. Hillier, *J. Chem. Soc. (Faraday II)*, **69**, 563 (1973); D T. Clark, I. W. Scanlan, and J. Muller, *Theoret. Chim. Acta*, **35**, 341 (1974).
58. L. Hedin and A. Johansson, *J. Phys. B: Atom Molec. Phys.*, **2**, 1336 (1969).
59. P. E. Cade, K. D. Sales, and A. C. Wahl, *J. Chem. Phys.*, **44**, 1973 (1966).
60. See, for example, R. K. Nesbet, *J. Chem. Phys.*, **36**, 1518 (1962); S. Meza and U. Wahlgren, *Theoret. Chim. Acta*, **21**, 323 (1971); T. E. H. Walker and J. A. Horsley, *Mol. Phys.* **21**, 939 (1971).

61. In principle, we could chose the inhomogeneous part in (4.10) to be the kernel arising from the self-consistent self-energy part of second order[27] instead of the "unrenormalized" zeroth-order kernel used here.

62. P. Noziéres, *Theory of Interacting Fermi Systems*, Benjamin, New York, 1964.

63. S. Ethofer and P. Schuck, *Z. Physik*, **228**, 264 (1969).

64. P. Schuck, F. Villars, and P. Ring, *Nucl. Phys.*, **A208**, 302 (1973).

65. D. J. Thouless, *Nucl. Phys.*, **22**, 78 (1961).

66. J. Schirmer and L. S. Cederbaum, to be published.

67. A. de Shalit and H. Feshbach, *Theoretical Nuclear Physics*, Wiley, New York, 1974.

68. A. M. Lane, *Nuclear Physics*, Benjamin, New York, 1964.

69. G. Ripka and R. Padjen, *Nucl. Phys.*, **A132**, 489 (1969).

70. K. Arita and H. Horie, *Nucl. Phys.*, **A173**, 97 (1971).

71. G. D. Purvis and Y. Öhrn, *Chem. Phys. Lett.*, **33**, 396 (1975).

72. O. Edqvist, E. Lindholm, L. E. Selin, and L. Åsbrink, *Physica Scripta*, **1**, 25 (1970).

73. R. N. Dixon and S. E. Hull, *Chem. Phys. Lett.*, **3**, 367 (1969).

74. a. G. Herzberg, *Molecular Spectra and Molecular Structure. I. Spectra of Diatomic Molecules*, Van Nostrand, New York, 1950.
   b. G. Herzberg, *Electronic Spectra and Electronic Structure of Polyatomic Molecules*, Van Nostrand, New York, 1966.

75. L. S. Cederbaum and W. Domcke, *J. Chem. Phys.*, **60**, 2878 (1974).

76. E. B. Wilson, J. C. Decius, and P. C. Cross, *Molecular Vibrations*, McGraw-Hill, New York, 1955.

77. The restriction to a HF potential and to a closed-shell system is a matter of convenience only. The extension to a non-HF picture and an open-shell system is straightforward.

78. M. Born and K. Huang, *Dynamical Theory of Crystal Lattices*, Oxford University Press, London, 1954, Appendix VII.

79. Equation 5.22 is obtained from (2.4a) by replacing a_i in (2.4b) by $\tau_{ei} a_i$. With this redefined Green's function, (2.4a) reads

$$P(\omega) = \frac{2e^2}{m^2 c^2 \hbar} \sum_{enl} \text{Im} \left\{ G_{ln}(\omega - \hbar\omega_0 - i\eta) \right\} \delta(\omega - E_e)$$

80. L. S. Cederbaum and W. Domcke, *J. Chem. Phys.*, **64**, 603 (1976).

81. H. Cramer, *Mathematical Methods of Statistics*, Princeton University Press, Princeton, 1946.

82. L. S. Cederbaum and W. Domcke, *Chem. Phys. Lett.*, **25**, 357 (1974).

83. C. B. Duke, N. O. Lipari, and L. Pietronero, *Chem. Phys. Lett.*, **30**, 415 (1975); N. O. Lipari and C. B. Duke, *J. Chem. Phys.*, **63**, 1748 (1975); C. B. Duke and N. O. Lipari, *Chem. Phys. Lett.*, **36**, 51 (1975).

84. H. A. Jahn and E. Teller, *Poc. Roy. Soc. (London)*, **A161**, 220 (1937).

85. H. C. Longuet-Higgins, *Advan. Spectrosc.*, **2**, 429 (1961).

86. M. D. Sturge, *Solid State Phys.*, **20**, 91 (1967).

87. R. Englman, *The Jahn-Teller Effect*, Wiley, New York, 1972.

88. R. N. Porter, R. M. Stevens, and M. Karplus, *J. Chem. Phys.*, **49**, 5163 (1968).

89. M. Gouterman, *J. Chem. Phys.*, **42**, 351 (1965).

90. M. Wagner, *Z. Physik*, **244**, 275 (1971); *Z. Physik*, **256**, 291 (1972).

91. C. S. Sloane and R. Silbey, *J. Chem. Phys.*, **56**, 6031 (1972).

92. M. Roche and H. H. Jaffé, *J. Chem. Phys.*, **60**, 1193 (1974); W. C. Johnson, Jr., and O. E. Weigang, Jr., *J. Chem. Phys.*, **63**, 2135 (1975); O. Atabek, A. Hardisson, and R. Lefebvre, *Chem. Phys. Lett.*, **20**, 40 (1973).

93. H. C. Longuet-Higgins, U. Öpik, M. H. L. Pryce, and H. Sack, *Proc. Roy. Soc. (London)*, **A244**, 1 (1958); W. Moffit and W. Thorson, in R. A. Daudel, Ed., *Calcul des functions d'onde moleculaires*, CNRS, Paris, 1958, p. 141; C. G. Rowland, *Chem. Phys. Lett.*, **9**, 169 (1971).

94. D. E. McCumber, *J. Math Phys.*, **5**, 508 (1964).

95. W. Benesch, J. T. Vanderslice, S. G. Tilford, and P. G. Wilkinson, *Astrophys. J.*, **143**, 236 (1966).

96. R. W. Nicholls, *J. Quant. Spectr. Radiat. Transfer*, **2**, 433 (1962).

97. J. B. Coon, R. E. DeWames, and C. M. Loyd, *J. Mol. Spectrosc.*, **8**, 285 (1962).

98. T. E. Sharp and H. M. Rosenstock, *J. Chem. Phys.*, **41**, 3453 (1964); R. Botter, V. H. Dibeler, J. A. Walker, and H. M. Rosenstock, *J. Chem. Phys.*, **44**, 1271 (1966).

99. P. A. Fraser and W. R. Jarmain, *Proc. Phys. Soc. (London)*, **A66**, 1145 (1953).

100. T. Y. Chang and M. Karplus, *J. Chem. Phys.*, **52**, 783 (1970).

101. A. Chutjian, D. C. Cartwright, and S. Trajmar, *Phys. Rev. Lett.*, **30**, 195 (1973).

102. A. B. Anderson and R. G. Parr, *Chem. Phys. Lett.*, **10**, 293 (1971).

103. R. Wallace, *Chem. Phys.*, **11**, 189 (1975).

104. S. Huzinaga and Y. Sakai, *J. Chem. Phys.*, **50**, 1371 (1969).

105. E. Clementi, *IBM J. Res. Develop. Suppl.*, **9**, 2 (1965).

106. P. S. Bagus and T. L. Gilbert, unpublished results, cited partially in A. D. McLean and M. Yoshimine, *IBM J. Res. Develop. Suppl.*, **12**, 206 (1968).

107. Ch. Froese-Fischer, Some Hartree-Fock Results for Atoms Helium to Radon, Special Report from the Department of Mathematics, University of British Columbia, Vancouver.

108. C. E. Moore, Natl. Bur. Std. (U. S.) Circular 467 (1949), Vol. I and corrections in Vol. II (1952) and Vol. III (1958).

109. M. W. Ribarsky, *Phys. Rev. A*, **12**, 1739 (1975).

110. S. F. Boys and N. C. Handy, *Proc. Roy. Soc. (London)*, **A310**, 43, 63 (1969).

111. C. M. Moser and R. K. Nesbet, *Phys. Rev. A*, **4**, 1336 (1971).

112. C. M. Moser and R. K. Nesbet, *Phys. Rev. A*, **6**, 1710 (1972).

113. A. W. Weiss, *Phys. Rev. A*, **3**, 126 (1971).

114. W. von Niessen, private communication. Here, as in all other calculations reported in this article, the program system MUNICH has been used for the SCF calculations; see Ref. 138.

115. L. S. Cederbaum and W. von Niessen, *J. Chem Phys.*, **62**, 3824 (1975).

116. D. P. Chong, F. G. Herring, and D. McWilliams, *J. Electr. Spectrosc.*, **7**, 429 (1975).

117. L. S. Cederbaum, Thesis, Munich 1972.

118. J. C. Lorquet and M. Desouter, *Chem. Phys. Lett.*, **16**, 136 (1972).

119. L. Åsbrink and C. Fridh, *Physica Scripta*, **9**, 338 (1974).

120. W. Domcke and L. S. Cederbaum, *J. Chem. Phys.*, **64**, 612 (1976).

121. C. Baker and D. W. Turner, *Proc. Roy. Soc. (London)*, **A308**, 19 (1968).

122. A. D. McLean and M. Yoshimine, *IBM J. Res. Develop. Suppl.* (1967).

123. L. S. Cederbaum, W. Domcke, and W. von Niessen, *Chem. Phys.*, **10**, 459 (1975).

124. L. H. Jones, *J. Mol. Spectr.*, **45**, 55 (1973).

125. C. R. Brundle and D. W. Turner, *Chem. Commun.*, **1967**, 314.

126. A. D. Baker, C. R. Brundle, and D. W. Turner, *Int. J. Mass Spectr. Ion Phys.* **1**, 285 (1968).

127. C. R. Brundle, M. B. Robin, N. A. Kuebler, and H. Basch, *J. Am. Chem. Soc.*, **94**, 1451 (1972).

128. I. P Batra and O. Robaux, *Chem Phys. Lett.*, **28**, 529 (1974).

129. L. S. Cederbaum, W. Domcke, and W. von Niessen, *Chem. Phys. Lett.*, **34**, 60 (1975).

130. B. J. Garrison, H. F. Schaefer, III, and W. A. Lester, *J. Chem. Phys.*, **61**, 3039 (1974).
131. T. Shimanouchi and I. Suzuki, *J. Chem. Phys.*, **42**, 296 (1965).
132. a. W. von Niessen, L. S. Cederbaum, and W. P. Kraemer, *J. Chem. Phys.*, **65**, 1378 (1976).
    b. W. von Niessen, G. H. F. Diercksen, and L. S. Cederbaum, *Chem. Phys.*, **10**, 345 (1975).
    c. W. von Niessen, W. P. Kraemer, and L. S Cederbaum, *J. Electr. Spectrosc.*, **8**, 179 (1976).
    d. W. von Niessen, L. S. Cederbaum, and G. H. F. Diercksen, *J. Am. Chem. Soc.*, **98**, 2066 (1976).
    e. W. von Niessen, W. P. Kraemer, and L. S. Cederbaum, *Chem. Phys.*, **11**, 385 (1975).
    f. W. von Niessen, L. S. Cederbaum, G. H. F. Diercksen, and G. Hohlneicher, *Chem. Phys.*, **11**, 399 (1975).
133. S. Huzinaga, *J. Chem Phys.*, **42**, 1293 (1965).
134. J. Almlöf, B. Roos, U. Wahlgren, and J. Johansen, *J. Electr. Spectrosc.*, **2**, 51 (1973).
135. J. D. Petke, J. L. Whitten, and J. A. Ryan, *J. Chem. Phys.*, **48**, 953 (1968).
136. E. Clementi, *J. Chem. Phys.*, **46**, 4736 (1967); **47**, 4485 (1967).
137. W. von Niessen, L. S. Cederbaum, W. Domcke, and G. H. F. Diercksen, *J. Chem. Phys.*, in press.
138. G. H. F. Diercksen and W. P. Kraemer, MUNICH, Molecular Program System, Reference Manual, Special Technical Report, Max-Planck Institut für Physik und Astrophysik, to be published; G. H. F. Diercksen, *Theoret. Chim. Acta*, **33**, 1 (1974).
139. A. Veillard, *Theoret. Chim. Acta*, **12**, 405 (1968).
140. S. Rothenberg, R. H. Young, and H. F. Schaefer, III, *J. Am. Chem. Soc.*, **92**, 3243 (1970).
141. U. Gelius, C. J. Allan, P. A. Allison, H. Siegbahn, and K. Siegbahn, *Chem. Phys. Lett.*, **11**, 224 (1971); U. Gelius, E. Basilier, S. Svensson, T. Bergmark, and K. Siegbahn, *J.Electr. Spectr.*, **2**, 405 (1974).
142. R. M. Friedman, J. Hudis, and M. L. Perlman, *Phys. Rev. Lett.*, **29**, 692 (1972); P. W. Shaw and T. D. Thomas, *Phys. Rev. Lett.*, **29**, 689 (1972).
143. U. Gelius, *J. Electr. Spectr.*, **5**, 985 (1974).
144. P. H. Citrin, P. Eisenberger, and D. R. Hamann, *Phys. Rev. Lett.*, **33**, 965 (1974).
145. U. Gelius, S. Svensson, H. Siegbahn, E. Basilier, A. Faxälv, and K. Siegbahn, *Chem. Phys. Lett.*, **28**, 1 (1974).
146. E. Clementi and H. Popkie, *J. Am. Chem. Soc.*, **94**, 4057 (1972).
147. W. Domcke and L. S. Cederbaum, *Chem. Phys. Lett.*, **31**, 582 (1975).
148. W. Meyer, *J. Chem. Phys.*, **58**, 1017 (1973).
149. C. R. Brundle, M. B. Robin, and H. Basch, *J. Chem. Phys.*, **53**, 2196 (1970).

APPLICATION OF DIAGRAMMATIC QUASIDEGENERATE RSPT IN QUANTUM MOLECULAR PHYSICS

VLADIMÍR KVASNIČKA

Department of Mathematics, Faculty of Chemistry
Slovak Technical University,
880 37 Bratislava Czechoslovakia

CONTENTS

I. INTRODUCTION

At present, the many-body diagrammatic (quasi-)degenerate Rayleigh-Schrödinger perturbation theory serves as a proper up-to-date theoretical tool, starting from first principles, to study finite inhomogeneous many-particle systems. This current technique was initially elaborated in microscopic nuclear physics[1-3] for theoretical foundation and justification of effective interactions in nuclei. Immediately after its originating in nuclear physics, it was transferred and successfully applied also in quantum theory

345

of many-electron systems.[4] The present work is intended primarily to give (1) a general approach for construction of the E-independent model Hamiltonian (which term is used here as a synonym for the Rayleigh–Schrödinger perturbation theory), and (2) also to demonstrate shortly its basic applications in quantum-molecular physics.

The model-Hamiltonian theory provides an answer to the following question. Let

$$H = H_0 + H_1 \tag{1.1}$$

be an arbitrary Hamiltonian defined in the entire Hilbert space $\mathcal{H}$. The calculation of its eigensystem is one of the basic problems of many-body quantum theory. If we attempt to solve this problem using the standard Ritz variational method, we meet some difficulties originating mainly from the diagonalization of a Hermitian matrix of enormous dimensionality. An alternative method for overcoming these difficulties is an idea of a model Hamiltonian defined in a finite-dimensional model space $D_0 \subset \mathcal{H}$. In passing from the entire Hilbert space $\mathcal{H}$ to a finite-dimensional space D_0, $\mathcal{H} \rightarrow D_0$, the total Hamiltonian $H = H_0 + H_1$ should be substituted by a model Hamiltonian

$$H_{RS} = H_0 + G \tag{1.2}$$

where G is called the model interaction. We require that if the model Hamiltonian H_{RS} is diagonalized in the model space D_0, it must yield some finite part of the eigenspectrum of H in the entire Hilbert space. Conversely, when increasing the dimension of the model space in such a way that it tends to the entire Hilbert space, $D_0 \rightarrow \mathcal{H}$, then the model Hamiltonian H_{RS} should tend to the original total Hamiltonian, or in other words, the model interaction G tends to the perturbation H_1,

$$\lim_{D_0 \rightarrow \mathcal{H}} G = H_1 \tag{1.3}$$

The concept of the model Hamiltonian may be theoretically established in many different ways, but at present only a perturbation approach seems to offer it in a manageable form that is immediately suitable for further theoretical as well as numerical applications. Therefore, in this work we use the term model Hamiltonian (and model interaction) as as equivalent to the perturbation theory. The model-Hamiltonian theory should be understood here as the most general version of the perturbation theory, extended to cover degenerate or quasidegenerate unperturbed levels. The model Hamiltonian can be classified into two types: Those that are dependent on the perturbed (exact) energy E, and those that are not dependent on E. This partitioning is made in accordance with general classification of the perturbation theory: (*1*) The Brillouin–Wigner perturbation theory producing the E-dependent model Hamiltonian, and (*2*) the Rayleigh–Schrödinger perturbation theory producing the E-independent model Hamiltonian. Although the structure of the E-dependent model Hamiltonian is usually not very complicated, its dependence on E causes difficulties in the solution of the model-space eigenvalue problem. This drawback is removed by the Rayleigh–Schrödinger perturbation theory, but the E independence of the model Hamiltonian is here paid for, probably, by a slower convergence of the perturbation expansions.

We study here two versions of the E-independent model Hamiltonian. First, the *formal* model Hamiltonian without any relation to many-body systems; this theory deals mainly with an algebraic structure of the perturbation theory. Second, the *many-body* model Hamiltonian representing, de facto, a many-body diagrammatic version of the formal model Hamiltonian. These two "different" levels of the model Hamiltonian are closely related, since their algebraic structure is quite similar. Some properties of the formal model Hamiltonian may be directly transferred also to the theory of many-body model Hamiltonian. Therefore, we believe that the many-body model Hamiltonian should be formulated concurrently with the formal model Hamiltonian. This fact provides some generalized insight into the algebraic structure of the many-body model Hamiltonian.

The many-body Hamiltonian can also be very profitable when it is used in quantum-molecular physics. Here it offers a powerful tool to study many-particle systems when the degeneracy or quasidegeneracy plays a basic role. Of course, these problems may be theoretically treated by the other well-known many-body techniques, for example, by Green's functions, random-phase approximation, and so on. The many-body version of the E-independent model Hamiltonian provides an alternative way to derive a series expansion for the many-body model interaction; this expansion is formally exact, energy independent, and free of the occurrence of the disconnected diagrams. Furthermore, it is a straightforward generalization of the nondegenerate many-body perturbation theory widely used in quantum theory of atoms and molecules.

The outline of the present work is as follows: In Section II we construct the Brillouin–Wigner perturbation theory, which serves in Section III, analogously to its use in Brandow's derivation,[2] as a starting point for construction of the Rayleigh–Schrödinger perturbation theory. In Section IV we present the many-body diagrammatic construction of the E-independent model Hamiltonian. Here, we use the so-called separability theorem, initially introduced in many-body perturbation theory by Hugen-holtz[5] and Löwdin.[6] In Section V we present some simple applications illustrating the fruitfulness of the many-body E-independent model Hamil-tonian in the quantum theory of atoms and molecules.

II. BRILLOUIN–WIGNER PERTURBATION THEORY

The basic concept of the perturbation theory is that the Hamiltonian H, defined in the entire Hilbert space $\mathcal{H}$, can be written as the sum of an "unperturbed" Hamiltonian H_0 and a "perturbation" term H_1,

$$H = H_0 + H_1 \tag{2.1}$$

The perturbed and unperturbed eigenproblems have the form

$$HP_i = E_i P_i \tag{2.2a}$$

$$H_0 P_0(\alpha) = E_\alpha^{(0)} P_0(\alpha) \tag{2.2b}$$

where P_i, $P_0(\alpha)$ and E_i, $E_\alpha^{(0)}$ are perturbed and unperturbed eigenprojectors and eigenenergies, respectively. We assume that the perturbed eigenen-ergies E_i (in contrast to $E_\alpha^{(0)}$) are nondegenerate, that is, $\text{Tr}(P_i) = 1$ and $\text{Tr}[P_0(\alpha)] = d_\alpha \geqslant 1$. Let us introduce a d-dimensional *model space* $D_0 \subset \mathcal{H}$ defined by the projector P_0,

$$P_0 = \sum_{\alpha \in M_0} P_0(\alpha) \tag{2.3a}$$

$$\text{Tr}(P_0) = \sum_{\alpha \in M_0} d_\alpha = d \tag{2.3b}$$

$$\overline{H}_0 = H_0 P_0 = P_0 H_0 = \sum_{\alpha \in M_0} E_\alpha^{(0)} P_0(\alpha) \tag{2.3c}$$

where the summations run over all unperturbed states α from the set M_0.

For further considerations it is appropriate to define a projector P into a d-dimensional space $D \subset \mathcal{H}$ spanned by perturbed eigenvectors,

$$P = \sum_{i \in M} P_i \tag{2.4a}$$

$$\text{Tr}(P) = d \tag{2.4b}$$

$$HP = PH = \sum_i E_i P_i \tag{2.4c}$$

where the set M of indices is determined by

$$\lim_{H_1 \to 0} P = P_0 \tag{2.5}$$

In other words, the space D tends to the model space D_0 when the perturbation H_1 is "switched-off," or there exist d perturbed eigenenergies $\{E_i; i \in M\}$ tending to $\{E_\alpha^{(0)}; \alpha \in M_0\}$ when $H_1 \to 0$.

After these introductory remarks we proceed to construct the Brillouin–Wigner perturbation theory (BWPT).[7-9] First of all, let us define a *wave operator* $U_{BW}(E_i)$ by

$$U_{BW}(E_i)P_0 = U_{BW}(E_i) \tag{2.6a}$$

$$P_i U_{BW}(E_i) = U_{BW}(E_i) \tag{2.6b}$$

$$P_0 U_{BW}(E_i) = P_0 \tag{2.6c}$$

for $i \in M$. The last formula expresses the so-called intermediate normalization of the perturbed state vector. From the first two defining conditions, (2.6a) and (2.6b), it follows that $U_{BW}(E_i)$ maps the model space D_0 into an eigensubspace of E_i. Multiplying the eigenproblem (2.2a) from the right side by $U_{BW}(E_i)$ and using (2.6b), the following formula is obtained:

$$H U_{BW}(E_i) = E_i U_{BW}(E_i) \tag{2.7}$$

which can be rewritten according to (2.1) in the form

$$(E_i - H_0) U_{BW}(E_i) = H_1 U_{BW}(E_i) \tag{2.8}$$

This equation may be uniquely solved,

$$(1 - P_0) U_{BW}(E_i) = \frac{1 - P_0}{E_i - H_0} H_1 U_{BW}(E_i) \tag{2.9}$$

Starting from the intermediate normalization (2.6c) we obtain $U_{BW}(E_i) = P_0 + (1 - P_0)U_{BW}(E_i)$, which together with (2.9) gives a nonlinear equation determining the wave operator,

$$U_{BW}(E_i) = P_0 + \frac{1 - P_0}{E_i - H_0} H_1 U_{BW}(E_i) \tag{2.10}$$

It can be simply verified that the wave operator determined in this way satisfies the all defining conditions (2.6a) to (2.6c). An iterative solution of (2.10) offers an infinite expansion,

$$U_{BW}(E_i) = P_0 + \sum_{n=1}^{\infty} \left(\frac{1 - P_0}{E_i - H_0} H_1 \right)^n P_0 \tag{2.11}$$

Let us define the *reaction operator* $T(E_i)$,

$$T(E_i) = H_1 U_{BW}(E_i) = T(E_i) P_0 \tag{2.12}$$

$$T(E_i) = H_1 P_0 + H_1 \frac{1 - P_0}{E_i - H_0} T(E_i) \tag{2.13}$$

Similarly, an iterative solution of (2.13) gives the reaction operator in the form of infinite expansion,

$$T(E_i) = H_1 P_0 + \sum_{n=1}^{\infty} H_1 \left(\frac{1 - P_0}{E_i - H_0} H_1 \right)^n P_0 \tag{2.14}$$

Introduction of (2.12) into the right-hand side of (2.8) gives

$$(E_i - H_0) U_{BW}(E_i) = T(E_i) P_0 \tag{2.15}$$

Multiplying this equation from the right side by the projector P_0 we obtain the crucial expression of the BWPT,

$$\overline{H}_0 + G_{BW}(E_i) = E_i P_0 \tag{2.16}$$

where the intermediate normalization (2.6c) was used. The operator $\overline{H}_0$ is an inner projection of H_0 into D_0 [see (2.3c)], and G_{BW} is an E-dependent *model interaction* defined by

$$G_{BW}(E_i) = P_0 H_1 U_{BW}(E_i) P_0 = P_0 T(E_i) P_0 \tag{2.17}$$

Introduction of (2.14) into (2.17) gives the model interaction in an infinite-expansion form,

$$G_{BW}(E_i) = P_0 H_1 P_0 + \sum_{n=1}^{\infty} P_0 H_1 \left(\frac{1 - P_0}{E_i - H_0} H_1 \right)^n P_0 \tag{2.18}$$

implying that G_{BW} is a Hermitian operator,

$$G_{BW}(E_i) = G_{BW}(E_i)^{\dagger} \tag{2.19}$$

According to (2.3a) and (2.3c), the expression (2.16) may be rewritten in the form

$$(E_i - E_{\alpha}^{(0)}) P_0(\alpha) = \begin{cases} P_0(\alpha) G_{BW}(E_i) P_0 \\ P_0 G_{BW}(E_i) P_0(\alpha) \end{cases} \tag{2.20}$$

Let us define in the model space D_0 a projector $q(E_i)$, then from (2.16) we immediately obtain a model eigenproblem specifying the projector $q(E_i)$ as well as the perturbed energy E_i,

$$H_{BW}(E)q(E) = Eq(E) \qquad (2.21a)$$

$$H_{BW}(E) = \overline{H}_0 + G_{BW}(E) \qquad (2.21b)$$

where we have used the abbreviation $E = E_i$ for $i \in M$. The Hermitian operator H_{BW} is the E-dependent *model Hamiltonian*. The model eigenproblem (2.21a) can be transformed in standard form if we introduce vectors $|\psi(E)\rangle \in D_0$ by $q(E)|\psi(E)\rangle = |\psi(E)\rangle$,

$$H_{BW}(E)|\psi(E)\rangle = E|\psi(E)\rangle \qquad (2.22)$$

According to the E dependence of the model Hamiltonian, the model eigenproblem (2.22) should be solved by an iterative technique. For instance, let us assume that E is a trial estimation of some perturbed energy from $\{E_i; i \in M\}$. Using (2.18) and (2.21) we calculate the model Hamiltonian, and on diagonalizing it we obtain a new estimation for the desired energy, and so on. The eigenvectors $\{|\psi(E_i)\rangle; i \in M\}$ are generally nonorthogonal, since the model Hamiltonian depends on the energy E.

A. Diagrammatic Interpretation

In order to visualize the BWPT we introduce a formal diagrammatic method.[10,11] It should be noted that this diagrammatic technique is here used only for formal (in contrast to many-body) perturbation theory, and therefore it cannot be confused with the many-body diagrammatic approaches,[12] although some common features do exist.

We first introduce the basic concepts of the present formal diagrammatic technique. The used vertices have the following diagrammatic interpretation:

$$\tau \!-\!\!\oslash\!\!-\!\omega = \bigvee\oslash = \cdots\cdots = \omega\!-\!\!\oslash\!\!-\!\tau = P_0(\tau)H_1P_0(\omega) \qquad (2.23a)$$

$$\alpha'\!\!\diagup\!\!O^{\diagup\alpha} = P_0(\alpha)GP_0(\alpha') \qquad (2.23b)$$

$$\overset{\mu}{\diagdown}\!\!\boxslash = P_0(\mu)UP_0(\alpha) \qquad (2.23c)$$

$$\boxslash\!\!\overset{\alpha'}{\diagdown}_{\mu'} = P_0(\alpha')U^+P_0(\mu') \qquad (2.23d)$$

where $\alpha, \alpha' \in M_0$, $\mu, \mu' \notin M_0$, and ω, τ are arbitrary indices. We use the convention that lines going from the left (right) to right (left) side correspond to the unperturbed states $\alpha, \alpha', \ldots \in M_0(\mu, \mu', \ldots \notin M_0)$, and they are called the *positive (negative) lines*.

At this point we interrupt our successive introduction of concepts and rules of the present formal diagrammatic technique and turn our attention to concrete examples. Let us introduce for the wave operator and model interaction the following notation [see (2.11) and (2.18)]:

$$U = U_{BW}(E) = \sum_{n=0}^{\infty} U^{[n]} \tag{2.24a}$$

$$G = G_{BW}(E) = \sum_{n=1}^{\infty} G^{[n]} \tag{2.24b}$$

$$U^{[n]} = \left(\frac{1 - P_0}{E - H_0} H_1 \right)^n P_0 \tag{2.24c}$$

$$G^{[n]} = P_0 H_1 \left(\frac{1 - P_0}{E - H_0} H_1 \right)^{n-1} P_0, \tag{2.24d}$$

where $U^{[0]} = P_0$ and $G^{[1]} = P_0 H_1 P_0$. The term $U^{[n]}$ for $n \geqslant 1$ can be "drawn" as follows:

$$U^{[1]} = \quad\text{} \tag{2.25a}$$

$$U^{[2]} = \quad\text{} \tag{2.25b}$$

$$U^{[3]} = \quad\text{} \tag{2.25c}$$

and generally,

$$U^{[n]} = \quad\text{} \tag{2.25d}$$

where the negative double lines correspond to energy denominator $(1 - P_0)/(E - H_0)$. The diagrammatic rules with help of which we may uniquely assign to any diagram its operator counterpart can be summarized as follows:

1. The lines of the given diagram are indexed following the convention introduced above, that is, the positive (negative) lines are indexed by indices $\alpha, \alpha', \ldots \in M_0$ ($\mu, \mu', \ldots \notin M_0$). For instance, the diagram from the right-hand side of (2.25b) is indexed by

$$(2.26)$$

2. The energy denominators are constructed in such a way that to each double negative line μ we assign the expression

$$\left(E - E_\mu^{(0)}\right)^{-1} \tag{2.27}$$

For instance, for (2.26) we have

$$\left(E - E_\mu^{(0)}\right)^{-1}\left(E - E_{\mu'}^{(0)}\right)^{-1} \tag{2.28}$$

3. The numerator (operator-valued entity) is constructed from the given diagram if we assign to each vertex its algebraic interpretation [see (2.23a) to (2.23d)]. Here, the double lines are treated as the single lines. The assignment is made step by step starting from the vertex having the external outgoing line [in (2.26) indexed by μ], and we continue in this process going from the one vertex to another one joined with the previous vertex by line. For instance, the diagram (2.26) has the numerator

$$P_0(\mu)H_1 P_0(\mu')H_1 P_0(\alpha) \tag{2.29}$$

4. Summing up over all lines.

Using these rules, the operator counterpart of the diagram (2.26) is equal to

$$\sum_{\alpha \in M_0} \sum_{\mu, \mu' \notin M_0} \frac{P_0(\mu)H_1 P_0(\mu')H_1 P_0(\alpha)}{\left(E - E_\mu^{(0)}\right)\left(E - E_{\mu'}^{(0)}\right)} \tag{2.30}$$

which can be directly rewritten in the standard form identical with $U^{[2]}$, when the spectral-resolution formula

$$\frac{1 - P_0}{E - H_0} = \sum_{\mu \notin M_0} \frac{P_0(\mu)}{E - E_\mu^{(0)}} \tag{2.31}$$

is applied.

The formal diagrammatic expression for the wave operator may be now presented as

$$U = P_0 + \text{[diagram]}$$ (2.32)

where the rectangular block represents [see (2.23c)] the operator $(1 - P_0)U$. From (2.10) after the first iteration we obtain

$$(1 - P_0)U = \frac{1 - P_0}{E - H_0} H_1 P_0 + \frac{1 - P_0}{E - H_0} H_1 \frac{1 - P_0}{E - H_0} U \qquad (2.33)$$

for which the diagrammatic picture

$$\text{[diagram]} = \text{[diagram]} + \text{[diagram]} \qquad (2.34)$$

serves us as the defining equation for the rectangular vertices. The Hermitian conjugate form of (2.32) may be "drawn" in this form:

$$U^+ = P_0 + \text{[diagram]} \qquad (2.35)$$

where the rectangular block from the right-hand side is defined by

$$\text{[diagram]} = \text{[diagram]} + \text{[diagram]} \qquad (2.36)$$

The diagrammatic interpretation of the model interaction and its individual terms $G^{[n]}$ can be introduced in the following form:

$$G - \text{[diagram]} = G^{[1]} + G^{[2]} + G^{[3]} + \cdots \qquad (2.37a)$$

$$G^{[1]} = \text{[diagram]} \qquad (2.37b)$$

$$G^{[2]} = \text{[diagram]} \qquad (2.37c)$$

$$G^{[3]} = \text{[diagram]} \qquad (2.37d)$$

and generally,

$$G^{[n]} = \text{[diagram]} \left(\text{[diagram]} \right)^{n-2} \text{[diagram]} \qquad (2.37e)$$

It follows from (2.17) that the model interaction is determined by $G = P_0 H_1 U P_0$, which may be simply rewritten as $G = P_0 H_1 P_0 + P_0 H_1 (1 - P_0) U P_0$, or diagrammatically

$$\tag{2.38a}$$

Furthermore, the model interaction can also be diagrammatically expressed in the following two alternative ways:

$$\tag{2.38b}$$

$$\tag{2.38c}$$

which can easily be verified by (2.34) and (2.36). Thus we have obtained three alternative diagrammatic expressions for the E-dependent model interaction, their operator are given by (2.17) and the following two formulas:

$$G = \begin{cases} P_0 H_1 P_0 + P_0 U^+ (1 - P_0) H_1 P_0 \\ P_0 H_1 P_0 + P_0 H_1 \dfrac{1 - P_0}{E - H_0} H_1 P_0 + P_0 U^+ (1 - P_0) H_1 (1 - P_0) U P_0 \end{cases}$$

$$\tag{2.39}$$

III. RAYLEIGH–SCHRÖDINGER PERTURBATION THEORY

The essence of the Rayleigh–Schrödinger perturbation theory (RSPT) is to get rid of the dependence on the exact energy, which characterize the BWPT. There are a number of ways of doing this. The standard method is to formally expand energy E and the corresponding state vector $|\Psi\rangle$ in powers of H_1, substitute in the Schrödinger equation, and equate terms of the same order. Although this method is widely used in standard textbooks, it is tedious, and it fails to provide any insight into general algebraic structure of the RSPT.

Another method[2,11] is to start from the BWPT, and expand the E-dependent model interaction $G = G_{BW}(E)$ as well as the wave operator $U = U_{BW}(E)$ in power series of $(E - E_\alpha^{(0)})$. This approach has been initially systematically used by Brandow[2] in his construction of the many-body degenerate RSPT.

In our further considerations we use the conventions (2.24b) and (2.24d),

$$G = G_{BW}(E) = \sum_{n=1}^{\infty} G^{[n]} \tag{3.1a}$$

$$G^{[n]} = P_0 H_1 \left(\frac{1 - P_0}{E - H_0} H_1 \right)^{n-1} P_0 \tag{3.1b}$$

where for simplicity the energy E is not explicitly displayed. The first term of (3.1a)

$$G^{[1]} = P_0 H_1 P_0 \tag{3.2}$$

is the E-independent operator. The E dependence is starting from the second term of (3.1a),

$$G^{[2]} = P_0 H_1 \frac{1 - P_0}{E - H_0} H_1 P_0 \tag{3.3}$$

on which we shall demonstrate the general features and properties of the present E-removing procedure. According to (2.3a) the term $G^{[2]}$ can be written as

$$G^{[2]} = \sum_{\alpha \in M_0} P_0 H_1 \frac{1 - P_0}{\left(E_\alpha^{(0)} - H_0 \right) + \left(E - E_\alpha^{(0)} \right)} H_1 P_0(\alpha) \tag{3.4}$$

where in the denominator we have simultaneously added and subtracted the unperturbed energy $E_\alpha^{(0)}$ for $\alpha \in M_0$. Applying the operator identity $(A + B)^{-1} = A^{-1} - A^{-1} B (A + B)^{-1}$ and (2.20), the expression (3.4) can be transformed to

$$G^{[2]} = \sum_{\alpha \in M_0} P_0 H_1 \frac{1 - P_0}{E_\alpha^{(0)} - H_0} H_1 P_0(\alpha)$$

$$- \sum_{\alpha \in M_0} P_0 H_1 \frac{1 - P_0}{\left(E_\alpha^{(0)} - H_0 \right)(E - H_0)} H_1 P_0(\alpha) G P_0 \tag{3.5}$$

The E dependence in the first term is completely removed, but the second term still contains a product of E-dependent and E-independent energy denominators. This last term may be transformed by a similar procedure,

$$- \sum_{\alpha \in M_0} P_0 H_1 \frac{1 - P_0}{\left(E_\alpha^{(0)} - H_0\right)\left(E - H_0\right)} H_1 P_0(\alpha) G P_0$$

$$= - \sum_{\alpha, \alpha' \in M_0} P_0 H_1 \frac{1 - P_0}{\left(E_\alpha^{(0)} - H_0\right)\left(E_{\alpha'}^{(0)} - H_0\right)} H_1 P_0(\alpha) G P_0(\alpha')$$

$$+ \sum_{\alpha, \alpha' \in M_0} P_0 H_1 \frac{1 - P_0}{\left(E_\alpha^{(0)} - H_0\right)\left(E_{\alpha'}^{(0)} - H_0\right)\left(E - H_0\right)} H_1 P_0(\alpha) G P_0(\alpha') G P_0$$

$$(3.6)$$

Here, analogously to the previous step (3.5), we have some freedom in selection of an unperturbed state with respect to which the E-removing procedure is carried out. Indeed, the term $G^{[2]}$ may be alternatively expressed by

$$G^{[2]} = \sum_{\alpha \in M_0} P_0(\alpha) H_1 \frac{1 - P_0}{E_\alpha^{(0)} - H_0} H_1 P_0$$

$$- \sum_{\alpha, \alpha' \in M_0} P_0 G P_0(\alpha) H_1 \frac{1 - P_0}{\left(E_\alpha^{(0)} - H_0\right)\left(E - H_0\right)} H_1 P_0 \qquad (3.7)$$

These two formulas (3.5) and (3.7) can be unified,

$$G^{[2]} = \sum_{\alpha \in M_0} \left[\lambda P_0 H_1 \frac{1 - P_0}{E_\alpha^{(0)} - H_0} H_1 P_0(\alpha) + (1 - \lambda) \text{H.c.} \right]$$

$$- \sum_{\alpha \in M_0} \left[\lambda P_0 H_1 \frac{1 - P_0}{\left(E_\alpha^{(0)} - H_0\right)\left(E - H_0\right)} H_1 P_0(\alpha) G P_0 + (1 - \lambda) \text{H.c.} \right] \qquad (3.8)$$

where H.c. means the Hermitian conjugate term, and λ is a factor taken $0 \leqslant \lambda \leqslant 1$. For $\lambda = 0$ and $\lambda = 1$ this last expression is equivalent to (3.7) and (3.5), respectively. For $\lambda = \frac{1}{2}$ it offers the operator $G^{[2]}$ in manifestly Hermitian form. This important general feature of the E-removing procedure can be kept also in the next E-removing steps by the above-indicated proper selection of the unperturbed states and the factor λ. Exactly the same method is also applicable for treating the terms $G^{[n]}$, $n \geqslant 3$.

In order to avoid complications arising from lengthy algebraic formulas, in further considerations we use the formal diagrammatic technique introduced in Section II.A in a slightly enlarged form. The diagrammatic rules 1 to 4 are completed by the following two rules:

5. The *sign* of the given diagram is equal to $(-1)^p$, where p is the total number of *internal* (i.e., running between two vertices) positive lines.

6. The energy denominators are constructed in such a way that to each *cut* (vertical line cutting one positive α line and one negative line μ) we assign the energy denominator

$$\left(E_\alpha^{(0)} - E_\mu^{(0)}\right)^{-1} \tag{3.9}$$

where $\alpha \in M_0$ and $\mu \notin M_0$.

In the framework of this formal diagrammatic technique, manipulations with complicated algebraic expressions are replaced by the "topological" considerations with diagrams. For instance, (3.5) and (3.6) may be "drawn" as follows:

$$\tag{3.10}$$

Similarly, the diagrammatic counterparts of (3.7) and (3.8) are

$$\tag{3.11a}$$

$$\tag{3.11b}$$

The *E*-removing procedure, illustrated by these simple examples, can be generalized in the following two alternative ways:

$$\tag{3.12a}$$

$$\tag{3.12b}$$

where the index $\alpha \in M_0$ is a preselected unperturbed state (called the *reference state*) from the model space D_0, with respect to which the E-removing procedure is carried out. The reference state α may be selected in any possible way from the positive lines of the given diagram. This freedom makes our E-removing procedure very flexible; by the proper successive selection of the reference states and parameters λ, we are able to construct an infinite class of the E-independent model interactions. For example, setting $\lambda = 1$ and assuming that the reference states are always external incoming positive lines, the expression (3.11b) may be expanded as follows:

$$G^{[2]} = \text{(diagram)} + \text{(diagram)} + \text{(diagram)} + \cdots \tag{3.13}$$

which is, de facto, equivalent to the right-hand side of (3.10), when the incoming external line was always used as the reference state. Similarly, from (3.11b) we obtain the manifestly Hermitian expression

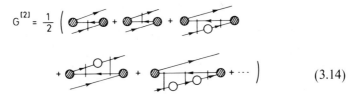

$$\tag{3.14}$$

where we set $\lambda = \frac{1}{2}$ and the reference states were selected either as incoming or outgoing external lines, which is already indicated in (3.11b).

Now, we can summarize some typical general features of the present E-removing procedure:

(*i*) The factor λ is kept fixed during all successive steps, and (*ii*) the cuts (or the successive selection of the reference states) may be done by an arbitrary way, while their "system" is determined by (3.12a) and (3.12b). This "system" of cuts for a diagram with p shaded and q unshaded vertices depends on the "system" of cuts of the preceding diagram with p shaded and $(q-1)$ unshaded vertices, genealogically joined with the first one by the E-removing procedure (3.12a) or (3.12b).

Then, the E-independent form of the $G^{[2]}$ may be written in the following formal way:

$$G^{[2]} = (R^{[2]})_{\Gamma_0} + (R^{[2]} \cdot G)_{\Gamma_1} + (R^{[2]} \cdot G \cdot G)_{\Gamma_2}$$

$$+ \cdots + (R^{[n]} \cdot G \cdot \cdots \cdot G)_{\Gamma_n} + \cdots \tag{3.15}$$

where $(R^{[2]})_{\Gamma_0}$ expresses those E-independent terms of $G^{[2]}$ which contain only two shaded vertices, and the "system" of cuts and the factor λ are abbreviated by the subscript Γ_0. Generally, the term $(R^{[n]} \cdot G \cdots \cdot G)_{\Gamma_n}$ expresses the diagrams with two shaded and n unshaded vertices, and the subscript Γ_n describes the "system" of cuts and the used factor λ. However, this subscript should depend on the subscript Γ_{n-1}; see (i) and (ii) above.

Up to now we have assumed that during all E-removing steps only one factor λ was used. This approach may be simply generalized by the following way: Let us have a diagram formally denoted by $(D)_\Gamma$, corresponding to a term with simultaneous appearance of E-dependent and E-independent energy denominators, and which was obtained from the $G^{[2]}$ after n applications of the E-removing procedure. That is, this diagram has one double line, n unshaded vertices, and $(n-1)$ internal positive lines; see, for example, the last diagram in (3.10). The subscript Γ describes the "system" of cuts and the used factor λ. As in (3.8), we may write

$$(D)_\Gamma = \lambda'(D)_\Gamma + (1-\lambda')(D)_\Gamma \tag{3.16}$$

where λ' is a new λ factor. Then, applying different E-removing procedures to the terms $\lambda'(D)_\Gamma$ and $(1-\lambda')(D)_\gamma$ in (3.16), we obtain

$$(D)_\Gamma = \lambda'(D_{10})_{\Gamma_{10}} + (1-\lambda')(D_{20})_{\Gamma_{20}}$$
$$+ \lambda'(D_{11})_{\Gamma_{11}} + (1-\lambda')(D_{21})_{\Gamma_{21}} \tag{3.17}$$

where $(D_{10})_{\Gamma_{10}}$ and $(D_{20})_{\Gamma_{20}}$ are E-independent diagrams (the negative double line was removed) with two shaded and n unshaded vertices and the subscripts Γ_{10} and Γ_{20} describe the "system" of cuts composed from the Γ and new cuts from the last E-removing step. The diagrams $(D_{11})_{\Gamma_{11}}$ and $(D_{21})_{\Gamma_{21}}$ with $(n+1)$ unshaded vertices have the simultaneous appearance of the E-dependent and E-independent energy denominators, and their "systems" of cuts (described by Γ_{11} and Γ_{21}) are topologically quite identical with the previous ones of the diagrams $(D_{10})_{\Gamma_{10}}$ and $(D_{20})_{\Gamma_{20}}$. Generally, the approach (3.17) can be used either for all or for some preselected E-removing steps; that is, we have a branching process when the E-independent terms of original E-dependent model interaction are constructed.

For a better understanding of the present diagrammatic method it is instructive to consider the "higher-order" terms from (3.1). For instance, expanding the second double line in the diagram (2.37d) we get:

$$\tag{3.18a}$$

or alternatively,

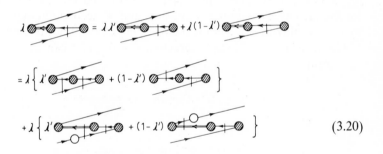

$$G^{[3]} = \quad = \quad + \qquad (3.18b)$$

Similarly, these two alternative diagrammatic formulas may be unified:

$$G^{[3]} = \lambda \quad + (1-\lambda) \quad$$

$$= \lambda \quad + (1-\lambda) \quad$$

$$+ \lambda \quad + (1-\lambda) \qquad (3.19)$$

The first diagram A from the right-hand side of (3.19) in the next E-removing step gives

$$\lambda \quad = \lambda\lambda' \quad + \lambda(1-\lambda') \quad$$

$$= \lambda \left\{ \lambda' \quad + (1-\lambda') \quad \right\}$$

$$+ \lambda \left\{ \lambda' \quad + (1-\lambda') \quad \right\} \qquad (3.20)$$

and analogously for other remaining diagrams in (3.19).

Generally, an approach completely analogous to the one above is applicable to all terms $G^{[n]}$, $n \geqslant 2$. Then the E-independent form of the model interaction G can be formally written as follows:

$$G = P_0 H_1 P_0 + (R)_{\Gamma_0} + (R \cdot G)_{\Gamma_1} + \cdots + (R \cdot G \cdots \cdot G)_{\Gamma_n} + \cdots \qquad (3.21)$$

where $(R)_{\Gamma_0}$ expresses the E-independent terms of $G^{[2]}, G^{[3]},\ldots$ containing only shaded vertices; this algebraic structure is uniquely determined by the "systems" of cuts and λ factors abbreviated by Γ_0. The nth term $(R \cdot G \cdots \cdot G)_{\Gamma_n}$ expresses the diagrams with n unshaded vertices; this algebraic structure is determined by Γ_n.

A final construction of the E-independent model interaction determined by the formal infinite expansion (3.21) can be done by an iterative method. For instance, an initial step of the model interaction can be chosen as follows:

$$G \stackrel{\circ}{=} G^{(0)} = P_0 H_1 P_0 + (R)_{\Gamma_0} \qquad (3.22)$$

that is, by all the diagrams without unshaded vertices. Substitution of (3.22) in the right-hand side of (3.21) gives the first step, $G^{(1)}$, of this simple iterative scheme, and so on. Finally, the E-independent model interaction containing merely *shaded* (perturbation) vertices is determined by

$$G = \lim_{k \to \infty} G^{(k)} \qquad (3.23)$$

where $G^{(k)}$ is the kth step of this iterative scheme.

A.   Simple Forms of the E-Independent Model Interaction

Let us assume, for simplicity, that (i) the λ factor is kept fixed during all E-removing steps, and (ii) the reference states are always either outgoing or incoming lines. Then the E-removing formula (3.12a) may be applied for the incoming external positive line, and its second alternative possibility (3.12b) may be applied for the outgoing external positive line.

The E-dependent model interaction $G = G_{\mathrm{BW}}(E)$ can be "drawn" in a split form,

$$+ (1-\lambda) \left(\ldots \right) \qquad (3.24)$$

for $0 \leqslant \lambda \leqslant 1$. Now we adopt the convention introduced above, that the reference states are always either incoming or outgoing external lines. In particular, we require that in the case of all diagrams with factor λ the reference states are merely the incoming external positive lines, and the diagrams with factor $(1-\lambda)$ use the reference states merely the outgoing external positive lines.

This idea was, for instance, used in (3.11b). Since the factor λ is kept fixed during all E-removing steps, the initial splitting of the model interaction into two "quasi-" independent series is preserved. Then the formal diagrammatic expression for the E-independent model interaction (depending on λ) can be formally written as follows:

$$G(\lambda) = \lambda \mathcal{G}(\lambda) + (1-\lambda) \mathcal{G}^+(\lambda) \qquad (3.25)$$

where the operators $\mathcal{G}(\lambda)$ and $\mathcal{G}^+(\lambda)$ represent all the possible diagrams with *uncut* outgoing and incoming, respectively, external positive lines. Diagrammatically,

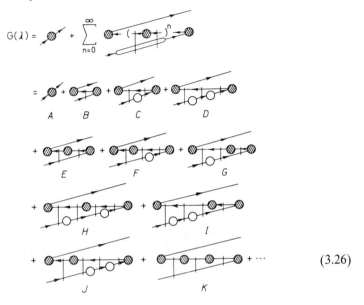

$$(3.26)$$

and similarly for the $\mathcal{G}^+(\lambda)$,

$$(3.27)$$

where we have presented, for simplicity, only a few first terms. In order to illustrate the formal expression (3.21) we relate its terms to the individual diagrams from (3.26): The diagrams A, B, E, and K belong to $(R)_{\Gamma_0}$, the diagrams C, F, and G belong to $(R \cdot G)_{\Gamma_1}$, and the diagrams D, H, I, and J belong to $(R \cdot G \cdot G)_{\Gamma_2}$.

Let us now distinguish three special cases of the factor λ, $\lambda = 1$, 0, and $\frac{1}{2}$.

Case A. $\lambda = 1$. Here the general expression (3.25) gives the model interaction in the following simple form:

$$G = G(1) = \mathcal{G}(1) \qquad (3.28)$$

where the operator $\mathcal{G}(1)$ is determined by the diagrammatic expression (3.26), the topological structure of which is invariant with respect to λ.

Finally, its formal counterpart, containing only perturbation shaded vertices, is constructed with the help of iterative method (3.22) and (3.23). Then the model interaction, expressed up to the fourth order, can be completely obtained after the first iterative step,

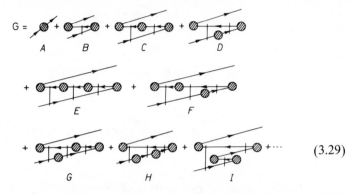

$$\tag{3.29}$$

where it is very instructive to assign a correspondence between these diagrams and the diagrams from (3.26). The diagrams A, B, C, and E are simply contained in $G^{(0)}$ [see (3.22)], that is, they are given by the zeroth iterative step. Following Brandow,[2] these "linear" diagrams are classified as the unfolded diagrams, the next iterative steps then give only folded diagrams. For instance, the folded diagrams D, F, G, and H are given by the first iterative step, when the unshaded vertices in the diagrams C, F, G, and D in (3.26) were simply changed by the shaded (perturbation) vertex. Especially, the diagram I from (3.29) is given by the first iterative step when the unshaded vertex in diagram C of (3.26) is changed by the total diagram B from (3.26).

The algebraic counterpart of the diagrammatic expression (3.29) can be simply obtained by using the diagrammatic rules 1 to 6 presented in Sections II.A and III. For instance, the model interaction expressed up to the third order (i.e., diagrams A, B, C, and D) from (3.29) is determined in algebraic form as follows:[13, 14]

$$G = P_0 H_1 P_0 + \sum_{\alpha \in M_0} P_0 H_1 \frac{1 - P_0}{E_\alpha^{(0)} - H_0} H_1 P_0 (\alpha)$$

$$+ \sum_{\alpha \in M_0} P_0 H_1 \frac{1 - P_0}{E_\alpha^{(0)} - H_0} H_1 \frac{1 - P_0}{E_\alpha^{(0)} - H_0} H_1 P_0 (\alpha)$$

$$- \sum_{\alpha, \alpha' \in M_0} P_0 H_1 \frac{1 - P_0}{\left(E_\alpha^{(0)} - H_0\right)\left(E_{\alpha'}^{(0)} - H_0\right)} H_1 P_0 (\alpha') H_1 P_0 (\alpha) \tag{3.30}$$

The model interaction determined by the expressions (3.28) and (3.29) is a non-Hermitian operator, and it is quite identical with the Bloch[15] model interaction extended for quasidegenerate perturbation theory.[13,14] This type of the model interaction is the main essence of Brandow's folded-diagram theory.[2] Recently, Kiselev,[16] and independently, following a different philosophy, the present author[10] suggested the direct diagrammatic construction of the non-Hermitian model interaction. In this latter approach the nth-order contributions of the model interaction are directly constructed by thhe diagrammatic technique without necessity to use any iterative technique.

Case B. $\lambda = 0$. We shall not study this particular case since the present model interaction,

$$G = G(0) = \mathcal{G}^+(0) \tag{3.31}$$

is simply related via the Hermitian conjugation,

$$G = G^+(1) \tag{3.32}$$

to the non-Hermitian model interaction studied just above. This model interaction can be obtained from (3.29) by carrying out the Hermitian-conjugation operation over all diagrams, which may be diagrammatically simply realized by (i) counterclockwise rotation about 180° of all diagrams, and (ii) by reverse reorientation of all lines. For instance, the present non-Hermitian model interaction, expressed up to the third order, can be "drawn" as follows:

$$G = \quad \text{(3.33)}$$

Its algebraic interpretation is simply constructed from (3.30) by the Hermitian conjugation.

Case C. $\lambda = \frac{1}{2}$. For this special intermediate value of the factor the model interaction is a manifestly Hermitian operator and can be written in the following form:

$$G = G\left(\tfrac{1}{2}\right) = \tfrac{1}{2}(\mathcal{G} + \mathcal{G}^+) \tag{3.34}$$

where $\mathcal{G} = \mathcal{G}\left(\frac{1}{2}\right)$ and $\mathcal{G}^+ = \mathcal{G}^+\left(\frac{1}{2}\right)$. Following (3.26), the operator $\mathcal{G}$ from (3.34) is formally determined by the same diagrammatic expression as the non-Hermitian model interaction (3.28), but now the unshaded vertices represent the model interaction defined by (3.34). Therefore, the operator $\mathcal{G}$ from (3.34) and the non-Hermitian model interaction (3.28) are not exactly the same operators; that is, the expression (3.34) does not represent a simple symmetrization of the non-Hermitian model interaction. The diagrammatic expansion containing merely shaded (perturbation) vertices can be constructed by the iterative procedure (3.22) and (3.23). The resulting operator $\mathcal{G}$ is up to the third order identical with the non-Hermitian model interaction (3.29). Only in the fourth order do the first differences emerge; in particular, diagram I from (3.29) is now split onto two diagrams,

$$\frac{1}{2} \left(\qquad + \qquad \right) \qquad (3.35)$$

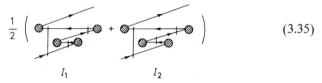

$$I_1 \qquad\qquad I_2$$

For the degenerate RSPT, where $P_0 = P_0(\alpha)$ for some fixed α, this splitting is removed, and the non-Hermitian model interaction (3.29) and the operator $\mathcal{G}$ are now identical up to the fourth order. Therefore we may assume that the operator $\mathcal{G}$ from (3.34) is well approximated by the Bloch non-Hermitian model interaction (3.28), now for uniqueness denoted by $G^{(B)}$,

$$\mathcal{G} \triangleq G^{(B)} \qquad (3.36)$$

Then the Hermitian model interaction is approximately determined by

$$G \triangleq \tfrac{1}{2}(G^{(B)} + G^{(B)+}) \qquad (3.37)$$

This formula offers a very simple possibility for constructing an approximate Hermitian model interaction. For the nondegenerate RSPT, that is, when $\mathrm{Tr}(P_0) = 1$, the expression (3.37) is exact.[18]

B.   Diagrammatic Construction of the Wave Operator

To complete the present diagrammatic construction of the formal RSPT it is necessary also to consider the diagrammatic construction of the E-independent wave operator. Similarly to the E-independent model interaction, there exists here some freedom in the construction of the E-independent wave operator, namely, the reference states from the model space may be selected in an arbitrary way. In our further considerations we do not elaborate the E-independent wave operator in its full generality, since the flexibility of the E-removing procedure was sufficiently illustrated for the model interaction. Therefore, we focus our attention on the simpler case in which the reference states are always the incoming external positive lines, in accordance with the convention used in Section III.A for the construction of model interactions with simple and transparent algebraic structure.

The E-dependent wave operator is determined by (2.24a) and (2.24c),

$$U = U_{BW}(E) = \sum_{n=0}^{\infty} U^{[n]} \qquad (3.38a)$$

$$U^{[n]} = \left(\frac{1-P_0}{E-H_0} H_1 \right)^{n-1} P_0 \qquad (3.38b)$$

or diagrammatically,

$$U^{[n]} = \ast\; (\;{=}\!\!\leftarrow\!\!\oslash\!\!\leftarrow\;)^{n-1}\!\!{=}\!\!\oslash \qquad (3.38c)$$

Applying the E-removing procedure (3.13a) together with the convention introduced above, we obtain

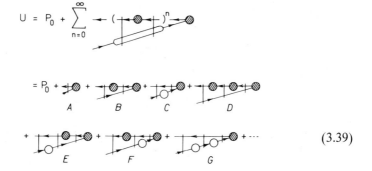

$$\qquad (3.39)$$

Generally, the E-independent wave operator may be formally written in the following form [cf. (3.21)]:

$$U = P_0 + (S_0) + (S_1 \cdot G) + (S_2 \cdot G \cdot G) + \cdots \qquad (3.40)$$

where the term (S_0) represents the E-independent terms of U containing only shaded vertices, that is, the diagrams A, B, and D in (3.39), and the term $(S_1 \cdot G)$ represents those contributions of U containing only one unshaded vertex, that is, the diagrams C, E, and F in (3.39), and so on.

The unshaded vertices from (3.39), representing the E-independent model interaction constructed in Section III.A, should be distinguished for the two alternative cases: The model interaction is (i) the non-Hermitian operator (3.28), or (ii) the Hermitian operator (3.34). In the first case the wave operator has, up to the third order, the following form:

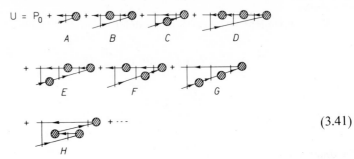

$$(3.41)$$

This is directly obtained from (3.39) by simple substitution of the un-shaded vertices by the shaded vertices, except that diagram H is constructed from diagram C in (3.39) when the unshaded vertex is changed by the second-order model interaction expressed by diagram B in (3.26). The algebraic interpretation of (3.41) is simply obtained by using the rules 1 to 6 presented in Sections II.A and III. For instance, the wave operator up to the second order is expressed by[13,14]

$$U = P_0 + \sum_{\alpha \in M_0} \frac{1 - P_0}{E_\alpha^{(0)} - H_0} H_1 P_0(\alpha)$$

$$+ \sum_{\alpha \in M_0} \frac{1 - P_0}{E_\alpha^{(0)} - H_0} H_1 \frac{1 - P_0}{E_\alpha^{(0)} - H_0} H_1 P_0(\alpha)$$

$$- \sum_{\alpha, \alpha' \in M_0} \frac{1 - P_0}{\left(E_\alpha^{(0)} - H_0\right)\left(E_{\alpha'}^{(0)} - H_0\right)} H_1 P_0(\alpha') H_1 P_0(\alpha) \qquad (3.42)$$

The wave operator determined by (3.41) is, de facto, the Bloch[15] wave operator generalized for quasidegenerate perturbation theory. Similarly to the method used for the non-Hermitian model interaction (3.28), the present wave operator can be constructed by direct formal diagrammatic technique.[16]

Now we turn our attention to the general expressions (2.38a) and (2.38b) determining the E-dependent model interaction by two alternative ways:

$$G = \begin{cases} P_0 H_1 U P_0 & (3.43a) \\ P_0 U^+ H_1 P_0 & (3.43b) \end{cases}$$

These formulas are also fulfilled in the present case of the non-Hermitian model interaction. Let us assume that a wave operator appearing in both cases (3.43a) and (3.43b) is determined by the diagrammatic expansion (3.41). Then, (*i*) the non-Hermitian model interaction (3.28) is determined by the first formula, (3.43a), and (*ii*) the non-Hermitian model interaction (3.31) is determined by the second formula, (3.43b). Generally, the non-Hermitian E-independent model interaction (3.28) may be written as follows:

$$G = P_0 H_1 U P_0 \tag{3.44}$$

where the wave operator is determined by (3.41).

The general expansion formula (3.39) for the wave operator is now studied for the case of the Hermitian model interaction (3.34). Introduction of the diagrammatic expansion for this type of the model interaction into (3.39) gives the diagrammatic expansion for the wave operator containing only perturbation vertices. The resulting expression is up to the third order almost identical with the previous one, (3.41), but now the diagram H is split onto two diagrams,

$$\frac{1}{2} \left(\quad \underset{H_1}{\text{}} \quad + \quad \underset{H_2}{\text{}} \quad \right) \tag{3.45}$$

This observation is closely related to (3.35) obtained for the Hermitian model interaction. As in (3.44), we may now express the Hermitian model interaction (3.34) in the following manifestly Hermitian form [see (3.34)]:

$$G = \tfrac{1}{2}(\mathcal{G} + \mathcal{G}^+) \tag{3.46a}$$

$$\mathcal{G} = P_0 H_1 U P_0 \tag{3.46b}$$

where the wave operator is determined by (3.39), when the unshaded vertices are changed by this type of the Hermitian model interaction.

To conclude this section, we now suggest an algebraic approach unifying the above three alternative cases of the model interaction. The general diagrammatic expression (3.39) can be algebraically expressed by[13, 14, 19] (see also Section III.C)

$$UP_0 = P_0 + \sum_{\alpha \in M_0} \frac{1 - P_0}{E_\alpha^{(0)} - H_0} (H_1 U - UG) P_0(\alpha) \tag{3.47}$$

Iterative solution of this expression gives just the wave operator in the form of (3.39). A simple "formalistic" proof of (3.47) can be presented as follows: Let us start from (2.10) determining the E-dependent wave operator; it may be rewritten in another form:

$$UP_0 = P_0 + \sum_{\alpha \in M_0} \frac{1 - P_0}{\left(E_\alpha^{(0)} - H_0\right) + \left(E - E_\alpha^{(0)}\right)} H_1 UP_0(\alpha) \qquad (3.48)$$

Applying the E-removing procedure, this expression becomes

$$UP_0 = P_0 + \sum_{\alpha \in M_0} \frac{1 - P_0}{E_\alpha^{(0)} - H_0} H_1 UP_0(\alpha)$$

$$- \sum_{\alpha \in M_0} \frac{1 - P_0}{\left(E_\alpha^{(0)} - H_0\right)\left(E - H_0\right)} H_1 UGP_0(\alpha) \qquad (3.49)$$

which is identical with (3.47), since $H_1 U = (H - H_0)U = (E - H_0)U$. Now, following the general idea (3.25), the E-independent model interaction can be introduced by

$$G = \lambda \mathcal{G} + (1 - \lambda)\mathcal{G}^+ \qquad (3.50a)$$

$$\mathcal{G} = P_0 H_1 UP_0 \qquad (3.50b)$$

for $0 \leqslant \lambda \leqslant 1$, and the wave operator is determined by (3.47). Thus we have obtained a system of coupled equations. The first, (3.47), is of a nonlinear nature and serves as an iterative determination of the wave operator. The model interaction is defined by the equations (3.50a) and (3.50b) via the wave operator. In (3.50a) we have used also the λ-factor approach; for $\lambda = 0$ and $\lambda = 1$ we obtain the non-Hermitian model interactions, which are mutually related by Hermitian conjugation. For $\lambda = \frac{1}{2}$ we obtain the Hermitian model interaction (3.46).

C. Algebraic Theory of RSPT

In the preceding sections we have used the formal diagrammatic approach for construction of the E-independent model interaction as well as the wave operator. We now outline its direct algebraic construction, where no reference to the BWPT is necessary. Let us start from the introductory mathematical preliminaries in the beginning of Section II. In the entire Hilbert space we have two eigenproblems,

$$H|\Psi_i\rangle = E_i|\Psi_i\rangle \qquad (3.51a)$$

$$H_{RS}|\psi_i\rangle = E_i|\psi_i\rangle, \qquad (3.51b)$$

where the first one is the perturbed eigenproblem, and the second one is the *model eigenproblem*. The operator H_{RS} from (3.51b) is the *model Hamiltonian*, and it is related to the perturbed Hamiltonian H through similarity transformation,

$$H_{RS} = U^{-1}HU \tag{3.52}$$

Then the perturbed state vector $|\Psi_i\rangle$ is determined by the *wave operator U*,

$$|\Psi_i\rangle = U|\psi_i\rangle \tag{3.53}$$

Let us assume that the model Hamiltonian H_{RS} may be expressed in the split form

$$H_{RS} = H_0 + G \tag{3.54}$$

where the *model interaction G* is constrained by

$$G = \overline{G} + \overline{\overline{G}} \tag{3.55a}$$

$$\overline{G} = P_0 G P_0 \tag{3.55b}$$

$$\overline{\overline{G}} = (1 - P_0)G(1 - P_0) \tag{3.55c}$$

That is, the model space D_0 should be stable with respect of H_{RS}, $H_{RS}D_0 = D_0$. The intermediate normalization of the perturbed state vectors can now be expressed by the two "boundary" conditions for inner projections of U,

$$\overline{U} = P_0 U P_0 = P_0 \tag{3.56a}$$

$$\overline{\overline{U}} = (1 - P_0)U(1 - P_0) = 1 - P_0 \tag{3.56b}$$

According to property (3.55a), the model eigenproblem (3.51b) can be considered only in the d-dimensional model space D_0,

$$\left(\overline{H}_0 + \overline{G}\right)|\psi_i\rangle = E_i|\psi_i\rangle \tag{3.57}$$

for all $i \in M$; that is, if we know the model interaction $\overline{G}$ then the subspectrum $\{E_i; i \in M\}$ of H is determined as the eigenvalues of a d-dimensional model eigenproblem (3.57).

Now we turn our attention to the construction of a commutator equation useful for the further determination of UP_0 and $\overline{G}$ in a perturbation-expansion form. Introducing (2.1) and (3.54) into (3.52) we obtain after simple algebraic manipulations the commutator equation[20, 21, 14, 11]

$$[U, H_0]_- = H_1 U - UG \tag{3.58}$$

playing a central role in the present algebraic construction of the RSPT. Following Primas[22, 23] (see also Ref. 13), this equation is a special case of the more general commutator equation,

$$[X, H_0]_- = A \tag{3.59}$$

where X is an operator to be constructed and A is a known operator. Applying the spectral-resolution theorem to H_0 we obtain from (3.59) two formulas determining "cross" projections of X,

$$(1 - P_0)XP_0 = \sum_{\alpha \in M_0} \frac{1 - P_0}{E_\alpha^{(0)} - H_0} AP_0(\alpha) \tag{3.60a}$$

$$P_0 X(1 - P_0) = -\sum_{\alpha \in M_0} P_0(\alpha)A \frac{1 - P_0}{E_\alpha^{(0)} - H_0} \tag{3.60b}$$

Then the general solution of (3.59) can be written as follows:

$$X = \bar{X} + \bar{\bar{X}} + \sum_{\alpha \in M_0} \left\{ \frac{1 - P_0}{E_\alpha^{(0)} - H_0} AP_0(\alpha) - P_0(\alpha)A \frac{1 - P_0}{E_\alpha^{(0)} - H_0} \right\} \tag{3.61}$$

A particular solution of (3.59) is obtained from (3.61) when the inner projections $\bar{X} = P_0 X P_0$ and $\bar{\bar{X}} = (1 - P_0)X(1 - P_0)$ are specified. For our further considerations it is sufficient to know only the right-side projection of U into D_0, that is, UP_0. The operator UP_0 is fully determined by the commutator equation (3.58) and by the constraint (3.56a). Thus, starting from (3.61) we arrive at the following nonlinear equation[13, 14]:

$$UP_0 = P_0 + \frac{1}{k}(H_1 U - UG) \tag{3.62}$$

where the "superoperator" $(1/k)(X)$ is defined by

$$\frac{1}{k}(X) = \sum_{\alpha \in M_0} \frac{1 - P_0}{E_\alpha^{(0)} - H_0} XP_0(\alpha) \tag{3.63}$$

This equation is identical to (3.47).

The model interaction $\bar{G}$ is determined by the commutator equation (3.58). Multiplying it from the right and left by the projector P_0 and using the intermediate normalization (3.56) we obtain

$$\bar{G} = P_0 H_1 U P_0 \qquad (3.64)$$

which is completely identical to the non-Hermitian model interaction (3.29). In order to obtain the model interaction in the general form (3.50a) and (3.50b), let us again turn our attention to the commutator equation (3.58) and its particular solution (3.62). The nonlinear equation (3.62) determining the operator UP_0 is invariant when to the right-hand side of the commutator equation is added an operator $A = \bar{A} + \bar{\bar{A}}$, since $(1/k)(A) = 0$. That is, the model interaction is determined by the commutator equation only up to the operator A,

$$\bar{G} \to \bar{G}' = \bar{G} + \bar{A} \qquad (3.65)$$

or in other words, the operator $\bar{A}$ specifies an actual form of the model interaction. Now, let us assume that the operator $\bar{A}$ is specified by

$$\bar{A} = (\lambda - 1)P_0 H_1 U P_0 + (1 - \lambda)P_0 U^+ H_1 P_0 \qquad (3.66)$$

where $0 \leqslant \lambda \leqslant 1$. Then, introducing (3.64) and (3.66) into the right-hand side of (3.65) we obtain

$$\bar{G} = \lambda P_0 H_1 U P_0 + (1 - \lambda)P_0 U^+ H_1 P_0 \qquad (3.67)$$

This result is identical to (3.50a) and (3.50b).

D. Separability Theorem

We now study general properties[6, 11] of the model interaction and wave operator for the case when they are constructed for a system composed from a collection of noninteracting subsystems $A, B, C, \dots$. Let us assume, for simplicity, that a system may be split into two noninteracting subsystems A and B. A perturbed Hamiltonian of such system can be written as follows:

$$H = H_A + H_B \qquad (3.68)$$

where $H_{A (B)}$ is a Hamiltonian of the isolated subsystem A (B). As we did above, we assume that these Hamiltonians are determined by

$$H_A = H_{0A} + H_{1A} \qquad (3.69a)$$

$$H_B = H_{0B} + H_{1B} \qquad (3.69b)$$

where H_{0A} and H_{1A} (H_{0B} and H_{1B}) are an unperturbed Hamiltonian and a perturbation of the isolated subsystem A (B). Then the total Hamiltonian (3.68) may be written in a split form,

$$H = H_0 + H_1 \tag{3.70a}$$

$$H_0 = H_{0A} + H_{0B} \tag{3.70b}$$

$$H_1 = H_{1A} + H_{1B} \tag{3.70c}$$

Now we specify the commutator equation (3.58) for the noninteracting subsystems,

$$[U_A, H_{0A}]_- = H_{1A} U_A - U_A G_A \tag{3.71a}$$

$$[U_B, H_{0B}]_- = H_{1B} U_B - U_B G_B \tag{3.71b}$$

where U_A and G_A (U_B and G_B) are the wave operator and model interaction of the subsystem A (B). Multiplying (3.71a) by U_B and (3.71b) by U_A, we obtain

$$[U_A U_B, H_{0A}]_- = H_{1A} U_A U_B - U_A U_B G_A \tag{3.72a}$$

$$[U_A U_B, H_{0B}]_- = H_{1B} U_A U_B - U_A U_B G_B \tag{3.72b}$$

where we have assumed that the operators for different subsystems automatically commute. Finally, addition of these two equations gives

$$[U_A U_B, H_0]_- = H_1 U_A U_B - U_A U_B (G_A + G_B) \tag{3.72c}$$

with the same formal structure as the general commutator equation (3.58). Therefore, if the total system is a collection of two noninteracting subsystems A and B, then the total model interaction and total wave operator should satisfy the separability theorem:

$$G = G_A + G_B \tag{3.73a}$$

$$U = U_A U_B \tag{3.73b}$$

This result can be directly rewritten in perturbation-expansion form,

$$G^{(n)} = G_A^{(n)} + G_B^{(n)} \tag{3.74a}$$

$$U^{(n)} = \sum_{p=0}^{n} U_A^{(p)} U_B^{(n-p)} \tag{3.74b}$$

where the superscripts describe the order of perturbation theory.

The separability theorem (3.73a) and (3.73b) can be simply generalized for the case when a total system is composed of more than two noninteracting subsystems $A, B, C, \ldots$:

$$G = G_A + G_B + G_C + \cdots \qquad (3.75a)$$

$$U = U_A U_B U_C \cdots \qquad (3.75b)$$

Let us consider two following instructive implications of the separability theorem:

A. The model interaction. Here, we introduce the following terminology which, of course, should not be confused with a similar one used in many-body diagrammatic technique:[5,24] The terms of total model interaction G containing merely the single model interactions of an isolated subsystem are called *connected*; in the opposite case they are called *disconnected*. Then, the nth-order contributions of the total model interaction G may be formally written as follows:

$$G^{(n)} = \{ G^{(n)} \}_C + \{ G^{(n)} \}_{DC} \qquad (3.76)$$

where the subscript C (DC) denotes the connected (disconnected) terms. The connected terms (cf. their definition) are expressible by

$$\{ G^{(n)} \}_C = G_A^{(n)} + G_B^{(n)} + G_C^{(n)} + \cdots \qquad (3.77)$$

Comparing (3.76) with the separability theorem (3.74a) we obtain

$$\{ G^{(n)} \}_{DC} = 0 \qquad (3.78)$$

that is, the nth-order disconnected terms with simultaneous appearance, at least, two different perturbations H_{1X} and H_{1Y}, should be mutually cancelled.

B. The wave operator. A discussion similar to the above may be also carried out for the total wave operator. Let us introduce the following terminology: The perturbation terms of the total wave operator containing merely the products [see (3.47b)] of the wave operators of subsystems are called *linked*; in the opposite case they are called *unlinked*. Then the nth-order contributions of the total wave operator may be formally written as follows:

$$U^{(n)} = \{ U^{(n)} \}_L + \{ U^{(n)} \}_{UL} \qquad (3.79)$$

where the subscript L (UL) denotes the linked (unlinked) terms. The linked terms (cf. their definition) are expressible as a sum of products,

$$\{U^{(n)}\}_L = U_A^{(n)}U_B^{(0)}U_C^{(0)}\cdots + U_A^{(n-1)}U_B^{(1)}U_C^{(0)}\cdots + \cdots$$

$$+ U_A^{(n-1)}U_B^{(0)}U_C^{(1)}\cdots + U_A^{(0)}U_B^{(n)}U_C^{(0)}\cdots + \cdots \qquad (3.80)$$

This sum for the case of the two isolated subsystems is identical to (3.74b). Then, comparing (3.79) with (3.75b) we obtain

$$\{U^{(n)}\}_{UL} = 0 \qquad (3.81)$$

that is, the nth-order unlinked terms of the total wave operator should be mutually cancelled.

The separability theorem and all its implications are satisfied for any type of the model interaction based on the commutator equation (3.58).

IV. MANY-BODY DIAGRAMMATIC RSPT

In order to apply the RSPT to many-body systems it is necessary, first of all, to elaborate its many-body version based on the following two techniques:

1. The second-quantization formalism and Wick's theorem,
2. Diagrammatic interpretation of individual perturbation contributions.

Both these techniques were initially used by Brueckner[9,25] for demonstration of the fact that up to the fifth order the nondegenerate ground-state energy calculated by the RSPT can be diagrammatically visualized in a simple way, and that the diagrams that appear should be of the connected type. This very serious observation was first generally proved by Goldstone[26] using the time-dependent S-matrix formalism and by Hugenholtz[5] in the time-independent formalism of resolvent-operator technique. The degenerate version of many-body perturbation theory, built up in the framework of the BWPT, was given by Bloch and Horowitz,[27] using an ingenious combination of a thermodynamic analog of the time-dependent technique of Goldstone, and the resolvent-operator technique of Hugenholtz. They showed that the total perturbed energy separates into two terms: The core part, determined by Goldstone result for the nondegenerate ground-state energy, and the E-dependent valence part containing merely linked diagrams. The similar result was obtained also by Hugenholtz[5] slightly earlier, when he calculated the nondegenerate excitation energies.

The first paper refering to the many-body quasidegenerate RSPT is probably that of Okubo.[28] He showed, in connection with an E-independent RSPT version of Tamm–Dancoff theory, that the many-body diagrams with the external lines overlapped by ground-state diagrams without external lines are cancelled by similar diagrams with different relative position of these disconnected parts. This was explicitly demonstrated in fourth order, and a sketchy argument was presented to prove that such a property should be true in general. The next attempt to derive the E-independent many-body model interaction RSPT together with "linked-cluster" theorem is due to Primas.[22,23] His argument was reproduced in a somewhat more detailed form by Klein.[29] Primas employed the unitary transformation technique and some refined arguments based on the fact that his algebraic expressions are of "Lie-group-structure" type. Recently, this approach was criticized by Brandow,[30] who gives some simple "counterexample" showing why the Primas rather formal arguments for "linked-cluster" structure of the model interaction are, to say the lease, insufficient. The first many-body diagrammatic construction of the E-independent model interaction was presented by Morita.[1] This work, based on the time-dependent formalism, contains all the fundamental ideas necessary for construction of many-body degenerate RSPT, namely, the folded diagrams were initially introduced here. Unfortunately, many Morita's arguments and mathematical operations are presented in a very intuitive form, and therefore it is usually unjustly cited in the current literature as an incomplete work. The Morita derivation was repeated in a more easy and transparent form by Oberlechner et al.[31]

The many-body diagrammatic degenerate RSPT in its full generality was constructed by Brandow.[2] He starts from the Bloch–Horowitz perturbation theory and uses the folded-diagram approach to eliminate the energy dependences as well as the disconnected diagrams. The E-independent many-body non-Hermitian model interaction he obtained is closely related to the Bloch[15] formal degenerate RSPT. A highly original time-dependent formalism for the construction of many-body model interaction was used by Johnson and Baranger.[32] They demonstrated that in the framework of their formalism one can derive an infinite variety of the many-body Hermitian and/or non-Hermitian E-independent model interactions. Here, the concrete form of the many-body model interaction is determined by the class of conditions applied during the "time-integration" process. It seems that this work represents, at present, the most general theoretical technique for the construction of many-body model interaction that we know.

A different approach directly based on the algebraic structure of Bloch's model interaction [15] was given by Sandars.[4] There is no discussion of the existence or nonexistence of the disconnected diagrams in his many-body model interaction. This drawback was recently removed by Lindgren,[14] who constructed the many-body model interaction defined in the quaside-generate model space. The work of Kuo et al.[33] follows the general adiabatic approach of Morita[1] with some small modifications. In related paper, Krenciglowa et al.[34] investigated a possible partial summation of the folded diagrams, namely, the "Q-box" summation. The resolvent-operator technique was used by Kvasnička for construction of the many-body model interaction defined in the degenerate[35] as well as quasidegenerate[36] model space, where in the first case a method quite identical with the "Q-box" summation of Kranciglowa et al.[34] was independently studied.

A. Many-Body Diagrammatic Technique

Let us assume that in the second-quantization formalism the unperturbed Hamiltonian H_0 and the perturbation H_1 have the form

$$H_0 = \sum_i \epsilon_i X_i^+ X_i, \tag{4.1a}$$

$$H_1 = \tfrac{1}{4} \sum_{ijkl} \langle ij|v|kl\rangle_A X_i^+ X_j^+ X_l X_k - \sum_{ij} \langle i|w|j\rangle X_i^+ X_j \tag{4.1b}$$

where X_i^+ and X_j are creation and annihilation operators defined on the orthonormal set of spin orbitals that are the solution of the one-particle Hermitian eigenproblem $(h+w)|i\rangle = \epsilon_i|i\rangle$, where h is a one-particle operator expressing the kinetic and external-field potential energy, and w is an effective (shell-model) Hermitian potential. The two-particle matrix elements $\langle ij|v|kl\rangle_A$ are antisymmetric with respect to the indices (i,j) and (k,l). The eigenstates of H_0 are generated by the creation operators: $X_i^+ X_j^+ \cdots |0\rangle$, where $|0\rangle$ is the normalized vacuum-state vector. The unperturbed eigenenergy corresponding to this state vector is equal to sum of one-particle energies: $\epsilon_i + \epsilon_j \cdots$.

The basic concept of the present many-body technique is the core-state vector (unperturbed nondegenerate) $|\Phi_0\rangle$, $H_0|\Phi_0\rangle = E_0^{(0)}|\Phi_0\rangle$, initially introduced in many-body diagrammatic technique by Bloch and Horowitz.[27] The main reasons for the use of the core-state vector $|\Phi_0\rangle$ can be summarized as follows:

1. The core-state vector serves as a new renormalized vacuum, that is, the particle-hole formalism is defined with respect to $|\Phi_0\rangle$. Then the one-particle states occupied (unoccupied) in $|\Phi_0\rangle$ are called *hole (particle)* states.

2. Following intuitive physical considerations, it is assumed that a total many-body system may be divided into two subsystems: (*a*) Inert core

covering a closed-shell subsystem, and (*b*) valence particles determining the low-energy physical phenomena of the given total system. The "frozen" core interacts with valence particles only through higher-order perturbation contributions (called core-polarization effects).

The unperturbed state vectors spanning the d-dimensional model space D_0 are generated through the core-state vector by $X_{p_1}^+ X_{p_2}^+ \cdots X_{h_1} X_{h_2} \cdots |\Phi\rangle$, where $(h_1, h_2, \cdots)$ and $(p_1, p_2, \cdots)$ are indices of hole and particle states, respectively. We assume that the model space is spanned by all the a-particle–b-hole (where a and b are kept fixed) unperturbed state vectors corresponding to some unperturbed eigenenergies forming the set $\{E_\alpha^{(0)}; \alpha \in M_0\}$. For a more precise determination of the model space it is necessary to introduce two sets of the particle and hole indices, $\{p_1, p_2, \ldots, p_{a'}\}$ and $\{h_1, h_2, \ldots, h_{b'}\}$, where $a \leq a'$ and $b \leq b'$. Then the unperturbed a-particle–b-hole state vectors spanning the model space D_0 are determined as all possible different vectors of the type $X_{p_1}^+ X_{p_2}^+ \cdots X_{p_a}^+ X_{h_1} X_{h_2} \cdots X_{h_b} |\Phi_0\rangle$ where the indices $(p_1, p_2, \ldots, p_a)$ and $(h_1, h_2, \ldots, h_b)$ are chosen from the sets of particle and hole indices introduced above.

Now we turn our attention to the problem of the many-body diagrammatic interpretation of individual perturbation contributions of the E-independent model interaction and wave operator. These operator entities were visualized by the formal diagrammatic technique; here we introduce their *many-body diagrammatic interpretation*. In all our further considerations we use the Hugenholtz diagrammatic technique[5,24] based on the following one- and two-particle vertices:

$$\text{\raisebox{-1em}{\Large$\times$}} \; = \; \langle ij \,|\, v \,|\, kl \rangle_A \; = \; \quad + \quad \qquad (4.2a)$$

$$\underline{\quad\bullet\quad} \; = -\langle i \,|\, w \,|\, j \rangle \; = \quad \qquad (4.2b)$$

where for completeness the corresponding Goldstone's vertices are also presented. The Hugenholtz classification of the many-body diagrams can be shortly summarized in this way:

1. A diagram that cannot be divided into parts without cutting any line is called *connected*; in the opposite case it is called *disconnected*. A disconnected diagram is made up of connected parts. Each connected part is called *component*.

2. A diagram that has no external lines is called a *ground-state diagram*. A component of a disconnected diagram without external lines is a *ground-state component*.

3. A diagram that may be generally disconnected, but does not contain ground-state components, is called *linked*; in the opposite case it is called *unlinked*. The term "linked" must not be confused with the term "connected."

A similar terminology has also been introduced in the formal perturbation theory when we studied some implications of the separability theorem. We demonstrate in Section IV.B that these two different concepts are completely overlapped.

Recall that we are interested now in the many-body diagrammatic interpretation of the individual formal perturbation contributions of the model interaction $\bar{G}$ determined by (3.25) and (3.50). In these formulas, each term of this model interaction can be expressed by formal diagrammatic technique, and its formal algebraic interpretation can be obtained by means of the rules *1* to *6* presented in Sections II.A and III.

Generally, let us consider an nth-order formal diagram $(D)_\Gamma$ with n shaded vertices; this "system" of cuts and λ factor is determined by the subscript Γ. For further considerations we have to keep in mind the following two points dealing with algebraic interpretation of the diagram $(D)_\Gamma$:

1. Its total multiplicative factor $w[(D)_\Gamma]$ is determined by

$$w\big[(D)_\Gamma\big] = (-1)^p \lambda \tag{4.3}$$

where p is the number of internal positive lines [see rule *5* in Section III], and λ is the factor determining the type of the model interaction.

2. The structure of energy denominators is uniquely determined by the system of cuts. This fact is extremely important for construction of the energy denominators in the framework of many-body diagrammatic technique, since they will be constructed with respect to formal algebraic interpretation of $(D)_\Gamma$.

The model interaction G is determined by matrix elements $\langle \Phi_\alpha | G | \Phi_{\alpha'} \rangle$, where $|\Phi_\alpha\rangle$ and $|\Phi_{\alpha'}\rangle$ are two arbitrary unperturbed state vectors spanning the model space D_0. Furthermore, these matrix elements should be equal to expansion over all possible distinct formal diagrams,

$$\langle \Phi_\alpha | G | \Phi_{\alpha'} \rangle = \sum_\Gamma \langle \Phi_\alpha | (D)_\Gamma | \Phi_{\alpha'} \rangle \tag{4.4}$$

where the matrix elements $\langle \Phi_\alpha | (D)_\Gamma | \Phi_{\alpha'} \rangle$ can be simply obtained from the formal algebraic interpretation of the diagram $(D)_\Gamma$. For instance, let us put

$$(D)_\Gamma = \frac{1}{2} \quad \text{[diagram]} \quad =$$

$$= -\frac{1}{2} \sum_{\alpha, \alpha' \in M_0} P_0 H_1 \frac{1 - P_0}{\left(E_\alpha^{(0)} - H_0 \right) \left(E_{\alpha'}^{(0)} - H_0 \right)} H_1 P_0 (\alpha') H_1 P_0 (\alpha) \quad (4.5)$$

which is the third-order contribution of the Hermitian model interaction defined by (3.34), and where the multiplicative factor (4.3) is equal to $(-\frac{1}{2})$. Then the matrix element $\langle \Phi_\alpha | (D)_\Gamma | \Phi_{\alpha'} \rangle$ is determined by

$$-\frac{1}{2} \sum_{\alpha'' \in M_0} \langle \Phi_\alpha | H_1 \frac{1 - P_0}{\left(E_{\alpha''}^{(0)} - H_0 \right) \left(E_{\alpha''}^{(0)} - H_0 \right)} H_1 P_0 (\alpha'') H_1 | \Phi_{\alpha'} \rangle \quad (4.6)$$

Following the Hugenholtz many-body diagrammatic technique,[5,24] each matrix element $\langle \Phi_\alpha | (D)_\Gamma | \Phi_{\alpha'} \rangle$ from the right-hand side of (4.4) can be expressed as a sum of all possible nonequivalent[37] many-body diagrams. Two many-body diagrams are topologically equivalent if one diagram is transformed onto another one by a topological deformation that preserves the orientation of lines as well as relative positions of many-body vertices. Then

$$\langle \Phi_\alpha | (D)_\Gamma | \Phi_{\alpha'} \rangle = \sum_K \langle \Phi_\alpha | (D)_\Gamma | \Phi_{\alpha'} \rangle_K \quad (4.7)$$

where the summation index K runs over all the pertinent many-body diagrams. We now give three remarks dealing with some slight modification of Hugenholtz's diagrammatic rules.[5,24] First, the numerator composed from a product of the one- and two-particle matrix elements from (4.1b) is constructed by the standard way, that is, following the expressions (4.2a) and (4.2b). Second, the total multiplicative factor of a given diagram corresponding to the formal diagram $(D)_\Gamma$ is determined by

$$(\pm 1) w \left[(D)_\Gamma \right] (-1)^{h + l + w} \left(\frac{1}{2} \right)^q \quad (4.8)$$

where $w[(D)_\Gamma]$ is the multiplicative factor (4.3) of the formal diagram $(D)_\Gamma$, h is the number of *internal* hole lines, l is the number of *closed loops*, w is the number of one-particle vertices (4.2b), and q is the number of *equivalent lines*.[5,24] Finally, the factor (± 1) in front of (4.8) is determined by a proper selection of the phase factors of incoming $|\Phi_{\alpha'}\rangle$ and outgoing $\langle\Phi_\alpha|$ unperturbed states.[32] Third, the energy denominators should be constructed according to the formal energy denominators from the algebraic interpretation of $(D)_\Gamma$. This last item can be formalized for the non-Hermitian model interaction (3.28) and (3.29) by the folded-diagram approach of Morita[1] and Brandow.[2]

For a better understanding of these general ideas let us consider, for example, a many-body counterpart of the formal term (4.5). For simplicity we assume that the model space D_0 is spanned by all independent two-particle states $X_p^+ X_{p'}^+ |\Phi_0\rangle$. A possible term from the right-hand side of (4.7) can be "drawn" as follows:

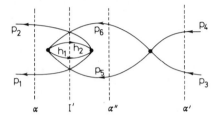

$$(4.9)$$

where $|\Phi_\alpha\rangle = X_{p_1}^+ X_{p_2}^+ |\Phi_0\rangle$, $|\Phi_{\alpha'}\rangle = X_{p_3}^+ X_{p_4}^+ |\Phi\rangle$, $|\Phi_{\alpha''}\rangle = X_{p_5}^+ X_{p_6}^+ |\Phi_0\rangle$, and $E_{\alpha'}^{(0)}$ $= E_0^{(0)} + \epsilon_{p_3} + \epsilon_{p_4} = E_0^{(0)} + e_{\alpha'}$, $E_{\alpha''}^{(0)} = E_0^{(0)} + \epsilon_{p_5} + \epsilon_{p_6} = E_0^{(0)} + e_{\alpha''}$. The intermediate state I between the first and second vertex has energy $E_I^{(0)} = E_0^{(0)}$ $+ \epsilon_{p_1} + \epsilon_{p_2} + \epsilon_{p_5} + \epsilon_{p_6} - \epsilon_{h_1} - \epsilon_{h_2} = E_0^{(0)} + e_I$. Then, using the rules of Hugenholtz's graphology together with its modifications presented above, we obtain an explicit algebraic interpretation of (4.9),

$$(\pm 1)\left(\tfrac{1}{2}\right)(-1)^{2+0+0}\left(\tfrac{1}{2}\right)^2$$

$$\times \sum_{h_1 h_2 \in \Phi_0} \sum_{p_5 p_6 \notin \Phi_0} \frac{\langle h_1 h_2 | v | p_5 p_6\rangle_A \langle p_1 p_2 | v | h_1 h_2\rangle_A \langle p_5 p_6 | v | p_3 p_4\rangle_A}{(e_{\alpha'} - e_I)(e_{\alpha''} - e_I)} \quad (4.10)$$

In a completely analogous way we may study the many-body diagrammatic interpretation of the wave operator $U P_0$. Here, similarly, this operator is determined by the formal diagrammatic expansion (3.39). The unshaded vertices that appear are changed by an actual form of the model interaction (3.50). Then

$$U P_0 = \sum_\Gamma (D)_\Gamma \quad (4.11)$$

where the summation runs over all distinct formal diagrams. Further, let us consider matrix elements $\langle \Phi_\mu | U | \Phi_\alpha \rangle$, where $|\Phi_\alpha \rangle$ is an unperturbed state from the model space D_0, and $|\Phi_\mu \rangle$ is an unperturbed state from its orthogonal complement. From (4.11) we obtain

$$\langle \Phi_\mu | U | \Phi_\alpha \rangle = \sum_\Gamma \langle \Phi_\mu | (D)_\Gamma | \Phi_\alpha \rangle \qquad (4.12)$$

The matrix elements from the right-hand side of (4.12) can be expressed as a sum of all possible nonequivalent many-body diagrams [cf. (4.7)],

$$\langle \Phi_\mu | (D)_\Gamma | \Phi_\alpha \rangle = \sum_K \langle \Phi_\mu | (D)_\Gamma | \Phi_\alpha \rangle_K \qquad (4.13)$$

with completely the same diagrammatic rules as have been presented for the many-body model interaction.

Summarizing, in the framework of the many-body diagrammatic technique the matrix elements of the model interaction $\bar{G}$ and the corresponding wave operator UP_0 may be expressed as

$$\langle \Phi_\alpha | G | \Phi_{\alpha'} \rangle = \sum_\Gamma \sum_K \langle \Phi_\alpha | (D)_\Gamma | \Phi_{\alpha'} \rangle_K \qquad (4.14a)$$

$$\Phi_\mu | U | \Phi_\alpha \rangle = \sum_\Gamma \sum_K \langle \Phi_\mu | (D)_\Gamma | \Phi_\alpha \rangle_K \qquad (4.14b)$$

where the first summations run over all formal diagrams, and the second ones over all many-body topologically nonequivalent diagrams.

B. "Linked-Cluster" Theorems

In Section IV.A we used the standard Hugenholtz "graphology" for many-body diagrammatic interpretation of the model interaction and the corresponding wave operator. The many-body diagrammatic contributions of these operators can be classified from the standpoint of Hugenholtz's classification scheme as connected and disconnected, linked and unlinked, and generally, as a combination of these terms. Therefore, on first sight it seems that such terminology does not facilitate our further considerations. But, in the next part of this section we demonstrate that such a classification scheme of the many-body diagrams can be used advantageously. On the basis of this classification we shall prove the following two useful statements called *"linked-cluster" theorems:*

1. The many-body model interaction contains only linked-connected diagrams.

2. The many-body wave operator contains only linked diagrams.

These two properties were initially formulated and proved by Brandow[2] for the non-Hermitian Bloch model interaction (3.28) using the folded-diagram approach.

In the framework of the many-body diagrammatic technique the matrix elements $\langle \Phi_\alpha | G | \Phi_{\alpha'} \rangle$ of the model interaction G can formally be written as follows:

$$\langle \Phi_\alpha | G | \Phi_{\alpha'} \rangle = \underset{A}{\alpha \overline{\boxed{///}} \alpha'} + \underset{B}{\alpha \overline{\boxed{///}} \alpha'} + \underset{C}{\alpha \quad \alpha'} \qquad (4.15)$$

where the first term A represents all linked diagrams (generally disconnected), the second term B represents all disconnected unlinked diagrams composed from the linked diagrams with incoming and outgoing lines (top block) and from ground-state diagrams (generally disconnected, bottom block). Finally, the third C represents the ground-state diagrams (connected or disconnected). This last term C is nonzero only for diagonal matrix elements $\langle \Phi_\alpha | G | \Phi_\alpha \rangle$. Following Hugenholtz[5] and Löwdin,[6] the many-body disconnected diagrams of the model interaction may be formally treated as the contributions of the model interaction built up for a system composed from two or more noninteracting subsystems. In order to illuminate this formal analogy let us consider a simple way in which the concept of noninteracting subsystems may be realized for the many-body systems:

1. The Hamiltonian H_x [see (3.68), (3.69a) and (3.69b)] for an isolated subsystem $X = A, B, C, \ldots$ is, in the second quantization formalism, written as follows:

$$H_X = H_{0X} + H_{1X} \qquad (4.16a)$$

$$H_{0X} = \sum_{i \in X} \epsilon_i X_i^+ X_i \qquad (4.16b)$$

$$H_{1X} = \tfrac{1}{4} \sum_{ijkl \in X} \langle ij|v|kl \rangle_A X_i^+ X_j^+ X_l X_k - \sum_{ij \in X} \langle i|w|j \rangle X_i^+ X_j \qquad (4.16c)$$

where the summations run over all possible indices from X. Let the total Hamiltonian be equal to sum of these sub-Hamiltonians, $H = H_A + H_B + \ldots$.

2. For calculation of the many-body diagrammatic contributions of the matrix elements $\langle \Phi_\alpha | G | \Phi_{\alpha'} \rangle$ by the Wick theorem,[38] we take into account only operator contractions from the same subsystem, and, furthermore, we

require that the many-body diagrammatic terms of a given subsystem be always connected, that is, it should contain only one component.

Now, following these two conventions we may state that the concept of connected and disconnected terms introduced in the formulation of the separability theorem is quite identical with the concept of connected and disconnected many-body diagrams of the model interaction. Generally, the matrix element $\langle \Phi_\alpha | G | \Phi_{\alpha'} \rangle$ of the model interaction can be factorized by the many-body diagrammatic technique factorized into two terms,

$$\langle \Phi_\alpha | G | \Phi_{\alpha'} \rangle = \langle \Phi_\alpha | G | \Phi_{\alpha'} \rangle_C + \langle \Phi_\alpha | G \Phi_{\alpha'} \rangle_{DC} \qquad (4.17)$$

where the subscript C (DC) denotes the connected (disconnected) many-body diagrammatic terms of the model interaction. Here, the connected diagrams are either linked-connected [term A in (4.15)] or ground-state connected [term C in (4.15)]. The disconnected diagrams from (4.17) contain, at least, two connected components, linked and ground-state. Therefore, applying the separability theorem (3.78) we obtain

$$\langle \Phi_\alpha | G | \Phi_{\alpha'} \rangle_{DC} = 0 \qquad (4.18)$$

that is, the disconnected many-body diagrams of the model interaction are cancelled. This cancellation is carried out separately for each order of perturbation theory; cf. (3.78). Then the matrix element $\langle \Phi_\alpha | G | \Phi_{\alpha'} \rangle$ can be written in the following form:

$$\langle \Phi_\alpha | G | \Phi_{\alpha'} \rangle = \langle \Phi_\alpha | G | \Phi_{\alpha'} \rangle_C$$
$$= \langle \Phi_\alpha | G | \Phi_{\alpha'} \rangle_C^{(A)} + \langle \Phi_\alpha | G | \Phi_{\alpha'} \rangle_C^{(C)} \cdot \delta_{\alpha\alpha'} \qquad (4.19)$$

where the first term represents those connected diagrammatic terms that interact with incoming and/or outgoing lines, that is, these terms should be linked and connected,

$$\langle \Phi_\alpha | G | \Phi_{\alpha'} \rangle_C^{(A)} = \langle \Phi_\alpha | G | \Phi_{\alpha'} \rangle_{LC} \qquad (4.20)$$

The second diagonal term from the right-hand side of (4.19) represents those connected diagrammatic terms that do not interact with incoming and outgoing lines; that is, their components should be connected and ground state. In this case, the free running external lines are fully inactive in the construction of the energy denominators. This fact can be simply verified, for instance, by the many-body diagram (4.9). Let us assume that this diagram also contains free running particle line indexed by p_7. Then, the unperturbed energies $E_\kappa^{(0)}$($\kappa = \alpha, \alpha', \alpha''$ and I) should be shifted by the factor $+\epsilon_{p_7}$, that is, $E_\kappa^{(3)} \to E_\kappa^{(0)} + \epsilon_{p_7}$. Since the energy denominators are formed from differences of two unperturbed energies, this modification is unimportant. Generally, the free running external lines are irrelevent for algebraic interpretation of the diagrammatic terms of the model interaction. According to this observation, the matrix element $\langle \Phi_\alpha | G | \Phi_\alpha \rangle_C^{(C)}$ from (4.19) can be rewritten as follows:

$$\langle \Phi_\alpha | G | \Phi_\alpha \rangle_C^{(C)} = \langle \Phi_0 | \sum_{n=0}^{\infty} H_1 \left(\frac{1}{E_0^{(0)} - H_0} H_1 \right)^n | \Phi_0 \rangle_C \qquad (4.21)$$

that is, as a sum of all ground-state connected diagrams. Here, the formal terms of the model interaction with intermediate states from the model space should be omitted, since they produce disconnected diagrams. Following Goldstone[26] and Hugenholtz,[5] (4.21) is the well-known many-body diagrammatic expression for the "correlation" energy ΔE_0 of the core subsystem,

$$\langle \Phi_\alpha | G | \Phi_\alpha \rangle_C^{(C)} = \Delta E_0 \qquad (4.22)$$

The total perturbed energy of the core is then equal to $E_0 = E_0^{(0)} + \Delta E_0$. Finally, introducing (4.20) and (4.22) in (4.19) we obtain

$$\langle \Phi_\alpha | G | \Phi_{\alpha'} \rangle = \Phi_\alpha | G | \Phi_{\alpha'} \rangle_{LC} + \delta_{\alpha\alpha'} \Delta E_0 \qquad (4.23a)$$

or formally,

$$\overline{G} = \{ \overline{G} \}_{LC} + \Delta E_0 P_0 \qquad (4.23b)$$

This represents, de facto, the "linked-cluster" theorem for the model interaction. Substitution of (4.23b) into the model eigenproblem (3.57) gives its many-body diagrammatic version,

$$\left(\overline{H}_0^1 + \{ \overline{G} \}_{LC} \right) | \psi_i \rangle = \Delta E_i | \psi_i \rangle \qquad (4.24a)$$

$$\overline{H}_0^1 = \overline{H}_0 - E_0^{(0)} P_0 = \sum_{\alpha \in M_0} (E_\alpha^{(0)} - E_0^{(0)}) P_0(\alpha) \qquad (4.24b)$$

where the eigenvalue $\Delta E_i = E_i - E_0$ can be interpreted as an a-particle-b-hole excitation energy.

The Pauli principle is properly taken into account. This is generally achieved by deriving the matrix elements $\langle \Phi_\alpha | G | \Phi_{\alpha'} \rangle$ in the notation of second quantization, where it is automatically fulfilled. However, when determining these matrix elements by a many-body diagrammatic technique, we obtain some diagrams that violate the Pauli principle. Such diagrams originate from (i) application of the separability theorem and also from (ii) the expression (4.22). Thus, the many-body diagrammatric terms of the matrix elements $\langle \Phi_\alpha | G | \Phi_{\alpha'} \rangle$ must contain all diagrams violating the Pauli principle.

For a better understanding of these general ideas leading to the many-body form of model interaction (4.23), let us consider the following simple examples:

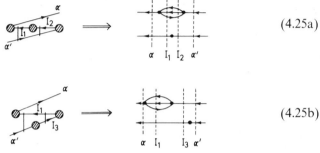

$$(4.25a)$$

$$(4.25b)$$

with the same two-particle model space as in (4.9). The states α, α', and I_3 are from the model space D_0, and the states I_1 and I_2 are from the orthogonal complement of D_0. The algebraic interpretation of these many-body diagrams is almost identical; there exist some differences only in the numerical factors (4.8), namely, the factor $w[(D)_\Gamma]$ is in the first (second) case equal to $+\lambda$ $(-\lambda)$ [cf. (4.3)]. Then the many-body diagrams (4.25a) and (4.25b) are exactly cancelled. Generally, the many-body disconnected diagram corresponding to a "linear" formal diagram (without internal positive lines) is cancelled by the proper disconnected diagrams of "nonlinear" formal diagrams. Of course, the "linear" as well as the "nonlinear" formal diagrams should be of the same order, that is, with the same number of shaded vertices. This observation may be illustrated by the formal diagrams from (3.29). The disconnected many-body diagram corresponding to the formal diagram C is cancelled by a similar one (with a different relative position of vertices) corresponding to the formal diagram D. Also, the disconnected diagram of E is cancelled by the disconnected diagrams of F to I.

The proof of the "linked-cluster" theorem (4.23) for the many-body version of the model interaction is completely based on the separability theorem. This powerful theoretical approach may be applied for any model interaction that is determined via the commutator equation (3.58). Therefore, the "linked-cluster" theorem is satisfied only for such a model interaction the formal counterpart of which is determined by the commutator equation. In Section III we have suggested the general method for construction of wide class of the E-independent formal model interactions. But, as it follows from Section III.C, only the interactions determined by (3.25) form a subclass of interactions which satisfy the commutator equation (3.58). Thus the "linked-cluster" theorem is satisfied only for a many-body model interaction, with formal counterpart (Hermitian or non-Hermitian) expressed by (3.25).

A similar approach is also applicable to the construction of a many-body expression determining the wave operator. We assume that the formal wave operator is defined by commutator equation (3.58). The matrix element $\langle \Phi_\mu | U | \Phi_\alpha \rangle$ can be formally written as follows:

$$\langle I_\mu | U | I_\alpha \rangle = \underset{A}{\overset{\mu}{\xleftarrow{}}\boxed{/\!/\!/\!/}\overset{\alpha}{\xleftarrow{}}} \quad \underset{B}{\overset{\mu}{\xleftarrow{}}\boxed{/\!/\!/\!/}\overset{\alpha}{\xleftarrow{}}} \qquad (4.26)$$

where the first term A represents linked many-body diagrammatic contributions. The second term B represents unlinked many-body diagrams composed, at least, from two components. Generally, in the framework of the many-body diagrammatic technique, the matrix element $\langle \Phi_\mu | U | \Phi_\alpha \rangle$ can be factorized into two terms,

$$\langle \Phi_\mu | U | \Phi_\alpha \rangle = \langle \Phi_\mu | U | \Phi_\alpha \rangle_L + \langle \Phi_\mu | U | \Phi_\alpha \rangle_{UL} \qquad (4.27)$$

Here, the subscript $L(UL)$ denotes the linked (unlinked) diagrammatic terms of the wave operator. Applying the separability theorem in the form (3.81) we obtain

$$\langle \Phi_\mu | U | \Phi_\alpha \rangle_{UL} = 0 \qquad (4.28)$$

that is, the unlinked many-body diagrams of the wave operator are mutually cancelled. Finally, introducing expression (4.28) into (4.27) we obtain

$$\langle \Phi_\mu | U | \Phi_\alpha \rangle = \langle \Phi_\mu | U | \Phi_\alpha \rangle_L \qquad (4.29a)$$

or formally,

$$UP_0 = P_0 + \{(1 - P_0)UP_0\}_L \qquad (4.29b)$$

The perturbed state vector $|\Psi_i\rangle$ determined by (3.53), can then be expressed by

$$|\Psi_i\rangle = |\psi_i\rangle + \{(1 - P_0)UP_0\}_L|\psi_i\rangle \qquad (4.30)$$

Here, the vector $|\psi_i\rangle$ is the eigenvector of the many-body eigenproblem (4.24a). The expressions (4.29a) and (4.29b) represent the "linked-cluster" theorem for the wave operator.

C. Mean Value of Observable

To complete our quasidegenerate many-body RSPT we turn our attention to the calculation of the mean values of an observable described by an operator Q. In order to realize this program we use so called double-perturbation approach.[2, 36] Let us introduce an auxiliary perturbed Hamiltonian

$$H(\omega) = H_0 + H_1(\omega) \qquad (4.31a)$$

$$H_1(\omega) = H_1 + \omega Q \qquad (4.31b)$$

Here, ω is a small parameter, and the operator $H_1(\omega)$ is taken as a new perturbation. Then the perturbed Schödinger equation (3.51a) and the many-body model eigenproblem (4.24) have the form

$$H(\omega)|\Psi_i(\omega)\rangle = E_i(\omega)|\psi_i(\omega)\rangle \qquad (4.32a)$$

$$\left(\overline{H_0'} + \{G(\omega)\}_{LC}\right)|\psi_i(\omega)\rangle = \Delta E_i(\omega)|\psi_i(\omega)\rangle \qquad (4.32b)$$

The many-body model interaction $\{G(\omega)\}_{LC}$ arises from replacing H_1 by $H_1(\omega)$ in (3.25). The perturbed eigenvalues $E_i(\omega)$ are connected with the "excitation" energies $\Delta E_i(\omega)$ by

$$E_i(\omega) = E_0(\omega) + \Delta E_i(\omega) \qquad (4.33)$$

or, in the matrix form,

$$\langle \Psi_i(\omega)|H(\omega)|\Psi_i(\omega)\rangle = \langle \Psi_0(\omega)|H(\omega)|\Psi_0(\omega)\rangle$$
$$+ \langle \varphi_i(\omega)|\{G(\omega)\}_{LC}|\psi_i(\omega)\rangle \qquad (4.34)$$

where $E_0(\omega) = \langle \Psi_0(\omega)|H(\omega)|\Psi_0(\omega)\rangle$ is the perturbed ground-state energy of the core subsystem described by the Hamiltonian (4.31a). The set $\{|\varphi_i(\omega)\rangle; i \in M\}$ is the basis of the left eigenvectors of (4.32b) normalized in such a way that

$$\langle \varphi_i(\omega)|\psi_{i'}(\omega)\rangle = \delta_{ii'} \qquad (4.35)$$

for $i, i' \in M$. Differentiating (4.34) with respect to ω and using (4.32a) and

(4.32b) we find

$$\langle \Psi_i(\omega)| \frac{dH(\omega)}{d\omega}|\Psi_i(\omega)\rangle = \langle \Psi_0(\omega)| \frac{dH(\omega)}{d\omega}|\Psi_0(\omega)\rangle$$

$$+ \langle \varphi_i(\omega)| \frac{d\{G(\omega)\}_{LC}}{d\omega}|\psi_i(\omega)\rangle \qquad (4.36)$$

Taking $\omega \to 0$ and $dH(\omega)/d\omega = Q$ [cf. (4.31b)], we obtain

$$\langle \Psi_i|Q|\Psi_i\rangle = \langle \Psi_0|Q|\Psi_0\rangle + \langle \varphi_i|Q_{RS}^{LC}|\psi_i\rangle \qquad (4.37)$$

where the operator Q_{RS}^{LC} defined in the model space D_0 is determined by

$$Q_{RS}^{LC} = \lim_{\omega \to 0} \frac{d\{G(\omega)\}_{LC}}{d\omega} \qquad (4.38)$$

which can be directly obtained from $\{G(\omega)\}_{LC}$ as the set of all many-body linked-connected diagrams that are linear in ω; that is, the many-body vertices corresponding to observable Q appear just once in all possible successive places. Following Thouless,[12] the core mean value $\langle \Psi_0|Q|\Psi_0\rangle$ is given by the sum of all ground-state diagrams with one Q vertex. Equation 4.37 is the principal result for this section; it represents a linked-connected expression for the evaluation of the mean values of an observable Q. Similarly to the model eigenproblem (4.24), the core mean value $\langle \Psi_0|Q|\Psi_0\rangle$ enters in (4.38) as an additive factor.

V. APPLICATIONS

In this section we apply the many-body diagrammatic theory of model Hamiltonian to some interesting problems of up-to-date quantum-molecular physics:

1. The diagrammatric perturbation theory of quantum-chemical effective Hamiltonians.

2. Direct calculation of low-lying ionization potentials, electron affinities, and excitation energies.

3. Calculation of Brueckner orbitals and generalized natural orbitals.

4. Many-body theory of rotation-vibration molecular spectra.

5. Fine and hyperfine interactions in atoms and molecules.

These simple examples demonstrate that the many-body diagrammatic theory of model Hamiltonian offers very fruitful ab initio approach, starting from first principles, for theoretical studies of many-electron molecular systems.

A. Effective Hamilitonians

A theoretical justification and foundation of quantum-chemical semiempirical parameters (or, generally, the quantum-chemical effective Hamiltonians) that would enable their complete a priori physical determination and understanding is a very interesting problem of present quantum chemistry. The first introductory steps toward solving this problem have been independently by Freed,[39, 40] Westhaus et al.,[41] and Kvasnička.[42] Freed used an energy-dependent model Hamiltonian determined by the use the BWPT, and the Sinanoğlu-type cluster functions.[43] The resulting effective matrix elements are dependent on energy as well as on the valence-bond configurations of the remaining electrons. On the basis of these two dependences, Freed deduced some general properties of the semiempirical parameters, for example, how these parameters change with the electronic state, or with degree of ionicity of the state. It seems that a many-body diagrammatic version of Freed's approach can be simply constructed when the Bloch-Horowitz[27] perturbation theory is applied. Westhaus et al.[41] tackled the same problem by a Van Vleck contact-transformation perturbation approach of the original full Hamiltonian. Its many-body realization together with some additional conditions produces an effective Hamiltonian with partitioning in one-, two-,..., body terms. Generally, the Van Vleck procedure gives the E-independent Hermitian model Hamiltonian,[45] and it can be rederived in the framework of the present algebraic approach (Section III.C), when it is assumed that the wave operator is an unitary operator.[21] This means[21] that the commutator equation (3.58) has its place in this special case of interest also; that is, the separability theorem and all its implications should be satisified. Indeed, Brandow[2] demonstrated that such a many-body model Hamiltonian contains only linked-connected terms, or in other words, that Westhaus' results may be uniquely diagrammatically interpreted.

Here, in order to construct the diagrammatic perturbation theory of the effective Hamiltonians we follow our recent publication,[42] where the present diagrammatic quasidegenerate RSPT with E-independent model interaction has been used. Let us have an orthonormal[46] set of atomic spinorbitals ASOs

$$\{|\varphi_i\rangle; \, i = 1, 2, \ldots\} \tag{5.1}$$

In the second-quantization formalism the original full Hamiltonian can be written in the form

$$H = \sum_{ij} \langle i|h|j\rangle X_i^+ X_j + \left(\tfrac{1}{4}\right) \sum_{ijkl} \langle ij|v|kl\rangle_A X_i^+ X_j^+ X_l X_k \tag{5.2}$$

where X_i^+ (X_j) are creation (annihilation) operators defined on the ortho-normal set of ASOs (5.1). Starting from intuitive quantum-chemical assumptions, the set (5.1) can be divided into three disjoint subsets, namely (1) the core orbitals $\{|\varphi_i\rangle; i \in C\}$, (2) the valence orbitals $\{|\varphi_i\rangle; i \in V\}$, and finally, (3) the excited orbitals $\{|\varphi\rangle; i \in E\rangle$. Unifying the first two subsets, the minimum basis set of ASOs is obtained; this is the basic concept of all semiempirical methods.[47] The core-state vector is now determined by

$$|\Phi_0\rangle = \prod_{i \in C} X_i^+ |0\rangle \tag{5.3}$$

where the product index i runs over all core ASOs. Then, using Wick's theorem,[38] the original full Hamiltonian (5.2) can rewritten in the normal form,

$$H = \mathcal{E}_0 + H_{(1)} + H_{(2)} \tag{5.4}$$

The scalar quantity $\mathcal{E}_0$ is determined by

$$\mathcal{E}_0 = \langle \Phi_0 | H | \Phi_0 \rangle = \sum_{i \in C} \langle i|h|i \rangle + \left(\tfrac{1}{2}\right) \sum_{ij \in C} \langle ij|v|ij \rangle_A \tag{5.5}$$

which can be interpreted as an unperturbed ground-state energy of the core subsystem. The one-particle term $H_{(1)}$ is given by

$$H_{(1)} = \sum_{ij} \langle i|f|j \rangle N\left[X_i^+ X_j \right] \tag{5.6a}$$

$$\langle i|f|j \rangle = \langle i|h|j \rangle + \sum_{k \in C} \langle ik|v|jk \rangle_A \tag{5.6b}$$

where $N[\cdots]$ is the normal product defined with respect to $|\Phi_0\rangle$, and the matrix elements $\langle i|f|j \rangle$ may be formally interpreted as matrix elements of the Hartree-Fock operator constructed for $|\Phi_0\rangle$. Finally, the two-particle term $H_{(2)}$ from (5.4) has the form

$$H_{(2)} = \left(\tfrac{1}{4}\right) \sum_{ijkl} \langle ij|v|kl \rangle_A N\left[X_i^+ X_j^+ X_l X_k \right] \tag{5.7}$$

Thus, the Hamiltonian (5.4) can be rewritten in the final form (4.1a) and (4.1b) appropriate for an application of the many-body diagrammatic perturbation theory:

$$H = \mathcal{E}_0 + H_0 + H_1 \tag{5.8}$$

where the operator H_0 (called the unperturbed Hamiltonian) is given by

$$H_0 = \sum_i \epsilon_i N[X_i^+ X_i] \tag{5.9a}$$

$$\epsilon_i = \langle i|f|i \rangle \tag{5.9b}$$

where the ϵ_i's play the rôle of one-particle energies. The operator H_1 (called the perturbation) has the form

$$H_1 = \sum_{ij} (1 - \delta_{ij}) \langle i|f|j \rangle N[X_i^+ X_j]$$

$$+ \left(\tfrac{1}{4}\right) \sum_{ijkl} \langle ij|v|kl \rangle_A N[X_i^+ X_j^+ X_l X_k] \tag{5.10}$$

with one-particle (first summation) and two-particle (second summation) terms. The one-particle core states from $|\Phi_0\rangle$ are called *hole states*, and either valence or excited one-particle states, *particle states*. The model space D_0 is determined as a subspace spanned by all possible p-particle (where p is the number of the occupied valence one-particle states) unperturbed state vectors,

$$D_0 \equiv \{ |\Phi_K\rangle = X_{k_1}^+ X_{k_2}^+ \cdots X_{k_p}^+ |\Phi_0\rangle;\ k_1 < k_i < \cdots < k_p \in V \} \tag{5.11}$$

That is, the model space D_0 is spanned by all N-electron Slater determinants with fixed core (q-particle subsystem), and the remaining p electrons ($N = p + q$) occupy merely the valence ASOs. In the Freed terminology,[40] this model space is equivalent to the "chemical sea." The projector P_0 onto the model space D_0 is then determined by

$$P_0 = \sum_K |\Phi_K\rangle\langle\Phi_K| \tag{5.12}$$

where the summation index K runs over all p-particle unperturbed states from (5.11).

Now, we are ready to apply the many-body diagrammatic model eigenproblem (4.24) in the following form:

$$H_{LC}|\psi_i\rangle = \Delta E_i |\psi_i\rangle \tag{5.13a}$$

$$H_{LC} = H_0 + \{ G \}_{LC} \tag{5.13b}$$

where the model Hamiltonian H_{LC} is expressed as the sum of the unperturbed Hamiltonian H_0, (5.9a) and (5.9b) and the many-body diagrammatic model interaction $\{G\}_{LC}$ with formal counterpart G determined by (3.34). The perturbed eigenenergy E_i is determined by $E_i = E_0^{core} + \Delta E_i$, where E_0^{core} is the perturbed (exact) energy of the core subsystem defined by (4.21). Since the model eigenproblem (5.13a) is fully defined in the model space D_0, the eigenfunctions $|\psi_i\rangle$ should be expressed as a linear combination of the unperturbed state vectors $|\Phi_K\rangle$ from (5.11),

$$|\psi_i\rangle = \sum_K C_{Ki} |\Phi_K\rangle \tag{5.14}$$

Then the model eigenproblem (5.13a) can be rewritten in the matrix form

$$\mathbf{H}_{LC}\mathbf{C}_i = \Delta E_i \mathbf{C}_i \tag{5.15}$$

where $\mathbf{H}_{LC}$ is a Hermitian matrix built up from the matrix elements $\langle \Phi_k | H_{LC} | \Phi_L \rangle$, and $\mathbf{C}_i$ is a column vector of C_{K_i} coefficients. Since the model Hamiltonian (5.13b) contains only linked-connected terms, it is possible to introduce an effective Hamiltonian H_{eff} by the following defining identity:

$$\langle \Phi_k | H_{LC} | \Phi_L \rangle \equiv \langle \Phi_k | H_{eff} | \Phi_L \rangle \tag{5.16}$$

In the second-quantization formalism H_{eff} can be expressed as follows:

$$H_{eff} = H_0 + \sum_{t=1}^{p} V_{eff}^{(t)} \tag{5.17a}$$

$$V_{eff}^{(t)} = (t!)^{-1} \sum_{\substack{i_1 i_2 \cdots i_t \in V \\ j_1 j_2 \cdots j_t \in V}} v_{eff}^{(t)}(i_1 i_2 \cdots i_t, j_1 j_2 \cdots j_t)$$

$$\times N\left[X_{i_1}^+ X_{i_2}^+ \cdots X_{i_t}^+ X_{j_t} \cdots X_{j_1} \right] \tag{5.17b}$$

Here, the term $V_{eff}^{(t)}$ is the t-particle effective interaction, and its matrix elements $v_{eff}^{(t)}(i_1 i_2 \cdots i_t, j_1 j_2 \cdots j_t)$ are determined by summation of all possible linked-connected diagrams (without free particle lines) with outgoing (incoming) valence particle lines indexed by $i_1, i_2, \ldots, i_t (j_1, j_2, \ldots, j_t)$,

$$v_{eff}^{(t)} = (i_1, i_2, \cdots i_t, j_1, j_2, \cdots j_t) = \tag{5.18}$$

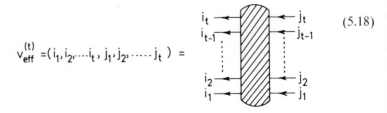

Finally, the present approach of the construction of the effective Hamiltonian H_{eff} can be very profitable used for an ab initio physical establishment of the semiempirical effective Hamiltonians.[42] We do not repeat the main reasons for it here, since they have been presented in an exhaustive form by Freed.[40] We note, for illustration, that the well-known three semiempirical parameters α_i, β_{ij}, and γ_{ij} appearing in almost all semiempirical methods, may be now determined as follows:

$$\alpha_i = v_{eff}^{(1)}(i,i) \qquad (5.19a)$$

$$\beta_{ij} = v_{eff}^{(1)}(i,j) \qquad (\text{for } i \neq j), \qquad (5.18b)$$

$$\gamma_{ij} = v_{eff}^{(2)}(ij,ij) \qquad (5.19c)$$

where $i, j \in V$. This means that in the semiempirical methods the remaining two-particle and all higher-particle effective matrix elements are neglected. Consequently, if we have selected some actual basis (5.1) of ASOs, then we may, by using (5.19), calculate the semiempirical parameters α_i, β_{ij}, and γ_{ij} directly without any reference to experimental values.

B. Ionization Potentials, Electron Affinities, and Excitation Energies

Let us assume that for a given many-electron atomic or molecular system an orthonormal set of the Hartree-Fock one-particle functions is known, and that in the zero-order approximation this system is described by a unperturbed state vector $|\Phi_0\rangle$, which formally serves also as the core vector. Then, in the second-quantization formalism the full Hamiltonian can be written in the form[37]

$$H = \langle \Phi_0 | H | \Phi_0 \rangle + H_0 + H_1 \qquad (5.20a)$$

$$H_0 = \sum_i \epsilon_i N[X_i^+ X_i] \qquad (5.20b)$$

$$H_1 = \left(\tfrac{1}{4}\right) \sum_{ijkl} \langle ij|v|kl \rangle_A N[X_i^+ X_j^+ X_l X_k] \qquad (5.20c)$$

where the matrix element $\langle \Phi_0 | H | \Phi_0 \rangle$ is the Hartree-Fock ground-state energy, and ϵ_i are orbital energies.

In order to calculate low-lying ionization potentials,[48] let us define the model space D_0 as follows:

$$D_0 \equiv \{ |\Phi_h\rangle = X_h|\Phi_0\rangle; \text{ for all occupied } h \} \qquad (5.21)$$

that is, the model space D_0 is spanned by all one-hole unperturbed state vectors. The projecter P_0 onto D_0 has the form

$$P_0 = \sum_h^{\text{occ}} |\Phi_h\rangle\langle\Phi_h| \tag{5.22}$$

Assuming that the formal model interaction G is determined by (3.24) in Hermitian form, the many-body model eigenproblem (4.24) can be now specified in the following form:

$$(H_0 + \{G\}_{LC})|\psi_i\rangle = (\text{I.P.})_i|\psi_i\rangle \tag{5.23}$$

where the eigenvalues $(\text{I.P.})_i$ are exact ionization potentials corresponding to states that are in the zero-order approximation described by one-hole states from (5.21). The model interaction $\{G\}_{LC}$ is determined in D_0 by the matrix elements $\langle\Phi_h|\{G\}_{LC}|\Phi_{h'}\rangle$, which have, up to the second order, this simple diagrammatic interpretation[48]:

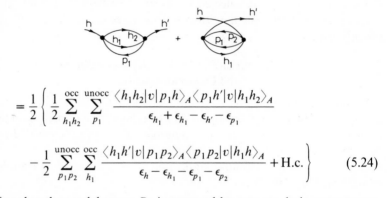

$$= \frac{1}{2}\left\{ \frac{1}{2}\sum_{h_1 h_2}^{\text{occ}}\sum_{p_1}^{\text{unocc}} \frac{\langle h_1 h_2|v|p_1 h\rangle_A \langle p_1 h'|v|h_1 h_2\rangle_A}{\epsilon_{h_1} + \epsilon_{h_1} - \epsilon_{h'} - \epsilon_{p_1}} \right.$$

$$\left. - \frac{1}{2}\sum_{p_1 p_2}^{\text{unocc}}\sum_{h_1}^{\text{occ}} \frac{\langle h_1 h'|v|p_1 p_2\rangle_A \langle p_1 p_2|v|h_1 h\rangle_A}{\epsilon_h - \epsilon_{h_1} - \epsilon_{p_1} - \epsilon_{p_2}} + \text{H.c.} \right\} \tag{5.24}$$

Assuming that the model space D_0 is spanned by one one-hole state vector $|\Phi_h\rangle$, that is, D_0 is a one-dimensional subspace, then the model eigenproblem (5.23) gives an expression determining directly[35] the ionization potential $(\text{I.P.})_h$,

$$(\text{I.P.})_h = -\epsilon_h + \langle\Phi_h|\{G\}_{LC}|\Phi_h\rangle \tag{5.25}$$

Here, the matrix elements $\langle\Phi_h|\{G\}_{LC}|\Phi_h\rangle$ also contain diagrammatic terms that are forbidden in (5.24) since the one-hole states $|\Phi_{h'}\rangle$ for $h' \neq h$ are now from the orthogonal complement of the model space D_0.

Electron affinities can be treated in an analogous way. Now, the model space D_0 should be spanned by all one-particle unperturbed states,

$$D_0 \equiv \{|\Phi_p\rangle = X_p^+|\Phi_0\rangle; \text{ for all unoccupied } p\} \tag{5.26}$$

and the projector P_0 onto the model space is

$$P_0 \sum_p^{\text{unocc}} |\Phi_p\rangle\langle\Phi_p| \tag{5.27}$$

Then the model eigenproblem (4.24) can be specified in the following form:

$$(H_0 + \{G\}_{LC})|\psi_i\rangle = (\text{E.A.})_i|\psi_i\rangle \tag{5.28}$$

where the eigenvalues $(\text{E. A.})_i$ are exact electron affinities corresponding to one-particle states (5.26). Similarly, the matrix elements $\langle\Phi_p|\{G\}_{LC}|\Phi_{p'}\rangle$ from (5.28) are diagrammatically interpreted up to the second order as follows:

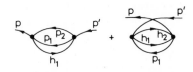

$$= \frac{1}{2} \left\{ \frac{1}{2} \sum_{p_1 p_2}^{\text{unocc}} \sum_{h_1}^{\text{occ}} \frac{\langle h_1 p|v|p_1 p_2\rangle_A \langle p_1 p_2|v|h_1 p'\rangle_A}{\varepsilon_{p'} + \varepsilon_{h_1} - \varepsilon_{p_1} - \varepsilon_{p_2}} \right.$$

$$\left. - \frac{1}{2} \sum_{h_1 h_2}^{\text{occ}} \sum_{p_1}^{\text{unocc}} \frac{\langle h_1 h_2|v|p_1 p'\rangle_A \langle p_1 p|v|h_1 h_2\rangle_A}{\varepsilon_{h_1} + \varepsilon_{h_2} - \varepsilon_{p_1} - \varepsilon_p} + \text{H.c.} \right\} \tag{5.29}$$

Assuming that the model space D_0 is spanned only by one one-particle vector $|\Phi p\rangle$, then from (5.28) we obtain

$$(\text{E.A.})_p = \varepsilon_p + \langle\Phi_p|\{G\}_{LC}|\Phi_p\rangle \tag{5.30}$$

Here we note that in these one-dimensional problems (5.25) and (5.30), the formal model interaction can be determined by (3.28), which is much more simple than its Hermitian form (3.34).

Let us now turn our attention to calculation of low-lying excitation energies,[49] corresponding to one-hole-one-particle (monoexcited) state vectors. This means that the model space should be defined as follows:

$$D_0 \equiv \{|\Phi_{ph}\rangle = X_p^+ X_h|\Phi_0\rangle; \text{ for all (un) occupied } h(p)\} \tag{5.31}$$

that is, by all monoexcited state vectors. The projector P_0 is then determined by

$$P_0 = \sum_h^{occ} \sum_p^{unocc} |\Phi_{ph}\rangle\langle\Phi_{ph}| \tag{5.32}$$

In this special case the model eigenproblem (4.24) gives

$$(H_0 + \{G\}_{LC})|\psi_i\rangle = \Delta E_i|\psi_i\rangle \tag{5.33}$$

where the formal model interaction G should be determined by (3.34) in the *Hermitian* form, and ΔE_i is an exact excitation energy corresponding to excitation process described by some monoexcited state vector from (5.31). Assuming that the model interaction is approximated only up to the first order, then the model eigenproblem (5.33) represents a simple configuration interaction realized over all monoexcited configurations; that is, the higher-order terms of $\{G\}_{LC}$ describe truncation effects of the model space D_0. The matrix elements $\langle\Phi_{ph}|\{G\}_{LC}|\Phi_{p'h'}\rangle$ from (5.33) can be partioned by the following way: (1) effective particle-particle interactions corresponding to diagrams with a free hole line, (2) effective hole-hole interactions corresponding to diagrams with a free particle line, and, finally, (3) effective two-particle-two-hole interactions corresponding to diagrams without free lines. Thus, the matrix elements $\langle\Phi_{ph}|\{G\}_{LC}|\Phi_{p'h'}\rangle$ may be written in the split form,

$$\langle\Phi_{ph}|\{G\}_{LC}|\Phi_{p'h'}\rangle = \delta_{pp'}j_{hh'}^{eff} + \delta_{hh'}h_{pp'}^{eff} + h_{ph,p'h'}^{eff} \tag{5.34}$$

where the effective matrix elements h^{eff} are related single effective interactions discussed above. In the second-quantization formalism the model interaction $\{G\}_{LC}$ from the right-hand side of (5.33) can be expressed as an "effective" interaction defined in the model space D_0 of all possible particle-hole states,

$$\{G\}_{LC} = \sum_{h_1 h_2} h_{h_1 h_2}^{eff} N[X_{h_1}^+ X_{h_2}] + \sum_{p_1 p_2} h_{p_1 p_2}^{eff} N[X_{p_1}^+ X_{p_2}^+]$$

$$+ \sum_{p_1 p_2} \sum_{h_1 h_2} h_{h_1 p_1, h_2 p_2}^{eff} N[X_{h_1}^+ X_{p_1}^+ X_{p_2} X_{h_2}] \tag{5.35}$$

where the effective matrix elements $h_{h_1 h_2}^{eff}$ ($h_{p_1 p_2}^{eff}$) are identical to the matrix elements $\langle\Phi_{h_1(p_1)}|\{G\}_{LC}|\Phi_{h_2(p_2)}\rangle$ from the calculation of the ionization potentials (electron affinities); cf. (5.24) and (5.29). The effective matrix elements $h_{h_1 p_1, h_2 p_2}^{eff}$ are diagrammatically interpreted, up to the second order, as follows:

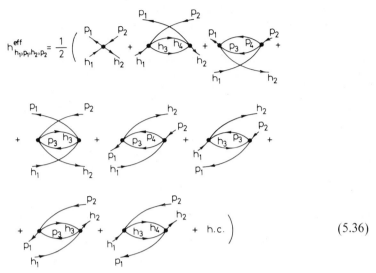

$$\tag{5.36}$$

Let us now assume that the model space D_0 is spanned by one hole-particle vector $|\Phi_{ph}\rangle$. Then the model eigenproblem (5.33) gives[35]

$$\Delta E_{ph} = \epsilon_p - \epsilon_h + \langle \Phi_{ph} | \{ G \}_{LC} | \Phi_{ph} \rangle \tag{5.37}$$

Applying the expression (5.34), this formula can be rewritten in the form[35]

$$\Delta E_{ph} = (\text{E.A.})_p - (\text{I.P.})_h + h^{\text{eff}}_{ph,ph} \tag{5.38}$$

where $(\text{E.A.})_p = \epsilon_p + h^{\text{eff}}_{pp}$ and $(\text{I.P.})_h = -\epsilon_h + h^{\text{eff}}_{hh}$ are the corresponding electron affinity and ionization potential, respectively.

Recently, Hubač and Urban[50] have used the expression (5.25) for direct calculation of ionization potentials of some small molecules by the ab initio method with Gausian AOs. They have observed that in order to calculate the ionization potentials with sufficient accuracy it is necessary to approximate the model interaction, at least up to the third order.

An interesting application of the present approach to the calculation of "excited" intermolecular interaction energies has been made by Kvasnička et al.[51] In this application the charge-transfer (A^+B^-) and local-excited (AB^*) interaction energies can be directly calculated using the many-body theory. Both these quantities are determined by $\Delta E_{XY} = \Delta E_{AB} + \Delta\Delta E_{XY}$, where $XY = A^+B^-$ or AB^*, and ΔE_{AB} is the ground-state interaction energy calculated by the many-body nondegenerate perturbation theory.

The $\Delta\Delta E_{XY}$ is determined by the linked-connected diagrammatic terms with two external hole and particle lines.

The many-body diagrammatic degenerate RSPT has been used by Kaldor[52, 53] for calculation of excited states of the H_2 molecule. For this simplest example containing only two electrons it is possible to identify the core state vector with the original vacuum state $|0\rangle$, that is, the many-body diagrammatic terms contain only particle lines; this reduces the total number of diagrams needed to be considered. Kaldor has included all diagrams up to the third order, and the calculated energies agree well with the exact values with an error of 1 to 2×10^{-3} a.u. This simple approach is also applicable for three- and four-particle systems. Generally, however, for many-particle systems, where the core-state vector should be for practical reasons identified with the Hartree-Fock ground-state vector, and where core-polarization effects enter into play, this straightforward approach is very impractical for an actual application.

C. Bruekner and Generalized Natural Orbitals

The independent-particle model[54] of the N-electron molecular systems can be well established by the Hartree-Fock theory, which is often used as its synonym. Of course, there are other possibilities how to choose the "best" independent-particle state-vector Slater determinant. The most important and useful alternative methods for determination of an orthonormal set of one-particle functions orbitals from which the independent-particle state vector is built-up are (1) the Brueckner (or maximum overlap) orbitals [55-59, 6] (BO) and the (2) the generalized natural orbitals[60-62] (GNO). These two alternative possibilities (or their combinations) have been extensively studied in the "microscopic" theory of nuclei[63-67] in order to accelerate the convergence of diagrammatic-perturbation series. There is introduced a special sort of one-particle functions that lie intermediate between BOs and GNOs and ensure that a maximal number of diagrams of the perscribed type is cancelled. Similarly, in many-body diagrammatic theory of molecular systems it can be also promising to turn one's attention to another type of one-particle functions, namely, to BOs and/or GNOs.

In our recent publication[68] we have suggested the one-particle pseudo-eigenvalue problem determining either the BOs or GNOs,

$$(f_0 + u)|i\rangle = \epsilon_i|i\rangle \tag{5.39}$$

where f_0 is standard Hartree-Fock operator and u is a Hermitian one-particle operator defined through its matrix elements $\langle i|u|j\rangle$. Starting from the defining conditions of the BOs and GNOs, respectively, we obtain an expression specifying matrix elements $\langle p|u|h \rangle = \langle h|u|p \rangle^*$, where $h(p)$ is an index of occupied (unoccupied) one-particle states. The remaining matrix elements $\langle p|u|p' \rangle$ and $\langle h|u|h' \rangle$ determine an additional factorization within the subspace of occupied and unoccupied one-particle functions, respectively. A similar situation also exists in the Hartree-Fock theory based on the Brillouin theorem.[69] This theorem can be rewritten in the form $\langle p|f_0|h \rangle = 0$, that is it serves only as a factorization procedure of the orbitals into two orthogonal subspaces of occupied and unoccupied orbitals. However, in the Hartree-Fock theory the situation is simpler than in the theory of BOs and GNOs, since the explicit form of the Hartree-Fock operator f_0 is known. Postulating the "canonical" Hartree-Fock orbitals,[69] that is, $\langle i|f_0|j \rangle = \delta_{ij}$, the Brillouin theorem is automatically satisfied. Unfortunately, in the present case this approach cannot be used because the one-particle operator u is defined only through its matrix elements $\langle i|u|j \rangle$.

For an additional specification of the matrix elements $\langle h|u|h' \rangle$ and $\langle p|u|p' \rangle$ we introduced[68] an "ideal" requirement[63–67] that the one-particle energies ϵ_i from (5.39) are exactly equal to minus ionization potential (for $i = h$) or to electron affinity (for $i = p$),

$$\epsilon_h = -(\text{I.P.})_h \qquad \text{(for all occupied } h) \qquad (5.40a)$$

$$\epsilon_p = (\text{E.A.})p \qquad \text{(for all unoccupied } p) \qquad (5.40b)$$

Both these conditions are a generalization of the well-known Koopmans's theorem,[70] which is valid in the Hartree-Fock theory up to the first order. The quantities $(\text{I.P.})_h$ and $(\text{E.A.})_p$ from (5.40a) and (5.40b) are exact ionization potentials and electron affinities respectively. According to the fact that the BOs and GNOs within the subspace of the occupied and unoccupied orbitals, respectively, are determined only up to a unitary transformation, the model eigenproblems (5.23) and (5.28) can be rewritten in the form

$$\left(H_0 + \{G^{(-)}\}_{LC}\right)|\psi_h\rangle = -\epsilon_h|\psi_h\rangle \qquad (5.41a)$$

$$\left(H_0 + \{G^{(+)}\}_{LC}\right)|\psi_p\rangle = \epsilon_p|\psi_p\rangle \qquad (5.41b)$$

where the eigenvalues $-\epsilon_h$ and ϵ_p satisfy the conditions (5.40a) and (5.40b), and the model interactions $\{G^{(-)}\}_{LC}$ and $\{G^{(+)}\}_{LC}$ are defined in the model spaces (5.21) and (5.26). Let us introduce the following partioning of the model interactions from (5.41a) and (5.41b),

$$\left\{ G^{(\pm)} \right\}_{LC} = \left\{ P_0^{(\pm)} H_1 P_0^{(\pm)} \right\}_{LC} + \left\{ \tilde{G}^{(\pm)} \right\}_{LC} \tag{5.42}$$

where we have explicitly separated the first-order term from the model interaction, and $P_0^{(\pm)}$ are the projectors onto the model spaces (5.26) and (5.21). Then, we obtain the final expression for the matrix elements,

$$\langle p|u|p' \rangle = \langle \Phi_p | \left\{ \tilde{G}^{(+)} \right\}_{LC} | \Phi_{p'} \rangle \tag{5.43a}$$

$$\langle h|u|h' \rangle = - \langle \Phi_{h'} | \left\{ G^{(-)} \right\}_{LC} | \Phi_h \rangle \tag{5.43b}$$

that is, the one-particle pseudoeigenvalue problem (5.39) is fully specified.

D. Rotation -Vibration Spectra

The purpose of this application is to call an attention to an interesting and potentially very useful possibility of formulating the many-body diagrammatic theory of rotation-vibration molecular spectra.[71] Although the many-body techniques are applied in wide branches of quantum-molecular physics, it is surprising that this powerful approach has never been applied in such a branch of molecular physics, for example, to the theory of rotation-vibration spectra.[72, 73] These many-body techniques, namely the theory of model Hamiltonian, have very great advantage in that they offer "microscopic" insight into the problem being studied. It seems, therefore, that even if the theory of rotation-vibration spectra is relatively closed, all basic concepts are already known,[72, 73] the many-body diagrammatic technique here opens new possibilities for theoreticians working in high-resolution molecular spectroscopy.

The basic concept of this theoretical attempt is a total rotation-vibration Hamiltonian expressed in second-quantization formalism,

$$H = e_0 + H_0 + H_1 \tag{5.44}$$

where e_0 is a scalar quantity (c-number) that can be omitted without loss of generality, H_0 is an unperturbed Hamiltonian, and H_1 is a perturbation. The unperturbed Hamiltonian is determined as a sum of two terms: the pure vibrational ($H_{0,V}$) and pure rotational ($H_{0,R}$) unperturbed Hamiltonains,

$$H_0 = H_{0,V} + H_{0,R} \tag{5.45a}$$

$$H_{0,V} = \sum_k \omega_k b_k^+ b_k \tag{5.45b}$$

$$H_{0,R} = \frac{1}{2} \left[\frac{J_x^2}{I_{xx}^{(e)}} + \frac{J_y^2}{I_{yy}^{(e)}} + \frac{J_z^2}{I_{zz}^{(e)}} \right] \tag{5.45c}$$

Here, ω_k's are harmonic wavenumbers of normal mode k, and b_k^+, b_k are corresponding creation and annihilation operators obeying the Bose-Einstein statistics. J_α's in (5.45c) are components of the total angular-momentum operator, and $I_{\alpha\alpha}^{(e)}$ are components of the equilibrium momentum of inertia. Let us now introduce the eigensystems of $H_{0,V}$ and $H_{0,R}$,

$$H_{0,V} |\mathbf{n}\rangle = E_{V,\mathbf{n}}^{(0)} |\mathbf{n}\rangle \tag{5.46a}$$

$$|\mathbf{n}\rangle = |n_1 n_2 \cdots n_k \cdots \rangle = \frac{(b_1^+)^{n_1} (b_2^+)^{n_2} \cdots (b_k^+)^{n_k} \cdots}{\left[(n_1!)(n_2!) \cdots (n_k!) \cdots \right]^{1/2}} |0\rangle \tag{5.46b}$$

$$E_{V,\mathbf{n}}^{(0)} = \sum_k n_k \omega_k \tag{5.46c}$$

and

$$H_{0,R} |j,\alpha\rangle = E_{R,j\alpha}^{(0)} |j,\alpha\rangle \tag{5.46d}$$

for $j = 0, 1, 2, \ldots$ and $\alpha = 0, 1, 2, \ldots, 2j+1$. The eigensystem of $H_{0,R}$ can be simply obtained by diagonalizing $H_{0,R}$ in finite-dimensional subspaces $\{|jm\rangle; |m| = 0, 1, \ldots, j\}$ spanned by eigenfunctions of J^2 with fixed quantum number j. Starting from these two eigensystems we may directly construct the eigensystem of the total unperturbed Hamiltonian H_0,

$$H_0 |\mathbf{n},j\alpha\rangle = E_{\mathbf{n},j\alpha}^{(0)} |\mathbf{n},j\alpha\rangle \tag{5.47a}$$

$$E_{\mathbf{n},j\alpha}^{(0)} = E_{V,\mathbf{n}}^{(0)} + E_{R,j\alpha}^{(0)} \tag{5.47b}$$

$$|\mathbf{n},j\alpha\rangle = |\mathbf{n}\rangle \otimes |j\alpha\rangle \tag{5.47c}$$

The perturbation H_1 from (5.44) should be generally determined as a sum of two terms,

$$H_1 = H_{1,V} + H_{1,R} \tag{5.48a}$$

Here, the "rotational" $H_{1,R}$ and "vibrational" $H_{1,V}$ part have, in the second-quantization formalism, the following form:

$$H_{1,X} = \sum_{n=1}^{\infty} \sum_{p+q=n} H_{1,X}^{(p;q)} \quad (\text{for } X = V, R) \tag{5.48b}$$

$$H_{1,V}^{(p;q)} = (p!q!)^{-1} \sum_{\substack{k_1 k_2 \cdots k_p \\ l_1 l_2 \cdots l_q}} A(k_1 k_2 \cdots k_p, l_1 l_2 \cdots l_q)$$

$$\times b_{k_1}^+ b_{k_2}^+ \cdots b_{k_p}^+ b_{l_1} b_{l_2} \cdots b_{l_q} \tag{5.48c}$$

$$H_{1,R}^{(p;q)} = (p!q!)^{-1} \sum_{\substack{k_1 k_2 \cdots k_p \\ l_1 l_2 \cdots l_q}} \sum_{j,\alpha\alpha'} A(j,\alpha\alpha'; k_1 k_2 \cdots k_p, l_1 l_2 \cdots l_q)$$

$$\times b_{k_1}^+ b_{k_2}^+ \cdots b_{k_p}^+ b_{l_1} b_{l_2} \cdots b_{l_q} P(j,\alpha\alpha') \tag{5.48d}$$

$$P(j,\alpha\alpha') = |j\alpha\rangle\langle j\alpha'|, P(j,\alpha\alpha')P(j,\alpha'',\alpha''') = \delta_{jj'}\delta_{\alpha'\alpha''}P(j,\alpha\alpha'') \tag{5.48e}$$

We assume that the coefficients $A(\cdots)$ are symmetric with respect to vibronic indices,

$$A(k_1 k_2 \cdots k_p, l_1 l_2 \cdots l_q) = A(k'_1, k'_2, \ldots, k'_p, l'_1, l'_2 \cdots l'_q) \tag{5.49a}$$

$$A(j,\alpha\alpha'; k_1 k_2 \cdots k_p, l_1 l_2 \cdots l_q) = A(j,\alpha\alpha'; k'_1 k'_2 \cdots k'_p, l'_1 l'_2 \cdots l'_q)$$

$$\tag{5.49b}$$

where $(k'_1 k'_2 \cdots k'_p)$ and $(l'_1 l'_2 \cdots l'_q)$ are arbitrary permutations of $(k_1 k_2 \cdots k_p)$ and $(l_1 l_2 \cdots l_q)$.

The rotation-vibration energies are determined as the eigenvalues of the Schrödinger equation,

$$H|\Psi\rangle = E|\Psi\rangle \tag{5.50}$$

where the total Hamiltonian H is defined in entire Hilbert space spanned by all unperturbed eigenvectors of H_0. This means that an eigenvector $|\Psi\rangle$ of (5.50) can be expressed as follows:

$$|\Psi\rangle = \sum_{\mathbf{n}} \sum_{\alpha=1}^{2j+1} c_{\mathbf{n},j\alpha} |\mathbf{n},j\alpha\rangle \tag{5.51a}$$

$$c_{\mathbf{n},j\alpha} = \langle \mathbf{n},j\alpha | \Psi \rangle \tag{5.51b}$$

for some fixed j (since $HJ^2 = J^2H$). In order to obtain the eigensystem of H an approximate method should be used. If we use the standard Ritz variational method, we meet some difficulties originating from the diagonalization of a Hermitian matrix of enormous dimensionality. An alternative method[71] outlined here is the many-body diagrammatic theory of model Hamiltonian. Before applying it, we have to introduce a proper diagrammatic interpretation of individual terms of H_1. We use the following diagrammatic convention:

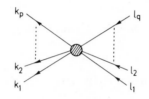

$$= A(k_1 k_2 \cdots k_p, l_1 l_2 \cdots l_q) b_{k_1}^+ b_{k_2}^+ \cdots b_{k_p}^+ b_{l_1} b_{l_2} \cdots b_{l_q} \tag{5.52a}$$

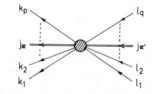

$$= A(j,\alpha\alpha'; k_1 k_2 \cdots k_p, l_1 l_2 \cdots l_q)$$

$$\times b_{k_1}^+ b_{k_2}^+ \cdots b_{k_p}^+ b_{l_1} b_{l_2} \cdots b_{l_q} P(j,\alpha\alpha') \tag{5.52b}$$

Here, the creation (annihilation) operator is represented by indexed outgoing (incoming) single line. The oriented double line represents the "projector" $P(j,\alpha\alpha')$.

Now we are ready to use the many-body diagrammatic theory of model Hamiltonian for calculation of excitation energies of a many-vibron system described by the Hamiltonian (5.44). First, we introduce the core-state vector $|\Phi_0\rangle$,

$$|\Phi_0\rangle = |0\rangle \otimes |00\rangle \tag{5.53a}$$

$$H_0|\Phi_0\rangle = 0 \tag{5.53b}$$

where $|0\rangle$ is the normalized vacuum-state vector from (5.46b), and $|00\rangle$ is the eigenfunction of (5.46d) for $j = \alpha = 0$. That is, in the present case we have only particle lines (running from right to left side); the hole lines cannot appear here, since the core-state vector (5.53a) is built-up from vacuum-state vector. Let $D_0(\mathbf{n},j)$ denotes the $(2j + 1)$-dimensional model space spanned by unperturbed state vectors $|\mathbf{n},j\alpha\rangle$ with fixed vibrational index $\mathbf{n}$ and the angular-momentum quantum number j,

$$D_0(\mathbf{n},j) \equiv \{|\mathbf{n},j\alpha\rangle; \alpha = 0, 1, \ldots, 2j + 1\} \tag{5.54}$$

The projector onto this model space is then determined by

$$P_0(\mathbf{n},j) = \sum_{\alpha=1}^{2j+1} P_0(\mathbf{n},\alpha) = \sum_{\alpha=1}^{2j+1} |\mathbf{n},j\alpha\rangle\langle\mathbf{n},\alpha| \tag{5.55a}$$

$$\overline{H}_0 = H_0 P_0(\mathbf{n},j) = P_0(\mathbf{n},j)H_0 = \sum_{\alpha=1}^{2j+1} E_{\mathbf{n},j\alpha}^{(0)} P_0(\mathbf{n},j\alpha) \tag{5.55b}$$

Finally, the general many-body model eigenproblem (4.24a) can be now specified as follows:

$$\left(\overline{H}_0 + \{G(\mathbf{n},j)\}_{LC}\right)|\varphi_{\mathbf{n},j\alpha}\rangle = \Delta E|\varphi_{\mathbf{n},j\alpha}\rangle \tag{5.56}$$

where the operator $\{G(\mathbf{n},j)\}_{LC}$ represents the many-body model interaction defined in the model space $D_0(\mathbf{n},j)$. The eigenvalue $\Delta E_{\mathbf{n},j\alpha}$ from (5.56) is the excitation energy between ground state (with energy $E_{0,00}$) and an excited state (with energy $E_{\mathbf{n},j\alpha}$). For $j = 0$ the model eigenproblem (5.56) is a one-dimensional problem, where the eigenfunction $|\varphi_{\mathbf{n},00}\rangle$ can be directly identified with $|\mathbf{n},00\rangle$,

$$\Delta E_{\mathbf{n},00} = \sum_k n_k \omega_k + \langle\mathbf{n},00|\{G(\mathbf{n},0)\}_{LC}|\mathbf{n},00\rangle \tag{5.57}$$

Since the vector $|\mathbf{n},00\rangle$ has zero total angular momentum, therefore the many-body model interaction $\{G(\mathbf{n},0)\}_{LC}$ contains only perturbation terms from $H_{1,v}$, and the terms from $H_{1,R}$ containing "projector" $P_0(j,\alpha\alpha')$

give in this special case zero contributions. In the diagrammatic language the model interaction $\{G(\mathbf{n},0)\}_{LC}$ is equal to a sum of all distinct linked-connected diagrams constructed merely from the vertices without angular-momentum double lines. As in Section V.A, the model interaction can be partioned into effective interactions,

$$\{G(\mathbf{n},0)\}_{LC} = \sum_{p \geqslant 1} \frac{1}{p!} \sum_{k_1 k_2 \cdots k_p} G_{k_1 k_2 \cdots k_p} b_{k_1}^+ b_{k_2}^+ \cdots b_{k_p}^+ b_{k_1} b_{k_2} \cdots b_{k_p} \quad (5.58)$$

where the symmetric coefficients $G_{k_1 \cdots k_p} (= G_{k'_1 \cdots k'_p})$ represent a sum of all distinct linked-connected diagrams (without free inactive lines) with p incoming and p outgoing lines indexed by $k_1, k_2, \ldots, k_p$. Introduction of (5.58) into (5.57) gives an alternative expression for the excitation energies,

$$\Delta E_{\mathbf{n},00} = \sum_k n_k \omega_k + \sum_{p \geqslant 1} \sum_{k_1 k_2 \cdots k_p} n_{k_1 k_2 \cdots k_p} G_{k_1 k_2 \cdots k_p} \quad (5.59a)$$

where the integer factors $n_{k_1 k_2 \cdots k_p}$ are determined by

$$n_{k_1 k_2 \cdots k_p} = (p!)^{-1} \langle \mathbf{n},00 | b_{k_1}^+ b_{k_2}^+ \cdots b_{k_p}^+ b_{k_1} b_{k_2} \cdots b_{k_p} | \mathbf{n},00 \rangle \quad (5.59b)$$

For a better understanding of these general results let us now study a simple example of an excitation energy $\Delta E_{\mathbf{n},00}$ corresponding to $\mathbf{n} = (1,0,\ldots,0)$. From (5.59a) and (5.59b) we obtain

$$\Delta E_{\mathbf{n},00} = \omega_1 + G_1 \quad (5.60)$$

where the lower-order diagrammatic contributions of G_1 are presented as follows:

$$(5.61)$$

Using the basic principles of the many-body diagrammatic technique we get from (5.61)

$$G_1 = A(1,1) + \frac{1}{2} \sum_{k_1 k_2} \frac{A(1,k_1 k_2)A(k_1 k_2,1)}{\omega_1 - \omega_{k_1} - \omega_{k_2}}$$

$$+ \frac{1}{2} \sum_{k_1 k_2} \frac{A(,1k_1 k_2)A(1k_1 k_2,)}{-\omega_1 - \omega_{k_1} - \omega_{k_2}} + \sum_{k_1} \frac{A(1,1k_1)A(k_1,)}{-\omega_{k_1}}$$

$$+ \sum_{k_1} \frac{A(,k_1)A(1k_1,1)}{-\omega_{k_1}} + \cdots, \tag{5.62}$$

where in the second and third term the Hugenholtz concept[5] of equivalent lines has been used. Let us introduce for our many-vibron diagrams a term of *p-fold equivalent lines*. Generally, p lines from a given diagram form the p-fold equivalent lines if they (*1*) all begin at the same vertex, and (*2*) all end at the same vertex (different from and not necessarily neighboring to the previous one). Then we include a factor $1/p!$ for each p-fold equivalent lines. Indeed, for the second and third diagram from the right-hand side of (5.61) we have $p = 2$.

The present many-body theory of rotation-vibration spectra can be related to the well-known technique based on the successive contact transformations[73] of the original full Hamiltonian. These successive transformations are made in order to diagonalize the Hamiltonian, namely, with respect to the pure vibrational unperturbed states. The basic difference that exists between the present many-body approach and contact-transformation method is that the model Hamiltonians built up in these two methods are different. In particular, in the contact-transformation method the intermediate normalization of perturbed state vectors is changed by the exact normalization, that is, it is assumed that the wave operator is a unitary operator.

E. Fine and Hyperfine Interactions in Atoms and Molecules

The many-body diagrammatic version of the E-independent model Hamiltonian can be profitably used for calculation of the relativistic[4,74] and electromagnetic[75] effects in atomic and/or molecular systems. In perturbative calculations of these effects, the original perturbation H_1, (4.1b), should be modified by additional one- and two-particle terms describing some preselected types of relativistic and electromagnetic interactions. Formally, these terms can be written as follows[76]:

$$H'_1 = \sum_\alpha H'_1(\alpha)a_\alpha + \sum_{\alpha\alpha'} H'_1(\alpha\alpha')a_\alpha a_{\alpha'} + \cdots \tag{5.63}$$

where the parameters a_α's denote (1) the components of an external electric or magnetic field, and (2) the components of total electronic spin, dipole and quadrupole nuclear (magnetic and/or electric) moments, and so on. Frequently, one is not primarily interested in the total energy of the system but rather in the effect of small additional perturbative effects (5.63). Let us assume that the total energy may be written as a power series of parameters a_α's,

$$E = E_0 + \sum_\alpha A_\alpha a_\alpha + \tfrac{1}{2} \sum_{\alpha\alpha'} B_{\alpha\alpha'} a_\alpha a_{\alpha'} + \cdots \tag{5.64}$$

Here, E_0 is the exact energy of the system described by the original Hamiltonian (4.1), and coefficients $A_\alpha, B_{\alpha\alpha'}$ are defined as follows:

$$A_\alpha = \frac{\partial E}{\partial a_\alpha}\bigg|_{a_\alpha = 0} \tag{5.65a}$$

$$B_{\alpha\alpha'} = \frac{\partial^2 E}{\partial a_\alpha \partial a_{\alpha'}}\bigg|_{a_\alpha = 0} \tag{5.65b}$$

These coefficients can be directly related to experiments dealing with fine and hyperfine structure, and with electric and magnetic properties of atoms and molecules.

The many-body diagrammatic theory of the model Hamiltonian is also convenient in this case. The coefficients A_α, $B_{\alpha\alpha'}$ can be calculated analogously to the calculation of the mean values (see Section IV.C). Assuming that the parameters a_α from (5.63) play formally the same role as the parameter ω from calculation of the mean values, we obtain for coefficients A_α and $B_{\alpha\alpha'}$ the following expressions:

$$A_\alpha = A_\alpha^{(0)} + \langle \varphi_i | Q_\alpha | \psi_i \rangle \tag{5.66a}$$

$$B_{\alpha\alpha'} = B_{\alpha\alpha'}^{(0)} + \langle \varphi_i | Q_{\alpha\alpha'} | \psi_i \rangle \tag{5.66b}$$

where $A_\alpha^{(0)}$ and $B_{\alpha\alpha'}^{(0)}$ are contributions to these coefficients arising from the core subsystem. They are determined by the Goldstone–Hugenholtz diagrammatic expansion[12] containing ground-state connected diagrams built up from original vertices, (4.2a) and (4.2b) and from (1) one $H'_1(\alpha)$ vertex (for calculation A_α), or (2) either two vertices $H'_1(\alpha)$, $H'_1(\alpha')$ or one vertex $H'_1(\alpha\alpha')$ (for calculation $B_{\alpha\alpha'}$). The operators Q_α and $Q_{\alpha\alpha'}$ from (5.66a) and

(5.66b) are formally defined as follows:

$$Q_\alpha = \frac{\partial \{ G \}_{LC}}{\partial a_\alpha} \bigg|_{a_\alpha = 0} \tag{5.67a}$$

$$Q_{\alpha\alpha'} = \frac{\partial^2 \{ G \}_{LC}}{\partial a_\alpha \partial a_{\alpha'}} \bigg|_{a_\alpha = 0} \tag{5.67b}$$

that is, they can be obtained from the original many-body model interactions as the set of all diagrams with new additional vertices appearing just once and twice, respectively, in all possible successive places [cf. (4.38)].

Acknowledgments

The author wishes to acknowledge the stimulation and help from friends and colleagues S. Biskupič, A. Holubec, I. Hubač, V. Laurinc, and J. Vojtik.

References

1. T. Morita, *Progr. Teoret. Phys.*, **29**, 351 (1963).
2. B. H. Brandow, *Rev. Mod. Phys.*, **39**, 771 (1967).
3. B. H. Brandow, in C. Bloch, Ed., *Proceedings of the International School of Physics "Enrico Fermi", Course 36*, Academic Press, New York, 1966.
4. P. G. H. Sandars, *Advan. Chem. Phys.*, **14**, 365 (1969).
5. N. M. Hugenholtz, *Physica*, **23**, 481 (1957).
6. P. O. Löwdin, *J. Math. Phys.*, **3**, 1171 (1962).
7. P. O. Löwdin, *J. Math. Phys.*, **3**, 969 (1962).
8. J. Koutecký and J. Čížek, *Czech. J. Phys. B*, **12**, 567 (1962).
9. Ka. A. Brueckner, in C. DeWitt, Ed., *The Many-Body Problem*, Dunod Cie, Paris, 1959.
10. V. Kvasnička, *Chem. Phys. lett.*, **32**, 167 (1975),
11. V. Kvasnička, *Czech. J. Phys. B* (to be published).
12. D. J. Thouless, *The Quantum Mechanics of Many-body Systems*, Academic Press, New York, 1961.
13. V. Kvasnička, *Czech. J. Phys. B*, **24** 605 (1974).
14. I. Lindgren, *J. Phys. B*, 7 2441 (1974).
15. C. Bloch, *Nucl. Phys.*, **6**, 329 (1958).
16. A. A. Kieselev and V. N. Popov, *Vestnik Leningradskogo Universitita, Fizika i Khimiya*, No. 22, 31 (1972); No. 4, 16 (1973).
17. This simple possibility for obtaining the approximate model interaction was initially suggested by Brandow[2] and Sanars.[4]
18. J. Soliverez, *J. Phys. C*, **2**, 2161 (1969).
19. F. Jørgensen, *Mol. Phys.*, **29**, 1137 (1975).
20. K. O. Fridrichs, *Perturbation of Spectra in Hilbert Space*, American Mathematical Society, Providence, Rhode Island, 1965, Chapter II, §6.
21. V. Kvasnička and A. Holubec, *Chem. Phys. Lett.*, **32**, 489 (1975).
22. H. Primas, *Rev. Mod. Phys.*, **35**, 710 (1963).

24. P. Roman, *Advanced Quantum Theory*, Addison-Wesley, Reading, Mass., 1965, Chapter 4.6.
25. K. A. Brueckner, *Phys. Rev.*, **97**, 1353 (1955).
26. J. Goldstone, *Proc. Roy. Soc. London, Ser. A*, **239**, 267 (1957).
27. C. Bloch and J. Horowitz, *Nucl. Phys.*, **8**, 91 (1958).
28. S. Okubo. *Progr. Theoret. Phys.*, **12**, 603 (1954).
29. D. J. Klein, *J. Chem. Phys.*, **61**, 786 (1974).
30. B. H. Brandow, in B. R. Barrett, Ed., *Effective Interactions and Operators in Nuclei, Lecture Notes in Physics*, Vol. 40, Springer-Verlag, Berlin, 1975.
31. G. Oberlechner, F. Owoni-N-Guema, and J. Richter, *Nuovo Cimento B*, **68**, 23 (1970).
32. M. B. Johnson and M. Baranger, *Annals of Physics*, **62**, 172 (1971).
33. T. T. S. Kuo, S. Y. Lee, and K. F. Ratcliff, *Nucl. Phys. A*, **176**, 65 (1971).
35. V. Kvasnička and I. Hubač, *J. Chem. Phys.*, **60**, 4483 (1974).
36. V. Kvasnička, *Czech. J. Phys. B*, **25**, 371 (1975).
37. J. Čížek, *Advan. Chem. Phys.*, **14**, 35 (1969).
38. S. S. Schweber, *An Introduction to Relativistic Quantum Field Theory*, Row-Peterson, Evanston, Ill., 1961, Chapter 4.
39. K. F. Freed, *Chem. Phys. Lett.*, **13**, 91 (1972); **17**, 331 (1972); **24**, 275 (1974).
40. K. F. Freed, *J. Chem. Phys.*, **60**, 1765 (1974).
41. P. Westhaus, E. G. Bradford, and D. Hall, *J. Chem Phys.*, **62**, 1607 (1975).
42. V. Kvasnička, *Phys. Rev. A*, **12**, 1159 (1975).
43. O. Sinanoğlu, *Advan. Chem. Phys.*, **6**, 315 (1964).
44. J. H. Van Vleck, *Phys. Rev.*, **33**, 467 (1929).
45. F. Jorgensen and J. Pedersen, *Mol. Phys.*, **27**, 33 (1974); **27**, 959 (1974).
46. The nonorthogonal set of ASOs has been used in Ref. 42.
47. J. N. Murrell and A. J. Harget, *Semiempirical Self-Consistent-Field Molecular Orbital Theory of Molecules*, Wiley, London, 1972.
48. I. Hubač, V. Kvasnička, and A. Holubec, *Chem. Phys. Lett.*, **23**, 381 (1973).
49. V. Kvasnička, A. Holubec, and I. Hubač, *Chem. Phys. Lett.*, **24**, 361 (1974).
50. I. Hubač and M. Urban, *Theoret. Chim Acta* (to be published).
51. V. Kvasnička, V. Laurinc, and I. Hubač, *Phys. Rev. A*, **10** 2016 (1974).
52. U. Kaldor, *Phys. Rev. Lett.*, **31**, 1338 (1973).
53. U. Kaldor, *J. Chem. Phys.*, **63**, 2199 (1975).
54. V. Kutzelnigg and V. H. Smith, *J. Chem. Phys.*, **41**, 896 (1964).
55. K. A. Brueckner and V. Wada, *Phys. Rev.*, **103**, 1008 (1956).
56. R. K. Nesbet, *Phys. Rev.*, **109**, 1632 (1958).
57. W. Brenig, *Nucl. Phys.*, **4**, 363 (1957).
58. H. Primas, in O. Sinanoğlu, Ed., *Modern Quantum Chemistry, Istanbul Lectures*, Part II, Academic Press, New York, 1965.
59. K. H. Kobe, *Phys. Rev. C*, **3**, 417 (1971).
60. P. O. Löwdin, *Phys. Rev.*, **97**, 1474 (1955).
61. D. H. Kobe, *J. Chem.Phys.*, **50**, 5183 (1969).
62. L. Schäfer and H. A. Weidenmüller, *Nucl. Phys. A*, **174**, 1 (1971).
63. B. H. Brandow, in K. T. Mahantheppa and W. E. Brittin, Eds. *Lectures in Theoretical Physics, Vol XIB*, Gordon and Breach, New York, 1969.
64. B. H. Brandow, *Annals of Physics*, **57**, 214 (1970).
65. M. W. Kirson, *Nucl. Phys. A*, **115**, 49 (1968).
66. M. W. Kirson, *The Structure of Nuclei*, IAEA, Vienna, 1972, p. 257.
67. B.. R. Barrett and M. W. Kirson, *Advan. Nucl. Phys.*, **6**, 219 (1973).

68. V. Kvasnička, *Theoret. Chim. Acta*, **36**, 297 (1975).
69. R. K. Nesbet, *Advan. Chem. Phys.*, **9**, 321 (1965).
70. C. C. J. Roothaan, *Rev. Mod. Phys.*, **23**, 69 (1951).
71. V. Kvasnička, *Mol. Phys.* (to be published).
72. H. H. Nielsen, *Rev. Mod. Phys.*, **23**, 90 (1951).
73. G. Amat, H. H. Nielsen, and G. Tarago, *Rotation-Vibration Spectra of Molecules*, Marcel Dekker, New York, 1971.
74. S. Garpman, I. Lindgren, J. Lindgren, and J. Morrison, *Phys. Rev. A*, **11**, 758 (1975).
75. S. Biskupič, Thesis, Department of Physical Chemistry, Slovak Technical University, 1976.
76. R. McWeeney and B. T. Sutcliff, *Methods of Molecular Quantum Mechanics*, Academic Press, New York, 1969, Appendix IV.

INTERACTIONS OF SLOW ELECTRONS WITH BENZENE AND BENZENE DERIVATIVES*

L. G. CHRISTOPHOROU[†] and M. W. GRANT

*Department of Physics, University of Tennessee,
Knoxville, Tennessee 37916*

and

D. L. McCORKLE

*Health Physics Division, Oak Ridge National Laboratory,
Oak Ridge, Tennessee 37830*

CONTENTS

*Sponsored, in part, by the U.S. Energy Research and Development Administration under Contract No. AT-(40-1)-4703.

†Also, Health Physics Division, Oak Ridge National Laboratory, Oak Ridge, Tennessee 37830.

413

I. INTRODUCTION

As part of our effort to extend our basic understanding of the funda-mental processes that accompany the interaction of radiation (ionizing and nonionizing) with matter from simple to complex molecules, we synthesize in this work knowledge on slow-electron interactions with benzene and its derivatives. Benzene and its derivatives remain the common building blocks of many biomolecules and relate to many biologically important structures. Hopefully, the synthesis of this knowledge will allow a better understanding of these basic organic molecules and their interactions with slow electrons, which, in turn, could serve as a basis for understanding bigger organic structures.

Benzene and the effect of substitution in the benzene ring have been the subject of many theoretical and experimental studies. However, the former studies have been almost exclusively confined to the neutral species and the latter have been conducted using mostly optical, rather than particle-impact, methods. Electron-impact studies of benzene and its derivatives are rather recent and still somewhat fragmentary in spite of their significance in providing supplementary information to photon-impact methods. On the other hand, most of the theoretical work has treated the π-electron system independently of the rest of the molecule. Although this is by no means always valid, it simplifies the task considerably. For the purpose of aiding the discussions in this chapter, the six π-molecular orbitals $\psi_1, \psi_2, \ldots, \psi_6$ in benzene (D_{6h} symmetry) are shown in Fig. 1. The ordering of these by orbital energy is $\varepsilon_1' < \varepsilon_2' = \varepsilon_3' < \varepsilon_4' = \varepsilon_5' < \varepsilon_6'$. In group terminology the orbitals corresponding to $\varepsilon_1' \ldots \varepsilon_6'$ are, respectively, a_{2u}, $e_{1g,1}$, $e_{1g,2}$, $e_{2u,1}$, $e_{2u,2}$, and b_{2g}. Thus the ground-state configuration of the π-electron system is $\psi_1^2 \psi_2^2 \psi_3^2$ or in group terminology $a_{2u}^2 e_{1g}^4$. This state is designated $^1A_{1g}$.

The interactions of slow electrons with benzene and its derivatives are treated coherently and comprehensively. Emphasis is, however, concentrated on direct and indirect excitation of benzene and its derivatives by slow electrons, negative-ion resonances, electron attachment and detachment processes, ionization and fragmentation under electron impact, and the motion of thermal and epithermal electrons through aromatic vapors. In Section VII a number of conclusions are drawn and some general observations are made.

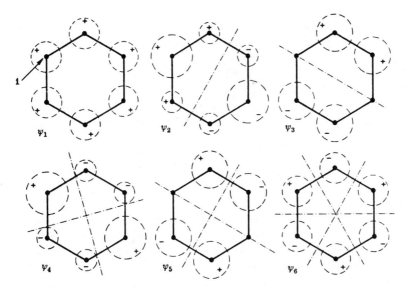

Fig. 1.   Shapes of orbital functions derived from six $2p_z$ orbitals of six carbon atoms in a hexagonal C_6 ring. Only the shape and sign of the functions above the C_6 plane are shown. The dot-dash lines indicate the nodal planes through the z-axis. (From G. Herzberg, *Molecular Spectra and Molecular Structure*, Vol. III, Litton Educational Publishing, Inc., 1966. Reprinted by permission of Van Nostrand Reinhold Company.)

II. ELECTRONIC EXCITATION BY ELECTRON IMPACT

A. The Electron-Scattering Process and the Nature of the Various Approximations

While the bulk of the theory on electron-molecule interactions has been confined largely to small molecules, some work has been reported on benzene. Before summarizing this latter work it is instructive to outline some of the approximations that are usually made in the theory of electron-scattering processes.

If, for convenience, we refer to electron-hydrogen atom scattering we may write[1] for the asymptotic form of the wave function $\phi_n(\mathbf{r})$ for an electron at large r following excitation of the nth state

$$\phi_n(\mathbf{r}) \cong e^{i\mathbf{k}_1 \cdot \mathbf{r}} \delta n_1 + \frac{1}{r} e^{ik_n r} f_n(\theta, \phi) \tag{1}$$

The quantity $f_n(\theta, \phi)$ is the scattering amplitude and has the form

$$f_n(\theta, \phi) = -\frac{2m}{4\pi\hbar^2} \int \int e^{-i\mathbf{k}_n \cdot \mathbf{r}_2} \psi_n^*(\mathbf{r}_1) \left| \frac{e^2}{r_{12}} - \frac{e^2}{r_2} \right| \Psi(\mathbf{r}_1, \mathbf{r}_2) \, d\mathbf{r}_1 \, d\mathbf{r}_2 \tag{2}$$

In (2) $\mathbf{r}_1$ and $\mathbf{r}_2$ are, respectively, the positions of the hydrogen-atom electron in the $1s$ state and the projectile electron, referred to the proton as origin; r_{12} is the distance between the two electrons; m is approximately the electron mass; $\mathbf{k}_1 = \mathbf{p}_1 \hbar$; $\mathbf{k}_n = \mathbf{p}_n \hbar$ are the wave vectors of the incident and scattered electrons; and (θ, ϕ) specify the direction of the scattered electron. The $\psi_n(\mathbf{r}_1)$ are hydrogen-atom wave functions and $\Psi(\mathbf{r}_1, \mathbf{r}_2)$, the total wave function characterizing the two-electron system, is the solution to

$$\left[-\frac{\hbar^2}{2m} (\nabla_1^2 + \nabla_2^2) - \frac{e^2}{r_1} - \frac{e^2}{r_2} + \frac{e^2}{r_{12}} - E_T \right] \Psi(\mathbf{r}_1, \mathbf{r}_2) = 0 \tag{3}$$

The quantity $f_n(\theta, \phi)$ is the probability amplitude for scattering the incident electron in the direction (θ, ϕ) with concomitant excitation of the target from the initial state 1 to an excited state n. The wave number k_n of the scattered electron is defined by

$$\frac{\hbar^2}{2m} k_n^2 = E_T - E_n \tag{4}$$

where E_n is the energy of the excited (final) state of the hydrogen atom.

The quantity

$$I_n(\theta,\phi) = \frac{k_n}{k_1} |f_n(\theta,\phi)|^2$$

is the corresponding differential excitation cross section and if $I_n(\theta,\phi)$ is integrated over all angles one obtains the total cross section, σ_n, for excitation into the state n,

$$\sigma_n = \frac{k_n}{k_1} \int \int |f_n(\theta,\phi)|^2 \sin\theta \, d\theta \, d\phi \tag{5}$$

The differential excitation cross section $I_n(\theta,\phi)$ is related to another quantity of interest in energy-loss experiments, namely, to the generalized oscillator strength $\hat{f}_n$, by[2,3] (in atomic units)

$$\hat{f}_n = \left(\frac{E_n}{2}\right)\left(\frac{k_1}{k_n}\right)(k_1 - k_n)^2 I_n(\theta,\phi) \tag{6}$$

The generalized oscillator strength approaches the value of the optical oscillator strength in the limit $(k_1 - k_n) \to 0$. This and other properties of the generalized oscillator strength have been discussed by Lassettre (see, for example, Refs. 3 and 4).

In order to calculate $f_n(\theta,\phi)$ various approximations for $\Psi(\mathbf{r}_1,\mathbf{r}_2)$ are introduced. In the Born approximation, $\Psi(\mathbf{r}_1,\mathbf{r}_2)$ is written as a product of the incident electron wave function $e^{i\mathbf{k}_1\cdot\mathbf{r}_2}$ and the hydrogenic wave function $\psi_1(\mathbf{r}_1)$:

$$\Psi(\mathbf{r}_1,\mathbf{r}_2) \cong e^{i\mathbf{k}_1\cdot\mathbf{r}_2}\psi_1(\mathbf{r}_1) \tag{7}$$

One then finds that the scattering amplitude is given by

$$f_n(\theta,\phi) = -\frac{2m}{4\pi\hbar^2} \int e^{i(\mathbf{k}_1 - \mathbf{k}_n)\cdot\mathbf{r}_2} V_{n1}(\mathbf{r}_2) \, d\mathbf{r}_2 \tag{8}$$

where

$$V_{n1}(\mathbf{r}_2) = \int \psi_n^*(\mathbf{r}_1)\left[\frac{e^2}{r_{12}} - \frac{e^2}{r_2}\right]\psi_1(\mathbf{r}_1) \, d\mathbf{r}_1 \tag{9}$$

In the Born approximation the excitation cross section for optically allowed transitions falls off as $\ln\varepsilon/\varepsilon$ for large incident electron energies ε. For optically forbidden transitions at large ε the excitation cross section decreases faster (as $1/\varepsilon$) than for optically allowed transitions.[5] The Born approximation is a high-energy approximation and is valid for large incident electron energies, usually at least 200 eV, but this value depends on the scattering target and the state involved.

A natural extension of the Born approximation is the Born–Oppenheimer approximation in which $\Psi(\mathbf{r}_1,\mathbf{r}_2)$ is the antisymmetric product function

$$\Psi^{\pm}(\mathbf{r}_1,\mathbf{r}_2) = N\left[e^{i\mathbf{k}_1\cdot\mathbf{r}_2}\psi_1(\mathbf{r}_1) \pm e^{i\mathbf{k}_1\cdot\mathbf{r}_1}\psi_1(\mathbf{r}_2)\right]\eta \tag{10}$$

in accord with the fact that electrons are fermions. The quantity η is the appropriate symmetric or antisymmetric spin function and N is a normalization constant. In this approximation the scattering amplitude is given by

$$f_n^{\pm}(\theta,\phi) = -\frac{2m}{4\pi\hbar^2}\left\{ \int\int e^{i(\mathbf{k}_1-\mathbf{k}_n)\cdot\mathbf{r}_2}\psi_n^*(\mathbf{r}_1)\left[\frac{e^2}{r_{12}} - \frac{e^2}{r_2}\right]\psi_1(\mathbf{r}_1)\,d\mathbf{r}_1\,d\mathbf{r}_2\right.$$

$$\left. \pm \int\int e^{i(\mathbf{k}_1\cdot\mathbf{r}_1-\mathbf{k}_n\cdot\mathbf{r}_2)}\psi_n^*(\mathbf{r}_1)\left[\frac{e^2}{r_{12}} - \frac{e^2}{r_1}\right]\psi_1(\mathbf{r}_2)\,d\mathbf{r}_1\,d\mathbf{r}_2\right\} \tag{11}$$

where the plus sign is for the singlet case and the minus sign is for the triplet case. The first term in (11) is the Born scattering amplitude, (8), and the second term corresponds to exchange scattering. As a consequence of this latter term the probability amplitude for spin-forbidden transitions is nonzero.

Ochkur[6] expanded the exchange amplitude in the Oppenheimer approximation in powers of k_1^{-1} and retained only the leading term, which behaves as k_1^{-2}. Furthermore, he assumed that the e^2/r_1 interaction decays faster than the k_1^{-2} term and can thus be neglected. Under these assumptions the exchange amplitude $g_n(\theta,\phi)$ is given by

$$g_n(\theta,\phi) \cong -\frac{2me^2}{\hbar^2 k_1^2}\int e^{i(\mathbf{k}_1-\mathbf{k}_n)\cdot\mathbf{r}_2}\psi_n^*(\mathbf{r}_2)\psi_1(\mathbf{r}_2)\,d\mathbf{r}_2 \tag{12}$$

and the scattering amplitude $f_n^{\pm}(\theta,\phi)$ by

$$f_n^{\pm}(\theta,\phi) = -\frac{2}{a_0}\left[\frac{1}{K^2} \pm \frac{1}{k_1^2}\right]\int\psi_n^*(\mathbf{r})e^{i(\mathbf{k}_1-\mathbf{k}_n)\cdot\mathbf{r}}\psi_1(\mathbf{r})\,d\mathbf{r} \tag{13}$$

where a_0 is the Bohr radius and $\mathbf{K}=\mathbf{k}_1-\mathbf{k}_n$. A variation of this approximation was obtained by Rudge.[7,8] The results of the Ochkur and the Ochkur–Rudge approximations as well as those of the Born approximation are presented below for the case of benzene.

B. Theoretical calculations of the excitation of benzene by electron impact

The calculation of $f_n(\theta,\phi)$ and $f_n^\pm(\theta,\phi)$ in the case of electron-impact excitation of benzene follows the development given above for the hydrogen atom except that the hydrogenic wave functions $\psi_n(\mathbf{r}_1)$ and $\psi_1(\mathbf{r}_1)$ are now replaced by multi-electron wave functions of benzene. These are usually some product function of occupied molecular orbitals. The simplest molecular orbital description of benzene is the free-electron model[9] in which the π-electron's motion is restricted to the perimeter of the molecule. Other descriptions in the LCAO (linear combination of atomic orbitals) formalism include Hückel theory, extended Hückel theory, and various SCF (self-consistent field) formulations.[10] Below we review some of the results obtained utilizing the free-electron model and the LCAO formalism.

Inokuti[11] and Matsuzawa[12] have applied the Born approximation to the scattering of fast electrons by the benzene molecule. They used free-electron molecular orbitals of the form $u_m(x)=(1/\sqrt{2\pi})e^{imx}$, $m=0,\pm1,\pm2,\ldots$ where x is the angular coordinate of the π electron. Inokuti[11] considered the direction of the incident electron to be along the axis perpendicular to the molecular plane and calculated total π-electron excitation cross sections at various incident electron energies. Matsuzawa[12] extended this work by averaging the cross sections over all incident angles. In Table I Matsuzawa's results for the $\pm1\to\pm2(\overline{Q_2^0})$ and the $0\to\pm3(\overline{Q_3^i})$

TABLE I
Free-Electron-Model-Based Electron-Impact Excitation
Functions for Benzene[12]

Incident Energy (eV)	$\overline{Q_2^0}$ (πa_0^2)	$\overline{Q_3^i}$ (πa_0^2)	$\overline{Q_3^0}+\overline{Q_3^i}$ (πa_0^2)
100	5.950	0.458	0.604
200	3.479	0.230	0.308
300	2.500	0.154	0.197
400	1.980	0.116	0.155
700	1.238	0.066	0.090
1000	0.919	0.046	0.063
1500	0.646	0.031	0.042
10000	0.124	4.6×10^{-3}	6.3×10^{-3}
30000	0.046	1.5×10^{-3}	2.1×10^{-3}

excitations are listed. From Table I it can be seen that the free-electron model predicts within the Born approximation that the total π-electron excitation cross sections are substantial and that they decrease as the incident electron energy increases.

Read and Whiterod[13] and Matsuzawa[14] have performed calculations on the electron impact excitation of benzene utilizing LCAO π-molecular orbitals. Using the Born approximation, Read and Whiterod calculated the total excitation cross sections for the singlet states $^1B_{1u}$, $^1B_{2u}$, and $^1E_{1u}$ of benzene. Similarly, Matsuzawa used the Ochkur and the Ochkur–Rudge approximations and calculated the total excitation cross sections for the triplet states $^3B_{1u}$, $^3B_{2u}$, and $^3E_{1u}$. These results are summarized in Table II. We have assumed that the expressions in column 2 of Table II hold down to the excitation thresholds and plotted in Fig. 2 the total excitation cross sections as a function of the incident electron energy ε. It is seen that the theory predicts increasing triplet-to-singlet ratios as the energy of the incident electron decreases. At $\varepsilon \gtrsim 100$ eV the singlet→triplet transitions are negligibly weak compared to singlet→singlet excitations. Close to the excitation thresholds, however, they are quite strong, owing to electron exchange. This, as is seen in the next section, is in agreement with the experimental results. In Fig. 3 the normalized angular distributions[14] for the three singlet-to-triplet transitions are plotted as a function of the scattering angle. It is seen that the maximum intensities for these transi-

TABLE II

LCAO Molecular Orbital Results for Benzene Excitation Functions

Excitation	Excitation Cross-section Function $(10^{-18}\text{cm}^2)^a$	Approximation	Remarks
$^1B_{1u}$	$1260/\varepsilon$	Born[b]	Symmetry-forbidden
$^1B_{2u}$	$90.3/\varepsilon$	Born[b]	Symmetry-forbidden
$^1E_{1u}$	$\dfrac{58,700}{\varepsilon}\ln\dfrac{13.24\varepsilon}{E^2}$ [c]	Born[b]	Optically-allowed
$^3B_{1u}$	$15.48\times10^4/\varepsilon^3$	O(OR)[d,e]	Spin-forbidden
$^3B_{2u}$	$3.78\times10^4/\varepsilon^3$	O(OR)[d,e]	Spin-forbidden
$^3E_{1u}$	$24.98\times10^4/\varepsilon^3$	O(OR)[d,e]	Spin-forbidden

[a] In the expressions below, E and ε are in eV.
[b] Reference 13; for $\varepsilon \geqslant 300$ eV.
[c] E is the energy difference between the excited state and the ground state.
[d] Reference 14.
[e] Above 100 eV the Ochkur (O) and Ochkur–Rudge (OR) approximations give approximately the same result.

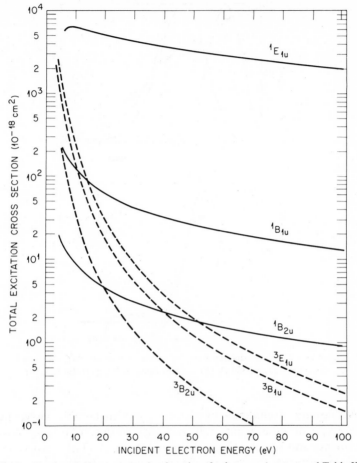

Fig. 2. Total excitation cross section functions for benzene (see text and Table II).

tions occur at large scattering angles and that the angles corresponding to the maximum intensities increase as the incident electron energy decreases. This is an interesting finding; it suggests that large-angle low-energy electron scattering may be used as a means for detecting weak transitions observed neither in forward scattering nor in optical absorption studies.

C. Experimental Studies

In general, the excitation of molecules by low-energy electrons has been studied by measuring either the scattered electron current or the radiation that is emitted following deexcitation of the excited molecule, or—for

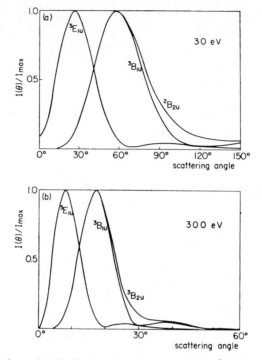

Fig. 3. Normalized angular distributions for excitations of the $^3B_{1u}$, $^3B_{2u}$, and $^3E_{1u}$ states for 30 eV (a) and 300 eV (b) incident electron energy (from Ref. 14).

metastable-state excitation—by detecting the metastable molecules. The former type of experiments fall into two groups: (*1*) energy-loss (EL) experiments and (*2*) threshold-electron excitation (TEE) experiments. In the EL experiments the primary electron energy is fixed and usually much larger than the excitation thresholds of the target. The energies of the scattered electrons in the forward direction or at some other angle are analyzed and the energy losses suffered by the primary electron beam in collisions with the target atoms or molecules are identified. The spectrum so obtained is referred to as the energy-loss spectrum (ELS). In the TEE experiments the energy of the primary electron beam is varied from 0 eV upward. When the energy of the primary beam matches the excitation thresholds of the target, the electron may lose essentially all of its energy in a single collision. When this happens the electrons are collected, trapped, and the resultant "trapped-electron" current reflects the excitation spectrum of the target, referred to as the threshold-electron excitation spectrum (TEES). The thermal electrons are trapped either by a scavenger gas with a large and sharply peaking capture cross section at ~ 0.0 eV such as SF_6 or by a small potential well depth such as in the "trapped-electron method." The TEE technique is uniquely suited for locating excited states and negative-ion resonances, but it yields very little information as to the magnitude and the energy dependence of the excitation cross sections.

From the data summarized by Christophorou[5] and those in this section on benzene it is apparent that at energies in excess of ~ 200 eV, the Born approximation satisfactorily accounts for the observed ELS. The electronic excitation spectra so obtained resemble the optical-dipole excitation (i.e., absorption) spectrum (see Section II.C.1). For sufficiently low-energy electrons, optical selection rules involving change in multiplicity, angular momentum, and symmetry are greatly relaxed, and a study of both optically allowed and optically forbidden transitions becomes possible. At primary electron energies less than ~ 50 to 60 eV, optically forbidden transitions are easily detectable in diatomic molecules in forward scattering, but even more easily at large scattering angles.[5] For polyatomic molecules, primary electron energies closer to the excitation thresholds are required to violate optical selection rules in forward scattering. Certain forbidden transitions not detectable in forward scattering can be studied with ease at large scattering angles. These features are clearly borne out by experiment.

1.   Energy-Loss Spectra of Benzene

The energy-loss spectrum of benzene obtained by Skerbele and Lassettre[15] at zero scattering angle and 300 eV incident electron energy is shown in Fig. 4. It is seen to resemble closely the optical excitation spectrum obtained by Koch and Otto[16] using synchrotron light, and hence optical selection rules hold at these electron energies and scattering angles. Energy-loss spectra of benzene taken with better energy resolution and at kinetic energies between 40 and 100 eV and scattering angles to 16° were also measured by Lassettre et al.,[17] in an effort to investigate electric-quadrupole and singlet→triplet transitions. Spectra taken with 40, 50, and 100 eV incident electrons and at small scattering angles (0° to 5°) were very similar to the one shown in Fig. 4. Vibrational structure due to excitation of the first singlet ($^1B_{2u}$) state was well resolved. Excitation of this state from an excited vibrational level in the ground state was also observed. However, no singlet→triplet transitions were found. Apparently, lower electron energies are required for such transitions (see discussion later in this section).

One of the most interesting findings in the work of Lassettre et al.[17] is the change in relative intensity of the ELS between 6.2 and 6.5 eV with scattering angle. This is shown in Figs. 5a and b for accelerating voltages

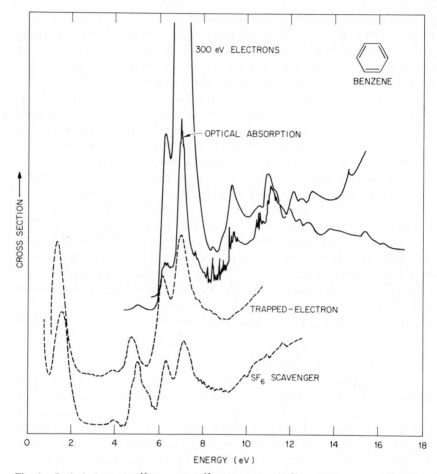

Fig. 4. Optical absorption,[16] energy-loss,[15] trapped-electron,[21] and SF_6 scavenger[22] spectra of benzene (see text).

50 and 90 V, respectively. It is further seen from Fig. 5c that the relative intensity of the ELS in this region is essentially independent of the incident electron energy. By comparing the data in Fig. 5 with similar ones on NH_3 and CO_2, Lassettre et al.[17] concluded that two different transitions are involved in this spectral region, one at 6.2 eV and another with vibrational structure at 6.31, 6.41, and 6.53 eV.

The ELS of benzene at even lower incident electron energies than those employed by Lassettre et al.[17] was obtained by Doering[18] with 13.6 and 20.0 eV incident electron energies. Doering identified, at large scattering

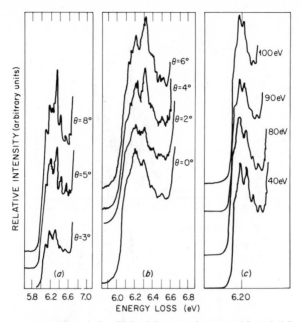

Fig. 5.   Relative intensity of the ELS of benzene between 6.2 and 6.5 eV at various scattering angles and 50 (*a*) and 90 (*b*) V accelerating voltages; relative intensity of the ELS of benzene at the indicated electron energies (*c*) (from Ref. 17).

angles, the three lowest excited triplet and singlet states of benzene. In Fig. 6 the ELS of benzene obtained at 75° and 14° scattering angles and for 13.6 eV incident electron energy are shown.[18] Five peaks in the 75° spectrum are seen at 3.9, 4.8, 5.6, 6.2, and 6.9 eV. The peaks at 3.9 and 5.6 eV were assigned to the first and third triplet states, T_1 and T_3; those at 6.2 and 6.9 eV were assigned to the second and third singlet states, S_2 and S_3. The 4.8-eV peak is a combination of excitation of the second triplet (T_2) and the first singlet (S_1) states. Lassettre et al.[17] had previously resolved the vibrational progression of the singlet state between 4.7 and 5.2 eV, but at higher incident electron energy and smaller scattering angles. Doering[18] separated T_2 from S_1 by noting the changes in the intensity of the energy-loss peak between 4.7 and 5.0 eV with increasing scattering angle, as is clearly seen in Fig. 7.

The assignment of the peak at 4.7 to 5.0 eV to a combination of T_2 and S_1 is further supported by the changes in relative intensities of the various peaks with scattering angle shown in Fig. 8. Although the ratio S_2/S_3 of the intensities of S_2 and S_3 is constant for 10° to 80° scattering angles, the ratios T_1/S_3 and T_3/S_3 of the intensities of T_1 and T_3 to S_3 decrease

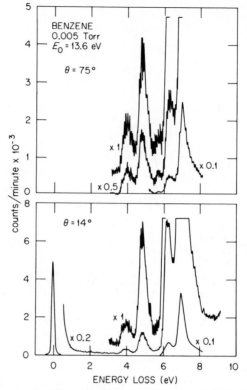

Fig. 6. ELS of benzene at 75° and 14° scattering angles and 13.6 eV incident electron energy (from Ref. 18).

sharply with decreasing scattering angle. The ratios T_1/S_3 and T_3/S_3 increase by an order of magnitude in the 10° to 80° scattering-angle range. Additionally, it was noted that although T_3/S_3 is proportional to T_1/S_3 in this scattering angle range, the ratio $(T_2+S_1)/S_3$ does not increase as rapidly as the triplet-to-singlet ratios. This is taken as confirmation of the earlier conclusion that the peak at 4.7 to 5.0 eV in the ELS is indeed a combination of T_2 and S_1. Further support for this is provided by the TEE spectra described in the next section.

The data in Figs. 7 and 8 clearly show that spin-forbidden transitions are more easily detectable at large scattering angles. Unfortunately, no ELS exist for benzene derivatives.

Finally, in a recent publication Azria and Schulz[19] reported cross-section functions for vibrational and triplet-state excitation of benzene. With 4.8-eV incident electrons, the ELS of benzene showed energy-loss

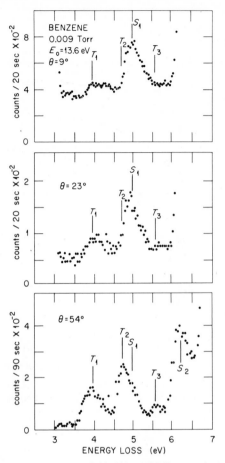

Fig. 7. ELS of benzene at 13.6 eV and 9°, 23°, and 54° scattering angles. Note the shift in the position of the maximum (4.7 to 5 eV) with increasing scattering angle (from Ref. 18).

processes of 0.125 and 0.38 eV, which correspond to the ν_2 and ν_1 vibrational modes of benzene. The differential cross section for the vibrational excitation of the ν_1 mode showed two peaks at 4.9 and ~8 eV. The 4.9-eV peak was attributed[19] to the third NIR state (see Section III) of benzene and that at ~8 eV to the decay of a shape resonance, which may also lead to dissociative attachment.[19]

The energy dependence of the differential cross section for excitation of the first triplet ($^3B_{1u}$) state of benzene at 60° is shown in Fig. 9. The curve exhibits three peaks at 6.0, 7.6, and 10.2 eV, which were interpreted[19] as due to the decay of NIR states at these energies. Also shown in Fig. 9 is

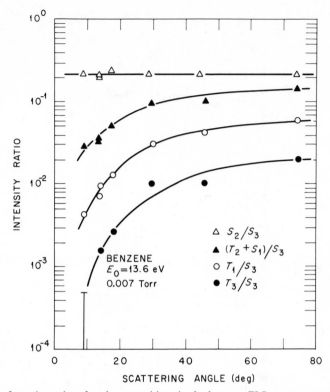

Fig. 8.   Intensity ratios of various transitions in the benzene ELS versus scattering angle. The bar for the ratio T_3/S_3 represents the upper limit that could be placed on the intensity of this transition from the 9° scattering spectrum. The incident energy used was 13.6 eV (from Ref. 18).

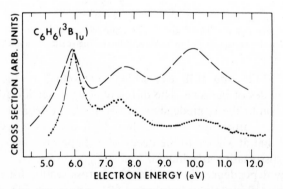

Fig. 9.   Energy dependence of the cross section for excitation of the first triplet state ($^3B_{1u}$) of benzene (-·-·-) (Ref. 19); (---) (Ref. 20). The two curves have been normalized to the same height at the 6-eV peak.

the total excitation function for the formation of metastable states of benzene measured by Smyth et al.[20] The latter investigators suggested, however, that only the first triplet state is detected in their experiment, although they did not rule out other possibilities (see further discussion in Section II.C.3).

2. Threshold-Electron Excitation of Benzene
And Benzene Derivatives

In Fig. 4 the threshold-electron excitation spectrum of benzene obtained using the trapped electron method[21] and that obtained using the SF_6 scavenger technique[22] is compared with the optical excitation spectrum of Koch and Otto[16] and the energy-loss spectrum of Skerbele and Lassettre[15] obtained with 300-eV primary electrons and in forward scattering. The EL spectrum is in general agreement with the optical absorption spectrum, but both differ drastically from the TEE spectra.

The most intense peak in the three electron-impact spectra in Fig. 4 is that due to excitation of the electric-dipole transition $^1A_{1g}^-(^1A) \rightarrow ^1E_{1u}^+(^1B)$[23] at 6.96 eV, which is symmetry allowed. While the intensity of this peak in both the optical absorption spectrum and the 300-eV ELS is much greater than those ascribed to excitation of the symmetry-forbidden transitions $^1A_{1g}^-(^1A) \rightarrow ^1B_{2u}(^1L_b)$ and $^1A_{1g}^-(^1A) \rightarrow ^1B_{1u}^+(^1L_a)$ at 4.9 and 6.2 eV, respectively, these peaks have about equal intensity in the TEES. The peak at 3.9 eV and the shoulder at $\sim$4.7 eV in the TEES have been attributed[22,24] to excitation of the first $^3B_{1u}^+(^3L_a)$ and the second $^3E_{1u}^+(^3B)$ π-triplet states. The energy loss peak at $\sim$1.40 eV in the TEES is due to the lowest two, degenerate, negative-ion resonant states of benzene. This energy-loss peak as well as others, at $\sim$4.9 and $\sim$5.9 eV, are attributed to negative-ion shape resonances and are discussed in Section III. In Table III we summarize the observed electronic transitions in benzene.

In Fig. 10 the TEE spectra—obtained using the trapped-electron method — are shown for the benzene derivatives benzaldehyde,[25,26] benzoic acid,[25] fluorobenzene,[25] acetophenone,[26] and hexafluorobenzene.[26] The solid arrows and accompanying numbers above the spectra indicate the positions of the maxima and the broken arrows indicate the positions of the onsets in the TEES of Christophorou, McCorkle, and Carter.[25] The broken and solid arrows perpendicular to the energy axis give, respectively, the positions of the band onsets (0$\rightarrow$0 transitions) and band maxima obtained with photophysical methods. There is good general agreement between the maxima and the onsets determined by the two methods. However, several optically forbidden as well as negative-ion resonant states observed in the TEES are naturally absent from the optical spectra. In the case of benzaldehyde, the maxima in the TEES of Christophorou, McCorkle, and

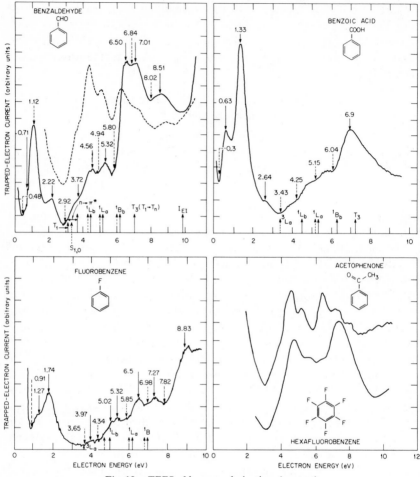

Fig. 10. TEES of benzene derivatives (see text).

Carter[25] are in good agreement with those of Brongersma et al.,[26] but the relative intensities are quite different—partly, perhaps, because of the different trapped depths employed—pointing out the large uncertainties in the intensity scale as determined in this type of experiment. The reader is referred to the original papers for a detailed discussion.

3.   Optical Emission from Benzene and Some of its Derivatives Excited by Low-Energy Electrons

In a controlled low-energy (0 to 300 eV) electron-impact experiment, Smyth et al.[27,28] measured the excitation functions for fluorescence ($S_1 \rightarrow S_0$) from benzene, toluene, and aniline. The fluorescence was observed at

TABLE III

Electronic Transitions in Benzene

Mode of excitation columns: Photons 0—0 energy[b] (eV) · Threshold electrons peak energy[c] (eV) · Energy loss via electron-impact peak energy (eV)

Assignment of observed transition[a]	Photons 0—0 energy[b] (eV)	Threshold electrons peak energy[c] (eV)	Energy loss via electron-impact peak energy (eV)	Comments	Ref.
Negative ion resonant state (shape resonance)			1.07 ± 0.07^d	Principal vibrational spacing $= 0.127 \pm 0.010$ eV	1
			1.14 ± 0.05^d	Principal vibrational spacing $= 0.123$ eV	2
			1.15 ± 0.05^d	Vibrational spacing $= 0.12 \pm 0.01$ eV	3
$^1A_{1g} \rightarrow$ Triplet		1.3			4
$\rightarrow {}^3B_{1u}$		1.35^e			5
$\rightarrow {}^3B_{1u}$		1.4			6
$\rightarrow {}^3B_{1u}$		1.55 ± 0.08			7
$\rightarrow {}^3B_{1u}$		(0.9 ± 0.3)			7
$\rightarrow ({}^3L_a)$	3.65			O$_2$ perturbation (solution)	8
$\rightarrow {}^3B_{1u}$	3.65			Phosphorescence	9
$\rightarrow T_1$	3.68			Phosphorescence	10
$\rightarrow {}^3E_{1u}$	3.68			O$_2$ perturbation (crystal)	11
		3.85^e			5
		3.9(3.6)			7
		3.9			4
			3.95		12
	4.58			O$_2$ perturbation (crystal)	11
	4.59			Fluorescence	13
$\rightarrow {}^1B_{2u} + {}^3B_{2u} + {}^3E_{1u}$	4.72(4.88)	4.7			4
$\rightarrow ({}^3B)$		~4.7(4.4)			7
$\rightarrow {}^1B_{2u}({}^1L_b)$		4.75^e			14
$\rightarrow {}^1B_{2u}$			4.75		5
$\rightarrow T_2$		4.80			12
$\rightarrow ({}^1L_b)$					7

431

TABLE III (Continued)

Assignment of observed transition[a]	Photons 0-0 energy[b] (eV)	Threshold electrons peak energy[c] (eV)	Energy loss via electron-impact peak energy (eV)	Comments	Ref.
→$^1B_{2u}$			4.902		15
→$^1B_{2u}$			4.93		16
→S_1			5.0		12
shape resonance[f]			~5		2
→$^3B_{2u}(^3L_b)$		~5.4			7
→T_3			5.6		12
→$^1B_{1u}(^1L_a)$	5.99(6.11)				14
→$^1B_{1u}$		6.0			4
→$^1B_{1u}$		6.15^e			5
→(^1L_a)		6.16			7
→$^1E_{2g}$ or $^1B_{1u}$			6.20		15
→S_2			6.2		12
→$^1B_{1u}$			6.21		16
Unidentified			6.31	Higher vibrational levels at 6.41 and 6.53 eV	15
→$^1E_{1u}(^1B)$	6.77(6.89)				14
→$^1E_{1u}$		6.9			4
→S_3			6.9		12
→$^1E_{1u}$			6.95		15
→$^1E_{1u}$			6.96		16
→$^1E_{1u}$		6.96^e			5
→(^1B)		6.96			7
→^3R			8.125		15
→$3R'$			8.358		15

432

→4R'	8.682	15
→5R'+5R"	8.853	15
→6R'+6R"	8.960	15

---

[a] This column lists the assignment as given in the respective reference; notation in parenthesis is that of J. R. Platt, *J. Chem. Phys.*, **17**, 484 (1949); T_1, T_2, and T_3 denote first, second, and third excited triplet states; S_1, S_2, and S_3 denote first, second, and third excited singlet states; R's refer to Rydberg states.

[b] Value in parenthesis refers to peak energy.

[c] Value in parenthesis refers to onset.

[d] Energy of lowest observed vibronic state.

[e] Calibration of energy scale referenced to the 6.96-eV peak in Ref. 16 below.

[f] Also, a higher-lying shape resonance at 6.9 and possibly another at 13.6 eV was reported by K. C. Smyth, J. A. Schiavone, and R. S. Freund, *J. Chem. Phys.*, **61**, 1782 (1974).

1. I. W. Larkin and J. B. Hasted, *J. Phys. B: Atom. Molec. Phys.*, **5**, 95 (1972).
2. L. Sanche and G. J. Schulz, *J. Chem. Phys.*, **58**, 479 (1973).
3. M. J. W. Boness, I. W. Larkin, J. B. Hasted, and L. Moore, *Chem. Phys. Lett*, **1**, 292 (1967).
4. H. H. Brongersma, J. A. v. d. Hart, and L. J. Oosterhoff, *Proceedings of the Nobel Symposium*, Vol. 5, Interscience, New York, 1967.
5. M. J. Hubin-Franskin and J. E. Collin, *Int. J. Mass Spectr. Ion Phys.*, **5**, 163 (1970).
6. R. N. Compton, L. G. Christophorou, and R. H. Huebner, *Phys. Lett.*, **23**, 656 (1966).
7. R. N. Compton, R. H. Huebner, P. W. Reinhardt, and L. G. Christophorou, *J. Chem. Phys.*, **48**, 901 (1968).
8. D. F. Evans, *J. Chem. Soc.*, 1351 (1957).
9. H. Shull, *J. Chem. Phys.*, **17**, 295 (1949).
10. G. C. Nieman, Thesis, California Institute of Technology, 1965.
11. S. D. Colson and E. R. Bernstein, *J. Chem. Phys.*, **43**, 2661 (1965).
12. J. P. Doering, *J. Chem. Phys.*, **51**, 2866 (1969).
13. J. B. Birks, C. L. Braga, and M. D. Lumb, *Proc. Roy. Soc. (London)*, **A283**, 83 (1965).
14. H. H. Jaffe and M. Orchin, *Theory and Applications of Ultraviolet Spectroscopy*, Wiley, New York, 1962. See also, J. B. Birks, L. G. Christophorou, and R. H. Huebner, *Nature*, **217**, 809 (1968).
15. E. N. Lassettre, A. Skerbele, M. A. Dillon, and K. J. Ross, *J. Chem. Phys.*, **48**, 5066 (1968).
16. A. Skerbele and E. N. Lassettre, *J. Chem. Phys.*, **42**, 395 (1965).

right angles with respect to the crossed molecular and electron beams. The excitation functions for each molecule were quite similar with main peaks at ~6.5, 13, and 150 eV. The excitation function for benzene exhibited shoulders at 5.2 and 17.5 eV and minor peaks at 9.5 and 11.3 eV. Since optical excitation yields fluorescence mainly from S_1 for these molecules, it can be argued that electron-impact-induced excitation also yields S_1 fluorescence. Smyth et al.[27,28] argued that at least the 6.5-eV peak and possibly the 13-eV peak are due to high-lying core-excited shape resonances that decay to produce S_1, although the latter is believed[28] to result from nonresonant scattering. The 150-eV peak was assigned to emission from dissociation fragments.[28]

Smyth et al.[20] also measured electron-impact excitation functions for the formation of metastable states in benzene, toluene, and aniline. The metastable molecules were detected in-line with the molecular beam by an Auger detector. The metastable excitation function for each of the three molecules exhibited similar features with peaks at 5.9, 7.7, and 10.1 eV for benzene, 5.7, 7.9, and 10 eV for toluene, and 5.6, 8.3, (shoulder) and 10.0 eV for aniline. The metastable species measured in each case was assumed to be in T_1. The peak at ~5.8 eV for each molecule was assigned to the decay of a temporary negative-ion resonance. The peak at ~7.8 eV was attributed to the maximum in the exchange-excitation cross section for T_1 and the maximum at ~10 eV to either a higher-lying NIR state or as the maximum in the exchange-excitation cross section for a higher-lying triplet state. A comparison of the electron-impact-induced fluorescence functions for each molecule with the corresponding metastable excitation function ruled out intersystem crossing as a possible channel for populating T_1 in this experiment.

Beenakker et al.[29] also measured the $S_1 \rightarrow S_0(^1B_{2u} \rightarrow ^1A_{1g})$ fluorescence from benzene produced by electron impact at incident energies of 0 to 30 eV. The measured emission cross section for the $S_1 \rightarrow S_0$ fluorescence exhibited a peak at ~7.3 eV incident electron energy. At this energy the apparent emission cross section was found to vary linearly with pressure, although at higher energies this pressure dependence disappeared. The pressure dependence was observed only at low energies and was interpreted as being due to indirect excitation of the $^1B_{2u}$ state by intersystem crossing mainly from the $^3E_{1u}$ state, which is approximately degenerate with $^1B_{2u}$ (see Table III). The intersystem crossing was thought to be induced through collisions between molecules in the $^3E_{1u}$ state and in the ground state.

The emission cross-section function for the $^1B_{2u} \rightarrow ^1A_{1g}$ fluorescence in benzene reported by Beenakker et al.[29] is at low energies similar to the excitation function for the S_1 fluorescence reported by Smyth et al.[20,28]

However, there is a disagreement as to the precursor of the S_1 emission. Beenakker et al. attributed the peak at $\sim$7.3 eV to population of S_1 via direct excitation of $T_2(^3E_{1u})$ [which is nearly degenerate with $S_1(^1B_{2u})$] followed by collisions with ground-state benzene molecules, while Smyth et al. assigned this peak (at $\sim$6.5 eV) to a core-excited shape resonance such as $(a_{2u})^2(e_{1g})^3(e_{2u})^2$.

Finally, emission spectra of the radical cations of hexa-, penta-, tetra-, and tri-fluorobenzenes excited in the gas phase by electron impact have been reported by Allan and Maier [A. Allan and J. P. Maier, *Chem. Phys. Lett.*, **34**, 442 (1975)]. These were attributed to the radiative relaxation of the parent radical cation from the excited state $\tilde{B}$ to the ground state $\tilde{X}$. The $\tilde{B}\rightarrow\tilde{X}$ transitions are dipole allowed. Interestingly enough, no parent cation emission was observed by these workers for benzene, fluoro-, and difluorobenzenes.

III. NEGATIVE-ION RESONANCES

A. Introduction: Classification of Resonances

A negative ion resonance (NIR), also referred to as a temporary-negative-ion or a negative-ion resonance, is formed when the incident electron attaches itself to a neutral molecule for a period of time that is longer than its normal transit time through the molecule. Such resonances are non-stationary states of the electron-molecule system; their probability densities are time dependent and they decay with a characteristic lifetime

$$\tau = \hbar/\Gamma \tag{14}$$

where Γ is the resonance width. Various types of resonances have been distinguished[30-32] depending on the mechanism by which the electron is trapped. The main types are listed in Table IV. In Fig. 11 a schematic illustration is shown for shape and nuclear-excited Feshbach resonances.

Shape or single-particle resonances result when the incident electron is trapped in a potential well arising from the interaction between the incoming electron and the neutral molecule in its electronic ground state. The neutral molecule exerts an attractive force on the incident electron while the relative motion of the two bodies gives rise to a repulsive centrifugal force. The combined effect of the attractive and the repulsive forces is an effective potential exhibiting a barrier as shown in Fig. 11a. A shape resonance arises when the incident electron is trapped in the attractive region by the barrier. To understand more concisely the origin of this type of resonance let us consider the Schrödinger equation for the electron-molecule system. If we separate the equation into angular and

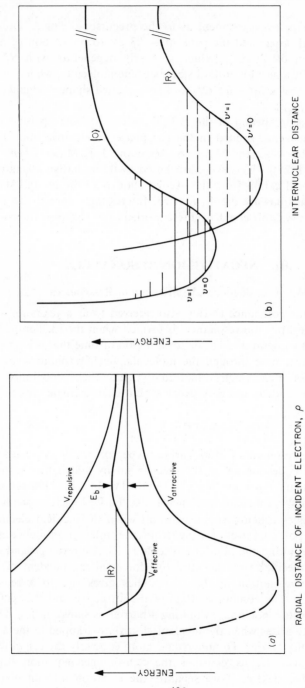

RADIAL DISTANCE OF INCIDENT ELECTRON, ρ

INTERNUCLEAR DISTANCE

Fig. 11. Schematic illustration of (a) shape and (b) nuclear-excited Feshbach resonances. The symbols $|0\rangle$ and $|R\rangle$ designate, respectively, the electronic ground state of the neutral and the negative-ion resonance.

436

TABLE IV

Types of Negative-Ion Resonances for Molecules
(Nomenclature Adopted by Various Authors)

Bardsley and Mandl[31] and present work	Taylor et al.[30] and Schulz[32]	Typical examples	
Shape	Single particle	N_2^-	2.3 eV[30,31,33]
		H_2^-	3.75 eV[30,31,34]
		$C_6H_6^-$	1.4 eV[31,35]
		$C_6H_6^-$	4.85 eV[36]
Electron-excited Feshbach	Core excited (type I)	H_2^-	10–13 eV[30,31,37]
		N_2^-	11.48 eV[30,38]
Core-excited shape	Core excited (type II)	H_2^-	8–12 eV[30]
Nuclear-excited Feshbach		O_2^-	~0.0 eV[31,39,a]
		SO_2^-	~0.0 eV[40]
		$C_6H_4NO_2X^-$	~0.0 eV[41]

[a] In contrast to this interpretation, the vibrational structure of this resonance has been ascribed [A. Herzenberg, *J. Chem. Phys.*, **51**, 4942 (1969)] to low-lying shape resonances.

radial parts, we obtain for the latter an equation of the form

$$\left[\frac{1}{\rho^2} \frac{d}{d\rho} \rho^2 \frac{d}{d\rho} + k^2 - \frac{l(l+1)}{\rho^2} - \frac{2\mu}{\hbar^2} V(\rho) \right] R_l(\rho) = 0 \qquad l = 0,1,2,3,\dots \quad (15)$$

where ρ is the radial distance of the incident electron from the molecule, $R_l(\rho)$ is the radial wave function for the incident electron, $V(\rho)$ is the attractive potential associated with the electronic ground state of the target, and $(\hbar^2 l / 2\mu)[(l+1)/\rho^2]$ is the repulsive centrifugal potential. In order for the repulsive term to be nonzero, l must be equal to or greater than one. Consequently, s waves will not lead to a shape resonance as do partial waves with higher l values.

A shape resonance decays back into the target in its ground state plus a free electron. It is possible, however, for the target to be left with vibrational and/or rotational energy. It is further possible that the shape resonance may decay via dissociative attachment as is schematically illustrated in Fig. 12.

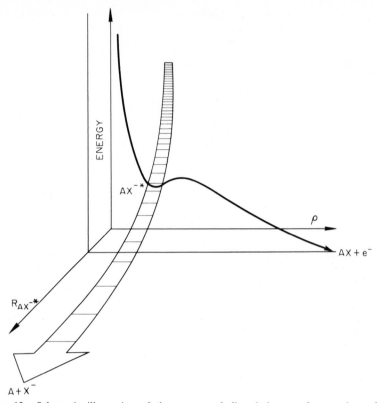

Fig. 12. Schematic illustration of the process of dissociative attachment via a shape resonance. It should be noted that the asymptote of $A + X^-$ (in the R_{AX^-*} direction) is in general different from that in the ρ direction.

Core-excited resonances are associated with excited electronic states of the target molecule. The incident electron is captured with the simultaneous excitation of one of the electrons of the target molecule. In the excited state the nuclei of the molecule are less well screened and the electron sees a slightly positive charge and becomes temporarily bound to the molecule. Such a negative-ion state decays via autodetachment or, if energetically possible, also via dissociative attachment. Although in these resonances the parent state is an excited electronic state of the neutral molecule, two types have been distinguished, namely, type I and type II, depending on whether the negative-ion resonance lies below (type I) or above (type II) the parent state. Because resonances are finite in width, the distinction between types I and II diminishes when molecular electronic levels lie close to one another. The type II core-excited resonances are also called core-excited shape resonances since their mode of formation is the same as the shape resonances except that $V(\rho)$ arises from the attractive interaction between the incident electron and an excited electronic state of the target molecule rather than the ground state.

Nuclear-excited Feshbach resonances (Fig. 11b) are similar to type I core-excited resonances in that the negative-ion state lies below the parent ground state exhibiting a positive electron affinity. Unless in vibrational levels v' higher than the lowest vibrational level $v = 0$ of the parent neutral state (see Fig. 11b), nuclear-excited Feshbach resonances cannot decay into the parent state. A similar situation may exist for type I core-excited resonances. In nuclear-excited Feshbach resonances the loss of energy by the captured electron is solely to vibrational modes of the molecule. This type of resonance entails the coupling of electronic and vibrational motions. For this reason any description of these resonances that utilizes potential-energy surfaces is artificial and the representation in Fig. 11b is simply a schematic illustration and not an actual physical description of the resonance.

The energies of the quasistationary negative-ion states are often approximately described in terms of the incident electron going into one of the unoccupied molecular orbitals of the target molecule. This description is adopted in the following sections.

B. Negative-Ion Resonances of Benzene

As stated in Section III.A, the incident electron may be temporarily captured into one of the unoccupied orbitals of the neutral molecule in its electronic ground state. In the case of benzene and its derivatives these are the unoccupied π-orbitals ψ_4, ψ_5, and ψ_6 (see Fig. 1), although the substituents themselves may introduce additional unoccupied orbital(s). Thus, the lowest benzene shape resonance would have the configuration $\psi_1^2\psi_2^2\psi_3^2\psi_4^1$, the configuration $\psi_1^2\psi_2^2\psi_3^2\psi_5^1$ being also possible. Furthermore, since ψ_4 and ψ_5 are degenerate in benzene, one would expect the resonances associated with ψ_4 and ψ_5 to be likewise degenerate.

The first observation of a NIR for benzene was made by Compton et al.[35] utilizing the SF_6-scavenger technique. The maximum of the energy-loss peak due to the NIR state was found at 1.4 eV (Fig. 13a). This resonance was subsequently attributed[25] to the degenerate $\psi_1^2\psi_2^2\psi_3^2\psi_4^1$ and $\psi_1^2\psi_2^2\psi_3^2\psi_5^1$ configurations. Boness et al.[42] observed vibrational structure in the energy-loss peak due to this resonance with a mean spacing of 0.12 ± 0.01 eV in a transmission experiment. The first vibrational peak was found at 1.15 ± 0.05 eV (Fig. 13b) and the observed structure was attributed to a C—C stretching mode. Larkin and Hasted[43] Fourier analyzed the transmission spectrum of benzene (Fig. 13*b*) and found the first vibrational peak of the resonance at 1.07 ± 0.07 eV. They deduced a mean vibrational spacing of 0.127 ± 0.010 eV. Furthermore, they observed the decay of the NIR state into the ground state and thus obtained a lower limit for the electron affinity $(EA)_B$ of benzene equal to -0.95 eV. The Fourier analysis of Larkin and Hasted[43] did not take into account the anharmonicity of the vibrations. In another transmission experiment Sanche and Schulz[44] observed the first vibrational peak of the resonance at 1.14 ± 0.05 eV with a first vibrational spacing ($v' = 0 \rightarrow v' = 1$) of 0.123 eV and with pronounced anharmonicity in the vibrational spacings (see Fig. 13*c*). Since the observed $v' = 0 \rightarrow v' = 1$ spacing is almost identical to that (0.124 eV) for the totally symmetric breathing vibrational mode of the $^1A_{1g}$ ground state of C_6H_6, Sanche and Schulz[44] concluded that the captured electron does not appreciably perturb the C—C bond.

Sanche and Schulz[44] also observed structure in the transmitted electron current in the 4 to 6-eV region (inset in Fig. 13*c*), which they attributed to a NIR with the configuration $\psi_1^2\psi_2^2\psi_3^2\psi_6^1$. Additional experimental work by Burrow and Sanche[45] indicated that this resonance decays into the ground state of benzene. Thus it appears that all of the shape resonances associated with the π-molecular orbitals ψ_4, ψ_5, and ψ_6 of benzene have been experimentally observed. In Fig. 14 similar results obtained by Nenner and Schulz[36] for C_6D_6 are shown. Both the $\psi_1^2\psi_2^2\psi_3^2\psi_4^1$ (or ψ_5^1) and the $\psi_1^2\psi_2^2\psi_3^2\psi_6^1$ resonances have been observed for C_6D_6.

Finally, as we have discussed in Section II.C.3, Smyth et al.[27,28] reported the first observations of core-excited shape resonances of benzene, toluene, and aniline. These results have been discussed in that section. The various experimentally observed shape resonances for benzene and its derivatives are summarized in Table V.*

*Recently D. Mathur and J. B. Hasted [*J. Phys. B: Atom. Mol. Phys.* **9**, L31 (1976)] have reported additional data on the resonances of benzene and some substituted benzenes from electron transmission experiments. For benzene they reported NIRs at 1.086 ± 0.03 and 4.93 ± 0.03 eV; they also attributed a broad resonance with its center at 8.85 ± 0.05 eV and a shoulder at 4.21 ± 0.03 eV to NIRs. For phenol, chlorobenzene, toluene and nitrobenzene, they observed resonances at $\varepsilon_1 \cdots \varepsilon_5$ as:

	ε_1	ε_2	$\varepsilon_3{}^a$	ε_4	ε_5
Phenol	1.03 ± 0.05	1.66 ± 0.05	3.3	4.78 ± 0.02	—
Chlorobenzene	0.90 ± 0.03	1.74 ± 0.04	3.1	4.68 ± 0.08	8.22 ± 0.08
Toluene	—	1.27 ± 0.03	3.0	4.9 ± 0.10	8.37 ± 0.20
Nitrobenzene	—	0.72 ± 0.03	3.3	4.62 ± 0.04	8.18 ± 0.08

^{a} Prominent shoulder.

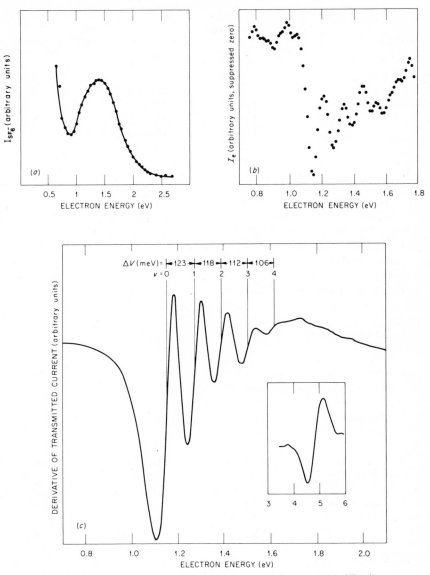

Fig. 13. Experimentally observed negative-ion states in benzene: (a) SF_6^- ion current versus incident electron energy (Ref. 35); (b) transmission function for electrons in benzene (Ref. 42); (c) derivative of the transmitted current versus incident electron energy in benzene; vertical lines indicate the center of each vibrational level of the first (degenerate) shape resonance in the 1 to 2 eV region; the inset is the "third" shape resonance near 5 eV (Ref. 44).

TABLE V

Shape Resonances Observed in Electron-Scattering Experiments[a]

Molecule	Peak energy (eV)	Onset (eV)	Method	Ref.
1. Benzene		1.07 ± 0.07[b]	f	1
C_6H_6	—; 5.0	1.14[c]; —	f	2
		1.15 ± 0.05[d]	f	3
	1.30	0.9	g	4
	1.35	0.8	h	5
	1.40	0.9	h	6
	1.55 ± 0.08	0.9 ± 0.3	h	7
	1.80	0.9	i	8
	2.1		j	9
2. Pentafluorobenzene C_6HF_5	~0		h	10
3. 1,2,3,4-Tetrafluorobenzene 1,2,3,4-$C_6H_2F_4$	~0		h	10
4. 1,3,5-Trifluorobenzene 1,3,5-$C_6H_3F_3$	0.3		h	10
5. 1,2-Dichlorobenzene o-$C_6H_4Cl_2$	0.4	0.0	h	6
6. 1,3-Difluorobenzene m-$C_6H_4F_2$	0.6		h	10
7. Bromobenzene C_6H_5Br	0.80	0.0	h	6
8. Chlorobenzene C_6H_5Cl	0.9	0.55	h	6
9. Fluorobenzene C_6H_5F	1.2	0.62	h	6
	1.27; 1.74	0.91; 1.4	g	11
	1.35		h	10
10. Phenol C_6H_5OH	0.61; 1.67	0.46; 1.2	g	11
	; 1.66 ± 0.05		f	12
	4.78 ± 0.07		f	12
11. Thiophenol C_6H_5SH	0.66; 1.10	0.48; 0.9	g	11

TABLE V   (*Continued*)

Molecule	Peak energy (eV)	Onset (eV)	Method	Ref.
12. Aniline	0.55; 1.88	0.33; 1.0	*g*	11
$C_6H_5NH_2$	1.8		*j*	9
13. Benzaldehyde	0.71; 1.12; 2.22(?)[e]	0.48; —; —	*g*	11
C_6H_5CHO	0.72	0.45	*h*	13
14. Benzoic Acid	0.63; 1.33; 2.64(?)[e]	0.30; 0.8; —	*g*	11
C_6H_5COOH				
15. Toluene	0.4; 1.6	—; 0.8	*g*	11
$C_6H_5CH_3$	1.3	0.8	*h*	6
	1.6		*j*	9
16. Anisole	1.9		*g*	11
$C_6H_5OCH_3$				
17. *N*-Methylaniline	1.30; 2.25	1.0 ; 2.0	*g*	11
$C_6H_5NHCH_3$				
18. Acetophenone	0.95	0.65	*h*	13
$C_6H_5COCH_3$				

[a] See section IV.B for shape resonances observed in the dissociative attachment channel. For nuclear-excited Feshbach resonances see Section IV.C and for electron-excited Feshbach resonances see Section III.B.

[b] This is the peak energy of the lowest vibrational level of the first NIR; vibrational spacing = 0.127 eV.

[c] This is the peak energy of the lowest vibrational level of the first NIR; vibrational spacing = 0.123 eV.

[d] This is the peak energy of the lowest vibrational level of the first NIR; vibrational spacing = 0.12 ± 0.01 eV.

[e] Position of an energy-loss process in the TEE spectrum,[11] possibly due to a NIR. In a recent paper on the application of molecular orbital theory to shape resonances of organic molecules (Chemical Physics, submitted), M. W. Grant and L. G. Christophorou pointed out that for substituted benzenes with an electron withdrawing substitutent (such as CHO) the carbonyl π^* orbital interacts in phase and out of phase with the symmetric ψ_s benzene orbital to form two orbitals ψ_s' and ψ_s''. Thus three NIRs can be observed in these cases due to capture into the ψ_s', ψ_s'' and the antisymmetric ψ_a'' orbital.

[f] Transmission in gas.

[g] Trapped electron.

[h] SF_6 scavenger.

[i] Electron impact at 11.6 eV incident energy.

[j] Electron energy loss in solid film.

1. I. W. Larkin and J. B. Hasted, *J. Phys. B: Atom. Molec. Phys.*, **5**, 95 (1972).
2. L. Sanche and G. J. Schulz, *J. Chem. Phys.*, **58**, 479 (1973).

TABLE V (*Continued*)

3. M. J. W. Boness, I. W. Larkin, J. B. Hasted, and L. Moore, *Chem. Phys. Lett.*, **1**, 292 (1967).
4. H. H. Brongersma, A. J. H. Boerboom, and J. Kistemaker, *Physica*, **44**, 449 (1969).
5. M.-J. Hubin-Franskin and J. E. Collin, *Int. J. Mass Spectr. Ion Phys.*, **5**, 163 (1970).
6. R. N. Compton, L. G. Christophorou, and R. H. Huebner, *Phys. Lett.*, **23**, 656 (1966).
7. R. N. Compton, R. H. Huebner, P. W. Reinhardt, and L. G. Christophorou, *J. Chem. Phys.*, **48**, 901 (1968).
8. J. P. Doering, *J. Chem. Phys.*, **51**, 2866 (1969).
9. K. Hiraoka and W. H. Hamill, *J. Chem. Phys.*, **57**, 3870 (1972).
10. W. T. Naff, C. D. Cooper, and R. N. Compton, *J. Chem. Phys.*, **49**, 2784 (1968).
11. L. G. Christophorou, D. L. McCorkle, and J. G. Carter, *J. Chem. Phys.*, **60**, 3779 (1974).
12. D. Mathur and J. B. Hasted, *Chem. Phys. Lett.*, **34**, 90 (1975).
13. W. T. Naff, R. N. Compton, and C. D. Cooper, *J. Chem. Phys.*, **57**, 1303 (1972).

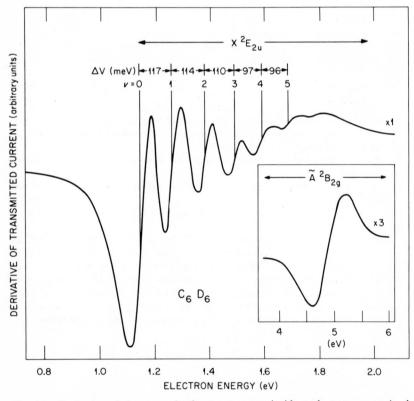

Fig. 14. Derivative of the transmitted current versus incident electron energy in deuterated benzene. The vertical lines indicate the center of each vibrational level of the first (degenerate) shape resonance in the 1 to 2 eV region. The inset is the "third" shape resonance near 5 eV (Ref. 36).

C.   Negative-Ion Resonances of Benzene Derivatives

The most obvious effect of substitution on the benzene NIR states is the removal of the degeneracy of the e_{2u} orbitals (ψ_4 and ψ_5) resulting in the appearance of two resonances below ~ 2.0 eV. In addition, the positions of these two resonances are shifted with respect to each other and the corresponding degenerate benzene resonance. These effects are brought about primarily by inductive and mesomeric interactions[46] between the ring and the substituent(s).

The first observation of double NIRs in benzene derivatives was made by Christophorou et al.,[25] although double NIRs for heterocyclic aromatics were reported[47,48] at a somewhat earlier date. The results of Christophorou et al. are reproduced in Figs. 15 and 16 and are summarized in Table V along with those of other investigators on shape resonances observed in electron-scattering experiments. The double NIRs for both the heterocyclic aromatics and the benzene derivatives were qualitatively explained using perturbation theory.[25,47] Below we elaborate further on these experimental and theoretical findings.

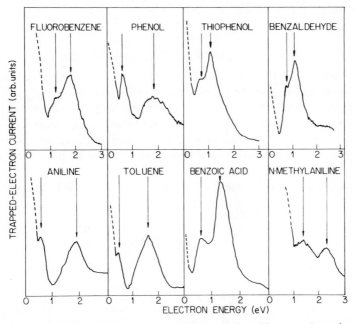

Fig. 15.   Double-shape resonances for several monosubstituted benzenes (reproduced from Ref. 25). The rapid increase in the trapped-electron current at energies approaching 0.0 eV is due to elastic scattering.

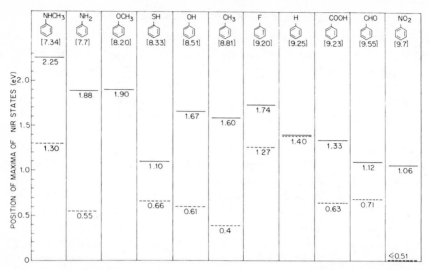

Fig. 16.   Position of the maxima of the two lowest negative-ion shape resonances in several monosubstituted benzenes arranged in ascending order of the ionization potential (reproduced from Ref. 25). The lower resonance for $C_6H_5OCH_3$ has also been observed[25] but it is not shown in the figure owing to uncertainty in its location.

1.   The Origin of the Double Negative-Ion Resonances

In benzene the NIR at 1 to 2 eV (see Fig. 13 and Table V) is attributed to the temporary capture of the electron into the degenerate $e_{2u,1}$ and $e_{2u,2}$ orbitals ψ_4 and ψ_5. The perturbation of the orbitals effected by the introduction of a substituent X onto the ring breaks this degeneracy, as has been discussed by Christophorou et al.[25,47] and is illustrated below by application of simple perturbation theory to the case of monosubstituted benzenes.

The wave functions for the $e_{2u,1}(=\psi_4)$ and $e_{2u,2}(=\psi_5)$ orbitals are

$$e_{2u,1} = \frac{(2p_1 - 2p_2 - p_3 + 2p_4 - p_5 - p_6)}{2[3(1 - S_1)]^{1/2}} \tag{16}$$

and

$$e_{2u,2} = \frac{(p_2 - p_3 + p_5 - p_6)}{2(1 - S_1)^{1/2}} \tag{17}$$

where p_i are the wave functions of the ith atomic p orbital perpendicular to the plane of the molecule and S_1 is the overlap integral between adjacent p orbitals. The perturbing Hamiltonian is assumed to be

$$H' = \sum_{i=1}^{6} H_i'$$

where H_i' corresponds to the perturbation at the ith carbon atom in the ring. The first-order energy corrections to the ψ_4 and ψ_5 orbitals are

$$\int \psi_4^* H' \psi_4 d\tau \quad \text{and} \quad \int \psi_5^* H' \psi_5 d\tau \tag{18}$$

To evaluate expressions (18) we used for ψ_4 and ψ_5 expressions (16) and (17), respectively, and assumed that

$$H_i' p_j = A_i p_j \tag{19}$$

where

$$A_i = \int p_i^* H_i' p_i d\tau$$

is the Coulomb integral involving the p orbital of the ith C atom; $H_i' p_j$ is taken equal to zero for $i \neq j$. If we now take

$$A_i \propto \Delta q_i \tag{20}$$

where Δq_i is the change in the π-electron density resulting from the substitution at position 1 (indicated by the arrow in Fig. 1) of X for H, the π-charge densities in various monosubstituted benzenes can be used to calculate Δq and consequently the relative magnitudes of the perturbations. These relative magnitudes can then be normalized with experimental data to obtain the position of the resonances. Some results of this procedure, as obtained by Christophorou et al.,[25] are shown in Fig. 17.

Within the simple theoretical treatment just outlined, the molecular orbital ψ_5 is not perturbed [see Fig. 1 and (17)]. This, however, is clearly seen from the experimental data on the double resonances in Table V not to be the case. It is a consequence of the fact that in this simple description the perturbation has been limited to position 1 ($H_i' p_j = 0$ for $i \neq j$) although inductive and mesomeric effects occur to a varying degree at all positions.

In Fig. 18, ε_2, the position of the maximum of the second shape NIR (see Fig. 15) is plotted[25] as a function of the first π-molecular ionization potential, I_M, which correlates with the energy of the highest occupied π-molecular orbital. It is seen that as I_M increases, ε_2 decreases (i.e., the vertical attachment energy decreases). This can be rationalized in a simple

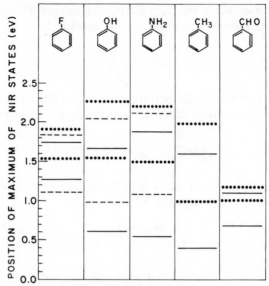

Fig. 17. Comparison of the experimental and theoretical values of the maxima of the double NIR states for fluorobenzene, phenol, aniline, toluene, and benzaldehyde. ——, experimental; ----, theoretical, based on the π-electron charge-density distributions of D. T. Clark and J. W. Emsley, *Mol. Phys.*, **12**, 365 (1967); $\cdots$, theoretical, based on the π-electron charge-density distributions of J. E. Bloor and D. L. Breen, *J. Phys. Chem.*, **72**, 716 (1968) (reproduced from Ref. 25).

fashion by inspecting the molecular orbitals in Fig. 1 (Section I). The orbitals ψ_3 and ψ_5 are not perturbed (in simple first-order theory) by the substituent in position 1, while the orbitals ψ_2 and ψ_4 are. The orbital energies of ψ_2 and ψ_4 should be increased relative to benzene in the case of electron-donating substituents and should decrease if the substituent is electron withdrawing. Thus in the case of electron-donating substituents, ψ_2 is the highest-lying occupied molecular orbital of the neutral molecule and the second resonance is correlated with ψ_4, hence the observed relation between ε_2 and I_M. In the case of electron-withdrawing substituents, ψ_3 and ψ_5 can be correlated with I_M and ε_2, respectively, and the simple theory would predict that the corresponding data points in Fig. 18 coincide with that of benzene. Evidently this is not the case.

2.   A Qualitative Discussion of the Effects of the Substituents On the Position of the Double NIR States

A number of interesting observations can be made by considering the results on the double NIRs in Table V and the inductive and mesomeric effects of the substituents.[49] Thus the variation of the lower resonance peak

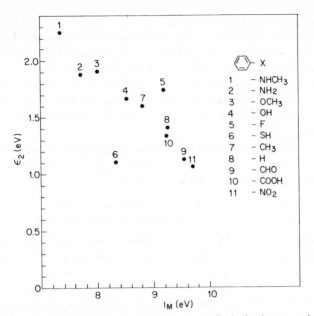

Fig. 18.   Second-shape resonance peak energy, ε_2, versus the ionization potential, I_M, of the neutral molecule (reproduced from Ref. 25).

energy ε_1 for substituents within the halogen family of the periodic table is

$$\varepsilon_1(F) > \varepsilon_1(Cl) > \varepsilon_1(Br) > \varepsilon_1(I)^{50}$$

that is, it decreases with decreasing electronegativity of the halogen atom. This decrease is opposite to what might be expected if ε_1 depended purely on the "classical" inductive effect and may be associated with the changes in the polarizability of the halogen atom series.

From the data in Table V it is seen also that

$$\varepsilon_1(OH) > \varepsilon_1(NH_2) > \varepsilon_1(CH_3)$$

that is, ε_1 decreases with decreasing electronegativity of the second period elements (O, N, C), and this sequence is again opposite to what would be expected in the case of a pure inductive effect.

It is interesting to compare the positions ε_1 and ε_2 of the double NIRs for the pairs of compounds phenol (OH) and anisole (OCH$_3$), aniline (NH$_2$) and N-methyl aniline (NHCH$_3$), and benzaldehyde (CHO) and acetophenone (COCH$_3$), which shows the effect of adding a methyl group to the substituent. As in many chemical phenomena, the methyl group appears to be electron releasing. Here the release is mainly onto that substituent atom bonded to the ring. For oxygen and nitrogen (phenol and aniline) this release of electrons by the methyl group increases the donating ability of the substituent. The carbonyl in the benzaldehyde exerts a classic —M effect. The release of electrons from the methyl onto the carbonyl carbon results in a reduction of the —M effect of the carbonyl. The effect of the methyl group in the compounds considered here is to increase the orbital energies, which in turn results in higher-lying negative-ion resonance states.

Finally, it is worth noting the decrease in ε_1 with increasing number of substituted fluorine atoms (see Table V) resulting in a more stable negative ion with increasing number of fluorine atoms.

3. Results of CNDO Calculations and The Need for a Rigorous Theoretical Approach

The discussion in the previous section—being itself a rationalization after the fact—stresses the need for a deeper theoretical understanding in this area. As an initial step in this direction we compare in Fig. 19 the experimental results on the position of the NIR states and the CNDO virtual π-orbital energies for a number of benzene derivatives. Using Koopmans' theorem[52] we assumed the resonance energy to be given by the energy of the corresponding virtual orbital. In spite of the limitations of Koopmans' theorem and the CNDO method, the results in Fig. 19 indicate a broad relation between the experimental positions of the NIRs and the theoretical orbital energies. It can be seen from Fig. 19 that for any given molecule, application of Koopmans' theorem to ψ_5 results in greater error than in its application to ψ_4.* This may be attributed to a greater reorganization energy upon the addition of an electron to the more localized ψ_5 orbital as compared with the less localized ψ_4 orbital. Furthermore, CNDO unrestricted Hartree-Fock calculations, which we have performed for the ground states of $C_6H_5F^-$, $C_6H_5NH_2^-$, $C_6H_5OH^-$, $C_6H_5CHO^-$, $C_6H_5COOH^-$, and $C_6H_5NO_2^-$ utilizing the neutral ground-state geometry, have shown that the extra electron is in the same orbital as predicted by Koopmans' theorem. However, when similar calculations were made for $C_6H_5SH^-$, it was found that the extra electron was in an orbital associated with the —SH group rather than the ψ_4 or ψ_5 π-orbitals.

The CNDO method is easy to use for a qualitative description of NIR states of large polyatomic molecules, but its results are semiquantitative at best. There is a need for both the further refinement of this and similar molecular orbital methods and for other descriptions of NIR states of large polyatomic molecules.

---

*It should be pointed out that CNDO calculations often give poor quantitative orbital energies and at times even qualitatively poor orderings.

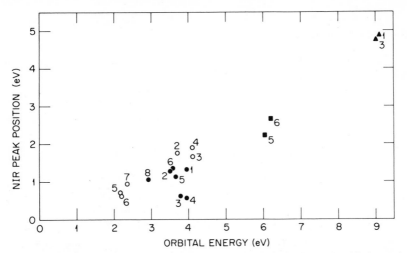

Fig. 19. Negative-ion resonance peak positions versus CNDO (virtual) orbital energies. (O) correlated with ψ_4; (●) correlated with ψ_5; (▲) correlated with ψ_6; (■) associated with the carbonyl group. The compounds are identified as follows: 1, benzene (1.3 eV, trapped electron; 4.85 eV (onset), transmission); 2, fluorobenzene (trapped electron); 3, phenol (trapped electron); 4, aniline (trapped electron); 5, benzaldehyde (trapped electron); 6, benzoic acid (trapped electron); 7, acetophenone (SF_6 scavenger); 8, nitrobenzene (dissociative attachment). The statements in the parentheses following each compound refer to the experimental method used to obtain the data plotted.

IV. NEGATIVE IONS

Negative ions of atoms and molecules are generally considered to have been formed when the incoming electron is associated with the atom or the molecule for a time longer than the transit time of the electron. This time may be as short as 10^{-15} sec, depending on the velocity of the electron and the size of the target. Negative-ion formation is thus viewed as proceeding via a negative-ion intermediate that can itself be formed either by electron capture in the field of the ground electronic state or by electron capture in the field of an excited electronic state.[53,54] The time the electron is retained by the molecule is referred to as the negative-ion (autodetachment) lifetime τ_a. Depending on the magnitude of τ_a—and thus on the way the negative ion can be observed experimentally—three classes of negative ions have been distinguished.[53]

1. Extremely short-lived ($10^{-15} \lesssim \tau_a \lesssim 10^{-12}$ sec)—these are observed as resonances in electron scattering and/or in dissociative attachment studies.

2. Moderately short-lived ($10^{-12} \lesssim \tau_a \lesssim 10^{-6}$ sec)—these can be stabilized in high- (1 to $\sim$1000 torr)- and very-high- ($>$1000 torr) pressure swarm experiments, and

3. Long-lived ($\tau_a > 10^{-6}$ sec)—these can be conveniently studied with conventional time-of-flight mass spectrometers.

The various channels through which a transient molecular negative ion AX^{-*} decays can be summarized as[53]

$$\xrightarrow{p_{sc}} AX \ (\text{or} \ AX^*) + e^{(\prime)} \tag{20a}$$

$$AX + e \xrightarrow{\sigma_0} AX^{-*} \xrightarrow{p_{da}} A \ (\text{or} \ A^*) + X^- \tag{20b}$$

$$\xrightarrow[p_{st}]{} AX^- + \text{energy} \tag{20c}$$

where σ_0 is the cross section for the formation of AX^{-*} and $p_{sc} = \sigma_{sc}/\sigma_0$, $p_{da} = \sigma_{da}/\sigma_0$, $p_{st} = \sigma_{st}/\sigma_0$ are, respectively, the probability that AX^{-*} will decay via autoionization (indirect elastic or inelastic electron scattering), will undergo dissociative attachment, or will be stabilized radiatively or collisionally; all of the above three decay processes can be in competition.

A. Short-Lived Negative Ions of Benzene and Benzene Derivatives

A number of short-lived negative ions have been detected as resonances by studying the elastically or inelastically scattered electrons (channel 20a) using electrostatic analyzers and/or threshold-electron excitation techniques. These negative ions have been the subject of much study recently and were discussed in Section III. They can, of course, be detected via channel 20b (see Section IV.B) and also in very-high-pressure gaseous media via channel 20c. Thus Christophorou and Goans,[56] using the electron-swarm method, found that benzene captures weakly thermal and epithermal electrons in mixtures with N_2, at N_2 pressures from 2000 to 15,000 torr. A representative set of their data is shown in Fig. 20. The attachment rate is seen to increase with increasing N_2 pressure at a fixed mean electron energy $\langle \varepsilon \rangle$, and to decrease with increasing $\langle \varepsilon \rangle$ at a fixed N_2 pressure. On the basis of the observed pressure and energy dependences of the attachment rate, Christophorou and Goans estimated a mean autodetachment lifetime for $C_6H_6^{-*}$ equal to $\sim 1 \times 10^{-12}$ sec at ~ 0.04 eV and 2×10^{-13} sec at ~ 0.18 eV. Obviously, these findings require that the electron affinity of benzene is positive (but small[56]), a fact that is at variance with the result of several experimental and theoretical estimates, as can be seen from Table VI. The fact that the maximum of the lowest $C_6H_6^{-*}$ resonance (at ~ 1.4 eV) in electron-scattering experiments coincides with the value (-1.4 eV) most theoretical calculations predict for the

TABLE VI
Literature Values for the "Electron Affinity"
of Benzene

Electron Affinity (eV)	Reference	Method
-1.63	1	a
-1.62	2	a
-1.59	3	b
-1.42	4	a
-1.4	5–7	a
$\geqslant -1.15 \pm 0.05^{c}$	8	d
-1.15	9	a
$\geqslant -1.14 \pm 0.05$	10	d
-1.1 ± 0.3	11	e
$\geqslant -1.07 \pm 0.07$	12	d
-1.06	19	a
$\geqslant -0.9^{f}$	13	g
$\geqslant -0.9^{f}$	14	h
$\geqslant -0.9^{f}$	15	h
$\geqslant -0.9^{f}$	16	i
$\geqslant -0.8^{f}$	17	h
-0.36	18	f
$\gtrsim 0$	20	k

^{a} Theory.

^{b} Magnetron method.

^{c} This is the peak energy of the lowest vibrational level of the lowest negative-ion resonance (NIR).

^{d} Electron transmission in gas.

^{e} Charge-transfer spectra.

^{f} This is the onset of the first (lowest) NIR.

^{g} Trapped electron.

^{h} SF_6 scavenger.

^{i} Electron impact at 11.6 eV incident energy.

^{j} Kinetics of electrode processes.

^{k} Electron swarm.

1. R. M. Hedges and F. A. Matsen, *J. Chem. Phys.*, **28**, 950 (1958).
2. J. Ehrenson, *J. Phys. Chem.*, **66**, 706 (1962).
3. A. F. Gaines and F. M. Page, *Trans. Faraday Soc.*, **59**, 1266 (1963).
4. D. R. Scott and R. S. Becker, *J. Phys. Chem.*, **66**, 2713 (1962).
5. N. S. Hush and J. A. Pople, *Trans. Faraday Soc.*, **51**, 600 (1955).
6. J. R. Hoyland and L. Goodman, *J. Chem. Phys.*, **36**, 21 (1962).
7. G. L. Caldow, *Mol. Phys.*, **18**, 383 (1970).
8. M. J. W. Boness, I. W. Larkin, J. B. Hasted, and L. Moore, *Chem. Phys. Lett.*, **1**, 292 (1967); I. Nenner and G. J. Schulz, *J. Chem. Phys.*, **62**, 1747 (1975).
9. B. J. McClelland, *J. Chem. Phys.*, **46**, 4158 (1967).
10. L. Sanche and G. J. Schulz, *J. Chem. Phys.*, **58**, 479 (1973).

TABLE VI   (*Continued*)

11. K. Kimura and S. Nagakura, *Mol. Phys.*, **9**, 117 (1967).
12. I. W. Larkin and J. B. Hasted, *J. Phys. B: Atom. Molec. Phys.*, **5**, 95 (1972).
13. H. H. Brongersma, A. J. H. Boerboom, and J. Kistemaker, *Physica*, **44**, 449 (1969).
14. R. N. Compton, L. G. Christophorou, and R. H. Huebner, *Phys. Lett.*, **23**, 656 (1966).
15. R. N. Compton, R. H. Huebner, P. W. Reinhardt, and L. G. Christophorou, *J. Chem. Phys.*, **48**, 901 (1968).
16. J. P. Doering, *J. Chem. Phys.*, **51**, 2866 (1969).
17. M.-J. Hubin-Franskin and J. E. Collin, *Int. J. Mass Spectr. Ion Phys.*, **5**, 163 (1970).
18. L. E. Lyons, *Nature (London)*, **166**, 193 (1950).
19. T. L. Kunii and H. Kuroda, *Theor. Chim. Acta*, **11**, 97 (1968).
20. L. G. Christophorou and R. E. Goans, *J. Chem. Phys.*, **60**, 4244 (1974).

electron affinity of benzene, shows the deficiency of most such calculations in that they yield the vertical attachment energy[53] rather than the adiabatic electron affinity. Furthermore, the fact that electron-swarm experiments indicate the existence of the benzene negative ion $C_6H_6^-$ in the gas phase although electron-scattering experiments show that for benzene the vertical attachment energy is ~1.4 eV may suggest that the potential-energy surface of $C_6H_6^-$ has a shallow minimum below and in a different geometry from that of the neutral benzene molecule. Although it appears that the swarm and the beam data are difficult to reconcile, it has to be noted that the experimental conditions in the two types of experiments are quite different.

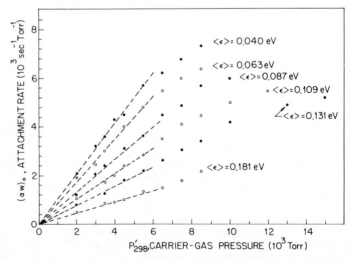

Fig. 20.   Attachment rate αw for benzene in mixtures with nitrogen as a function of the nitrogen pressure, P_{N_2}, at the indicated mean electron energies, $\langle \varepsilon \rangle$. The broken lines are a linear least-squares fit to the data for $P_{N_2} \leqslant 6000$ torr (from Ref. 56).

B.   Dissociative Electron Attachment to Benzene Derivatives

1.   General Considerations

Reaction (20b) is the resonant dissociative electron-attachment process. Electrons in a restricted energy range are captured by the molecule forming a compound negative-ion intermediate, which subsequently dissociates into neutral and negative-ion fragments.

On the basis of reactions (20a) and (20b) one may visualize the formation of AX^{-*} as proceeding via a Franck–Condon transition from the initial state—neutral molecule and electron separated an infinite distance —to the final state representing the compound negative ion. For a diatomic or a "diatomic-like" system, this is schematically illustrated in Fig. 21. The cross section for dissociative attachment, σ_{da}, is a product of the cross section for formation of AX^{-*}, σ_0, and the probability, p_{da}, that

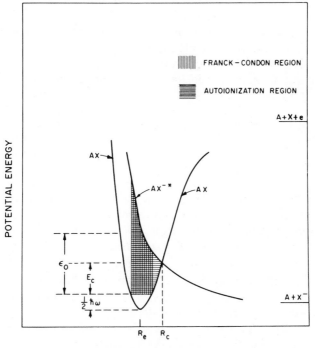

Fig. 21.   Schematic potential-energy diagram illustrating the dissociative electron attachment process for a diatomic or a "diatomic-like" molecule. Note that the ground state shown is for the neutral molecule AX plus the electron at rest at infinity.

AX^-* will dissociate, once formed, rather than decay by autodetachment. More precisely,[53,55]

$$\sigma_{da} = \sigma_0 \exp\left[-\int_{R_\varepsilon}^{R_c} \frac{\Gamma_a(R)}{\hbar} \frac{dR}{v(R)} \right] = \sigma_0 \exp\left[-\frac{\bar{\Gamma}_a}{\hbar} \tau_s \right] = \sigma_0 \exp\left(-\frac{\tau_s}{\bar{\tau}_a} \right) \quad (21)$$

where $e^{-\tau_s/\bar{\tau}_a}$ is the survival probability of AX^-*, that is, the probability that it will not autoionize while separating from the point of formation (at an internuclear distance R_ε) to the point (at an internuclear distance R_c) beyond which the negative-ion curve lies below that of the $AX + e$ system and thus AX^-* will dissociate with unit efficiency. The time, τ_s, taken by $A - X^-$ to separate from R_ε to R_c is

$$\tau_s = \int_{R_\varepsilon}^{R_c} \frac{dR}{v(R)}$$

where $v(R)$ is the velocity of separation of A and X^-, itself a function of the internuclear separation R. $\bar{\tau}_a (= \hbar/\bar{\Gamma}_a)$ is the mean autoionization lifetime; τ_a is a function of R. For a diatomic molecule initially in the lowest vibrational level $\nu = 0$ of the electronic ground state and with a purely repulsive potential energy curve for AX^-* (such as is shown in Fig. 21), the cross section for dissociative attachment can be expressed as[55]

$$\sigma_{da}(\varepsilon)_{\nu=0} = \underbrace{\frac{4\pi^{3/2}}{(2m/\hbar^2)\varepsilon} \bar{g} \frac{\Gamma_{\bar{a}}}{\Gamma_d} \exp\left[\frac{\Gamma_a^2 - 4(\bar{\varepsilon}_0 - \varepsilon)^2}{\Gamma_d^2} \right]}_{\sigma_0} e^{-\rho(\varepsilon)} \quad (22)$$

where m is the electron mass, ε is the incident electron energy, $\bar{g}$ is a statistical factor, $\Gamma_{\bar{a}}$ is the partial autoionization width, Γ_a is the total autoionization width, Γ_d is the experimentally determined dissociative attachment cross-section width, $\bar{\varepsilon}_0 = \varepsilon_0 + \frac{1}{2}\hbar\omega$, ε_0 is the electron energy at the peak of $\sigma_{da}(\varepsilon)$, $\frac{1}{2}\hbar\omega$ is the zero-point energy (see Fig. 21), and $e^{-\rho(\varepsilon)}$ is the survival probability. Although (22) is limited to diatomic molecules and to situations as pictured in Fig. 21, it is of great significance. It provides an explicit expression for σ_0, an explicit dependence of σ_0 on ε, and predicts explicitly the dependence of σ_{da} on the reduced mass of the $A - X^-$ system. It has been discussed by, among others, Christophorou and Stockdale,[57] Bardsley and Mandl,[31] Chen,[58] Christophorou,[53] Christodoulides and Christophorou,[59] Schulz,[32] and Fiquet-Fayard.[60]

Before discussing the experimental results on dissociative attachment to benzene and benzene derivatives, it should be noted that the negative-ion potential-energy curve may not be purely repulsive in the Franck–Condon region and that higher-lying negative-ion states exist abundantly with a variety of shapes of potential-energy curves or surfaces. Often, such states can be reached via the decay of higher-lying states. When the negative-ion state lies at or above known electronic states of the neutral molecule, the magnitude of σ_{da} decreases because the negative ion can now decay to an electronically excited neutral molecule in addition to a rotationally and/or vibrationally excited neutral molecule. The importance of this process has been stressed by Christophorou and Stockdale.[57] When, however, dissociative electron attachment proceeds via a purely repulsive negative-ion state in the Franck–Condon region with the peak energy, ε_{max}, of $\sigma_{da}(\varepsilon)$ occurring at energies below that of the lowest excited electronic state of the neutral molecule, the magnitude of σ_{da} is primarily determined by σ_0 since autoionization becomes relatively unimportant. In such a situation σ_{da} increases with decreasing ε_{max}. It is, finally, noted that as far as dissociative attachment to polyatomic molecules is concerned, in certain cases this may be treated as though the polyatomic molecule is "diatomic-like," as has been indicated by the work of Christodoulides and Christophorou[59] on the production of Br^- from $n\text{-}C_nH_{2n+1}$ Br molecules. In such cases, (22) may serve as a guide in understanding the variation of the magnitude of σ_{da} with respect to ε_{max}.

2. Experimental Results

Dissociative electron attachment cross sections (producing halogen negative ions) have been reported by Christophorou et al.[61] for the halogenated benzenes C_6H_5Cl, $o\text{-}C_6H_4Cl_2$, $o\text{-}C_6H_4CH_3Cl$, C_6H_5Br, C_6D_5Br, and $o\text{-}C_6H_4CH_3Br$ using the swarm-beam technique.[62] These are shown in Fig. 22. The overall widths of the resonances are large and their peak energies, ε_{max}, lie in the range $(\frac{3}{2})kT$ to 1 eV, that is, well below the energy of the lowest known excited electronic states of these molecules. The magnitude of the cross section at ε_{max} is seen to increase with decreasing value of ε_{max} [i.e., in the order Cl, Br, (I)[61]], in agreement with (22).

Electron-beam data for dissociative attachment have been reported[61] also for C_6H_5I, $C_6H_5NO_2$, o- and $m\text{-}C_6H_4CH_3NO_2$, and m- and $p\text{-}C_6H_4Cl_2$. These data as well as data on other benzene derivatives have been collected in Table VII. It is noted that $C_6H_5NO_2$ and o- and m-$C_6H_4CH_3NO_2$ are known to form long-lived parent negative ions at thermal energies (see Section IV.C). Also, for a number of molecules in Table VII,

$$D(A-X)-EA(X)<0$$

[$D(A-X)$ is the dissociation energy of the halogen substituent], and thus

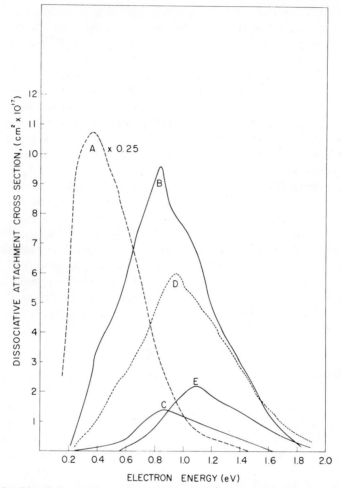

Fig. 22.   Dissociative attachment cross sections for halogenated benzene derivatives. Curve A, $Cl^-/o\text{-}C_6H_4Cl_2$; B, Br^-/C_6H_5Br; C, Cl^-/C_6H_5Cl; D, $Br^-/C_6H_4CH_3Br$; E, $Cl^-/o\text{-}C_6H_4CH_3Cl$ (from Ref. 61).

dissociative electron attachment is possible for these molecules even with "zero" energy electrons. Although for most of these benzene derivatives σ_{da} was not measured, it is expected to be large in the majority of the cases owing to the low value of ε_{max}.

In Fig. 23, the data of Christophorou et al.[61] on the NO_2^- current produced by dissociative attachment to $C_6H_5NO_2$, $o\text{-}C_6H_4CH_3NO_2$, and $m\text{-}C_6H_4CH_3NO_2$ are plotted as a function of the incident electron energy.

TABLE VII
Negative Ions Formed via Dissociative Electron Capture by Benzene Derivatives

Compound	Negative ion	Onset[a] (eV)	Position of maximum[a] (eV)	Cross section (cm²)	Ref.
C_6H_6	H^-	7.9 ± 0.3			1
		~ 10			1
	C_2H^-	8.3 ± 0.3			1
		10.5			1
	$C_4H_3^-$	8.5 ± 0.3			1
	$C_6H_5^-$	5.8 ± 0.3			1
C_6D_5Br	Br^-		0.80	1.04×10^{-16}	2
C_6F_5Br	$C_6F_5^-$	~ 0	0.06		3
	Br^-		0.1		3
C_6F_5Cl	$C_6F_5^-$	~ 0	0.2		3
	Cl^-	~ 0	0.2		3
C_6F_5I	$C_6F_5^-$	~ 0	0.0		3
	I^-		0.0		3
C_6F_6	F^-		?[b]		4
C_6F_5OH	$C_6F_4O^-$	~ 0	0.11		3
			0.73		3
	$C_6F_5O^-$	~ 0	0.5		3
$C_6F_5NH_2$	$C_6F_4NH_2^-$	0	1.1		3
$o\text{-}C_6H_4Cl_2$	Cl^-		0.36	4.3×10^{-16}	2
$m\text{-}C_6H_4BrCl$	Cl^-	~ 0	0.31		3
	Br^-		~ 0.0		3
$o\text{-}C_6H_4ClF$	Cl^-	~ 0	0.24		3
$m\text{-}C_6H_4ClF$	Cl^-	~ 0	0.26		3
$m\text{-}C_6H_4ClI$	Cl^-	~ 0	0.10		3
	I^-		~ 0.0		3
$m\text{-}C_6H_4ClNO_2$	Cl^-	~ 0	0.9		3
			3.4		3
	NO_2^-	~ 0	0.9		3
			3.3		3
$m\text{-}C_6H_4INO_2$	I^-		0.0		3
			0.8		3
	NO_2^-	~ 2.0	4.1		3
C_6H_5Br	Br^-		0.84	9.6×10^{-17}	2
C_6H_5Cl	Cl^-		0.86	1.4×10^{-17}	2
C_6H_5I	I^-		0.0		2
$C_6H_5SiCl_3$	Cl^-	0.5			5
		3.8			5
		6.1			5
	$SiCl_3^-$	3.7			5
		5.6			5
		6.1			5
	$C_6H_5SiCl_2^-$	5.9			5
		6.3			5
	$C_6H_4SiCl_3^-$	6.4			5
		7.2			5

TABLE VII   (*Continued*)

Compound	Negative ion	Onset[a] (eV)	Position of maximum[a] (eV)	Cross section (cm²)	Ref.
$C_6H_5NO_2$	O^-	3.5	5		6
	$OH^-(?)^c$	3.5	4.2		6
		6.0	6.7		6
	$CN^-(?)^d$	2.8	4		6
	C_2H^-	4.2	4.5		6
		4.7	5		6
		6	7		6
	NO_2^-	0.42	1.06		2
		1.0	1.5		6
		1.9	3.53		2
		2.5	4		6
	$C_6H_4NO_2^-$	2.8	3.8		6
		~6	7.2		6
C_6H_5SH	$C_6H_5S^-$	0.04	0.24		7
		0.34	0.75		7
		0.6	1.32		7
	SH^-	0.40	0.9		7
		1.4	2.3		7
		3.9	4.4		7
		4.6	5.0		7
		5.4	6.3		7
		7.5	8.9		7
	S^-	4.9	5.4		7
		7.5	8.7		7
C_6F_5CHO	$C_5F_5^-$	~0	0.4		3
		2.8	3.8		3
$o\text{-}C_6H_4BrCH_3$	Br^-		0.95	6.0×10^{-17}	2
$o\text{-}C_6H_4ClCH_3$	Cl^-		1.10	2.2×10^{-17}	2
$C_6F_5OCH_3$	$C_6F_4O^-$	~0	0.1		3
			0.8		3
	$C_6F_5O^-$	~0	1.1		3
		~3	4.4		3
$o\text{-}C_6H_4CH_3NO_2$	NO_2^-	0.5	0.62		2
		2.1	3.22		2
$m\text{-}C_6H_4CH_3NO_2$	NO_2^-		1.06		2
		1.8	3.50		2
$C_6H_5CH_2SH$	S^-	4.8	5.8		7
		7.5	8.5		7
	SH^-	0.41	0.92		7
	$C_6H_5S^-$	0.15	0.92		7
		0.42	0.97		7
	$C_6H_5CHS^-$	0.0	0.2		7
			0.7		7
			1.1		7

TABLE VII   (*Continued*)

Compound	Negative ion	Onset[a] (eV)	Position of maximum[a] (eV)	Cross section (cm^2)	Ref.
	$C_6H_5CH_2S^-$	0.35	0.9		7
	$CH_2=CH—CH_2S^-$	0.42	0.97		7
o-$C_6H_4OHNO_2$	$C_6H_4NO_2^-$	~0	1.1		8
		1.8	3.5		8
	$C_6H_4OH^-$	2.4	4.4		8
	NO_2^-				8
o-$C_6H_4NH_2NO_2$	$C_6H_4NO_2^-$	1.8	3.2		8
	$C_6H_4NH_2^-$				8
	NO_2^-	2.0	3.2		8

[a] These are either given by the respective authors or taken from their reported graphical data.

[b] See discussion in R. N. Compton, D. R. Nelson, and P. W. Reinhardt, *Int. J. Mass Spectr. Ion Phys.*, **6**, 117 (1971).

[c] Negative ion with mass of 17 a. m. u.

[d] Negative ion with mass of 26 a. m. u.

1. Von L. V. Trepka and H. Neuert, *Z. Naturforsch.*, **18a**, 1295 (1963).
2. L. G. Christophorou, R. N. Compton, G. S. Hurst, and P. W. Reinhardt, *J. Chem. Phys.*, **45**, 536 (1966).
3. W. T. Naff, R. N. Compton, and C. D. Cooper, *J. Chem. Phys.*, **54**, 212 (1971).
4. M. M. Bibby and G. Carter, *Trans. Faraday Soc.*, **62**, 2637 (1966).
5. K. Jäger and A. Henglein, *Z. Naturforsch.*, **23a**, 1122 (1968).
6. K. Jäger and A. Henglein, *Z. Naturforsch.*, **22a**, 700 (1967).
7. K. Jäger and A. Henglein, *Z. Naturforsch.*, **21a**, 1251 (1966).
8. A. Hadjiantoniou, L. G. Christophorou, and J. G. Carter, Oak Ridge National Laboratory Report ORNL-TM-3990 (1973).

The negative-ion currents for the three compounds have been normalized to the intensity at ~3.5 eV. The peak energy of the NO_2^- current for nitrobenzene and *m*-nitrotoluene are seen to occur at about the same energy, that is, 1.06 and ~3.5 eV. However, ortho substitution of the CH_3 group to nitrobenzene lowers the positions of both NO_2^- peaks by ~0.3 eV. The relative intensity of the first peak for *m*-$C_6H_4CH_3NO_2$ is lower than for the other two molecules indicating increased autoionization. All three molecules form long-lived parent negative ions at ~0.0 eV, and thus all three possess (at least) three negative-ion states below ~4 eV.

In Section II the compound negative-ion resonances observed for C_6H_5Cl, C_6H_5Br, o-$C_6H_4CH_3Cl$, o-$C_6H_4Cl_2$, $C_6H_5NO_2$, and C_6H_5SH are listed. The peak energies and shapes of these resonances are found to coincide, within experimental error, with those observed in dissociative attachment studies, suggesting, as was noticed earlier,[35,53] that both processes, namely, indirect inelastic electron scattering and dissociative

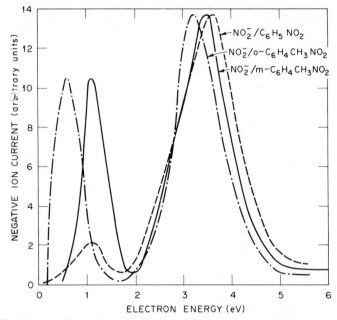

Fig. 23. Negative-ion yield as a function of electron energy for NO_2^- produced by dissociative electron attachment to $C_6H_5NO_2$ (----), $o\text{-}C_6H_4CH_3NO_2$ (-··-··-), and $m\text{-}C_6H_4CH_3NO_2$ (——) (based on data in Ref. 61).

attachment, proceed via the same intermediate negative-ion state. This coincidence, however, may not be exact since p_{sc} and p_{da} may depend differently on ε. A further example is shown in Fig. 24, where the negative-ion current for $C_6H_5S^-$ and SH^- from C_6H_5SH and the trapped-electron current (a measure of the inelastically scattered electrons) are plotted as a function of the electron energy. The maxima in the $C_6H_5S^-$ current and those of the TEE spectrum at 0.75 and 1.3 eV compare well. The peak at thermal energy seen in the $C_6H_5S^-$ ion current could not be observed in the TEE spectrum of C_6H_5SH because it is masked by the large contribution to the trapped-electron current at thermal and epithermal energies resulting from elastic energy losses. Above ~ 2 eV the SH^- current indicates the existence of other negative-ion states, as is to be expected for this and other benzene derivatives (see Table VII).

3. Dissociative Excitation of Benzene by Electron Impact

Dissociative excitation of benzene by electron impact has been investigated recently by Beenakker and De Heer[63] (see also Vroom and De Heer[64]). These authors determined the threshold energies and measured

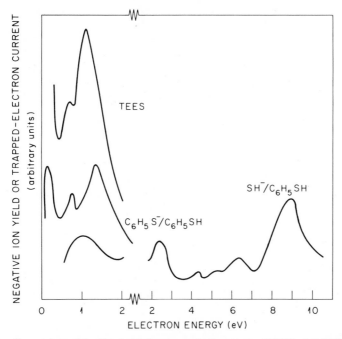

Fig. 24.   Comparison of the threshold-electron excitation spectra (TEES) of C_6H_5SH (from Ref. 25) and the yield of $C_6H_5S^-$ and SH^- ions produced from C_6H_5SH by dissociative attachment [from K. Jäger and A. Henglein, *Z. Naturforsch.*, **21a**, 1251 (1966)]. The relative intensities of the $C_6H_5S^-$ negative-ion peaks at 0.24, 0.75, and 1.32 eV, are, respectively 1060, 810, and 1080. Those for SH^- at 0.9, 2.3, 4.4, 5.0, 6.3, and 8.9 eV are 1.6, 1.4, 0.3, 0.3, 0.8, and 3.0, respectively.

the emission cross sections, σ_{em}, for the Balmer series of H atoms and the $A^2\Delta \rightarrow X^2\Pi$ emission of the CH fragments, which are proportional to the excitation cross sections of the molecular states of benzene that subsequently dissociate into these excited fragments. The emission cross sections from the excited H atoms and CH fragments were measured as functions of incident electron energy, ε, and were plotted as $\sigma_{em}\varepsilon/4\pi a_0^2 R$ versus $\ln\varepsilon$, where a_0 is the Bohr radius and R is the Rydberg energy. Had the production of excited H and CH fragments proceeded via optically allowed transitions in benzene, such a plot according to the Bethe theory[65] would yield a straight line with a positive slope. The results of Beenakker and De Heer shown in Fig. 25 are not consistent with this, and the observed profound divergence from the straight line is taken to suggest that the excited H and CH fragments are formed via optically forbidden excitation processes (initially produced by electrons) in benzene.

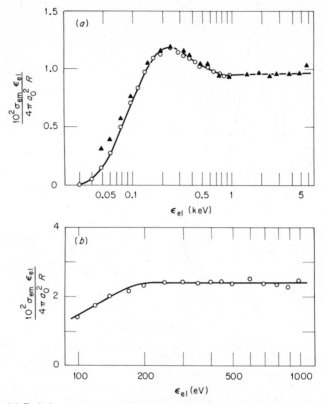

Fig. 25.   (a) Emission cross sections for Balmer β radiation: ($\blacktriangle$) Ref. 64; (O) Ref. 63. (b) Emission cross sections for $CH(A^2\triangle\text{-}X^2\Pi)$ radiation (Ref. 63).

C.   Long-Lived Parent Negative Ions of Benzene Derivatives

1.   Long-Lived Parent Negative Ions

Although no long-lived parent negative ions of diatomic molecules formed by electron capture under isolated-molecule conditions have yet been discovered—they are expected to fly apart within $\lesssim 10^{-12}$ sec—a large number of complex polyatomic molecules including many benzene derivatives with positive (>0 eV) electron affinity have been found to capture slow electrons (in the preponderance of known cases thermal and epithermal) with large cross sections, and to form long-lived parent negative ions. The long lifetimes ($>10^{-6}$ sec) have been thought of as resulting from the fact that for a complex negative ion the extra energy (comprised primarily of the electron affinity of the molecule and the kinetic energy of the incident electron) is shared with the many vibrational degrees of freedom so that electron ejection is drastically delayed. Using this hypothesis Compton et al.[66] considered the reaction

$$(AB \cdots CD)^{-*} \underset{\tau_a^{-1}}{\overset{v\sigma_a}{\rightleftharpoons}} AB \cdots CD + e \qquad (23)$$

where $v\sigma_a$ is the rate of the forward (capture) and τ_a^{-1} the rate of the backward (autodetachment) reaction. The two rates are related through the principle of detailed balancing

$$\tau_a = \left(\frac{\rho^-}{\rho^0}\right)\left(\frac{1}{v\sigma_a}\right) \qquad (24)$$

where ρ^- is the density of states of the negative ion, ρ^0 that of the electron plus the molecule, v is the velocity of the incident electron, and σ_a is the attachment cross section.

Compton et al. assumed that the molecule is left in its ground state after electron ejection, and they thus used for ρ^0 the density of states of the free electron ρ_e given by

$$\rho_e = \frac{m^2 v}{\pi^2 \hbar^3} \qquad (25)$$

For the density of states of the negative ion, ρ^-, they used the expression

$$\rho^- = \frac{(\varepsilon' + a\varepsilon_z)^{N-1}}{\left[\Gamma(N) \prod_{i=1}^{N} h\nu_i\right]} \qquad (26)$$

which was given earlier by Rabinovitch and Diesen[67] for a molecule with N vibrational degrees of freedom and ν_i vibrational frequencies. In (26), $\Gamma(N)$ is the gamma function of N, $\prod_{i=1}^{N} h\nu_i$ is the product of all vibrational frequencies $h\nu_i$, a is a constant, $\varepsilon_z = \frac{1}{2}\Sigma_i h\nu_i$ is the "zero-point" energy, and ε' is the internal energy of the metastable ion in excess of ε_z, that is, the sum of the kinetic energy of the incident electron, the electron affinity EA of the molecule, and the vibrational energy of the molecule above the zero-point energy before electron attachment. Using expressions (25) and (26) for ρ^0 and ρ^-, respectively, and taking the spin degeneracy of the negative ion to be 2, we have

$$\tau_a = \frac{2\pi^2 \hbar^3}{m^2 v} \left[\frac{(\varepsilon' + a\varepsilon_z)^{N-1}}{\Gamma(N) \prod\limits_{i=1}^{N} h\nu_i} \right] \left[v\sigma_a(v) \right]^{-1} \tag{27}$$

which relates τ_a to $\sigma_a(v)$, EA, N and ν_i.

The above treatment has been extended by Christophorou et al.[68] in an effort to explain the experimentally observed[41, 68–72] large dependences of τ_a on ε_i, the energy of the incident electron. Christophorou et al.[68] considered the reaction

$$e(\varepsilon_i) + AX(\varepsilon_z + \varepsilon_t) \xrightarrow{\sigma_a(\varepsilon_i)} AX^{-*}(\varepsilon_z + \varepsilon_t + \varepsilon_i + EA)$$

$$\xrightarrow{k = 1/\tau_a} AX^{(*)} \left[\varepsilon_z + \varepsilon_t + (\varepsilon_i - \varepsilon_f) \right] + e(\varepsilon_f) \tag{28}$$

where ε_t is the translational energy of the neutral molecule, ε_i and ε_f are, respectively, the energies of the incident and autodetached electrons, ε_z is the zero-point energy (assumed to be the same for the neutral molecule and the negative ion), and the rest of the symbols are as have been defined earlier. They used for the density of states of the negative ion ρ^-, the density of states of the molecule to which the compound negative ion state decays ρ_M, and the density of states of the products (molecule + e) to which the compound negative ion decays ρ^0, the expressions

$$\rho^- = \frac{\left[\varepsilon_T + (1 - \beta\omega')\varepsilon_z \right]^{N-1}}{\Gamma(N) \prod\limits_{i=1}^{N} h\nu_i} \tag{29}$$

$$\rho_M = \frac{\left[\varepsilon_t + (\varepsilon_i - \varepsilon_f) + (1 - \beta\omega'')\varepsilon_z \right]^{N-1}}{\Gamma(N) \prod\limits_{i=1}^{N} h\nu_i} \tag{30}$$

$$\rho^0 = \rho_e N_M = \int_0^{\varepsilon_i} \frac{m^2 v_f}{\pi^2 \hbar^3} \frac{\left[\varepsilon_i - \varepsilon_f + (1 - \beta\omega'')\varepsilon_z \right]^{N-1}}{\Gamma(N) \prod\limits_{i=1}^{N} h\nu_i} \, d\varepsilon_f \tag{31}$$

In the above expressions $\varepsilon_T = \varepsilon_i + \varepsilon_t + EA$, and $(1 - \beta\omega')$ and $(1 - \beta\omega'')$ are correction factors.[66,67] It is to be noted that in deriving (29) and (31) it was assumed that $\varepsilon_t = 0$, and that ρ^0 was taken equal to the density of states of the ejected electron, (25), times the number of molecular states N_M. It is stressed that ρ^0 is the number of states per unit energy per unit volume for the final products $AX^{(*)}$ and $e(\varepsilon_f)$, and ρ^- is the number of states per unit energy of the negative ion AX^{-*}.

From (24), and (29) to (31) we have[68]

$$
\tau_a^{-1} = \frac{\Gamma(N) \prod\limits_{i=1}^{N} h\nu_i}{\left[\varepsilon_i + EA + a\varepsilon_z\right]^{N-1}} \left(\frac{2\varepsilon_i}{m}\right)^{1/2} \sigma_a(\varepsilon_i) \int_0^{\varepsilon_i} \frac{m^2}{\pi^2\hbar^3} \left(\frac{2\varepsilon_f}{m}\right)^{1/2}
$$

$$
\times \frac{\left[\varepsilon_i - \varepsilon_f + a'\varepsilon_z\right]^{N-1}}{\Gamma(N) \prod\limits_{i=1}^{N} h\nu_i} d\varepsilon_f \tag{32}
$$

where a, a', and ε are $1 - \beta\omega'$, $1 - \beta\omega''$, and $\frac{1}{2}mv^2$, respectively.

If we assume a twofold degeneracy for the negative ion and multiply and divide (32) by $(\varepsilon_i + a'\varepsilon_z)^{(N+1/2)}$, we have[68]

$$
\tau_a^{-1} = \frac{\sigma_a(\varepsilon_i)\varepsilon_i^{1/2}}{\left[\varepsilon_i + EA + a\varepsilon_z\right]^{N-1}} \frac{m}{\pi^2\hbar^3} (\varepsilon_i + a'\varepsilon_z)^{(N+1/2)} I \tag{33}
$$

where

$$
I = \int_0^A X^{1/2} (1 - X)^{N-1} dX
$$

$$
A = \frac{\varepsilon_i}{(\varepsilon_i + a'\varepsilon_z)} \quad \text{and} \quad X = \frac{\varepsilon_f}{(\varepsilon_i + a'\varepsilon_z)}
$$

The quantity I can be evaluated numerically for any value of A. In spite of the many well-recognized[68] limitations of the above treatment, (33) is an important equation; it relates $\tau_a(\varepsilon_i)$, $\sigma_a(\varepsilon_i)$, N, $h\nu_i$, EA, and also the energies ε_i and ε_f of the incoming and outgoing electrons. It allows a qualitative and at times a semiquantitative understanding of long-lived negative ions as well as of the energy dependence of τ_a on ε_i. These features are brought out in the discussion that follows where the experimental work on τ_a and σ_a

and their energy dependences are summarized and discussed for a number of benzene derivatives.*

2. Experimental Results

The lifetimes of long-lived parent negative ions of many benzene derivatives are listed in Table VIII. The existing limited data on the respective thermal attachment rates and cross sections are given as footnotes to the table. It is interesting to note that among the many monosubstituted benzene derivatives investigated, so far only nitrobenzene and cyanobenzene, for which the substituents are strongly electron withdrawing, have been found to form long-lived parent negative ions. From their comprehensive studies of long-lived parent negative ions formed via nuclear-excited Feshbach resonances, Christophorou and co-workers[41,68-72] concluded that all NO_2-containing benzene derivatives attach thermal and epithermal electrons very efficiently and form long-lived parent negative ions, except for those cases where a fast dissociative electron attachment process destroys the parent negative ion in a time very much shorter than 1 μsec (e.g., m-$C_6H_4NO_2I$).

With minor exceptions [e.g., when, as in the case of benzil,[68,71] geometric changes may take place concomitantly with electron attachment requiring electrons with higher (still < 1 eV) energies than thermal], the cross sections were found to attain their highest values at thermal energy. A representative example of the dependence of the current for parent negative-ion

---

*It should be noted, of course, that process (28) can be written, more appropriately, as

$$e(\epsilon_i) + AX(\epsilon_z + \epsilon_t) \overset{\sigma_a(\epsilon_i)}{\to} AX^{-*}(\epsilon_z' + \epsilon_t + \epsilon_i + EA) \overset{1/\tau_a}{\underset{\sigma_a^{(\cdot)}(\epsilon_f)}{\rightleftarrows}}$$

$$\text{(a)} \qquad\qquad \text{(b)}$$

$$AX^{(\cdot)}\big(\epsilon_z + \epsilon_t + (\epsilon_i - \epsilon_f)\big) + e(\epsilon_f)$$

$$\text{(c)}$$

where ϵ_z, ϵ_z' are, respectively, the zero-point energies for the neutral molecule and the negative ion and $\sigma_a(\epsilon_i)$ and $\sigma_a^{(\cdot)}(\epsilon_f)$ are, respectively, the attachment cross sections for $a \to b$ and for $c \to b$. Similarly, (24) can be expressed as

$$\tau_a = \frac{\rho^-}{\rho^0}\left(\frac{1}{v_f \sigma_a^{(\cdot)}(v_f)}\right)$$

In deriving expression (32) in the text it was assumed that $v_f \sigma^{(\cdot)}(v_f) = v_i \sigma(v_i)$, but, of course, this assumption can be removed in which case $\left(\dfrac{2\epsilon_i}{m}\right)^{1/2} \sigma_a(\epsilon_i)$ in (32) should be replaced by $\left(\dfrac{2\epsilon_f}{m}\right)^{1/2} \sigma_a^{(\cdot)}(\epsilon_f)$ and be placed inside the integral.

TABLE VIII
Long-Lived Negative Ions of Benzene Derivatives[a]

Compound	Negative Ion	Lifetime (μsec)	Reference
8.1 NO$_2$-containing benzene derivatives			
Deuterated nitrobenzene		22	1
o-Bromonitrobenzene		18	2
m-Bromonitrobenzene		21	2
p-Bromonitrobenzene		10	2
o-Chloronitrobenzene		17	2
m-Chloronitrobenzene		47	1

469

TABLE VIII   (*Continued*)

Compound	Negative Ion	Lifetime (µsec)	Reference

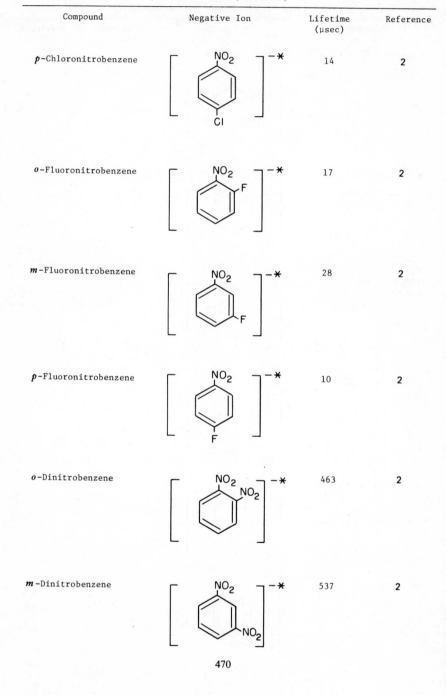

p-Chloronitrobenzene		14	2
o-Fluoronitrobenzene		17	2
m-Fluoronitrobenzene		28	2
p-Fluoronitrobenzene		10	2
o-Dinitrobenzene		463	2
m-Dinitrobenzene		537	2

470

TABLE VIII   (*Continued*)

Compound	Negative Ion	Lifetime (μsec)	Reference
p-Dinitrobenzene		421	2
Nitrobenzene		17.5;47.3	1, 3
o-Nitrophenol		460[b]	4
m-Nitrophenol		31	4
p-Nitrophenol		14	4
p-Nitrothiophenol		23	2

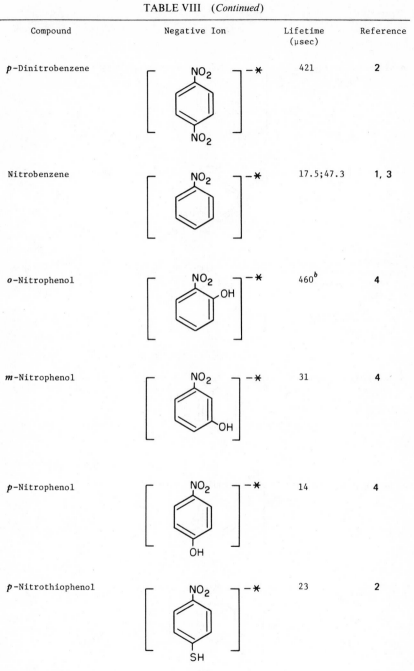

471

TABLE VIII (*Continued*)

Compound	Negative Ion	Lifetime (μsec)	Reference
o-Nitro-α,α,α-tri-fluorotoluene		200	2
m-Nitro-α,α,α-tri-fluorotoluene		187[b]	2
p-Nitro-α,α,α-tri-fluorotoluene		143	2
o-Cyanonitrobenzene		209	2
m-Cyanonitrobenzene		315[b]	2
p-Cyanonitrobenzene		205	2

TABLE VIII   (*Continued*)

Compound	Negative Ion	Lifetime (μsec)	Reference
o-Nitrobenzoic Acid		44	2
m-Nitrobenzoic Acid		338	2
p-Nitrobenzoic Acid		142	2
o-Nitrobenzaldehyde		395	2
m-Nitrobenzaldehyde		205	2
p-Nitrobenzaldehyde		47	2

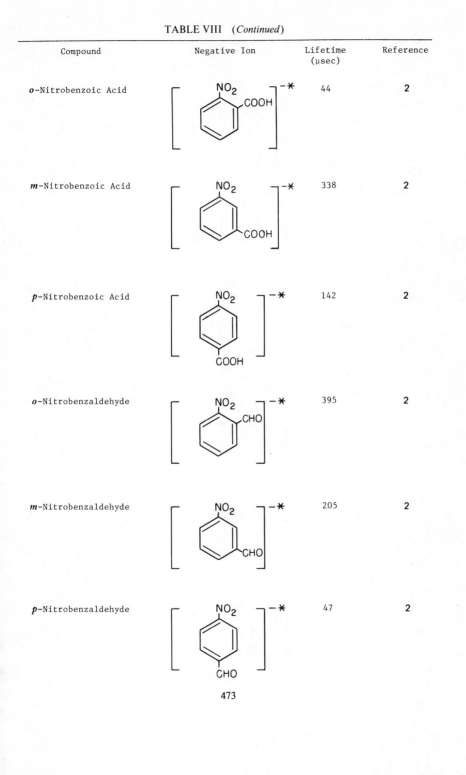

473

TABLE VIII   (*Continued*)

Compound	Negative Ion	Lifetime (μsec)	Reference
o-Nitroaniline	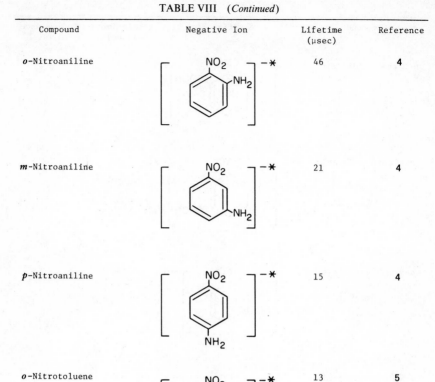	46	**4**
m-Nitroaniline		21	**4**
p-Nitroaniline		15	**4**
o-Nitrotoluene		13	**5**
m-Nitrotoluene	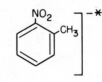	19	**5**
p-Nitrotoluene	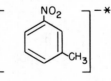 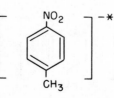	14	**5**

TABLE VIII   (*Continued*)

Compound	Negative Ion	Lifetime (μsec)	Reference
o-Nitroanisole		16	2
m-Nitroanisole		52	2
p-Nitroanisole		10	2
o-Nitroacetophenone		189	2
m-Nitroacetophenone		310[b]	2
p-Nitroacetophenone		196	2

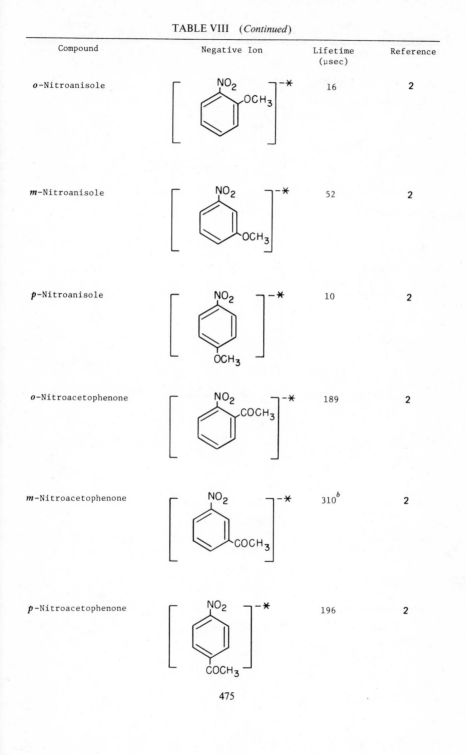

TABLE VIII (*Continued*)

Compound	Negative Ion	Lifetime (μsec)	Reference

8.2 <u>CN-containing benzene derivatives</u>

Cyanopentafluorobenzene	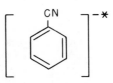	17	1
o−, *m*−, *p*-Cyanonitro-benzene	see above (8.1)	209; 315[b]; 205	2
Cyanobenzene	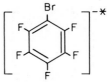	∿5	1

8.3 <u>Other benzene derivatives</u>

Bromopentafluorobenzene	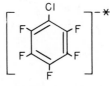	21	1
Chloropentafluoro-benzene	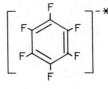	17.6	1
Hexafluorobenzene		12	6, 7

TABLE VIII (*Continued*)

Compound	Negative Ion	Lifetime (μsec)	Reference
Octafluorotoluene		12.2	6
Pentafluorobenzaldehyde		36	1
Cinnamaldehyde		12	5

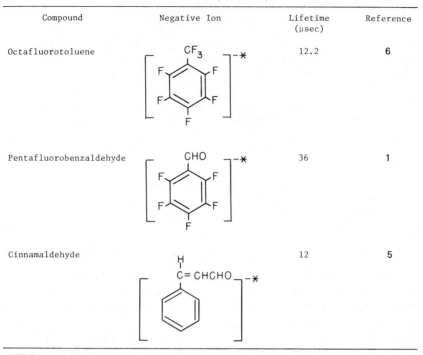

---

[a]Unfortunately, there are only very few reported values for the thermal attachment rate and the electron affinity of the benzene derivatives that form long-lived parent negative ions. The thermal attachment rates for nitrobenzene, cinnamaldehyde and hexafluorobenzene are 2.1×10^7 sec^{-1} torr^{-1} [L. G. Christophorou and R. P. Blaunstein, *Rad. Res.* **37**, 229 (1969)]; 2×10^8 sec^{-1} torr^{-1} [W. E. Wentworth and E. Chen, *J. Phys. Chem.* **71**, 1929 (1967)]; and 3.3×10^9 sec^{-1} torr^{-1} [K. S. Gant and L. G. Christophorou, *J. Chem. Phys.* (in press)], respectively. Electron affinity values have been reported for nitrobenzene: $\geqslant 0.4$ eV [R. N. Compton, L. G. Christophorou, G. S. Hurst and P. W. Reinhardt, *J. Chem. Phys.* **45**, 4634 (1966)]; $\geqslant 0.51$ eV [E. L. Chaney and L. G. Christophorou, *Oak Ridge National Laboratory Report* ORNL-TM-2613 (1970)]; $\geqslant 0.70$ eV [(C. Lifshitz, T. O. Tiernan and B. M. Hughes, *J. Chem. Phys.* **59**, 3182 (1973)]; 1.19 eV, 1.34 eV [E. C. M. Chen and W. E. Wentworth, *J. Chem. Phys.* **63**, 3183 (1975)]; for *o*-dinitrobenzene: 1.07 eV, 1.66 eV [E. C. M. Chen and W. E. Wentworth, *J. Chem. Phys.* **63**, 3183 (1975)]; for *m*-dinitrobenzene: 1.26 eV, 1.58 eV [E. C. M. Chen and W. E. Wentworth, *J. Chem. Phys.* **63**, 3183 (1975)]; 1.45 eV [M. Batley and L. E. Lyons, *Nature* **196**, 573 (1962)]; for *p*-chloronitrobenzene: 1.19 eV, 1.34 eV [E. C. M. Chen and W. E. Wentworth, *J. Chem. Phys.* **63**, 3183 (1975)]; for benzonitrile: 0.25 eV [W. E. Wentworth, L. W. Kao and R. S. Becker, *J. Phys. Chem.* **79**, 1161 (1975)]; for *p*-nitrobenzonitrile: 1.82 eV [E. C. M. Chen and W. E. Wentworth, *J. Chem. Phys.* **63**, 3183 (1975)]; for benzaldehyde: 0.42 eV [W. E. Wentworth, L. W. Kao, and R. S. Becker, *J. Phys. Chem.* **79**, 1161 (1975)]; 0.43 eV [W. E. Wentworth and E. Chen, *J. Phys. Chem.* **71**, 1929 (1967)]; for cinnamaldehyde: 0.82 eV [W. E. Wentworth and E. Chen, *J. Phys. Chem.* **71**, 1929 (1967)]; for hexafluorobenzene: $\geqslant 1.8 \pm 0.3$ eV [C. Lifshitz, T. O. Tiernan, and B. M. Hughes,

TABLE VIII   (*Continued*)

J. Chem. Phys. **59**, 3182 (1973)]; 1.2 ± 0.1 eV [F. M. Page and G. L. Goode, *Negative Ions and the Magnetron*, Wiley-Interscience, New York, 1969)]; and for octafluorotoluene: ≥ 1.7 ± 0.3 eV [C. Lifshitz, T. O. Tiernan, and B. M. Hughes, *J. Chem. Phys.* **59**, 3182 (1973)].

   b Measured lifetime is found to decrease with increasing electron energy.

1.   W. T. Naff, R. N. Compton, and C. D. Cooper, *J. Chem. Phys.*, **54**, 212 (1971).
2.   J. P. Johnson, L. G. Christophorou, D. L. McCorkle, and J. G. Carter, *J. Chem. Soc., Faraday Trans. II*, **71**, 1742 (1975).
3.   P. W. Harland and J. C. J. Thynne, *J. Phys. Chem.*, **75**, 3517 (1971).
4.   A. Hadjiantoniou, L. G. Christophorou, and J. G. Carter, *J. Chem. Soc., Faraday Trans. II*, **69**, 1691 (1973).
5.   L. G. Christophorou, J. G. Carter, E. L. Chaney, and P. M. Collins, in J. F. Duplan and A. Chapiro, Eds., *Proceedings of the IVth International Congress of Radiation Research, Evian, France (July 1970), Vol. 1, Physics and Chemistry*, Gordon and Breach, London, 1973, pp. 145–157.
6.   W. T. Naff, C. D. Cooper, and R. N. Compton, *J. Chem. Phys.*, **49**, 2784 (1968).
7.   J. P. Johnson and L. G. Christophorou (1975, unpublished result).

formation on the incident energy is shown in Fig. 26 for *o*-, *m*-, and *p*-$C_6H_4OHNO_2^-$*. The parent negative-ion current is seen to occur across a narrow resonance whose maximum coincides with that of SF_6^-* at ~0.0 eV[71] and which appears at energies lower than the maximum of the SF_5^- from SF_6 resonance at ~0.37 eV.[73] The widths (≲0.2 eV) of all the negative-ion resonances of the substituted nitrobenzenes at thermal energies are similar to that for SF_6^-* and instrumental.[41,53,71] The energy scale in Fig. 26 for the production of the *o*-$C_6H_4OHNO_2^-$* ion was also established by using the Cl^- from HCl resonance which peaks at 0.8 eV.[74]

For a number of NO_2-containing benzene derivatives, indicated by the superscript *b* in Table VIII, τ_a was found to decrease greatly with increasing incident electron energy above thermal. The lifetime values listed in Table VIII are those that coincide with the peak of the respective parent negative-ion resonance, that is, at ~0.0 eV. These, naturally, represent averages over the electron-beam energy spread, and might have been found to be much longer had it been possible to use a truly monoenergetic electron beam. This can easily be seen from Fig. 27, where the variation of τ_a with ε is shown for *m*-$C_6H_4CF_3NO_2^-$*. Had the electron beam been much narrower than that represented in the figure by the negative-ion resonance (x), the τ_a values for all molecules in Table VIII would be much larger and the decrease of τ_a with ε much faster. It has been argued,[41,68,71,72] however, that since the values of τ_a listed in Table VIII for the NO_2-containing disubstituted benzene derivatives were all taken under more or less identical experimental conditions, they can be compared

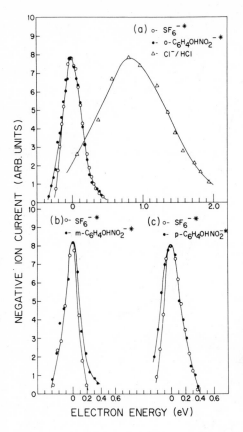

Fig. 26. Negative-ion current versus electron energy for (*a*) *o*-nitrophenol (●), SF₆ (O), and Cl⁻ from HCl (△); (*b*) *m*-nitrophenol (●) and SF₆ (O); and (*c*) *p*-nitrophenol (●) and SF₆ (O). The shapes of $SF_6^-{}^*$ and *o*-, *m*-, and *p*-$C_6H_4OHNO_2^-{}^*$ resonances are instrumental, that is, the actual resonances are narrower than shown (from Ref. 71).

directly and the observed differences can be attributed to basic differences in molecular structure as is discussed below.

3. Comparison with the Theory

Unfortunately, for none of the compounds in Table VIII for which there is an observed variation of τ_a with electron energy are the necessary data available to calculate τ_a using (33). However, such data do exist for 1,4-naphthoquinone whose parent negative ion is long-lived[69] with a lifetime that decreases with ε as shown in Fig. 28 (curve *A*). For this molecule Christophorou et al.[68] treated the experimental data as suggested by (33) and calculated the variation of τ_a with ε, which is also shown in Fig. 28 (curves *B* and *C*). In spite of the limitations intrinsic in such calculations,

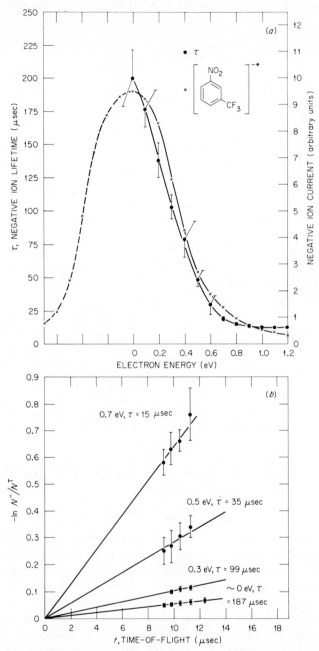

Fig. 27. (a) Dependence of the negative-ion lifetime (●) and the negative-ion current (x) for the m-$C_6H_4NO_2CF_3^-$* metastable ion on the incident electron energy. (b) Determination of τ_a for m-$C_6H_4NO_2CF_3^-$* at ∼0.0, 0.3, 0.5, and 0.7 eV (from Ref. 41).

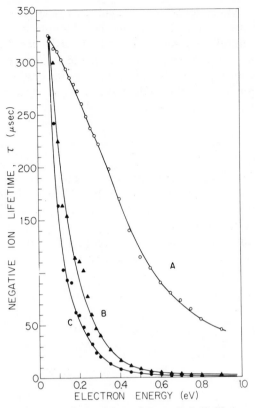

Fig. 28.   Experimental (A) and calculated (B and C; see Ref. 68) lifetimes for $1,4\text{-}C_{10}H_6O_2^-{}^*$ as a function of electron energy.

it is interesting to observe the much faster decrease of the calculated τ_a compared with that which is experimentally observed. This has been attributed to the poor energy resolution in such experiments and points to the fact that not only is the lifetime a very strong function of the incident energy, but it could also be much longer if one had truly thermal electrons, a situation that could perhaps be better approximated under certain conditions in ion-cyclotron-resonance experiments[75,76] (see further discussion in Ref. 68).

The comprehensive work of Johnson et al.[41] on over 40 substituted nitrobenzenes indicated that for all these compounds the capture cross sections attain their maximum value at thermal energies and that they decrease rapidly thereafter with increasing energy close to thermal. John-son et al.[41] pointed out that since all their lifetime measurements were made under identical experimental conditions and since these compounds are similar in structure, it might be reasonable to assume that the vibra-tional frequencies $h\nu_i$ and the attachment cross sections $\sigma_a(\varepsilon)$ for the substituted nitrobenzenes they studied were roughly the same. Under these conditions (33) would predict that

$$\tau_a^{1/(N-1)} \propto EA \tag{34}$$

Electron affinities for the nitrobenzenes are virtually nonexistent. How-ever, for most of these compounds polarographic half-wave potentials, $E_{1/2}$, which should be proportional to EA,[77] are available. In Fig. 29 $\tau^{1/(N-1)}$ is plotted as a function of $E_{1/2}$ and the results are seen to be in reasonable agreement with (34).

As has been discussed earlier in this section, τ_a is affected by N, internal energy of the ion, molecular geometry, and configurational changes that

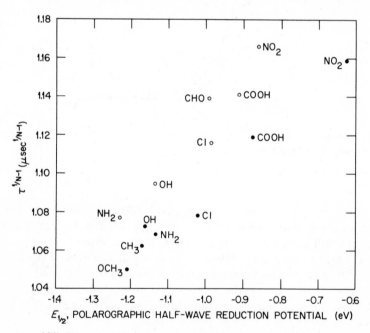

Fig. 29. $\tau_a^{1/(N-1)}$ versus $E_{1/2}$, the polarographic half-wave potential for substituted nitrobenzenes. (O) meta-isomers; (●) para-isomers. The compounds are identified by the second substituent. The number of vibrational degrees of freedom N was taken equal to $3n-6$ where n is the number of atoms in the molecule (from Ref. 41).

may take place concomitantly with electron capture. If for a given series of molecules it were possible to keep these factors except N approximately the same, it would be possible to test the prediction of (27) [or (33)] that τ_a, actually $\ln \tau_a$, increases with N. A rough linear increase of $\ln \tau_a$ with N has been reported[78] for a number of alicyclic and aromatic fluorocarbons.

It is, finally, noted that (27) has been employed to provide a rough estimate of EA from a measurement of $\tau_a(\varepsilon)$ and $\sigma_a(\varepsilon)$ and a knowledge of N and $h\nu_i$. Such EA estimates have been made for some systems.[66,69,70,71]

4. Dependence of the Lifetime on the Electron Donor-Acceptor Properties of the Substituents and the π-Electron Charge-Density Distributions in the Neutral Molecule and the Respective Negative Ion

The lifetimes of the parent negative ions of NO_2-containing disubstituted benzene derivatives were found[41] to depend strongly on the π-electron donor-acceptor properties of the substituent groups (NO_2 and X): strong electron-withdrawing substituents X (NO_2 is a strongly electron-withdrawing group as well) greatly increase τ_a, while τ_a is little affected by π-electron donating substituents. Furthermore, intramolecular interaction (complexing) between the substituents was shown[41,71] to lead to enhanced lifetimes for ortho substituted nitrobenzenes (see Refs. 41 and 71 and Table VIII).

CNDO-2 molecular orbital calculations on substituted nitrobenzenes with the two substituents NO_2 and X in the para position to avoid intramolecular complexing between them, have shown[41] that when X is an electron acceptor, the magnitude of τ_a correlates with the amount of π-electron charge withdrawn from the ring, while τ_a is little affected by the amount of π-electron charge donated to the ring by X when X is an electron donor. This is shown in Fig. 30a. If the ring π-electron charge for the neutral molecule is subtracted from that for the negative ion, the net increase in ring π-electron charge can be estimated for the parent negative ion. It was found[41] that the ring π-electron charge increased by 0.2 to 0.3 electron units when X is an electron donor and by 0.4 to 0.7 electron units when X is an electron acceptor. This large change in the latter case is due to the greater Coulomb attraction between the attached (extra) electron and the positively charged ring, and is responsible for the increased stability of the negative ion. In Fig. 30b the increase in τ_a with increasing ring π-electron charge for the negative ion is seen.

The comprehensive work of Christophorou and co-workers on substituted nitrobenzenes shows that the lifetime is not only affected by the nature of X, but also by the relative position of X with respect to NO_2. It indicates that it might be possible to infer intramolecular complexing

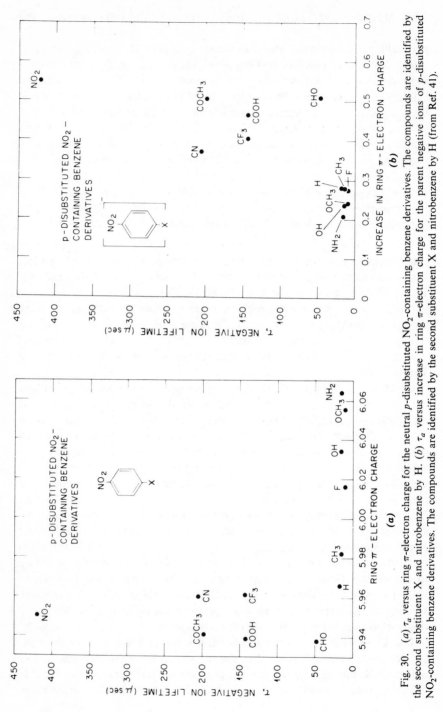

Fig. 30. (a) τ_a versus ring π-electron charge for the neutral p-disubstituted NO_2-containing benzene derivatives. The compounds are identified by the second substituent X and nitrobenzene by H. (b) τ_a versus increase in ring π-electron charge for the parent negative ions of p-disubstituted NO_2-containing benzene derivatives. The compounds are identified by the second substituent X and nitrobenzene by H (from Ref. 41).

between X and NO_2 and intramolecular charge transfer from X to the ring and vice versa from a measurement of the negative-ion lifetime.

V. IONIZATION BY ELECTRON IMPACT

A. Ionization Cross-Section Functions

Cross sections for single ionization of atoms and molecules begin from a value of zero at the ionization threshold, rise more or less linearly close to the threshold, reach a broad maximum usually in the region of ~ 100 eV, and decline monotonically thereafter.[5] Although, to our knowledge, no such cross-section measurements have been made for benzene, its single ionization cross section, σ_i, is expected to behave in a similar fashion. For sufficiently high impact energies ($\gtrsim 200$ eV) the Born approximation expression for the excitation cross section in Section II can be extended to the case of single ionization.[5] In this case, σ_i is expected to decrease asymptotically as

$$\sigma_i \propto \varepsilon^{-1} \ln B\varepsilon \qquad (35)$$

where ε is the incident electron energy and B^{-1} is a constant on the order of the binding energy of the ejected electron.

For double electron ejection the double-ionization cross section, σ_{2i}, is expected[5] to be smaller in magnitude than σ_i and to have a higher threshold. The asymptotic form of σ_{2i} in the Born approximation is

$$\sigma_{2i} \propto \varepsilon^{-1} \qquad (36)$$

Electron impact ionization cross sections have been calculated for benzene by Sen and Basu,[80] who used a modification of the free-electron model[81] in conjunction with the Born approximation. The result of this calculation is shown in Fig. 31.

Relative geometric charge cross sections, $\langle r^2 \rangle$, where r is the distance from the center of charge, for benzene and several methyl substituted benzene cations have been calculated by King[82] from Hückel charge distributions. Some of King's results are presented in Table IX. The relative geometric charge cross sections in column 3 have been determined by assuming a point charge distribution for the π electrons, while those in column 4 were determined by assuming the π electrons to be distributed in $2p$ Slater orbitals and the methyl electrons to be in sp^3 hybrids constructed from Slater $2s$ and $2p$ atomic orbitals. In column 5 the relative $\langle r^2 \rangle$ listed were obtained by weighting the methyl charge densities by a factor of 3.3 in order to achieve a better fit between experimental and theoretical results

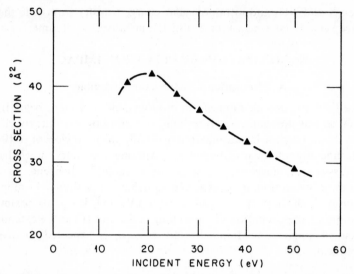

Fig. 31. Cross section versus incident electron energy for benzene π-electron ionization (from Ref. 80).

TABLE IX

Comparison of Relative Ionization Probabilities and Geometric Charge Cross Sections $\langle r^2 \rangle$ for the Polymethyl Benzenes[82]

	Relative ionization probability[a]	Relative $\langle r^2 \rangle$		
		Point charges	sp^3 for methyls $2p$ for π orbitals	Methyls weighted with a factor = 3.3
Benzene	(1.00)	(1.000)	(1.000)	(1.000)
Toluene	1.44	1.278	1.229	1.926
p-Xylene	1.74	1.537	1.436	2.807
m-Xylene	1.80	1.410	1.332	2.364
o-Xylene	1.58	1.374	1.308	2.300
Hemimellitene	2.49	1.369	1.306	2.383
Pseudocumene	2.46	1.588	1.478	2.992
Mesitylene	2.52	1.411	1.334	2.383
Durene	3.34[b]	1.705	1.573	3.372
Isodurene	3.08[b]	1.604	1.493	3.098
Prehnitene	2.82[b]	1.618	1.502	3.077
Pentamethylbenzene	3.34	1.705	1.573	3.372
Hexamethylbenzene	3.39	1.705	1.573	3.372

[a] Source, unless otherwise indicated, is Ref. 84.
[b] Reference 82.

486

for the more highly substituted compounds.[83] When the results of these calculations are compared with the experimental relative ionization probabilities near threshold in column 2, it appears that a rough correlation exists between the $\langle r^2 \rangle$ values and the ionization probability near threshold. Thus it would seem that the ionization probabilities near threshold increase with increasing size of the π-electron system. Furthermore, from measurements on a large number of substituted benzenes, Deverse and King[84] found that the ionization probabilities of these compounds near threshold increase with increasing electron-donating ability of the substituents.

On the experimental side, Harrison et al.[85] measured total cross sections for ionization of benzene and alkyl benzenes by 75-eV incident electrons. Their results are listed in Table X. The cross sections are seen to be significantly large and for the alkyl benzenes they seem to increase with increasing polarizability.[85] It is interesting to note that the experimental cross section, 13.4×10^{-16} cm^2, for benzene is comparable to that, $\sim 25 \times 10^{-16}$ cm^2, estimated from Basu's calculation (see Fig. 31) for 75 eV energy.

TABLE X
Electron-Impact Ionization Cross Sections for Several
Alkylbenzenes (Incident Energy: 75 eV)[85]

Molecule	σ_i (10^{-16} cm^2)
Benzene	13.4
Toluene	16.1
o-Xylene	18.7
Ethylbenzene	19.0
n-Propylbenzene	22.2
Cumene	22.2
p-Ethyltoluene	22.1
1,2,3-Trimethylbenzene	21.6
1,2,4-Trimethylbenzene	21.8
n-Butylbenzene	24.1
t-Butylbenzene	24.1
1,2,3,5-Tetramethylbenzene	23.6

B. Threshold Behavior and Structure Near Threshold In the Ionization Cross Section

While exceptions do seem to exist, many electron and photon impact processes seem to follow simple "laws" in the threshold region. Based on the theoretical work of Wigner[86] and Wannier[87] and the experimental

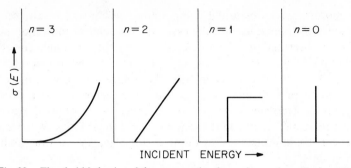

Fig. 32.   Threshold behavior of the cross section for various values of n (see text).

results of Morrison,[88,89] it has been postulated[90,91] that the energy dependence of the total cross section, $\sigma(E)$, near threshold is of the form

$$\sigma(E) \sim (\Delta E)^{n-1} \qquad (37)$$

where $\Delta E = E - E_{th}$ is the energy above threshold, E_{th} is the threshold energy, and n is the number of electrons leaving the collision complex.

In Fig. 32 the threshold behavior of $\sigma(E)$ as exemplified by (37) is shown for $n = 3$, $n = 2$, $n = 1$, and $n = 0$. The $n = 3$ case corresponds to double ionization by electron impact or triple ionization by photon impact. For $n = 2$ one has single ionization by electron impact or double ionization by photon impact. The case $n = 1$ corresponds to electron impact excitation or single photoionization while the $n = 0$ case corresponds to electron attachment or photoexcitation.

However useful the threshold behavior of the cross section is in yielding information about the nature of the process involved, it is very difficult to discern the threshold behavior of the processes mentioned above from the experimental data. Usually, two or more processes take place near threshold and as a consequence the total cross section does not exhibit a simple type of threshold behavior as exemplified by (37). Actually, in certain instances single ionization cross sections do not agree with an integer power law.[92] A major problem in identifying the threshold behavior of the cross section is the appearance of structure near threshold. This is clearly seen in Fig. 33 where the electron impact ionization efficiency curve for benzene is shown.[93] The first straight-line segment of the curve is due to the production of $C_6H_6^+$ in the lowest possible energy configuration, which requires ~ 9.24 eV. The breaks in the ionization efficiency curve can be due to a number of processes such as vibrational excitation of $C_6H_6^+$, autoionizing states of neutral C_6H_6, or ion-pair processes. Since the photoelectron spectra of benzene in this energy region exhibit structure that has

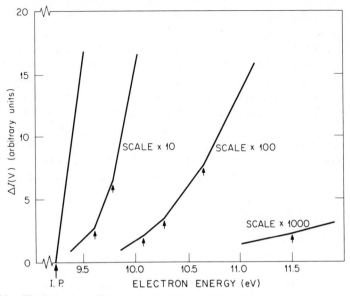

Fig. 33.   The ionization efficiency curve of benzene (Ref. 93). The spacings of the breaks are (left to right) 0.37 ± 0.03, 0.54 ± 0.03, 0.80 ± 0.03, 1.00 ± 0.03, 1.37 ± 0.03, and 2.27 ± 0.09 eV above the ionization threshold at 9.24 eV.

been attributed[94] to vibrational modes, it would seem reasonable to attribute the breaks in Fig. 33 to vibrationally excited $C_6H_6^+$ ions.

C.   First Ionization Potentials

The study of the first ionization potential of benzene and its derivatives has been carried out by a number of experimental methods other than the electron impact, including photoionization, photoabsorption, photoelectron spectroscopy, and charge-transfer spectra. While in this chapter we are mainly concerned with electron-impact phenomena, many of the results discussed in this section are from experiments utilizing these other techniques.

In electron-impact studies a quasimonoenergetic beam of electrons traverses the collision chamber containing the gas under investigation at pressures low enough to approximate single-collision conditions. The energy of the electron beam is increased until ions are formed. The positive ions, drawn out of the collision chamber by an electric field, are mass-analyzed and their current is plotted as a function of the incident electron energy. The potential required to accelerate the electrons to the lowest value of electron energy for a particular positive ion to appear is referred to as the appearance potential for that ion. (The appearance potential of the parent molecular ion represents an upper limit to the first molecular ionization potential.)

In Table XI the first ionization potentials, IP, for benzene and a number of its derivatives are listed. In each case the technique by which a particular value of IP was obtained and the appropriate references are given. In Fig. 34 the electron-impact value of IP is plotted versus the photoabsorption, photoionization, or photoelectron spectroscopy value of IP. As can be seen from Fig. 34, the electron-impact values are consistently higher than those obtained by photon-impact methods. Although part of this difference may be due to experimental uncertainties, expecially in the electron-impact method, the higher values of IP for electrons may be

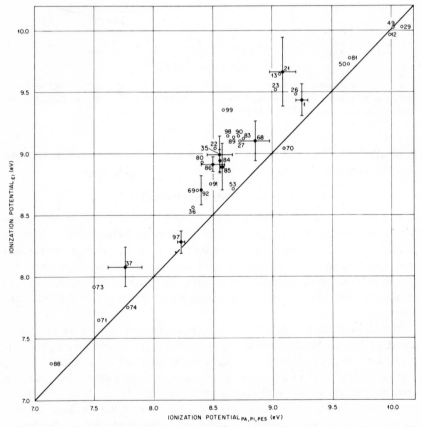

Fig. 34.   Electron-impact values of the ionization potential versus ionization potential values determined by photoabsorption, photoionization, or photoelectron spectroscopy.

TABLE XI
First Ionization Potentials of Benzene and Benzene Derivatives*

	Molecule	Formula	I (eV)	Ref.	Method
1.	Benzene $[9.25 \pm 0.05]_{a,b,c}$ $[9.43 \pm 0.13]_d$	C_6H_6	9.24	1, 2	a
			9.25	3, 28	a
			9.19	4	b
			9.24	5	b
			9.25	6	b
			9.24	7, 8	c
			9.25	9, 10	c
			9.40	30	c
			9.21	11, 12	d
			9.38	13	d
			9.5	14	d
			9.52	15, 16, 17	d
			9.56	18	d
			9.25 ± 0.03	19	e
			9.21	20	f
			9.29	21	f
			$9.35^\S; 9.22^\dagger$	42	f
			9.37	22	f
			9.45	23	f
			9.74	27	f
2.	Deuterated Benzene $[9.23 \pm 0.03]_{a,b}$	C_6D_6	9.24	2	a
			9.19	4	b
			9.25	6	b
3.	Hexafluorobenzene	C_6F_6	9.88 ± 0.05	10	c
4.	1,2,4-Trifluoro-benzene	$1,2,4\text{-}C_6H_3F_3$	9.30 ± 0.05	10	c
5.	1,4-Bromochloro-benzene	$p\text{-}C_6H_4BrCl$	9.04	30	c
6.	1,3-Bromonitro-benzene	$m\text{-}C_6H_4BrNO_2$	9.82 ± 0.1	24	d
7.	1,4-Bromonitro-benzene	$p\text{-}C_6H_4BrNO_2$	9.76 ± 0.1	24	d
8.	1,4-Dibromo-benzene	$p\text{-}C_6H_4Br_2$	8.97	30	c
9.	1,4-Chlorofluoro-benzene	$p\text{-}C_6H_4ClF$	9.26	30	c
10.	1,4-Chloronitroso-benzene	$p\text{-}C_6H_4ClNO$	9.74	30	c

TABLE XI (*Continued*)

	Molecule	Formula	I (eV)	Ref.	Method
11.	1,3-Chloronitro-benzene	m-C$_6$H$_4$ClNO$_2$	9.92 ± 0.1	24	d
12.	1,4-Chloronitro-benzene	p-C$_6$H$_4$ClNO$_2$	9.96 ± 0.1	24	d
			9.99	30	c
13.	1,2-Dichloro-benzene	o-C$_6$H$_4$Cl$_2$	9.07	25	a
			9.06	26	b
			9.64	18	d
14.	1,3-Dichloro-benzene	m-C$_6$H$_4$Cl$_2$	9.12	25	a
15.	1,4-Dichloro-benzene	p-C$_6$H$_4$Cl$_2$	8.94	25	a
			8.95	26	b
			9.17	30	c
16.	1,3-Fluoronitro-benzene	m-C$_6$H$_4$FNO$_2$	9.93	24	d
17.	1,4-Fluoronitro-benzene	p-C$_6$H$_4$FNO$_2$	10.00	24	d
18.	1,4-Difluoro-benzene	p-C$_6$H$_4$F$_2$	9.15 ± 0.06	10	c
			9.50	30	c
19.	1,3-Dinitrobenzene	m-C$_6$H$_4$(NO$_2$)$_2$	10.62	24	d
20.	1,4-Dinitrobenzene	p-C$_6$H$_4$(NO$_2$)$_2$	10.63	24	d
21.	Bromobenzene $[9.09 \pm 0.11]_{a,c}$ $[9.66 \pm 0.28]_d$	C$_6$H$_5$Br	8.98	28	a
			9.05	29	c
			9.25	30	c
			9.41	31	d
			9.52	18	d
			10.05	32	d
22.	1,4-Bromophenol	p-C$_6$H$_4$BrOH	8.52	30	c
			9.04	18	d
23.	Chlorobenzene $[9.03 \pm 0.25]_{a,b,c}$	C$_6$H$_5$Cl	9.07	28	a
			8.7	33	b
			9.31	30	c
			9.42	31	d
			9.60	18	d
24.	1,2-Chlorophenol	o-C$_6$H$_4$ClOH	9.28	18	d
25.	1,4-Chlorophenol	p-C$_6$H$_4$ClOH	8.69	30	c

TABLE XI (*Continued*)

	Molecule	Formula	I (eV)	Ref.	Method
26.	Fluorobenzene	C_6H_5F	9.19	28	a
	$[9.20 \pm 0.01]_{a,b,c}$		9.20	5, 26, 34	b
			9.21 ± 0.04	10	c
			9.3	32	d
			9.67	31	d
27.	Iodobenzene	C_6H_5I	8.73	28	a
			8.78	30	c
			9.10	31	d
28.	Nitrosobenzene	C_6H_5NO	9.97	30	c
29.	Nitrobenzene	$C_6H_5NO_2$	9.92	35	a
	$[10.02 \pm 0.21]_d$		10.26	30	c
			9.65 ± 0.10	24	d
			9.94	36	d
			10.15	37	d
			10.18	18, 38	d
30.	1,4-Nitrophenol	$p\text{-}C_6H_4NO_2OH$	7.38	39	d
	$[8.58 \pm 0.89]_d$		8.84 ± 0.1	24	d
			9.52	18	d
31.	1,4-Chloroaniline	$p\text{-}C_6H_4ClNH_2$	8.18	30	c
32.	1,2-Nitroaniline	$o\text{-}C_6H_4NO_2NH_2$	8.66	18	d
33.	1,3-Nitroaniline	$m\text{-}C_6H_4NO_2NH_2$	8.73 ± 0.1	24	d
			8.80	18	d
34.	1,4-Nitroaniline	$p\text{-}C_6H_4NO_2NH_2$	8.43	39	d
	$[8.63 \pm 0.17]_d$		8.62 ± 0.1	24	d
			8.85	18	d
35.	Phenol	C_6H_5OH	8.50	28	a
	$[8.56 \pm 0.11]_{a,c}$		8.52	1	a
	$[8.99 \pm 0.15]_d$		$8.48 \pm 0.05^\dagger$	40	c
			$8.74^\S$	40	c
			8.75	30	c
			9.01	15	d
			9.03	31	d
			9.16	18	d
36.	Phenyl mercaptan	C_6H_5SH	8.33	25	a
			8.56	15	d

TABLE XI   (*Continued*)

Molecule	Formula	I (eV)	Ref.	Method
37. Aniline	$C_6H_5NH_2$	7.69	1	*a*
$[7.76 \pm 0.14]_{a,c}$		7.70	28	*a*
$[8.08 \pm 0.16]_d$		$7.68 \pm 0.05^{\dagger}$	40	*c*
		$7.71^{\dagger}$	41	*c*
		$8.02^{\S}$	41	*c*
		8.04	30	*c*
		$8.08^{\S}$	40	*c*
		7.84	12	*d*
		8.0	14	*d*
		8.23	15	*d*
		8.32	18	*d*
38. 1,2-Diamino-benzene	o-$C_6H_4(NH_2)_2$	8.00	18	*d*
39. 1,3-Diamino-benzene	m-$C_6H_4(NH_2)_2$	7.96	18	*d*
40. 1,4-Diamino-benzene	p-$C_6H_4(NH_2)_2$	7.16	39	*d*
		7.58	18	*d*
41. p-Trifluoromethyl-bromobenzene	p-$C_6H_4BrCF_3$	9.55	30	*c*
42. 1,4-Bromobenzo-nitrile	p-C_6H_4BrCN	9.54	30	*c*
43. p-Trifluoromethyl-chlorobenzene	p-$C_6H_4ClCF_3$	9.80	30	*c*
44. 1,3-Nitroben-zonitrile	m-$C_6H_4NO_2CN$	10.29	24	*d*
45. 1,4-Nitroben-zonitrile	p-$C_6H_4NO_2CN$	10.23	24	*d*
46. 1,4-Chlorobenzal-dehyde	p-C_6H_4ClCHO	9.59	30	*c*
47. α,α,α-Trifluoro-toluene	$C_6H_5CF_3$	9.90	30	*c*
48. Phenyl-trifluoro-methyl ether	$C_6H_5OCF_3$	10.00	30	*c*
49. Benzonitrile	C_6H_5CN	10.02	30	*c*
		9.95	31	*d*
		10.09	18	*d*

TABLE XI (*Continued*)

Molecule	Formula	I (eV)	Ref.	Method
50. Benzaldehyde $[9.64 \pm 0.12]_{a,c}$	C_6H_5CHO	9.51	28	*a*
		9.60	1	*a*
		9.80	30	*c*
		9.63	43	*d*
		9.82	31	*d*
51. 1,2-Bromotoluene	o-$C_6H_4BrCH_3$	8.78	28	*a*
52. 1,3-Bromotoluene	m-$C_6H_4BrCH_3$	8.81	46	*a*
53. 1,4-Bromotoluene	p-$C_6H_4BrCH_3$	8.67	28	*a*
		8.71	30	*c*
54. 1,4-Bromoanisole	p-$C_6H_4BrOCH_3$	8.49	30	*c*
55. 1,2-Chlorotoluene	o-$C_6H_4ClCH_3$	8.83	25	*a*
56. 1,3-Chlorotoluene	m-$C_6H_4ClCH_3$	8.83	25	*a*
		8.93	27	*a*
57. 1,4-Chlorotoluene $[8.75 \pm 0.09]_{a,c}$	p-$C_6H_4ClCH_3$	8.69	26, 28	*a*
		8.70	25	*a*
		8.90	30	*c*
58. 1,2-Fluorotoluene	o-$C_6H_4FCH_3$	8.92	44	*a*
59. 1,3-Fluorotoluene	m-$C_6H_4FCH_3$	8.92	44	*a*
60. 1,4-Fluorotoluene	p-$C_6H_4FCH_3$	8.79	44	*a*
61. 1,2-Iodotoluene	o-$C_6H_4ICH_3$	8.62	44	*a*
62. 1,3-Iodotoluene	m-$C_6H_4ICH_3$	8.61	44	*a*
63. 1,4-Iodotoluene	p-$C_6H_4ICH_3$	8.50	44	*a*
64. 1,3-Nitrotoluene	m-$C_6H_4NO_2CH_3$	9.48	24	*d*
65. 1,4-Nitrotoluene $[9.63 \pm 0.14]_d$	p-$C_6H_4NO_2CH_3$	9.50 ± 0.1	24	*d*
		9.56	39	*d*
		9.82	18	*d*
66. 1,3-Nitroanisole	m-$C_6H_4NO_2OCH_3$	9.09	24	*d*
67. 1,4-Nitroanisole	p-$C_6H_4NO_2OCH_3$	9.04	24	*d*
68. Toluene $[8.86 \pm 0.12]_{a,b,c}$ $[9.10 \pm 0.16]_d$	$C_6H_5CH_3$	8.81	1	*a*
		8.82	3	*a*
		8.77	33	*b*
		8.82	5	*b*
		8.82	29	*c*
		9.13	30	*c*

TABLE XI   (*Continued*)

Molecule	Formula	I (eV)	Ref.	Method
		8.80	45	*d*
		9.1	14	*d*
		9.18	18	*d*
		9.20	15	*d*
		9.23	37	*d*
69. Anisole	$C_6H_5OCH_3$	8.20	28	*a*
		8.54	30	*c*
		8.56	15	*d*
		8.83	18	*d*
70. α-Aminotoluene	$C_6H_5CH_2NH_2$	9.10	41	*c*
		9.04	12	*d*
71. *N*-methylaniline	$C_6H_5NHCH_3$	7.34	1	*a*
		7.73	30	*c*
		7.65	12	*d*
72. 1,2-Aminotoluene	o-$C_6H_4CH_3NH_2$	7.68	12	*d*
		8.38	18	*d*
73. 1,3-Aminotoluene	m-$C_6H_4CH_3NH_2$	7.50	1	*a*
		7.57	12	*d*
		8.27	18	*d*
74. 1,4-Aminotoluene	p-$C_6H_4CH_3NH_2$	7.78	30	*c*
[7.75 ± 0.28]$_d$		7.5	14	*d*
		7.60	12	*d*
		8.14	18	*d*
75. 1,4-Methoxyaniline	p-$C_6H_4NH_2OCH_3$	9.39	39	*d*
76. Phenyl-pentafluoro ether	$C_6H_5OC_2F_5$	9.97	30	*c*
77. *N*,*N*-Diethylaniline	$C_6H_5N(C_2H_5)_2$	7.51	30	*c*
78. Ethynylbenzene	$C_6H_5C{\equiv}CH$	9.15	31	*d*
79. 1,4-Cyanotoluene	p-$C_6H_4CH_3CN$	9.31	39	*d*
		9.76	18	*d*
80. Styrene	$C_6H_5CH{=}CH_2$	8.47	46	*a*
		8.35	47	*b*
		8.86	37	*d*
		9.00	18	*d*
81. Acetophenone	$C_6H_5COCH_3$	9.65	48	*a*
		9.77	31	*d*

TABLE XI   (*Continued*)

Molecule	Formula	I (eV)	Ref.	Method
82. 1,4-Methoxybenzaldehyde	$p\text{-}C_6H_4(CHO)OCH_3$	8.87	30	c
83. Ethylbenzene	$C_6H_5C_2H_5$	8.75	49	b
		8.77	5	b
		9.12	17,31	d
84. 1,2-Dimethylbenzene $[8.56 \pm 0.01]_{a,b}$ $[8.94 \pm 0.09]_d$	$o\text{-}C_6H_4(CH_3)_2$	8.56	1,28	a
		8.56	26	b
		8.58	5	b
		8.8	14	d
		8.96	15	d
		8.97	16	d
		9.04	18	d
85. 1,3-Dimethylbenzene $[8.58 \pm 0.01]_{a,b}$ $[8.89 \pm 0.10]_d$	$m\text{-}C_6H_4(CH_3)_2$	8.56	28	a
		8.59	1	a
		8.58	5	b
		8.8	14	d
		9.01	15	d
		9.02	16	d
		9.05	18	d
86. 1,4-Dimethylbenzene $[8.50 \pm 0.10]_{a,b,c}$ $[8.91 \pm 0.06]_d$	$p\text{-}C_6H_4(CH_3)_2$	8.44	1,28	a
		8.45	3	a
		8.45	26	b
		8.48	5	b
		8.71	30	c
		8.4	14	d
		8.86	15	d
		8.88	16	d
		8.99	18	d
87. 1,4-Dimethoxybenzene	$p\text{-}C_6H_4(OCH_3)_2$	7.90	30	c
88. N,N-Dimethylaniline	$C_6H_5N(CH_3)_2$	7.14	1	a
		7.30	12	d
89. Isopropylbenzene $[8.68 \pm 0.07]_{a,b}$	$C_6H_5\text{-iso-}C_3H_7$	8.69	28	a
		8.60	49	b
		8.76	5	b
		9.13	31	d
90. Propylbenzene	$C_6H_5C_3H_7$	8.72	28	a
		9.14	31	d

TABLE XI (*Continued*)

Molecule	Formula	I (eV)	Ref.	Method
91. 1,2,3-Trimethyl-benzene	$C_6H_3(CH_3)_3$	8.48	50	*a*
		8.75	16	*d*
92. 1,3,5-Trimethyl-benzene	$C_6H_3(CH_3)_3$	8.39	28	*a*
		8.41	1	*a*
$[8.40 \pm 0.01]_{a,b}$		8.39	26	*b*
$[8.70 \pm 0.12]_d$		8.5	14	*d*
		8.74	18	*d*
		8.76	15	*d*
		8.79	16	*d*
93. *p*-Methyl-*N,N*-di-methyl aniline	$p\text{-}C_6H_4(N(CH_3)_2)CH_3$	7.48	30	*c*
94. *p*-Chlorophenyl-*t*-butyl ether	$p\text{-}C_6H_4(OC(CH_3)_3)Cl$	8.72	30	*c*
95. Phenyl-*t*-butyl ether	$C_6H_5OC(CH_3)_3$	8.75	30	*c*
96. *N,N*-Diethyl-aniline	$C_6H_5N(C_2H_5)_2$	7.51	30	*c*
97. Biphenyl	$C_6H_5C_6H_5$	8.27	25	*a*
$[8.23 \pm 0.03]_{a,c}$		$8.20 \pm 0.05^{\dagger}$	51	*c*
$[8.28 \pm 0.09]_d$		8.23	8	*c*
		$8.41 \pm 0.05^{\S}$	52	*c*
		8.22	12,51	*d*
		8.22 ± 0.03	51	*d*
98. Butylbenzene	$C_6H_5C_4H_9$	8.69	28,50	*a*
$[8.63 \pm 0.09]_{a,b}$		8.5	28	*b*
		9.14	17,31	*d*
99. **tert**-Butylbenzene	$C_6H_5C_4H_9$	8.68	50	*a*
		8.5	49	*b*
		9.35	31	*d*
100. 1,2-Diethylbenzene	$o\text{-}C_6H_4(C_2H_5)_2$	8.91	16	*d*
101. 1,3-Diethylbenzene	$m\text{-}C_6H_4(C_2H_5)_2$	8.99	16	*d*
102. 1,4-Diethylbenzene	$p\text{-}C_6H_4(C_2H_5)_2$	8.93	16	*d*

*The molecules in this table are ordered according to increasing number of their carbon atoms. Molecules are subordered according to the number of hydrogen atoms, and are further subordered alphabetically according to the numbers of other atoms. The parent molecule benzene, which occupies the first position, is deliberately placed out of sequence. Only experimental values are listed except for benzene, for which results of typical theoretical calculations are included.

TABLE XI   (*Continued*)

---

†Quoted by authors as adiabatic IP value.

§Quoted by authors as vertical IP value.

^aPhotoionization.

^bPhotoabsorption.

^cPhotoelectron spectroscopy.

^dElectron impact.

^eModified energy distribution difference method (see Ref. 19 below).

^fCalculation.

1. F. I. Vilesov and A. N. Terenin, *Dokl. Akad. Nauk SSSR*, **115**, 744 (1957).
2. V. H. Dibeler, R. M. Reese, and M. Krauss, *Advances in Mass Spectrometry*, Vol. III, The Chaucer Press, London, 1966.
3. K. Watanabe, *J. Chem. Phys.*, **22**, 1564 (1954).
4. W. C. Price and R. W. Wood, *J. Chem. Phys.*, **3**, 439 (1935).
5. V. J. Hammond, W. C. Price, J. P. Teegan, and A. D. Walsh, *Discus. Faraday Soc.*, **9**, 53 (1950).
6. M. F. A. El-Sayed, M. Kasha, and V. Tanaka, *J. Chem. Phys.*, **34**, 334 (1961).
7. M. J. S. Dewar and S. D. Worley, *J. Chem. Phys.*, **51**, 263 (1969).
8. M. J. S. Dewar, E. Haselbach, and S. D. Worley, *Proc. Roy. Soc. London*, **315A**, 431 (1970).
9. M. I. Al-Joboury and D. W. Turner, *J. Chem. Soc. (London)*, 4434 (1964).
10. I. D. Clark and D. C. Frost, *J. Am. Chem. Soc.*, **89**, 244 (1967).
11. R. E. Fox and W. M. Hickam, *J. Chem. Phys.*, **22**, 2059 (1954).
12. J. H. D. Eland, P. J. Shepherd, and C. J. Danby, *Z. Naturforsch.*, **21a**, 1580 (1966).
13. M. E. Wacks and V. H. Dibeler, *J. Chem. Phys.*, **31**, 1557 (1959).
14. H. Hartmann and M. B. Svendsen, *Z. Physik Chem. Neue Folge*, **11**, 16 (1957).
15. H. Baba, I. Omura, and K. Higasi, *Bull. Chem. Soc. Jap.*, **29**, 521 (1956).
16. F. H. Field and J. L. Franklin, *J. Chem. Phys.*, **22**, 1895 (1954).
17. J. D. Morrison, *J. Chem. Phys.*, **19**, 1305 (1951).
18. G. F. Crable and G. L. Kearns, *J. Phys. Chem.*, **66**, 436 (1962).
19. R. E. Ellefson, Ph.D. Thesis, University of Wyoming, 1972.
20. P. J. Hay and I. Shavitt, *J. Chem. Phys.*, **60**, 2865 (1974).
21. T. L. Kunii and H. Kuroda, *Theor. Chim. Acta*, **11**, 97 (1968).
22. R. M. Hedges and F. A. Matsen, *J. Chem. Phys.*, **28**, 950 (1958).
23. K. Nishimoto, *Theor. Chim. Acta*, **7**, 207 (1967).
24. P. Brown, *Org. Mass Spectr.*, **4**, 533 (1970).
25. R. I. Reed, *Ion Production by Electron Impact*, Academic Press, London, 1962.
26. R. Bralsford, P. V. Harris, and W. C. Price, *Proc. Roy. Soc. London*, **A258**, 459 (1960).
27. J. M. Schulman and J. W. Moskowitz, *J. Chem. Phys.*, **47**, 3491 (1967).
28. K. Watanabe, *J. Chem. Phys.*, **26**, 542 (1957).
29. D. W. Turner, C. Baker, A. D. Baker, and C. R. Brundle, *Molecular Photoelectron Spectroscopy*, Wiley-Interscience, London, 1970.
30. A. D. Baker, D. P. May, and D. W. Turner, *J. Chem. Soc. (London)*, B22 (1968).
31. J. D. Morrison and A. J. C. Nicholson, *J. Chem. Phys.*, **20**, 1021 (1952).
32. J. R. Majer and C. R. Patrick, *Trans. Faraday Soc.*, **58**, 17 (1958).
33. W. C. Price and A. D. Walsh, *Proc. Roy. Soc. (London)*, **A191**, 22 (1947).
34. R. Gilbert and C. Sandorfy, *Chem. Phys. Lett.*, **9**, 121 (1971).
35. R. W. Kiser, *Introduction to Mass Spectrometry and Its Applications*, Prentice-Hall, Englewood Cliffs, N.J., 1965.

TABLE XI   (*Continued*)

36. R. A. W. Johnstone and F. A. Mellon, *J. Chem. Soc. Faraday II*, **68**, 1209 (1972).
37. S. Nagakura and J. Tanaka, *J. Chem. Phys.*, **22**, 236 (1954).
38. I. Howe and D. H. Williams, *J. Am. Chem. Soc.*, **91**, 7137 (1969).
39. R. A. W. Johnstone and F. A. Mellon, *J. Chem. Soc. (London)*, **69**, 36 (1973).
40. J. H. D. Eland, *Int. J. Mass Spectr. Ion Phys.*, **2**, 471 (1969).
41. T. P. Debies and J. W. Rabalais, *Inorg. Chem.*, **13**, 308 (1974).
42. M. J. S. Dewar, J. A. Hashmall, and C. G. Venier, *J. Am. Chem. Soc.*, **90**, 1953 (1968).
43. R. I. Reed and M. B. Thornley, *Trans. Faraday Soc.*, **54**, 949 (1958).
44. K. Watanabe, T. Nakayama, and J. Mottl, Final Report on the Ionization Potential of Molecules by a Photoionization Method, No. DA-04-200-ORD480 (1959).
45. C. E. Melton and W. H. Hamill, *J. Chem. Phys.*, **41**, 3464 (1964).
46. K. Watanabe, T. Nakayama, and J. Mottl, *J. Quant. Spectr. Radiat. Transfer*, **2**, 369 (1962).
47. L. Issaacs, W. C. Price, and R. Ridley, "On the Threshold of Space," Proceedings of the Conference on Chemical Aeronomy, Cambridge, 1956, p. 143.
48. F. I. Vilesov, *Dokl. Akad. Nauk SSSR*, **132**, 632, 1332 (1960).
49. W. C. Price, *Chem. Rev.*, **41**, 257 (1947).
50. W. C. Price, R. Bralsford, P. V. Harris, and R. G. Ridley, *Spectrochim. Acta*, **14**, 45 (1959).
51. J. H. D. Eland and C. J. Danby, *Z. Naturforsch.*, **23a**, 355 (1968).
52. J. H. D. Eland, *Int. J. Mass Spectr. Ion Phys.*, **9**, 214 (1972).

understood by considering the two physical processes involved, that is, ionization by electron and by photon impact. The wave packet associated with the incident electrons is considerably narrower than that associated with a photon of the same energy. Consequently, the photon interacts with the molecule for a longer period of time than does the electron. Therefore, the transition from the neutral molecule to the positive ion is adiabatic in the former and vertical (Franck–Condon) in the latter case. This is illustrated in Fig. 35.

Often, the parent molecular positive ion is unstable and it undergoes fragmentation or isomerization. In the former case one is, of course, interested in the fragments (neutral and charged) produced.

A number of monosubstituted and *para*-disubstituted benzenes were investigated by Baker et al.[96] using photoelectron spectroscopy. While ionization in this work is by photon rather than by electron impact, the results of Baker et al. are relevant to the present discussion. Baker et al., for example, found that substituents with a large $+M$ effect split the first band (~ 9.3 eV) into two parts, reflecting the breaking of the degeneracy of the e_{1g} orbitals by substitution. In this case the energy of the e_{1g} orbital with a node at the 1 and 4 positions (Fig. 1) is little affected, whereas the

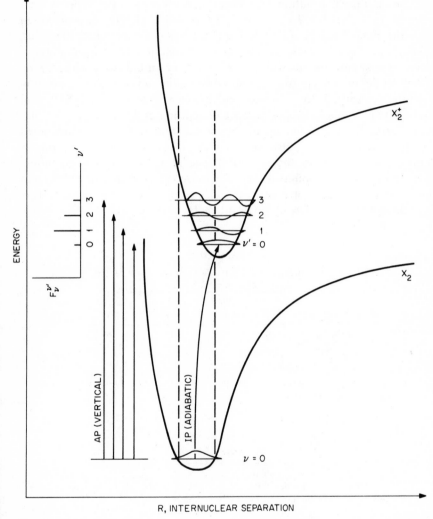

Fig. 35.   Vertical and adiabatic transitions for diatomic X_2 and X_2^+ systems (from Ref. 95).

energy of the other e_{1g} orbital with a finite electron density at the 1 and 4 positions is raised, resulting in a lower ionization potential. This is indicated in Table XII for the case of C_6H_5OH and $C_6H_5NH_2$. It is also seen from the data listed in Table XII for the halogen-substituted benzenes, although in this case both inductive and mesomeric effects are important. In the case of substituents with a $-M$ effect (or an inductive withdrawing effect), little or no splitting is observed, although the first ionization potential is shifted to a higher energy (see data on $C_6H_5NO_2$ and C_6H_5CHO in Table XII). The absence of splitting in this case suggests that the e_{1g} orbitals are affected equally by the substituent. For para-substituted benzenes, Baker et al.[96] observed that the magnitude of the splitting is approximately equal to the sum of the splittings of the corresponding monosubstituted compounds except for those cases where one substituent produced a very large change in the charge density in the benzene ring [e.g., Br and OCH_3, CH_3 and $N(CH_3)_2$, OCH_3 and OCH_3].

TABLE XII

Effect of Substituents on the First Two π-Ionization Potentials of Benzene[96]

Compound	$(I.P.)_1{}^a$ (eV)	$(I.P.)_2{}^b$ (eV)	$\Delta(I.P.)$ (eV)
C_6H_6	9.40	9.40	0.00
$+M$ $\left[\begin{array}{l} C_6H_5OH \\ C_6H_5NH_2 \end{array}\right.$	8.75	9.45	0.70
	8.04	9.11	1.07
$-M$ $\left[\begin{array}{l} C_6H_5NO_2 \\ C_6H_5CHO \end{array}\right.$	10.26	c	~ 0
	9.80	c	~ 0
C_6H_5F	~ 9.5	9.86	~ 0.3
C_6H_5Cl	9.31	9.71	0.40
C_6H_5Br	9.25	9.78	0.53
C_6H_5I	8.78	9.75	0.97

^{a} First π-ionization potential.

^{b} Second π-ionization potential.

^{c} Could not be resolved due to its proximity to $(IP)_1$.

D. Higher Ionization Potentials and Their Assignments

A number of investigators (see, for example, Refs. 95 to 98) have measured higher ionization potentials of the benzene molecule. Despite this, however, the assignments of the benzene ionization potentials have not been conclusive. The only exception seems to be the assignment of the first ionization potential to the degenerate $e_{1g}\pi$ orbital. In Table XIII the complete set of assignments of Jonsson and Lindholm[97] is given. To make their assignments these authors utilized photoelectron, electron-impact, mass-spectral, ultraviolet, and theoretical data (see also discussion in Eland[94] and Hay and Shavitt[99]).

TABLE XIII

Assignment of Benzene Ionization Potentials (After Ref. 97)[a]

Orbital		I.P. (eV)	Orbital		I. P. (eV)
$1e_{1g}$	π	9.3	$1b_{2u}$ σ		14.7
$1e'_{1g}$			$2b_{1u}$ σ		15.4
			$3a_{1g}$ σ		16.9
$3e_{2g}$	σ	11.4[b]	$2e_{2g}$	σ	19.2
$3e'_{2g}$			$2e'_{2g}$		
$1a_{2u}$	π	12.1	$2e_{1u}$	σ	≈ 26
$3e_{1u}$	σ	13.8	$2e'_{1u}$		
$3e'_{1u}$			$2a_{1g}$ σ		≈ 30

[a] Orbitals formed from $1s$ electrons of the carbon atoms are not considered.

[b] See a recent confirmation of this assignment by Price et al. [W. C. Price, A. W. Potts, and T. A. Williams, *Chem. Phys. Lett.*, **37**, 17 (1976)].

There are only scattered electron-impact data on the higher ionization potentials of benzene derivatives.[100] However, a number of benzene derivatives have been studied by Baker et al.[96] using photoelectron spectroscopy, and the reader is referred to this work for further discussion.

E. Multiple Ionization

Dorman and Morrison[101] reported on the double and triple ionization of C_6H_5D and $C_6H_5CH_3$ by electron impact with appearance potentials of 26.0 ± 0.2, 44 ± 5, 24.5 ± 0.2, and 42 ± 5 eV for $C_6H_5D^{++}$, $C_6H_5D^{+++}$, $C_6H_5CH_3^{++}$, and $C_6H_5CH_3^{+++}$, respectively. Earlier, Hustrulid et al.[102] obtained a value of 27.0 ± 1 eV and Wacks and Dibeler[103] a value of 25.4 eV for $C_6H_6^{++}$. Doubly and triply ionized species are more common in compounds with olefinic or aromatic bonds than in compounds with only aliphatic bonds.[104, 105]

The effect of substitution on multiple ionization was investigated by Engel et al.,[104] who reported that electron-withdrawing groups suppress the yield of multiply-charged ions whereas electron-donating groups enhance it. An exception is benzonitrile, in which dipositive ion formation is significant.[104] It was noted[104] that if the electron-withdrawing group is cleaved from the parent molecule, then multiply-charged ions are observed. The ratio of the intensities of doubly to singly ionized parent molecules varied from ~ 0 for strongly electron-withdrawing substituents such as NO_2 to ~ 0.25 for strongly electron-donating substituents such as NH_2. Finally, Biemann[105] has suggested that multiple ionization may be facilitated if there are regions of high electron density in the molecule.

F. The Fragmentation of Benzene Under Electron Impact

In Fig. 36 the abundances of selected ions of benzene fragments as measured by Ellefson[95] are plotted as a function of the incident electron energy. Such a plot is referred to as a clastolog. It is seen that the parent ion $C_6H_6^+$ is the predominant one not only close to the ionization threshold but for considerably higher electron energies. The abundance of the parent molecular ion decreases uniformly with increasing energy above threshold, while the abundances of the fragment ions have different types of energy dependences. The experimental data on benzene fragmentation have been somewhat confusing, but recently a clear picture has begun to emerge. Momigny et al.[106] studied the mass spectra and thermochemistry of benzene and its linear isomers and concluded that the reactive $C_6H_6^+$ ion is acyclic. A similar conclusion has been reached by Keough et al.[107] from their charge-transfer experiments; they concluded that this acyclic structure accounts for all fragmentation paths. Andlauer and Ottinger[108] ascertained that those processes leading to $C_6H_5^+$ and $C_6H_4^+$ are not competitive with those leading to $C_3H_3^+$ and $C_4H_4^+$. (The appearance potential of the $C_6H_5^+$ and $C_6H_4^+$ ions was reported[97] at 13.8 eV and that of the $C_4H_4^+$ and $C_3H_3^+$ fragments at ~ 14 eV.) Reexamining the benzene mass spectrum and metastable peak intensities within the standard quasiequilibrium description of ionic fragmentation,[109] Rosenstock et al.[110] concluded that $C_6H_5^+$ and $C_6H_4^+$ are formed from the $C_6H_6^+$ ground state whereas $C_4H_4^+$ and $C_3H_3^+$ are formed from either the first excited state of $C_6H_6^+$ at 2.25 eV above the ground state or from an open-chain isomer. Recent photoelectron-photoion coincidence studies by Eland[111] and Eland and Schulte[112] are in agreement with the conclusion of Rosenstock et al.[110] (see also Chupka[113]).

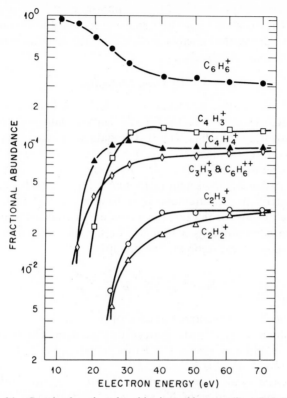

Fig. 36.   Castolog for selected positive ions of benzene (from Ref. 95).

VI.   ELECTRON TRANSPORT THROUGH
AROMATIC VAPORS

The transport of slow electrons through a number of aromatic vapors has been studied by Christophorou and co-workers[114–120] in their efforts to understand the role of microscopic and macroscopic molecular characteristics on the scattering of slow electrons by molecules. In particular, Christophorou, Blaunstein, and Pittman[118] measured the drift velocity, w (w is related to the electron mobility μ by $w = \mu E$ where E is the applied uniform electric field in V/cm), as a function of the pressure-reduced electric field, E/P, for three groups of π-electron containing molecules—linear, cyclic, and aromatic hydrocarbons—for which the electric dipole moment D is zero or very small and the static polarizability α is more or less the same, in order to ascertain the effect of induced π-orbital polarization on the scattering process. The results of Christophorou et al.[118,121] are elaborated upon here. They convincingly show that the mobility of thermal electrons decreases and the scattering cross section increases with increasing number of doubly-occupied π-orbitals.

A.   Electron Mobilities

In Fig. 37 w is plotted as a function of E/P for benzene and benzene derivatives for which $D \cong 0$. In Fig. 38 similar data are presented for benzene derivatives for which $D > 0$. The data were taken from Refs. 118 and 121 and are for temperature $T = 298°K$. Measurements of w at elevated temperatures ($298 \leqslant T \leqslant 450°K$) were made[116] for C_6H_5Cl and have shown that w increases with increasing T for this molecule.

Unfortunately, no measurements have been made of the lateral electron diffusion coefficients for the gaseous media under consideration and hence μ cannot be determined as a function of the mean electron energy. At thermal energies we can estimate the quantity

$$\mu_{\text{torr}} = \frac{w}{E} = \frac{w}{E/P} \frac{1}{P_1} = \frac{S}{P_1} \tag{38}$$

where S is the slope of the linear portion of the w versus E/P plots in Figs. 37 and 38 obtained by a least-squares fit to the data. The quantity μ_{torr} is the electron mobility normalized to 1 torr pressure (P_1). Values of μ_{torr} are listed in column 4 of Table XIV.

Christophorou et al.[118] compared the data on μ_{torr} in Table XIV with similar data on a number of linear, cyclic, and aromatic hydrocarbons containing varying numbers of doubly occupied π-orbitals and concluded that the scattering of thermal and near-thermal electrons is strongly affected by the number of doubly occupied π-orbitals contained in each molecule. They have found, for example, that for the linear hydrocarbons hexane, 1-hexene, 1,2-hexadiene, or 1,3-hexadiene with 0, 1, and 2 doubly occupied π-orbitals, respectively, μ_{torr} decreases from 20.6 to 10 to ~5 × 10^5 cm²/(V sec), and for the cyclic nonaromatic molecules cyclohexane, cyclohexene, 1,4-cyclohexadiene, and 1,3,5-cycloheptatriene with 0, 1, 2 and 3, doubly occupied π-orbitals, respectively, μ_{torr} decreases from 14.7 to 6.8, 5.5, and 4.5 × 10^5 cm²/(V sec). This indicates the direct influence of the microscopically induced π-orbital polarization on μ_{torr}. Christophorou et al.[118] also noted that the effect of π-orbital polarization is not additive, which may suggest that the individual π-orbital polarization is influenced by that of the rest. It is noted (see Table XIV) that the value of μ_{torr} for 1,4-dimethyl benzene is larger than that for benzene although the value of

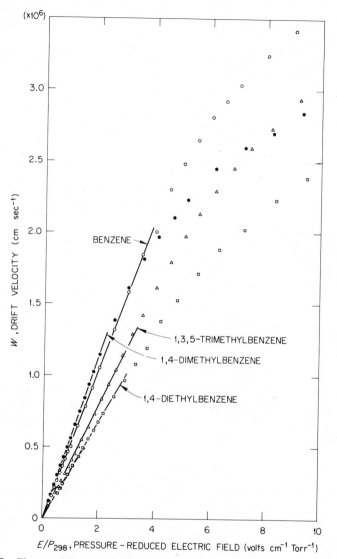

Fig. 37. Electron drift velocity, w, as a function of the pressure-reduced electric field, E/P, for benzene and benzene derivatives for which $D \cong 0$.

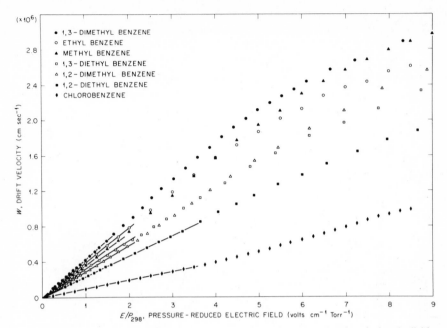

Fig. 38.   Electron drift velocity, w, as a function of the pressure-reduced electric field, E/P, for benzene derivatives for which $D > 0$.

α for 1,4-dimethyl benzene is somewhat higher than for benzene. This, as was pointed out earlier,[118] may indicate that the CH_3 groups in 1,4-dimethyl benzene interfere with the polarization of the π-orbitals and decrease the overall polarization compared to benzene.

B.   Cross Sections

It would be desirable to use the data in Table XIV and determine directly the scattering cross sections for these molecules at thermal and epithermal energies. There is no straightforward way to do this, mainly because the velocity dependence of the scattering cross section is not known and electron diffusion coefficient data are lacking. An estimate of the average scattering cross section at thermal energies can, however, be obtained by using the data in Table XIV.

For elastic electron scattering and a Maxwellian distribution of electron velocities, μ_{torr} is related to the momentum transfer cross section, $\sigma_m(v)[= \int d\sigma(v)(1 - \cos\theta)$, where $d\sigma(v)$ is the differential scattering cross section and v the electron velocity] by[117, 118, 124]

$$\mu_{\text{torr}} = \frac{S}{P_1} = \frac{w}{E} = \left(\frac{2}{\pi}\right)^{1/2} \frac{e}{3NP} \frac{m^{3/2}}{(kT)^{5/2}}$$

$$\times \int_0^\infty \left\{\frac{v^3}{\sigma_m(v)}\right\} \exp\left(-\frac{mv^2}{2kT}\right) dv \tag{39}$$

In (39) e, m, and v are, respectively, the electron charge, mass, and velocity, k is the Boltzmann constant, P is the pressure in torr, N is the number of molecules per cm^3 per torr at the temperature T, and the rest of the symbols are as defined earlier. On the basis of (39), relative changes in μ reflect relative changes in σ_m. If we now define an average cross section as[114, 117]

$$\langle \sigma_m(v) \rangle \equiv 4\pi \int_0^\infty \sigma_m(v) f_0(v) v^2 \, dv \tag{40}$$

where $4\pi f_0(v)v^2$ is the normalized Maxwellian distribution and assume a form for the v dependence of σ_m, an estimate of $\langle \sigma_m(v) \rangle$ can be obtained. The functional dependence of σ_m on v is, of course, determined by the scattering potential.

1. Nonpolar Molecules

A pure R^{-4} (R is the electron-molecule interaction distance) polarization scattering potential would be consistent with a cross section that varies inversely with the electron velocity. If we assume such a velocity dependence for σ_m, we obtain through (39) and (40)

$$\langle \sigma_m(v) \rangle_{\text{exp}} = \left(\frac{2}{\pi mkT}\right)^{1/2} \frac{e}{N} \frac{1}{\mu_{\text{torr}}} = 6.42 \times 10^{-9} \frac{1}{\mu_{\text{torr}}} \qquad \text{(for } T = 298°\text{K)} \tag{41}$$

This quantity has been determined and is listed in column 5 of Table XIV. From similar data on a number of other π-electron containing organic molecules, it was concluded[118] that the scattering cross sections increase with increasing number of doubly occupied π-electron orbitals. It was found,[118] for example, that $\langle \sigma_m(v) \rangle_{\text{exp}}$ is equal to 4.4, 9.4, 11.7, and 14.4×10^{-15} cm^2 for cyclohexane (zero doubly occupied π orbitals), cyclohexene (one doubly-occupied π orbital), 1,4-cyclohexadiene (two

TABLE XIV

Electric Dipole Moment D, Static Polarizability α, Thermal Electron Mobility μ_{torr} ($\equiv S/P$; $P = 1$ torr), and Average Momentum Transfer Cross Sections, $\langle \sigma_m(v) \rangle_{exp}$ and $\langle \sigma_m(v) \rangle_{theor}$

Molecule	D (Debye)	α (10^{-24} cm^3)	μ_{Torr} (10^5 cm^2 V^{-1} sec^{-1})	$\langle \sigma_m(v) \rangle_{exp}$ (10^{-15} cm^2)	$\langle \sigma_m(v) \rangle_{theor}$ (10^{-15} cm^2)
(Benzene)	0.0[a] 0.0[b]	10.3[c]	5.2	12.5[d]	76.2[f]
(1,4-Dimethyl benzene)	~ 0.0[b]	14.2[c]	5.8	11.1[d]	89.4[f]
(1,4-Diethyl benzene)	0.0[b]		3.4	18.9[d]	
(1,3,5-Trimethyl benzene)	0.0[b]		3.9	16.7[d]	

510

Compound					
CH$_3$ / CH$_3$ (1,3-Dimethyl benzene)	0.32[b]	14.2[c]	4.3	15[d]	89.4[f]
C$_2$H$_5$ (Ethyl benzene)	0.58[a] / 0.36[b]		3.9	16.4[d]	
C$_2$H$_5$ / C$_2$H$_5$ (1,3-Diethyl benzene)	0.36[b]		3.3	19.5[d]	
CH$_3$ (Methyl benzene)	0.37[a] / 0.34[b]	12.3[c]	3.6	17.7[d]	83.2[f]
CH$_3$ / CH$_3$ (1,2-Dimethyl benzene)	0.62[a] / 0.49[b]	14.1[c]	2.9	22.1[d]	89.1[f]

TABLE XIV (Continued)

Molecule	D (Debye)	α (10^{-24} cm^3)	μ_{Torr} (10^5 cm^2 V^{-1} sec^{-1})	$\langle\sigma_m(v)\rangle_{exp}$ (10^{-15} cm^2)	$\langle\sigma_m(v)\rangle_{theor}$ (10^{-15} cm^2)
C₂H₅—C₂H₅ (benzene) (1,2-Diethyl benzene)	0.59[b]		2.36	27.2[d]	
Cl (benzene) (Chlorobenzene)	1.70[a] 1.47[b]	12:3	0.96	180[e]	99[g]

(a) Average of gaseous values in Ref. 122.

(b) Average of liquid phase values given in Ref. 122.

(c) Gaseous value given in Ref. 123.

(d) Average experimental cross section using Equation (41).

(e) Average experimental cross section using Equation (43).

(f) Average theoretical cross section using Equation (42).

(g) Average theoretical cross section using Equation (44).

doubly occupied π orbitals), and 1,3,5-cycloheptatriene (three doubly occupied π orbitals), respectively, and that it increases from 12.5×10^{-15} cm^2 for benzene to 20.7×10^{-15} cm^2 for naphthalene.

The work of Christophorou et al.[118] and the results in Table XIV clearly show that although for these molecules $D \approx 0$, the scattering cross sections are large. This would indicate that double-bonded systems can effectively thermalize slow electrons, a property increasing with increasing number of doubly occupied π orbitals.

Theoretically, there is no satisfactory way of calculating $\sigma_m(v)$ when $D = 0$ (see discussion in Ref. 117). An R^{-4} polarization scattering potential would predict (for $T = 298°$K)

$$\langle \sigma_m(v) \rangle_{\text{theor}} = 4e \left(\frac{2\pi\alpha}{kT} \right)^{1/2} = 2.373 \times 10^{-2} \alpha^{1/2} \tag{42}$$

where α is the static polarizability. Values of $\langle \sigma_m(v) \rangle_{\text{theor}}$ are listed in column 6 of Table XIV and are seen to be very much larger than experimentally found.[125] This is in accord with the earlier conclusion[114,117] (see also recent work on n-alkanes[126]) that although an R^{-4} polarization potential can successfully describe low-energy ion-molecule collision processes, it is inappropriate for electron-molecule scattering.

2.   Polar Molecules

The compounds listed in Table XIV are weakly polar with the exception of chlorobenzene (C$_6$H$_5$Cl) for which $D \cong 1.70$ Debye (see Table XIV). For strongly polar molecules the electric-dipole term in the scattering potential predominates and in this case it has been found[114,115,117] that the experimentally determined cross sections are in comfortable agreement with those calculated theoretically on the basis of an electron-electric dipole interaction potential. An R^{-2} scattering potential is consistent with a v^{-2} dependence of the scattering cross section. Assuming such a dependence of σ_m on v, we find that ($T = 298°$K)

$$\langle \sigma_m(v) \rangle_{\text{exp}} = 17.176 \times 10^{-9} \mu_{\text{torr}}^{-1} \tag{43}$$

For C$_6$H$_5$Cl, (43) yields $\langle \sigma_m(v) \rangle_{\text{exp}} = 180 \times 10^{-14}$ cm^2, which is significantly higher than the theoretical estimate,

$$\langle \sigma_m(v) \rangle_{\text{theor}} = 3.8033 \times 10^{-14} D^2 = 99 \times 10^{-14} \text{ cm}^2 \tag{44}$$

one obtains using for $\sigma_m(v)$ the expression

$$\frac{8\pi}{3} \left(\frac{eD}{\hbar v} \right)^2 = 1.72 \frac{D^2}{v^2} \qquad (D \text{ in debye units})$$

determined by Altshuler[127] for electron scattering by a point dipole in the Born approximation. The higher experimental value has been attributed[115,117] to the presence of π electrons and to the possibility that temporary electron trapping takes place since the value of D for C_6H_5Cl slightly exceeds that (1.625 debye) which is necessary to bind an electron to a finite, fixed dipole (see discussion in Refs. 115 and 117).

Finally, Christophorou and Pittman[116] have observed a significant increase in w with increasing T in the range 298 to 450°K for C_6H_5Cl and 13 other polar molecules. They used these data to determine both the magnitude and the v dependence of σ_m, having assumed $\sigma_m = A_b/v^b$, where A_b is a constant. Although for C_6H_5Cl they found $b = 2.375$, the average value of b for 14 polar molecules was 2.12 ± 0.17, which is not too dissimilar from the value of 2 that is predicted theoretically for an R^{-2} scattering potential.

C. Electron Mobilities in Gases and Liquids

In their effort to relate studies on slow electron-molecule interaction processes in low pressure gases with studies on the same processes in condensed media, Christophorou et al.[120] measured thermal ($T \cong 298°K$) electron mobilities in a number of organic vapors that they related to those in the corresponding liquids. They defined a quantity

$$\mu_G \equiv S\frac{N_{\text{torr}}}{N_L} \tag{45}$$

which may be regarded as the gaseous electron mobility adjusted for the change in density between the gas and the liquid. In (45), N_{torr} is the number of molecules per cm^3 per torr at the specified gas temperature and N_L is the number density of the corresponding liquid at the temperature of the liquid.

Although for certain highly symmetric (spherical) molecules and for the heavier rare-gas atoms μ_L was found[119,120] to exceed μ_G often by a large factor (i.e., in this case, electrons are more mobile in the liquid than in the gas[119,120]) for benzene and benzene derivatives μ_L is much less than μ_G, as can be seen from the data in Table XV. This can be attributed to temporary electron trapping in the liquid, the repeated electron capture and loss process causing a delay in the electron motion. Of course, other energy-loss processes characteristic of the condensed phase can increase the scattering cross section and thus reduce the electron mobility (see further discussion in Refs. 119 and 120).

TABLE XV

μ_L, μ_G, and μ_L/μ_G for Benzene and Benzene Derivatives

Molecule	$\mu_L{}^a$ [cm^2/(V/sec)]	$\mu_G{}^b$ [cm^2/(V/sec)]	$\mu_L/\mu_G{}^c$
Benzene	0.114[d]; 0.6[e]	2.5	0.046
Methylbenzene	0.063[d]; 0.54[f]	2.1	0.030
1,2-Dimethylbenzene	0.018[d]	2.2	0.008
1,3-Dimethylbenzene	0.057[d]	3.3	0.017
1,4-Dimethylbenzene	0.062[d]	4.5	0.014
1,3,5-Trimethylbenzene	0.16[d]	2.9	0.055

[a] Measurements at $T = 292$ to $293°$K except those of Refs. e and f below, which were at $T = 300°$K.

[b] Measurements at $T \cong 298°$K.

[c] Using the μ_L data of Ref. d below.

[d] K. Shinsaka and G. R. Freeman, Can. J. Chem., 52, 3495 (1974).

[e] R. M. Minday, L. D. Schmidt, and H. T. Davis, J. Chem. Phys., 54, 3112 (1971).

[f] R. M. Minday, L. D. Schmidt, and H. T. Davis, J. Chem. Phys., 76, 442 (1972).

VII. GENERAL REMARKS AND CONCLUSIONS

Excitation of benzene and some of its derivatives by threshold electrons provided useful supplementary information to photon-impact data, as did the energy-loss spectra which, unfortunately, have been limited to benzene. The experimental as well as the theoretical work on the excitation of benzene and its derivatives by electron impact is, however, limited. As a rule, the theory is crude and the experimental techniques (TEE) are hindered by poor energy resolution. It appears that there are no data on the energy dependence of the excitation cross section for any of these systems, other than the cross-section functions obtained indirectly by observing the optical emission from excited fragments of benzene following excitation by electron impact.

With regard to negative-ion resonances, benzene and its derivatives are among the best-studied polyatomics. Although the phenomena are reasonably well understood, there is a need for further experimental work, especially with regard to measurements of absolute cross sections, electron affinities, and the resolving of vibrational structure. A deeper theoretical understanding based on a comprehensive approach is also needed.

The processes of dissociative and nondissociative electron attachment in benzene and a number of its derivatives have been widely investigated and are generally understood within the theoretical treatment of such processes in simple molecules. There is, however, a need for further absolute cross-section measurements, for investigation of dissociative attachment at energies to ~ 10 to 15 eV, and also of ion-pair processes. As far as long-lived parent negative-ion formation is concerned, the substituted benzenes are the most extensively studied group of molecules to date. A reasonable understanding of this phenomenon has been reached and the general conclusion has been drawn that all NO_2-, CN-containing, and perfluorinated benzenes form long-lived parent negative ions unless a fast dissociative attachment process is competing. The effects of intramolecular interaction between substituents on the benzene periphery, intramolecular π-charge transfer between the substituent(s) and the benzene ring, internal energy of the superexcited ion and the wider effect of molecular structure in general, on the autodetachment lifetime have been reasonably well understood. There is, however, still a need for electron-affinity measurements and also of electron attachment cross sections as a function of electron energy.

Determinations of the first ionization potential of benzene and many of its derivatives with both photon- and electron-impact techniques have been abundant. The values obtained by electron-impact methods are, as a rule, higher than those obtained by photon-impact methods. There have been, also, scattered reports on higher ionization potentials, multiple ionization, and fragmentation. The assignments of the higher ionization potentials and the fragmentation patterns seem to be somewhat uncertain. There are no experimental data on the energy dependence of the ionization cross sections for these molecules, and the theoretical treatments are rudimentary. Extensive experimental and theoretical work is needed.

Finally, electron transport through vapors of benzene and some benzene derivatives indicated the role of the π-electron system in slowing-down subexcitation electrons. The scattering cross section for both polar and nonpolar aromatics are higher than for systems without π electrons.

In conclusion, although much work is still needed in all the main areas (excitation, ionization, resonances, negative-ion formation and fragmentation, and electron transport) outlined in this chapter, an appreciable amount of knowledge does exist that allows a reasonable understanding of the nature and the magnitude of the phenomena taking place when slow electrons collide with benzene and its derivatives. This knowledge will aid the understanding of similar processes in more complex organic structures.

VIII. REFERENCES

1. B. L. Moiseiwitsch and S. J. Smith, *Rev. Mod. Phys.*, **40**, 238 (1968).
2. H. A. Bethe, *Ann. Physik*, **5**, 325 (1930).
3. E. N. Lassettre, *Can. J. Chem.*, **47**, 1733 (1969); E. N. Lassettre, A. Skerbele, and M. A. Dillon, *J. Chem. Phys.*, **50**, 1829 (1969).
4. E. N. Lassettre, in C. Sandorfy, P. J. Ausloos, and M. B. Robin, Eds., *Chemical*

Spectroscopy and Photochemistry in the Vacuum-Ultraviolet, D. Reidel Publishing, Dordrecht-Holland, 1974, pp. 43–73.

5. L. G. Christophorou, *Atomic and Molecular Radiation Physics*, Wiley-Interscience, New York, 1971, Chapter 5.

6. V. I. Ochkur, *Zh. Eksp. Teor. Fiz.*, **45**, 734 (1963) [*Soviet Phys.—JETP*, **18**, 503 (1964)].

7. M. R. H. Rudge, *Proc. Phys. Soc. (London)*, **85**, 607 (1965).

8. M. R. H. Rudge, *Proc. Phys. Soc. (London)*, **86**, 763 (1965).

9. J. R. Platt, *J. Chem. Phys.*, **17**, 484 (1949).

10. See, for example, M. Orchin and H. H. Jaffé, *Symmetry, Orbitals, and Spectra*, Wiley-Interscience, New York, 1971.

11. M. Inokuti, *J. Phys. Soc. Japan*, **13**, 537 (1958).

12. M. Matsuzawa, *J. Phys. Soc. Japan*, **18**, 1473 (1963).

13. F. H. Read and G. L. Whiterod, *Proc. Phys. Soc. (London)*, **85**, 71 (1965).

14. M. Matsuzawa, *J. Chem. Phys.*, **51**, 4705 (1969).

15. A. Skerbele and E. N. Lassettre, *J. Chem. Phys.*, **42**, 395 (1965).

16. E. E. Koch and A. Otto, *Chem. Phys. Lett.*, **12**, 476 (1972).

17. E. N. Lassettre, A. Skerbele, M. A. Dillon, and K. J. Ross, *J. Chem. Phys.*, **48**, 5066 (1968).

18. J. P. Doering, *J. Chem. Phys.*, **51**, 2866 (1969).

19. R. Azria and G. J. Schulz, *J. Chem. Phys.*, **62**, 573 (1975).

20. K. C. Smyth, J. A. Schiavone, and R. S. Freund, *J. Chem. Phys.*, **61**, 1789 (1974).

21. H. H. Brongersma, J. A. v. d. Hart, and L. J. Oosterhoff, *Proceedings of the Nobel Symposium*, Vol. 5, Interscience, New York, 1967, p. 211.

22. R. N. Compton, R. H. Huebner, P. W. Reinhardt, and L. G. Christophorou, *J. Chem. Phys.*, **48**, 901 (1968).

23. Notation in parentheses is that due to Platt.[9]

24. J. B. Birks, L. G. Christophorou, and R. H. Huebner, *Nature*, **217**, 809 (1968).

25. L. G. Christophorou, D. L. McCorkle, and J. G. Carter, *J. Chem. Phys.*, **60**, 3779 (1974).

26. H. H. Brongersma, Ph.D. Thesis, Leiden, 1968.

27. K. C. Smyth, J. A. Schiavone, and R. S. Freund, *J. Chem. Phys.*, **61**, 1782 (1974).

28. K. C. Smyth, J. A. Schiavone, and R. S. Freund, *J. Chem. Phys.*, **61**, 4747 (1974).

29. C. I. M. Beenakker, F. J. De Heer, and L. J. Oosterhoff, *Chem. Phys. Lett.*, **28**, 324 (1974).

30. H. S. Taylor, G. V. Nazaroff, and A. Golebiewski, *J. Chem. Phys.*, **45**, 2872 (1966).

31. J. N. Bardsley and F. Mandl, *Rept. Progr. Phys.*, **31**, 471 (1968).

32. G. J. Schulz, *Rev. Mod. Phys.*, **45**, 378, 423 (1973).

33. G. J. Schulz, Phys. Rev., **125**, 229 (1962); **135**, A988 (1964).

34. G. J. Schulz and R. K. Asundi, *Phys. Rev. Lett.*, **15**, 946 (1965).

35. R. N. Compton, L. G. Christophorou, and R. H. Huebner, *Phys. Lett.*, **23**, 656 (1966).

36. I. Nenner and G. J. Schulz, *J. Chem. Phys.*, **62**, 1747 (1975).

37. D. E. Golden and H. W. Bandel, *Phys. Rev. Lett.*, **14**, 1010 (1965); C. E. Kuyatt, J. A. Simpson, and S. R. Mielczarek, *J. Chem. Phys.*, **44**, 437 (1966); H. G. M. Heideman, C. E. Kuyatt, and G. E. Chamberlain, *J. Chem. Phys.*, **44**, 440 (1966).

38. H. G. M. Heideman, C. E. Kuyatt, and G. E. Chamberlain, *J. Chem. Phys.*, **44**, 355 (1966).

39. See, for example, D. L. McCorkle, L. G. Christophorou, and V. E. Anderson, *J. Phys. B: Atom. Molec. Phys.*, **5**, 1211 (1972).

40. J. Rademacher, L. G. Christophorou, and R. P. Blaunsteun, *J. Chem. Soc. Faraday Trans. II*, **71**, 1212 (1975).

41. J. P. Johnson, D. L. McCorkle, L. G. Christophorou, and J. G. Carter, *J. Chem. Soc. Faraday Trans. II*, **71**, 1742 (1975).

42. M. J. W. Boness, I. W. Larkin, J. B. Hasted, and L. Moore, *Chem. Phys. Lett.*, **1**, 292 (1967).

43. I. W. Larkin and J. B. Hasted, *J. Phys. B: Atom. Molec. Phys.*, **5**, 95 (1972).

44. L. Sanche and G. J. Schulz, *J. Chem. Phys.*, **58**, 479 (1973).

45. P. D. Burrow and L. Sanche, quoted in Ref. 44.

46. Mesomeric interactions are also called resonance interactions. To avoid any confusion of the latter term with the negative-ion resonances discussed in this chapter the term "mesomeric" is adopted.

47. M. N. Pisanias, L. G. Christophorou, J. G. Carter, and D. L. McCorkle, *J. Chem. Phys.*, **48**, 2110 (1973).

48. M. N. Pisanias, L. G. Christophorou, and J. G. Carter, *Chem. Phys. Lett.*, **13**, 433 (1972).

49. In aromatic molecules the inductive effect is often divided into σ and π components. [For a brief discussion see A. R. Katritzky and R. D. Topsom, *Angew. Chem. (Int. Ed.)*, **9**, 87 (1970); *J. Chem. Ed.*, **48**, 427 (1971).] The former results when the electronegativity of the carbon to which the substituent is attached is altered by the substituent and the latter involves the repulsion between the ring π electrons and the electrons in the corresponding p orbital of the substituent. The σ and π inductive effects are sometimes subdivided further. The mesomeric effect arises from the overlap of the π orbitals with similarly aligned p-atomic orbitals (or π-molecular orbitals) on the substituent. If electrons are donated to the ring by the substituent via this mechanism, we speak of a $+ M$ effect and if electrons are withdrawn from the ring we speak of a $- M$ effect. In addition to the inductive and mesomeric effects other factors such as the polarizability of the substituent may be important in certain cases.

50. I^- from C_6H_5I peaks at ~ 0.0 eV [L. G. Christophorou, R. N. Compton, G. S. Hurst, and P. W. Reinhardt, *J. Chem. Phys.*, **45**, 536 (1966)].

51. J. A. Pople and D. L. Beveridge, *Approximate Molecular Orbital Theory*, McGraw-Hill, New York, 1970.

52. T. Koopmans, *Physica*, **1**, 104 (1933).

53. L. G. Christophorou, *Atomic and Molecular Radiation Physics*, Wiley-Interscience, New York, 1971, Chapters 6 and 7.

54. Recently, O. H. Crawford and B. J. D. Koch [*J. Chem. Phys.*, **60**, 4512 (1974)] argued that diabatic electron capture is responsible for the electron-attachment processes in alkyl bromides and hydrogen halides below ~ 2 eV and that compound negative-ion states do not exist in these systems at these energies.

55. T. F. O'Malley, *Phys. Rev.*, **150**, 14 (1966); **156**, 230 (1967); see also J. N. Bardsley, A. Herzenberg, and F. Mandl, in M. R. C. McDowell, Ed., *Atomic Collision Processes*, North Holland, Amsterdam, 1964; J. N. Bardsley, H. Herzenberg, and F. Mandl, *Proc. Phys. Soc. (London)*, **89**, 321 (1966); J. C. Y. Chen, *Phys. Rev.*, **148**, 66 (1966); and J. C. Y. Chen and J. L. Peacher, *Phys. Rev.*, **163**, 103 (1967).

56. L. G. Christophorou and R. E. Goans, *J. Chem. Phys.*, **60**, 4244 (1974).

57. L. G. Christophorou and J. A. D. Stockdale, *J. Chem. Phys.*, **48**, 1956 (1968).

58. J. C. Y. Chen, in M. Burton and J. L. Magee, Eds., *Advances in Radiation Chemistry*, Vol. 1, Wiley-Interscience, New York, 1969.

59. A. A. Christodoulides and L. G. Christophorou, *J. Chem. Phys.*, **54**, 4691 (1971).

60. F. Fiquet-Fayard, *Vacuum*, **24**, 533 (1974).

61. L. G. Christophorou, R. N. Compton, G. S. Hurst, and P. W. Reinhardt, *J. Chem. Phys.*, **45**, 536 (1966).

62. L. G. Christophorou, R. N. Compton, G. S. Hurst, and P. W. Reinhardt, *J. Chem.*

Phys., **43**, 4273 (1965).
63. C. I. M. Beenakker and F. J. De Heer, *Chem. Phys. Lett.,* **29**, 89 (1974).
64. D. A. Vroom and F. J. De Heer, *J. Chem. Phys.,* **50**, 573 (1969).
65. H. A. Bethe, *Ann. Physik,* **5**, 325 (1930).
66. R. N. Compton, L. G. Christophorou, G. S. Hurst, and P. W. Reinhardt, *J. Chem. Phys.,* **45**, 4634 (1966).
67. B. S. Rabinovitch and R. W. Diesen, *J. Chem. Phys.,* **30**, 735 (1959); G. Z. Whitten and B. S. Rabinovitch, *J. Chem. Phys.,* **38**, 2466 (1963).
68. L. G. Christophorou, A. Hadjiantoniou, and J. G. Carter, *J. Chem. Soc. Faraday Trans. II,* **69**, 1713 (1973).
69. P. M. Collins, L. G. Christophorou, E. L. Chaney, and J. G. Carter, *Chem. Phys. Lett.,* **4**, 646 (1970).
70. L. G. Christophorou, J. G. Carter, E. L. Chaney, and P. M. Collins, in J. F. Duplan and A. Chapiro, Eds., *Proceedings of the IVth International Conference of Radiation Research, (July 1970), Vol. 1, Physics and Chemistry,* Gordon and Breach, London, 1973, pp. 145–157.
71. A. Hadjiantoniou, L. G. Christophorou, and J. G. Carter, *J. Chem. Soc. Faraday Trans. II,* **69**, 1691 (1973).
72. A. Hadjiantoniou, L. G. Christophorou, and J. G. Carter, *J. Chem. Soc. Faraday Trans. II,* **69** 1704 (1973).
73. L. G. Christophorou, D. L. McCorkle, and J. G. Carter, *J. Chem. Phys.,* **54**, 253 (1971).
74. L. G. Christophorou, R. N. Compton, and H. W. Dickson, *J. Chem. Phys.,* **48**, 1949 (1968).
75. J. M. S. Henis and C. A. Mabie, *J. Chem. Phys.,* **53**, 2999 (1970).
76. R. W. Odom, D. L. Smith, and J. H. Futrell, *J. Phys. B: Atom. Molec. Phys.,* **8**, 1349 (1975).
77. See, for example, F. A. Matsen, *J. Chem. Phys.,* **24**, 602 (1956).
78. W. T. Naff, R. N. Compton, and C. D. Cooper, *J. Chem. Phys.,* **54**, 212 (1971).
79. E. L. Chaney, L. G. Christophorou, P. M. Collins, and J. G. Carter, *J. Chem. Phys.,* **52**, 4413 (1970).
80. P. N. Sen and S. Basu, *Int. J. Quantum Chem.,* **1**, 591 (1967).
81. S. Basu, *J. Chem. Phys.,* **41**, 1453 (1964); S. Basu, *Theoret. Chim. Acta,* **3**, 156 (1965); **3**, 238 (1965).
82. A. B. King, *J. Chem. Phys.,* **47**, 2701 (1967).
83. According to Ref. 82 the factor 3.3 can be considered as a group-ionization probability for the methyl group as compared to the aromatic carbon atom.
84. F. T. Deverse and A. B. King, *J. Chem. Phys.,* **41**, 3833 (1964).
85. A. G. Harrison, E. G. Jones, S. K. Gupta, and G. P. Nagy, *Can. J. Chem.,* **44**, 1967 (1966).
86. E. Wigner, *Phys. Rev.,* **73**, 1002 (1948).
87. G. H. Wannier, *Phys. Rev.,* **100**, 1180 (1955).
88. J. D. Morrison and A. J. C. Nicholson, *J. Chem. Phys.,* **31**, 1320 (1959).
89. F. H. Dorman and J. D. Morrison, *J. Chem. Phys.,* **34**, 1407 (1961).
90. J. D. Morrison, *J. Appl. Phys.,* **28**, 1409 (1957).
91. J. D. Morrison, in R. Stoop, Ed., *Energy Transfer in Gases,* Interscience, New York, 1962, p. 397.
92. See, for example, G. H. Wannier, *Phys. Rev.,* **90**, 817 (1953); K. Omidvar, *Phys. Rev.,* **140**, A26 (1965); K. Omidvar, *Phys. Rev. Lett.,* **18**, 153 (1967); A. Temkin, *Phys. Rev. Lett.,* **16**, 835 (1966).
93. R. A. W. Johnstone, F. A. Mellon, and S. D. Ward, *Int. J. Mass Spectr. Ion Phys.,* **5**, 241 (1970).

94. J. H. D. Eland, *Photoelectron Spectroscopy*, Butterworths, London, 1974, p. 96.

95. R. Ellefson, Ph.D. Thesis, University of Wyoming, 1972.

96. A. D. Baker, D. P. May, and D. W. Turner, *J. Chem. Soc. (B)*, 22 (1968).

97. B. O. Jonsson and E. Lindholm, *Arkiv für Fysik*, **39**, 65 (1969).

98. a. R. E. Fox and W. M. Hickam, *J. Chem. Phys.*, **22**, 2059 (1954).
    b. P. Natalis and J. L. Franklin, *J. Phys. Chem.*, **69**, 2935 (1965).
    c. J. A. R. Samson, *Chem. Phys. Lett.*, **4**, 257 (1969).
    d. D. G. Streets and A. W. Potts, *J. Chem. Soc. Faraday Trans. II*, **70**, 1505 (1974).

99. P. J. Hay and I. Shavitt, *J. Chem. Phys.*, **60**, 2865 (1974).

100. Refs. 96, 98b, and C. E. Melton and W. H. Hamill, *J. Chem. Phys.*, **41**, 3464 (1964).

101. F. H. Dorman and J. D. Morrison, *J. Chem. Phys.*, **35**, 575 (1961).

102. A. Hustrulid, P. Kusch, and J. T. Tate, *Phys. Rev.*, **54**, 1037 (1938).

103. M. E. Wacks and V. H. Dibeler, *J. Chem. Phys.*, **31**, 1557 (1959).

104. R. Engel, D. Halpern, and B.-A. Funk, *Org. Mass Spectr.*, **7**, 177 (1973).

105. K. Biemann, *Mass Spectrometry, Organic Chemical Applications*, McGraw-Hill, New York, 1962.

106. J. Momigny, L. Brakier, and L. d'Or, *Bull. Classe Sci. Acad. Roy. Belg.*, **48**, 1002 (1962) (quoted in Ref. 107).

107. T. Keough, T. Ast, J. H. Beynon, and R. G. Cooks, *Org. Mass Spectr.*, **7**, 245 (1973).

108. B. Andlauer and Ch. Ottinger, *Z. Naturforsch.*, **27a**, 293 (1972).

109. H. M. Rosenstock, M. B. Wallenstein, A. L. Wahrhaftig, and H. Eyring, *Proc. Natl. Acad. Sci. U.S.*, **38**, 667 (1952).

110. H. M. Rosenstock, J. T. Larkins, and J. A. Walker, *Int. J. Mass Spectr. Ion Phys.*, **11**, 309 (1973).

111. J. H. D. Eland, *Int. J. Mass Spectr. Ion Phys.*, **13**, 457 (1974).

112. J. H. D. Eland and H. Schulte, *J. Chem. Phys.*, **62**, 3835 (1975).

113. W. A. Chupka, in C. Sandorfy, P. J. Ausloos, and M. B. Robin, Eds., *Chemical Spectroscopy and Photochemistry in the Vacuum-Ultraviolet*, D. Reidel Publishing, Dordrecht-Holland, 1974.

114. L. G. Christophorou, G. S. Hurst, and A. Hadjiantoniou, *J. Chem. Phys.*, **44**, 3506 (1966); **47**, 1883 (1967).

115. L. G. Christophorou and A. A. Christodoulides, *J. Phys. B: Atom Molec. Phys.*, **2**, 71 (1969).

116. L. G. Christophorou and D. Pittman, *J. Phys. B: Atom. Molec. Phys.*, **3**, 1252 (1970).

117. L. G. Christophorou, *Atomic and Molecular Radiation Physics*, Wiley-Interscience, New York, 1971, Chapter 4.

118. L. G. Christophorou, R. P. Blaunstein, and D. Pittman, *Chem. Phys. Lett.*, **22**, 41 (1973).

119. L. G. Christophorou, *Int. J. Rad. Phys. Chem.*, **7**, 205 (1975).

120. L. G. Christophorou, R. P. Blaunstein, and D. Pittman, *Chem. Phys. Lett.*, **18**, 509 (1973).

121. L. G. Christophorou and A. A. Christodoulides, unpublished results.

122. A. L. McClellan, *Tables of Experimental Dipole Moments*, W. H. Freeman, San Francisco, 1963.

123. H. H. Landolt, *Zahlenwerte und Funktionen aus Physik, Chemie, Astronomie, Geophysik und Technik*, Bd 1/3, Springer, Berlin, 1951, p. 511.

124. W. P. Allis, *Hand. Phys.*, **21**, 413 (1956).

125. It should be noted that part of this difference arises from the fact that $\langle \sigma_m(v) \rangle_{theor}$ is not weighted, as $\langle \sigma_m(v) \rangle_{exp}$ is, by the factor $(1 - \cos\theta)$, where θ is the scattering angle.

126. L. G. Christophorou, M. W. Grant, and D. Pittman, *Chem. Phys. Lett.*, **38**, 100 (1976).

127. S. Altshuler, *Phys. Rev.*, **107**, 114 (1957).

AUTHOR INDEX

Numbers in parentheses are reference numbers and show that an author's work is referred to although his name is not mentioned in the text. Numbers in *italics* indicate the pages on which the full references appear.

Aarous, L. J., 239(57), *341*
Ablow, C. M., *191, 204*
Abrikosov, A., 213(38), 215(38), *224, 341*
Abrines, R., *77, 135*
Alexander, M. H., *118–20, 133, 135, 136*
Al-Joboury, M. I., 491(9), *499*
Allan, A., *435*
Allan, C. J., 334(141), 336(141), *344*
Allis, W. P., 509(124), *520*
Allison, A. C., *65, 135, 181, 199*
Allison, P. A., 334(141), 336(141), *344*
Almhöf, J., *330, 344*
Altshuler, S., *514, 520*
Amat, G., 402(73), 408(73), *412*
Amdur, I., *174, 175, 193, 199, 201*
Anderson, A. B., *165, 199, 304, 343*
Anderson, J. B., *163, 202*
Anderson, V. E., 437(39), *517*
Andlauer, B., *504, 520*
Appleton, J. P., *170, 203*
Argyres, P. N., 3(5), *60, 70, 135*
Arita, K., 259(70), *342*
Arnol'd, V. I., *37, 39, 60, 61*
Aroeste, H., *143, 166, 172, 172, 199, 201*
Arthurs, A. M., *106, 135*
Asbrink, L., 268(72), 314(119), *342, 343*
Ast, T., 504(107), *520*
Asundi, R. K., *517*
Atabek, O., 293(92), *342*
Audibert, M. M., *65, 135*
Augustin, S. D., *123, 124, 128, 135*
Avez, A., 37(23), *60*
Azria, R., *426,* 427(19), 428(19), *517*

Baba, H., 491(15), 493(15), 494(15), 496–98(15), *499*
Bacskay, G. B., *147, 148, 159, 199*

Baer, M., *202*
Baer, Y., 206(1), 212(1b), 266(1b), 307(1b), 314(1b), 334(1b), *339*
Bagus, P. S., 239(57), *306, 341, 343*
Baker, A. D., 206(1), 308(1c), 309(1c), 311(1c), 316(1c), 318(1c), 319(1c), *321, 322,* 324(1c), 325(1c, 126), 326-28(1c), 330(1c), 331(1c), 333(1c), *339, 343,* 491(30), 492(29, 30), 493(30), 494(30), 495(29, 30), 496–98(30), *499, 500, 501, 503, 520*
Baker, C., 206(1), 308(1c), 309(1c), 311(1c), 316(1c, 121), 318(1c), 319(1c), 322(1c), 324–28(1c), 330(1c), 331(1c), 333(1c), *339, 343,* 492(29), *499*
Baker, D. A., *199*
Balian, R., 3(3), *53, 59, 61*
Balint-Kurti, G. G., *65, 81, 114, 118, 132, 136, 138*
Bandel, H. W., 427(37), *517*
Baranger, M., *377,* 382(32), *411*
Barbanis, B., *49, 61*
Bardsley, J. N., 435(31), *437, 456, 517, 518*
Barg, G. D., *123, 126, 136*
Barker, R. S., *149, 150, 199*
Barnett, C. F., *197, 199*
Barrett, B. R., 400(67), 401(67), *411*
Barua, A. K., *138*
Basch, H., 321(127), 322(127), 325(127), 334(149), *343, 344*
Basilier, C., 212(29), 334(141, 145), 335(145), 336(141), *341, 344*
Baskin, C. P., *156, 199*
Basu, S., *485,* 486(80), *519*
Batley, M., *477*
Batra, I. P., 322(128), 323(128), *343*
Batter, R., 293(98), 294(98), *343*

521

SUBJECT INDEX